OXFORD STATISTICAL SCIE

OXFORD STATISTICAL SCIENCE SERIES

For a full list of titles please visit

http://www.oup.co.uk/academic/science/maths/series/osss/

10. J.K. Lindsey: *Models for repeated measurements*
11. N.T. Longford: *Random coefficient models*
12. P.J. Brown: *Measurement, regression, and calibration*
13. Peter J. Diggle, Kung-Yee Liang, and Scott L. Zeger: *Analysis of longitudinal data*
14. J.I. Ansell and M.J. Phillips: *Practical methods for reliability data analysis*
15. J.K. Lindsey: *Modelling frequency and count data*
16. J.L. Jensen: *Saddlepoint approximations*
17. Steffen L. Lauritzen: *Graphical models*
18. A.W. Bowman and A. Azzalini: *Applied smoothing methods for data analysis*
19. J.K. Lindsey: *Models for repeated measurements, second edition*
20. Michael Evans and Tim Swartz: *Approximating integrals via Monte Carlo and deterministic methods*
21. D.F. Andrews and J.E. Stafford: *Symbolic computation for statistical inference*
22. T.A. Severini: Likelihood methods in statistics
23. W.J.Krzanowski: Principles of multivariate analysis: *A user's perspective, revised edition*
24. J.Durbin and S.J.Koopman: *Time series analysis by state space models*
25. Peter J. Diggle, Patrick Heagerty, Kung-Yee Liang, and Scott L. Zeger: *Analysis of longitudinal data, second edition*
26. J.K. Lindsey: *Nonlinear models in medical statistics*
27. Peter J. Green, Nils L. Hjort, and Sylvia Richardson: *Highly structured stochastic systems*
28. Margaret S. Pepe: *The statistical evaluation of medical tests for classification and prediction*
29. Christopher G. Small and Jinfang Wang: *Numerical methods for nonlinear estimating equations*
30. John C. Gower and Garmt B. Dijksterhuis: *Procrustes problems*
31. Margaret S. Pepe: *The statistical evaluation of medical tests for classification and prediction, paperback*
32. Murray Aitkin, Brian Francis and John Hinde: *Generalized Linear Models: Statistical modelling with GLIM4*
33. Anthony C. Davison, Yadolah Dodge, N. Wermuth: *Celebrating Statistics: Papers in honour of Sir David Cox on his 80th birthday*
34. Anthony Atkinson, Alexander Donev, and Randall Tobias: *Optimum Experimental Designs, with SAS*
35. M. Aitkin, B. Francis, J. Hinde, and R. Darnell: *Statistical Modelling in R*

Statistical Modelling in R

MURRAY AITKIN
Department of Psychology, University of Melbourne

BRIAN FRANCIS
Centre for Applied Statistics, Lancaster University

JOHN HINDE
School of Mathematics, Statistics and Applied Mathematics
National University of Ireland Galway Ireland

ROSS DARNELL
CSIRO Mathematical and Information Sciences, Australia

OXFORD
UNIVERSITY PRESS

Great Clarendon Street, Oxford OX2 6DP

Oxford University Press is a department of the University of Oxford.
It furthers the University's objective of excellence in research, scholarship,
and education by publishing worldwide in

Oxford New York

Auckland Cape Town Dar es Salaam Hong Kong Karachi
Kuala Lumpur Madrid Melbourne Mexico City Nairobi
New Delhi Shanghai Taipei Toronto

With offices in

Argentina Austria Brazil Chile Czech Republic France Greece
Guatemala Hungary Italy Japan Poland Portugal Singapore
South Korea Switzerland Thailand Turkey Ukraine Vietnam

Published in the United States
by Oxford University Press Inc., New York

First published 2009

British Library Cataloguing in Publication Data
Data available

Library of Congress Cataloging in Publication Data
Data available

Typeset by Newgen Imaging Systems (P) Ltd., Chennai, India
Printed in Great Britain
on acid-free paper by
CPI Antony Rowe, Chippenham, Wiltshire

ISBN 978–0–19–921914–8
978–0–19–921913–1 (pbk)

1 3 5 7 9 10 8 6 4 2

Preface

This book is an adaptation to R of the book *Statistical Modelling in GLIM4* by Aitkin, Francis, and Hinde. The need for an R version became clear while the GLIM4 book was being developed, and the GLIM4 authors invited Ross Darnell, who did the detailed text and example checking for the GLIM4 book, to join in the preparation of the R version.

R is approaching the status of the default statistical package for Universities and research organizations, and is making steady inroads into the commercial market. Its base in S and its object-oriented design make it both more powerful and more complicated to use than simpler command-line or menu-driven statistical packages. In adapting the book for R we have therefore extended the discussion of the package structure and facilities, and the more detailed output possible has increased the page content.

Chapter 1 serves as an introduction to the R language and we purposely leave discussion of more specialized statistical and graphical functions to later chapters. Chapter 2 gives a detailed account of the population modelling process, with a full discussion of frequentist, Bayesian and likelihood inferential approaches to simple models. Chapters 3–6 discuss the normal, binomial, Poisson/multinomial and survival models. Chapter 7 discusses continuous and finite mixtures, Chapter 8 overdispersed models and Chapter 9 variance component models. The data sets and custom functions are held in the SMIR package downloadable from a CRAN mirror.

The book was written over the period 2005–2008. We thank the Science Foundation of Ireland and the Australian Research Council for support, and we are grateful to Nick Sofroniou for very detailed suggestions for improvements to the procedures first used in the GLIM4 book, and to Jochen Einbeck for the development, with John Hinde and Ross Darnell, of the npmlreg function. Any errors in the text and procedures are entirely the authors' responsibility.

Melbourne, Lancaster, Galway, and Brisbane
2008

Contents

1
Introducing R

1.1 Statistical packages and statistical modelling

Statistical modelling is an iterative process. We have a research question of interest, the experimental or observational context, the sampling scheme and the data. Based on these data and the other information, we develop an initial model, fit it to the data and examine the model output. We then plot the residuals, check the model assumptions and examine the parameter estimates. Based on this model output, we may modify the first model possibly through some transformation or the addition or subtraction of model terms. Fitting the second model may in turn lead to further models for the data, with the form of the current model being based on the information provided by the previous models. This *interactivity* between the data and the data modeller is a crucial feature of the process of statistical modelling. While automatic model selection procedures provided in many software packages can provide guidance in large problems, we take a different route in this book.

We focus on generalized linear models (GLMs) and extensions of these models, which can be fitted using an iterative re-weighted least squares (IRLS) algorithm. The computational extensions which we cover here (into such areas as survival analysis and random effects models) are developed by building on the standard IRLS algorithm, incorporating it into more complex *functions*.

There are numerous interactive and programmable statistical packages on the market which can be used to fit GLMs. In no particular order, these include R, S-PLUS, GENSTAT, GLIM, SAS, STATA, and XLISPSTAT. With appropriate procedure development, all could be used to fit the models introduced in this book. All can be used interactively and all allow the user to develop models in the way outlined above. In addition, there are other packages such as SPSS and MINITAB which can fit specific types of GLM although not in a unified framework.

R is well supported with an excellent homepage located at `http://www.r-project.org/` with mirrors worldwide that constitute the Comprehensive R Archive Network (CRAN).

1.2 Getting started in R

To start R from a UNIX terminal window or a Microsoft command window, the command 'R' is sufficient provided the systems PATH variable includes its pathname. Under Windows the R installation typically generates an R shortcut which runs a console-based graphical user interface (GUI). Although this shortcut makes starting R very easy it is more convenient to change the shortcut properties so that R runs in the directory in which you would like the current dataset kept.

We would also encourage the new user to read the *R-intro.pdf* file at `http://cran.r-project.org/manuals.html` and installed with R in the `doc/manual` subdirectory.

After invoking R, we see

```
R : Copyright 2005, The R Foundation for Statistical Computing
Version 2.1.1  (2005-06-20), ISBN 3-900051-07-0

R is free software and comes with ABSOLUTELY NO WARRANTY.
You are welcome to redistribute it under certain conditions.
Type 'license' or 'licence' for distribution details.

  Natural language support but running in an English locale

R is a collaborative project with many contributors.
Type 'contributors' for more information and
'citation' on how to cite R or R packages in publications.

Type 'demo' for some demos, 'help' for on-line help, or
'help.start' for a HTML browser interface to help.
Type 'q' to quit R.

> options(STERM='iESS', editor='emacsclient')
>
```

The '>' character which follows the message is the R prompt, and tells you that R is waiting to receive instructions or commands from you. If a command line is entered and not complete, R will respond with the '+' character as a prompt to signify that the user needs to complete the current command.

A command line system like R benefits from a help system and a good editor. R has several on-line help methods. To search the help system for documentation using fuzzy or regular expression matching, use the *help.search* function. This is a good starting point to search using a keyword. Once the correct package and function are found, function documentation is retrieved by inserting the function name as the argument of the *help* function. A more succinct alternative is to prepend a question mark to the function name. For example, try *?help*. Many functions come with examples which the user can run by using the *example* function. The user might like to try *example(image)*. Source code of the examples provides useful templates for the user. There are some R examples that can be run using the *demo* function. R's *source* function allows input from a named file.

R's console supports several of the basic Unix type command line functions, such as CTL-A, CTL-E to help edit the command line. It also keeps a command history which can be accessed by the UP and DOWN arrow keys. There are several source editors available that provide the user with many more features and these are described in Section 1.13.

With the MS-windows operating systems, a graphical interface for R can be in invoked by the `Rgui.exe` command. The resulting screen appears as

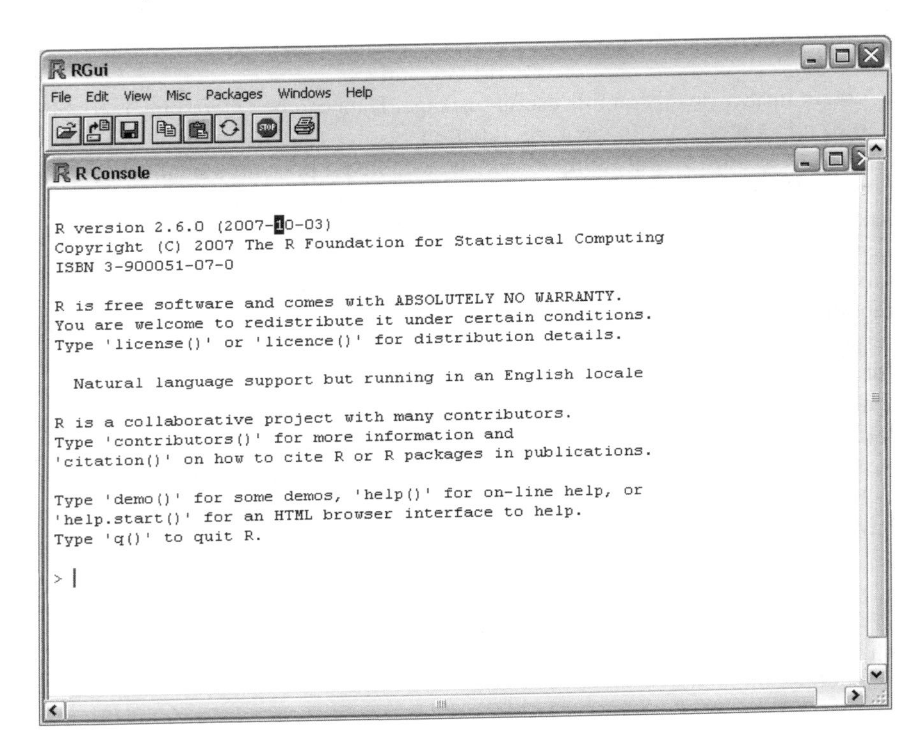

Note that R does not have the drop-down menu system seen in other packages, however the RCMDR package, a platform-independent package, does provide some of these features.

Information about downloading and using packages is given later in Section 1.14.

1.3 Reading data into R

Data can be entered into R in many ways. A simple and robust approach is by reading column-separated values in which rows represent cases or experimental units and columns represent variables with variable names included as the first row. Often data are entered using spreadsheet programs like Excel. Saving the spreadsheet as a `csv` type allows the data to be easily read by R although the *gdata* package includes the `read.xls` function which reads the Excel file format directly. Use the `search.help` function to discover other ways to read data. The *foreign* package provides functions to input data files in formats used by other statistical packages including Minitab, S, SAS, SPSS, and Stata.

Enter

```
> library(help = "foreign")
```

to obtain further details on this useful package. Data can be entered interactively into R, for example, using the `scan` function.

We start by introducing a simple set of data. Table 1.1 presents a small set of data, given in Erickson and Nosanchuk (1979), extracted from a larger set collected by Atkinson and Polivy (1976) to examine the effects of unprovoked verbal attack. Ten male and nine female subjects were asked to fill out a questionnaire which mixed innocuous questions with questions attempting to assess the subject's self-reported hostility. A hostility score for each individual was calculated from these responses. After completing the questionnaire, the subjects were then left waiting for a long time, and were subjected to insults and verbal abuse by the experimenter when the questionnaire was eventually collected. All subjects were told that they had filled out the questionnaire incorrectly, and were instructed to fill it out again. A second hostility score was then calculated from these later responses.

For each of the 19 individuals involved in this psychological experiment we have three items of information, or variables:

(1) the hostility score of the individual before being insulted;
(2) the hostility score of the individual after being insulted; and
(3) the sex of the individual.

Table 1.1. The insult data set

Hostility		Sex	Hostility		Sex
Before	After		Before	After	
51	58	Female	86	82	Male
54	65	Female	28	37	Male
61	86	Female	45	51	Male
54	77	Female	59	56	Male
49	74	Female	49	53	Male
54	59	Female	56	90	Male
46	46	Female	69	80	Male
47	50	Female	51	71	Male
43	37	Female	74	88	Male
			42	43	Male

One of the variables (the sex of the individual) is a categorical variable with only two values: male and female. The remaining two variables are continuous – that is, variables which are measured on an underlying continuous scale to a certain measurement accuracy (Section 2.2 contains a further discussion of types of variables). The data are arranged in a data matrix, with the rows of the matrix representing individuals, units or cases and the columns representing the variables recorded for the individuals. There are no missing data; we have complete information on all the variables for each individual.

This data is to be read and stored in R as a *dataframe*. Dataframes can be thought of as matrices with columns of mixed variable types. Many of the modelling functions used in R expect data to be stored as dataframes. Dataframes also exhibit *list* properties. Character variables are by default converted to a factor. Factor levels can be defined to represent ordered levels and model-fitting functions will respond differently to these different classes.

```
> data(insult, package = "SMIR")
```

Typing the name of the dataframe, or of any object, at the prompt will cause R to print the object using the default print method appropriate for the object class. A more succinct form to view the contents of a dataframe is to use the `str` function.

```
> str(insult)

'data.frame':        19 obs. of  3 variables:
 $ hbefore: int  51 54 61 54 49 54 46 47 43 86 ...
 $ hafter : int  58 65 86 77 74 59 46 50 37 82 ...
 $ sex    : Factor w/ 2 levels "female","male": 1 1 1 1 1 1 1 1
            1 2 ...
```

So the dataframe 'insult' has three variables, the first two being integer variables and the third a factor since the `read.csv` function converts character columns into factors by default. The `str` function lists the first few values of each variable. Factors are stored as a set of integer codes to save space.

Note that if sex could have been recorded as a numeric code in the original data file, for example, a 1 for a female and a 2 for a male, R would have treated this as an integer variable as well. The user would need to convert this to a factor and optionally explicitly define the labels by using the `factor` function.

```
> insult$sex.factor <- factor(insult$sex, labels = c("female",
+       "male"))
```

Individual columns of the dataframe can be extracted by using the dollar character separating the dataframe name and the column name, or by using square brackets and treating the dataframe as a matrix or even by the extraction operator '`[`'.

```
> insult$hbefore
> insult[,1]
> "["(insult,1)
```

It is also important to remember that names for objects, functions etc. in R are case-sensitive. The R help files state that identifiers (including object names) consist of a sequence of letters, digits, the period ('.') and the underscore. They must not start with a digit nor underscore, nor with a period followed by a digit.

The actual R commands may also be stored in a 'script' file using one of the text editors discussed in Section 1.13. Each line of code is read using the *source* function and 'interpreted' until the end of the file is reached and then the expressions are evaluated. The editors also allow lines of code to be directly copied to the R processor using 'cut-and-paste' type parsing.

1.4 Assignment and data generation

An alternative to the read and scan functions is to combine values using the *c* function and assign the resulting vector to a variate by use of the assignment operator '<-'.

```
> x <- c(51, 54, 62, 54, 49, 54, 46, 47, 43, 86,
+      28, 45, 59, 49, 56, 69, 51, 74, 42)
> ls()
[1] "insult" "x"
```

The vector x has been written to the user's workspace along with the `insult` dataframe. The function *ls* returns a vector of object names in the user's workspace. To remove an object from the workspace, *rm* is used but to remove a column from a dataframe you need to 'overwrite' the column with the *NULL* object. For example,

```
> insult$x <- x
> names(insult)

[1] "hbefore"    "hafter"     "sex"         "sex.factor"
[5] "x"

> rm(x)
> insult$x <- NULL
> names(insult)

[1] "hbefore"    "hafter"     "sex"         "sex.factor"
```

The *names* function gets the names of the dataframe columns but can also be used to set the names of the columns. For example,

```
insult$sex.factor <- NULL    # remove unwanted variable
names(insult) <- c('hb4','harfter','gender')
```

The operator '<-' is not the only assignment operator in R. The operators '<<-' (or the reverse '->') and '=' assign into the local environment, usually the user's workspace, but use of the '=' often leads to confusion as to what is the variable name and what is the value to be assigned to the variable or object.

Examples of assigning values to variables are

```
> x <- c(1, 4)
```

will define a variate called *x* of length 2, and

```
> x <- c(x, 3, 6, x)
```

will increase the length of *x* to now have values *1,4,3,6,1,4*. Number sequences are available, and take the form *from.value:to.value*. The increment is one, however a better method is by using *seq*. So, for example,

```
> z <- c(4:13)
> z

 [1]  4  5  6  7  8  9 10 11 12 13
```

or alternatively

```
> z <- seq(from = 4, to = 13, by = 1)
> z

 [1]  4  5  6  7  8  9 10 11 12 13
```

Factors which have a regular structure can be generated automatically using the *gl* function. So, for example, we might have a table of counts as follows:

	B					
	1			2		
	C			C		
A	1	2	3	1	2	3
1	10	0	0	2	1	0
2	4	6	2	3	7	1
3	0	3	8	2	4	8
4	3	4	11	1	7	12

We read the values of the table into a variable called *count*, and then use the *gl* function to generate the cross-classifying factors.

```
> count <- c(10, 0, 0, 2, 1, 0, 4, 6, 2, 3, 7, 1,
+      0, 3, 8, 2, 4, 8, 3, 4, 11, 1, 7, 12)
> count.df <- data.frame(a = gl(4, 6), b = gl(2,
+      3, 24), c = gl(3, 1, 24), count)
> str(count.df)
```

```
'data.frame':  24 obs. of  4 variables:
 $ a    : Factor w/ 4 levels "1","2","3","4": 1 1 1 1 1 1 2 2 2 2 ...
 $ b    : Factor w/ 2 levels "1","2": 1 1 1 2 2 2 1 1 1 2 ...
 $ c    : Factor w/ 3 levels "1","2","3": 1 2 3 1 2 3 1 2 3 1 ...
 $ count: num  10 0 0 2 1 0 4 6 2 3 ...
```

The first argument of the *gl* function defines the number of levels of the factor, the second argument the number of repeat values and the third represents the length. To force the lengths of factors *b* and *c* to be the correct length, it was necessary to define the third argument explicitly.

Another method that achieves the same result is to use the *expand.grid* function to generate the factor structure then add the column of counts using the *cbind* function.

```
> count.df1 <- cbind(expand.grid(c = factor(1:3),
+     b = factor(1:2), a = factor(1:4)), count)
```

1.5 Displaying data

Vectors can be displayed using the default print method. Arguments to the function can be set to modify the minimum number of significant digits but the same number of decimal places is used for all values in the vector. The number of digits to print can be set for a session by setting the *digits* argument in the *options* function. Alternatively *round* can be used.

```
> insult$hbefore

 [1] 51 54 61 54 49 54 46 47 43 86 28 45 59 49 56 69 51 74
[19] 42

> insult$hbefore/100

 [1] 0.51 0.54 0.61 0.54 0.49 0.54 0.46 0.47 0.43 0.86 0.28
[12] 0.45 0.59 0.49 0.56 0.69 0.51 0.74 0.42
```

The data values are printed across the screen, to the accuracy as defined by *getOption('digits')*.

Dataframes as well as 'sub-matrices' are printed by default in a 'sensible' way by entering the name of the dataframe;

```
> insult$sex.factor <- NULL
> insult
  hbefore hafter    sex
1      51     58 female
2      54     65 female
3      61     86 female
4      54     77 female
5      49     74 female
6      54     59 female
7      46     46 female
```

```
8         47      50 female
9         43      37 female
10        86      82   male
11        28      37   male
12        45      51   male
13        59      56   male
14        49      53   male
15        56      90   male
16        69      80   male
17        51      71   male
18        74      88   male
19        42      43   male
```

```
> insult[1:10, 1:2]
```

```
   hbefore hafter
1       51     58
2       54     65
3       61     86
4       54     77
5       49     74
6       54     59
7       46     46
8       47     50
9       43     37
10      86     82
```

A negative index can be used to remove particular rows or columns.

```
> insult[-c(1:10), 3]
```

```
 [1] male male male male male male male male male
 Levels: female male
```

Apart from the using the default print method, dataframes can be displayed and edited using either the `edit` or `fix` functions.

1.6 Data structures and the workspace

R has a fully featured programming language and so includes many types of objects. These include vectors which can have logical, integer, numeric, complex and character modes. Other object types include matrices, arrays, dataframes and functions. All these objects can be combined into lists. Components of lists can be extracted using the '`[`' operator while elements are extracted using the '`[[`' operator. To extract the first ten values of the second component of a list we first construct the list and then extract the required values. Suppose we use the vectors from the insults dataframe to compose a list called `my.list` containing the factor `gender` and the numeric vector `hbefore`.

```
> my.list <- list(insult[, 3], insult[, 2])
> my.list[[1]][1:10]

 [1] female female female female female female female female
 [9] female male
Levels: female male

> is.list(my.list[[1]])

[1] FALSE

> is.list(my.list[1])

 [1] TRUE
```

The command *my.list[[1]]* extracts the first component of the list which in this case is the gender factor and the first ten values are extracted using the '*[[*' operator. Note the difference between extracting the first component as a list using the operator '*[*' as opposed to extracting the first component as a vector using the '*[[*' operator. In the first extraction, the result is a vector (as reported by *is.list*), while in the second the result is a list of one component.

R has several special constants including *pi* and the character strings *LETTERS*, *letters*, *month.abb*, *Month.name*.

```
> LETTERS[1:10]

 [1] "A" "B" "C" "D" "E" "F" "G" "H" "I" "J"

> month.name

 [1] "January"   "February" "March"     "April"
 [5] "May"       "June"     "July"      "August"
 [9] "September" "October"  "November"  "December"
```

R allows the user to tailor procedures and output to their requirements. Although many of the objects generated by R like the linear modelling function *lm*, are often large lists, only those elements of the lists that are relevant for a particular purpose would be extracted. Most functions that generate these objects also come with a print method that prints only the minimal amount of information. As well many of the model objects have accompanying functions like *anova* and *summary* that produce ANOVA tables and regression parameters with standard errors.

Functions are single lines or blocks of R code that operate on arguments to produce objects. Functions make repetitive tasks more convenient and powerful.

The R-intro document describes a function: 'A function is defined by an assignment of the form

```
> name <- function(arg_1, arg_2,  ...) expression
```

The expression is an R expression, (usually a grouped expression), that uses the arguments, arg_i, to calculate a value.'

Many functions are available in the base R package as well as from the large number of packages available from the CRAN site. Many are written using R so are useful as templates and guides to develop your own functions.

The workspace can quickly increase in size, and this can increase the time it takes for R to be invoked and closed. Unwanted objects can be removed from the workspace using the *rm* function. For example, we can remove the object *my.list* by typing

```
> rm(my.list)
```

Dataframe columns are removed by assigning the constant *NULL* to the existing column;

```
> insult$hafter <- NULL
```

The R session is completely ended by using *q*. However, by default, the *save.image* function is run first so the user will be asked whether they want the workspace saved or to cancel the quit. If the user responds 'Yes', the workspace is saved in a file (*.RData* by default), in the working directory and R will close.

When run from the operating system command line, R does not keep a log of any output or a journal file. It is prudent to run R commands from a source file and cut-and-pasted into the R command window. Output from R can be redirected to a file by using the *sink(filename)* function which needs to be closed using *sink* again but with no argument, to redirect the output back to the console.

1.7 Transformations and data modification

We now describe the functions which are used for general calculation, transformation and modification of vectors. The R command

```
> 5 + 6

 [1] 11
```

returns the value 11; the command

```
> insult$hafter - insult$hbefore

 [1]  7 11 25 23 25  5  0  3 -6 -4  9  6 -3  4 34 11 20 14
[19]  1
```

returns the change in hostility sores after being insulted.

The symbols +, –, *, /, ^, %%, and %/% when used in arithmetic expressions, are known as arithmetic operators, and represent addition, subtraction, multiplication, division, exponentiation, modulus and integer division operations, respectively. The order of precedence of unitary and binary operators are defined in the help listing for `Syntax`. Within an expression operators of equal precedence are evaluated from left to right except where altered by parenthesis.

The results of calculations, instead of being displayed on the screen, may be assigned to a vector. Note that R does not specify a scalar type, only a vector of unitary length.

Transformations of existing vectors, matrices and arrays can also be carried out. For example,

```
> x <- 1:10
> y <- 2^(x - 1)
```

creates a vector `y` with length 10, with each element of `y` containing the transformation 2^{x-1} of the corresponding element of `x`.

Vectors may be transformed `y <- (y + 1) / 2` and reassigned `y <- x^ 3`. Two vectors appearing in the same expression need not have the same length. The shorter vector will be recycled until it matches the length of the longer vector. Constants such as `pi` will be repeated. So in the expression

```
> 2 * pi * seq(1:10)/5

 [1]  1.2566  2.5133  3.7699  5.0265  6.2832  7.5398  8.7965
 [8] 10.0531 11.3097 12.5664
```

the constants 2, `pi` and 5 will be repeated 10 times. To display a complete list of arithmetic operators try `help(Arithmetic)`.

When transforming many components of a dataframe, the `transform` function is particularly useful to reduce the amount of typing. For example, to calculate the change and relative change in hostility scores we can use the following code:

```
> insult <- transform(insult, hchange = hafter -
+       hbefore, hrelchange = (hafter - hbefore)/hbefore)
```

1.8 Functions and suffixing

A range of functions is provided in R and a full list of those supplied by the base library alone can be found through

```
> library(help = base)
```

Help on individual functions is available by issuing a question mark, `?`, followed by the function name, for example, `?log`.

1.8.1 *Structure functions*

Structure functions return the characteristics of an identifier, for example, its length (`length`) and number of levels of a factor (`nlevels`).

```
> x <- seq(1, 19, by = 3)
> lenx <- length(x)
> print(lenx)

[1] 7
```

will store the value 7 in the unitary length vector `lenx`.

Structure functions pertaining to arrays include `NROW`, `NCOL`, `dim` and `dimnames` which return the number of rows, the number of columns, the dimension and, if defined, the the dimension labels of an array.

Similarly for dataframes, `dim` displays the number of rows and columns in the dataframe, while `names` lists the labels given to columns and `rownames` lists the row labels.

1.8.2 *Mathematical functions*

R provides a comprehensive set of mathematical functions, including the natural log (`log`), exponential (`exp`), absolute value (`abs`), truncation (`tr`) and those related to the beta, gamma and trigonometric functions. Thus

```
> x <- seq(-1, 4)
> lx <- log(x)
```

will assign the natural log of all units of `x` to the equivalent units of `lx`. Where `x` $= 0$ or -1, R returns `-Inf` and `NaN`, respectively. A warning message is printed. This behaviour applies to all other functions and expressions that give an invalid result, for example, taking the square root of a negative number. R returns the number `NaN` which stands for 'Not a Number'. Dividing by zero returns the number `Inf` but no warning.

To list all the functions relating to the beta and gamma functions use the help command for `Special`, and for trigonometric functions, see help for `Trig`.

Another useful function is `cumsum`. The statement `sum(x)`:

```
> x <- seq(1, 6)
> print(sum(x))

[1] 21
```

will calculate the total of the vector `x`, while the function `cumsum`

```
> print(cumsum(x))

[1]  1  3  6 10 15 21
```

provides functions to calculate cumulative products and minima or maxima of a vector. For more information try help on `Arithmetic`.

1.8.3 *Logical operators*

Binary comparison operators return the logical (global) vectors `TRUE` or `FALSE`. The functions are '<', '>', '<=', '>=', '==', and '! =' representing 'less than', 'greater than', 'greater than or equal to', 'less than or equal to', 'equal to' and 'not equal to', respectively. If you want the logical operators to return zeroes and ones instead of `FALSE` and `TRUE` you can include the logical constant in a mathematical expression or an `ifelse` clause.

```
> x <- c(13, 21, 11, 36, 15, 19)
> print((x > 20) * 1)

[1] 0 1 0 1 0 0

> ifelse(x > 20, 1, 0)

[1] 0 1 0 1 0 0
```

A vector of ones and zeroes is returned based on the test given as the first argument in the `ifelse` statement.

Logical operators are

! x	logical negation
x & y	logical 'AND' elementwise
x && y	logical 'AND' uses only the first element of each vector
x | y	logical 'OR' elementwise
x || y	logical 'OR' using the first element of each vector
xor(x,y)	elementwise exclusive 'OR'
isTRUE(x)	an abbreviation of `identical(TRUE,x)`

See help for `Logic` to obtain more information.

1.8.4 *Control functions*

The R programming language includes several control-flow constructs. There is a basic `if` and `ifelse`, `for`, `while`, `repeat`, `break` and `next`. For more information regarding these topics see help for `Control`.

1.8.5 *Statistical functions*

There are over three hundred statistical functions available in R's basic `stat` package installed by default. Many other packages available from CRAN provide the user with additional features.

Statistical functions include the cumulative distribution and inverse cumulative distribution (or probability) functions for a range of distributions. For example, *qnorm* gives the deviate of a standard normal given a probability value, and *pnorm* gives the cumulative distribution value for a given deviate value. Other distributions provided include the chisquared (*qchisq*, *pchisq*); Poisson, (*qpois*, *ppois*); Student's t (*qt* and *pt*), binomial and beta and many others. Type *help.search('distribution')* to get a full list. Many of the functions have other arguments for distributional parameters and degrees of freedom. Thus *qchisq(0.95,1)* returns 3.84; *pnorm(1.96)* returns 0.98, when rounded to two decimal places.

1.8.6 *Random numbers*

Random numbers may be generated for many statistical distributions mentioned in Section 1.8.5. For example, sequences of uniform random numbers in the range (0, 1) can be generated by

```
> options(digits = 3)
> runif(20)

 [1] 0.4649 0.7294 0.1281 0.9343 0.3935 0.2491 0.9474 0.8417
 [9] 0.4152 0.8917 0.2067 0.9502 0.4785 0.1103 0.6294 0.2935
[17] 0.8303 0.3280 0.8737 0.0276

> runif(20)

 [1] 0.54004 0.47894 0.87790 0.48683 0.22236 0.50914 0.00953
 [8] 0.08849 0.31858 0.54531 0.30298 0.73988 0.01028 0.59864
[15] 0.71310 0.46599 0.21659 0.11643 0.41058 0.63919
```

Note these two sequences are not the same since R uses different seeds for each use of the random number function. The sequence is repeatable by setting the same random seed using the *set.seed* function. So the two sequences below are the same since the same seed is used.

```
> options(digits = 3)
> set.seed(1234)
> runif(20)

 [1] 0.1137 0.6223 0.6093 0.6234 0.8609 0.6403 0.0095 0.2326
 [9] 0.6661 0.5143 0.6936 0.5450 0.2827 0.9234 0.2923 0.8373
[17] 0.2862 0.2668 0.1867 0.2322

> set.seed(1234)
> runif(20)

 [1] 0.1137 0.6223 0.6093 0.6234 0.8609 0.6403 0.0095 0.2326
 [9] 0.6661 0.5143 0.6936 0.5450 0.2827 0.9234 0.2923 0.8373
[17] 0.2862 0.2668 0.1867 0.2322
```

1.8.7 *Suffixes in expressions*

All vectors, dataframes, matrices, arrays and lists may be suffixed. For example, sub-matrices can be extracted simply,

```
> (A <- matrix(1:12, nrow = 4, ncol = 3))
     [,1] [,2] [,3]
[1,]    1    5    9
[2,]    2    6   10
[3,]    3    7   11
[4,]    4    8   12

> A[1:2, 1:2]

     [,1] [,2]
[1,]    1    5
[2,]    2    6
```

The suffix itself can be a vector, so we can permute the rows of the matrix A by

```
> row.ord <- c(2, 1, 4, 3)
> A[row.ord, ]

     [,1] [,2] [,3]
[1,]    2    6   10
[2,]    1    5    9
[3,]    4    8   12
[4,]    3    7   11
```

reorders the rows of `A`.

Suffixes can also be used to expand objects. If we want to duplicate each row of the matrix `A` three times, then this could be achieved as follows:

```
> A[rep(1:4, 3), ]

      [,1] [,2] [,3]
 [1,]    1    5    9
 [2,]    2    6   10
 [3,]    3    7   11
 [4,]    4    8   12
 [5,]    1    5    9
 [6,]    2    6   10
 [7,]    3    7   11
 [8,]    4    8   12
 [9,]    1    5    9
[10,]    2    6   10
[11,]    3    7   11
[12,]    4    8   12
```

1.8.8 *Extracting subsets of data*

A common requirement is to extract subsets of data from a larger dataset. Using the insult data as an example, these subsets will usually have particular characteristics, such as 'female only' or those with a before hostility score of 65 or more.

The indexing methods described in Section 1.8.7 can be used for this purposes For example, to extract a subset of the insult dataset for those values of the 'before' hostility 65 or greater, we use

```
> insult[insult$hbefore >= 65, ]

   hbefore hafter  sex hchange hrelchange
10      86     82 male      -4    -0.0465
16      69     80 male      11     0.1594
18      74     88 male      14     0.1892
```

Many of the regression model functions include an argument, `subset`, which allows certain cases to be included in the analysis. Alternatively, the user could define a vector of 0, 1's to be used as weights in a regression model.

An alternative method is to use the `subset` function. To generate a new dataframe of males only from the `insult` dataframe,

```
> male.insult <- subset(insult, sex == "male")
```

See help for `Extract` for more information.

1.8.9 *Recoding variates and factors into new factors*

The `cut` function is a useful way to convert continuous variables into a factor. The `breaks` argument defines the cut points (if given as a vector) or the number of intervals (if a scalar). Labels for the levels can also be stipulated.

We use as an example the insult data, and recode the `hbefore` score into a four category factor `hb` defined as follows; less than 50, 50, less than 65, 65, less than 75, 75, or greater. We then generate a new factor which combines the extreme levels `(-Inf,50]` and `(75,Inf]`.

```
> insult <- transform(insult, hb = cut(hbefore, +
breaks = c(-Inf, 50, 65, 75, Inf)))
> insult$hbnew <- insult$hb
> levels(insult$hbnew) <- c(3, 1, 2, 3)
> with(insult, table(hbnew, hb))

     hb
hbnew (-Inf,50] (50,65] (65,75] (75, Inf]
    3         8       0       0         1
    1         0       8       0         0
    2         0       0       2         0
```

Note the use of the function *with* which obviates the need to include the dataframe name when referencing any of the variables within it.

1.9 Graphical facilities

Statistical modelling of data is aided considerably by the graphical display of data. R provides excellent graphical capabilities. R's object-oriented approach means that the plot method depends on the object being 'plotted'. The reader is encouraged to generate examples of various plotting methods. Currently there are two plotting engines in R, the traditional plot routines and those used by the *lattice* package. The latter generates better plots and can be extended to generate Trellis (Cleveland, 1985) plots.

The basic plotting functions like *plot* and *hist* exist; however, there are many variations of these two basic functions.

We return to the insult dataset, and illustrate the directive with a scatter-plot of the before and after hostility scores. The simplest way to produce this plot is

```
plot(hafter~hbefore,data=insult).
```

This will produce a scatter-plot on the active graphics window.

To obtain a hard copy of the graphics window use *dev.print*. If you need to copy the graph to a file, there are several device types available for the output file. Using help on *device* lists these devices which includes 'postscript', 'jpeg', 'png' and 'pdf'. For Microsoft operating systems, 'wmf' and 'emf' device types are available as well.

For simple scatter plots, *plot.default* will be used. However, there are 'plot' methods for many R objects. Use *methods(plot)* and the documentation for these. Users who want to control how axes, symbols, line types and other aspects of the plot are drawn should use the *par* function. See the help on *par* for details. So the code

```
> par(mfrow = c(2, 2))
> plot(hafter ~ hbefore, data = insult, xlab = "hostility before",
+      ylab = "hostility after", pch = 4, main = "(a)")
> plot(hafter ~ hbefore, data = insult, xlab = "hostility before",
+      ylab = "hostility after", pch = 4, main = "(b)")
> abline(a = 0, b = 1, lty = 2)
```

produces scatter-plots with axis labelling and specifying the plotting symbol $\times$ (*pch=4*).

We can see from Fig. 1.1(a) that there is a strong relationship between the hostility scores before and after the insults. We may want to superimpose a line of no change on the graph. This is most easily produced by using the *abline* function and specifying the intercept as zero and the slope to be one.

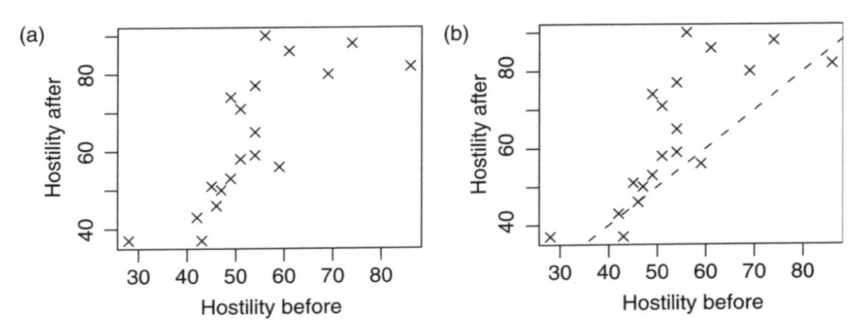

Fig. 1.1. (a) Scatter-plot of *hafter* (y-axis) plotted against *hbefore* (x-axis) (b) with, superimposed line of no change

Figure 1.1(b) shows that most individuals increase their hostility score after the experiment, with only three individuals showing slight decreases. There is also some evidence of increasing spread in the plot as the hbefore score increases. This may be due to different behaviour of the two sexes. However, the plot does not distinguish between male and female subjects; both are plotted with the same symbol. If a factor has been defined with a number of levels, we can choose to identify each of these levels with a different plotting symbol. Here, the factor *sex* has two levels, and we use different plotting symbols for females and males. Use the *demo* function for *points* to see what plotting symbols are available.

Convenient methods to generate different plots for various levels of a factor in R include using conditional plots or lattice graphics. So in this example the conditioning factor is *sex* and the appropriate function call is

```
> coplot(hafter ~ hbefore | sex, data = insult,
+      panel = function(x, y, ...) {
+           points(x, y, ...)
+           abline(lm(y ~ x), ...)
+      })
```

The intercept and slope values are supplied by the *lm* function embedded in the *abline* function.

The lattice package provides the user with greater flexibility in producing plots and will by default produce better plots than the traditional R graphics functions.

```
> library(lattice)
> print(xyplot(hafter ~ hbefore | sex, data = insult,
+      panel = function(...) {
+           panel.xyplot(...)
+           panel.lmline(...)
+     }))
```

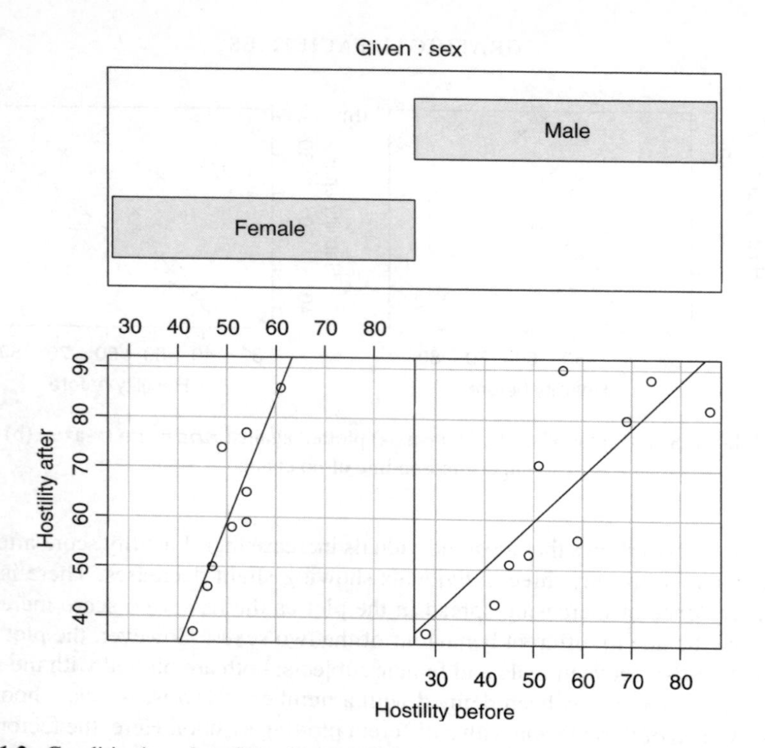

Fig. 1.2. Conditioning plot of hostility levels before and after insult for each of the sexes

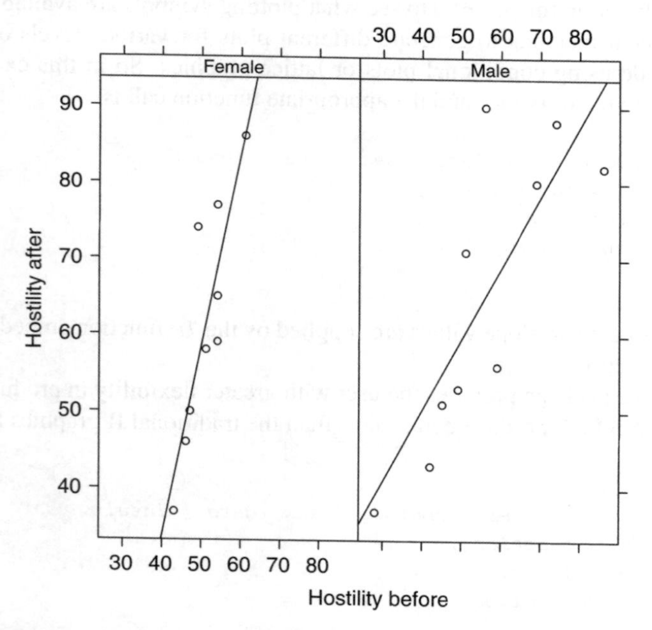

Fig. 1.3. Lattice panel plot of hostility scores before and after insult for each of the sexes

In Figs. 1.2 and 1.3 we see that the females appear to have a steeper relationship between before and after scores compared to the males. We have included fitted regression lines to highlight this difference, although no test for statistical significance of the interaction has been carried out.

1.10 Text functions

There are several functions related to text or string handling in R. These include *substr*, *paste*, *nchar*, *strsplit* and *strtrim*.

We may want to label the output from similar analyses with a specific heading. The repeated string can be pasted to another that contains the counter.

```
> line.of.stars <- paste(paste(rep("*", 50), collapse = ""),
+     "\n\n", sep = "")
> heading <- "ANALYSIS OF INSULT DATA LEAVING OUT CASE NUMBER "
> nrecs <- dim(insult)[[1]]
> insult$raised <- factor(insult$hbefore >= 65)
> for (i in seq(1, nrecs)) {
+     cat(paste(line.of.stars, heading, i, " \n\n",
+     line.of.stars, sep = ""))
+ print(anova(lm(hafter ~ sex * raised, data = insult[-i,
+     ])))
+ }
```

A part of the output appears as

```
**************************************************

ANALYSIS OF INSULT DATA LEAVING OUT CASE NUMBER 1

**************************************************

Analysis of Variance Table

Response: hafter
          Df Sum Sq Mean Sq F value Pr(>F)
sex        1     50      50    0.19  0.670
raised     1   1425    1425    5.40  0.035
Residuals 15   3956     264

**************************************************

ANALYSIS OF INSULT DATA LEAVING OUT CASE NUMBER 2

**************************************************
```

```
Analysis of Variance Table

Response: hafter
          Df Sum Sq Mean Sq F value Pr(>F)
sex        1     79      79    0.30  0.591
raised     1   1425    1425    5.41  0.034
Residuals 15   3953     264
```

Look for `help` on `character` for more information.

1.11 Writing your own functions

More often a function is written to store sequences of commands which the user can then use as required.

We illustrate with an example to calculate the correlation coefficient r of two variables X and Y with values x_i and y_i, for $i = 1, 2, \ldots, n$.

We use the formula

$$r = \frac{\sum(x_i - \bar{x})(y_i - \bar{y})}{\sqrt{\{\sum(x_i - \bar{x})^2 \sum(y_i - \bar{y})^2\}}}$$

```
> correl <- function(x, y) {
+     lenx <- length(x)
+     leny <- length(y)
+     xname <- deparse(substitute(x))
+     yname <- deparse(substitute(y))
+     if (lenx != leny)
+         stop("lengths of input vectors not equal")
+     meanx <- sum(x)/lenx
+     meany <- sum(y)/leny
+     xp <- sum((x - meanx) * (y - meany))
+     ssx <- sum((x - meanx)^2)
+     ssy <- sum((y - meany)^2)
+     corr <- xp/sqrt(ssx * ssy)
+     cat("The correlation coefficient of ", xname,
+         " and ", yname, " is \ n")
+     corr
+ }
```

The function is assigned to '`correl`' and accepts two arguments. The function defines two local variables, '`x`' and '`y`'. The length of each is calculated and compared to ensure they are of equal length. The function stops if this is not true giving an error message. Otherwise it continues calculating the means and sums of squares and cross-products, and eventually returns the value of the correlation estimate. The final line parses the value from `correl` back to the calling environment. To use the function, the order of the arguments must match that defined in the function itself unless explicitly labelled.

```
> correl(insult$hbefore, insult$hafter)

The correlation coefficient of  insult$hbefore  and  insult$hafter
  is
[1] 0.768
```

1.12 Sorting and tabulation

We conclude the chapter by a short discussion of the facilities for sorting and tabulation. There are two functions associated with sorting, `sort` and `order`. The former will sort a vector or factor into ascending order, returning the sorted vector, while `order` returns a permutation of indices. So `x[order(x)]` is equivalent to `sort(x)`:

```
> x <- c(20, 50, 35, 34, 55, 36, 92)
> sort(x)

[1] 20 34 35 36 50 55 92

> order(x)

[1] 1 4 3 6 2 5 7

> x[order(x)]

[1] 20 34 35 36 50 55 92
```

The `order` function is commonly used to sort dataframes into another new set according to the values of a particular vector. We may use this to create a new insult dataset ordered according to the `hbefore` hostility score. This is achieved by:

```
> insult <- insult[order(insult$hbefore), ]
```

We have now rearranged the rows of the `insult` dataframe to be ordered by the ascending `hbefore` values.

These functions can be used to find the order statistics of a vector by extracting the relevant elements of the sorted vector. For example,

```
> len <- length(insult$hbefore)
> sort(insult$hbefore)[c(1, (len + 1)/2, len)]

[1] 28 51 86
```

produces the minimum, median and maximum values of `hbefore`.

However, such functions are already included in the base R installation as well as *summary* which produces the five-number summary for numeric vectors along with the mean value. Tukey's five-number summary can be generated using *fivenum* which calculates the minimum, lower-hinge, median, upper-hinge, and maximum of a vector.

```
> summary(insult$hbefore)

   Min. 1st Qu.  Median    Mean 3rd Qu.    Max.
   28.0    46.5    51.0    53.6    57.5    86.0

> fivenum(insult$hbefore)

[1] 28.0 46.5 51.0 57.5 86.0
```

Other functions which generate basic sample statistics are *min*, *max*, *range*, *median*, *mean*, *sd*, *var*, *IQR* (the inter-quartile range) and *quantile*. In R the default action to handle vectors with missing data (signified by *NA*) for these functions is not to remove the missing data, therefore when *NA*'s exist, the resultant statistic will also be *NA*. To strip missing values before calculating the statistic, set the argument *na.rm=T*.

```
> c <- c(1:10, NA, 12:20)
> mean(c)

[1] NA

> mean(c, na.rm = T)

[1] 10.5
```

Tables of counts are produced using either the *table* or *xtabs* functions; the latter includes the dimension labels.

```
> table(insult$sex)

female   male
     9     10

> xtabs(~sex, data = insult)

sex
female   male
     9     10
```

The number of non-missing values for the vector entered as the argument are displayed – we see that there are nine subjects with *sex=="female"* and ten subjects with *sex=="male"*. Contingency tables can also be generated using *xtabs* which uses a formula interface to specify the cross-classifying factors, each factor separated by a '+' on the right-hand side of the formula.

To determine how many subjects of each sex have a raised initial hostility score we use the results of a logical statement as one of the arguments of *table*.

```
> with(insult, table(sex, raised = hbefore >= 65))
```

```
        raised
sex      FALSE TRUE
  female     9    0
  male       7    3
```

A table with four counts is produced. Three of the males have raised initial hostility score, and none of the females.

We can generate and assign this table to a dataframe named `cells` by combined use of the *data.frame* and *table* (or *xtabs*) functions.

```
> cells <- data.frame(with(insult, table(sex = sex,
+     raised = hbefore >= 65)))
> str(cells)
```

```
'data.frame':      4 obs. of  3 variables:
 $ sex   : Factor w/ 2 levels "female","male": 1 2 1 2
 $ raised: Factor w/ 2 levels "FALSE","TRUE": 1 1 2 2
 $ Freq  : int  9 7 0 3
```

The dataframe *cells* has 4 rows and 3 columns. The columns are labelled explicitly as *sex* and *raised* and are classed as factors.

Tables of other statistics such as the mean for combinations of factor levels can be produced using the *aggregate* function. For example, to produce a table of means of the variate *hafter* rather than cell counts, we use

```
> with(insult, aggregate(hafter, list(sex = sex,
+     raised = hbefore >= 65), mean, na.rm = T))
```

```
     sex raised    x
1 female  FALSE 61.3
2   male  FALSE 57.3
3   male   TRUE 83.3
```

Since there are no females with raised initial hostility scores, R does not report a mean for this group.

Two-way tables of combinations of summary statistics can be produced using the functions *aggregate.table* and *interleave* available in the *gdata* package (Warnes, 2005). For example, to produce a table of means of the column *hafter*, standard deviations and cell counts cross-classified by sex and raised initial hostility score, we produce the three tables and then interleave them as follows:

```
> library("gdata")
> count.tab <- with(insult, table(sex, raised = hbefore >=
+     65))
> mean.tab <- with(insult, tapply(hafter, list(sex,
+     raised = hbefore >= 65), mean, na.rm = TRUE))
> sd.tab <- with(insult, tapply(hafter, list(sex,
+     raised = hbefore >= 65), sd), na.rm = TRUE)
> interleave(Mean = round(mean.tab, 2), "Std Dev" = round(sd.tab,
+     2), N = count.tab, sep = " ")

                FALSE  TRUE
female Mean      61.3    NA
female Std Dev   15.8    NA
female N          9.0  0.00
male   Mean      57.3 83.33
male   Std Dev   17.9  4.16
male   N          7.0  3.00
```

To produce multidimensional tables in a neat format, use the `ftable` function (see examples in Chapter 5).

1.13 Editing R code

The effort spent on learning to use source editors like TINN-R (`http://www.sciviews.org/Tinn-R/`), R-WinEdt (`http://www.winedt.com/`) or (X) Emacs quickly return benefits to the R user. All three run on the Microsoft Windows operating systems while only (X)Emacs is available for multiple platforms.

To quote the TINN-R web-page

> *'Tinn-R is a small, free and simple, yet efficient, replacement for the basic code editor provided by Rgui or SciViews-R.'*

R-WinEdt is a 'plug in' for the shareware program WinEdt which was specifically written as an editor for the TEX system. Details can be found on the CRAN.

Emacs and XEmacs are large applications that run on many computers and operating systems. Their learning curve is steep but the flexibility of these programs allow for very sophisticated support. The ESS user interface provides a common editing interface for many statistical packages not just R. Emacs with the ESS interface, AUCTEX and the R package *Sweave*, provide a very powerful environment for generating beautifully typeset reports all from the one source file by combining R code with LATEX code. Applications that use the OpenDocument Format such as OpenOffice, have a similar facility with the *odfWeave* package (Kuhn and Weaston, 2007).

1.14 Installing and using packages

There are many purpose-written functions available in R. Most can be found in packages stored at a CRAN site and mirrored worldwide at

```
> read.csv(file.path(R.home(), "doc", "CRAN_mirrors.csv"),
+       as.is = TRUE)[, 1:4]
```

Packages available at CRAN mirror sites can be downloaded by using the `install.packages` function. To install a particular package include the package name as an argument:

```
> install.packages("MASS")
```

These may fail if you do not have write permission to the library location. Try installing these to a directory to which you have write access, and define this as `lib=directory.name` argument in the `install.packages` function where `directory.name` is a character vector giving the library directory. For example, to install the `MASS` package to the Windows `USERPROFILE` directory:

```
> install.packages("MASS", lib = Sys.getenv("USERPROFILE"))
```

Since we have not specified the source repository, the user will be asked to choose a site. If no package is specified, a list of available packages names will be presented for the user to choose from.

Once installed the package is loaded ready for use by

```
> library(MASS)
```

All functions and data-sets supplied with the package are then available to the user.

All datasets and functions defined in this monograph are available from CRAN in the package `SMIR`.

2
Statistical modelling and inference

2.1 Statistical models

In this book, we will be concerned with the statistical analysis of data from experimental and observational studies. In experimental studies randomization or random assignment of individual experimental units (e.g. human or animal subjects, agricultural plots) to the experimental treatments plays a fundamentally important role.

The experimental units may themselves be sampled randomly according to some sample design from a larger population, in which case conclusions from the experiment can be drawn about the larger population, or they may be all that is available for the experiment (e.g. all patients with a particular disease being assigned to treatments in a hospital clinical trial), in which case conclusions may not be generalizable to a broader population.

In observational studies, observational units are drawn from a population according to some random sample design (which may be the complete examination or *census* of the entire population), and conclusions are to be drawn about the complete population.

In both kinds of study, the random selection or allocation of observations is critical to the analysis and interpretation of the data. Studies which are not based on sample or experimental designs, that is, in which the data collected are accidental, or are subjectively chosen, are particularly hazardous to interpret because of the possibility that inclusion in the study is systematically related to important variables, so that the data are not representative of the population about which conclusions are to be drawn.

The object of statistical modelling is to present a simplified or smoothed representation of the underlying population. This is done by separating systematic features of the data (e.g. sex differences in height or weight) from random variation (natural variability in height or weight within sex). The systematic features are represented by a regression function involving parameters which can be simply related to the structure of the population and to important variables measured on each experimental or observational unit. The random variation unrelated to important variables is represented by a probability distribution depending on a small number of parameters, typically one or two. Interpretation of the data, and conclusions about the population, can then be based on the regression function, with no attempt to interpret random variation.

How do we distinguish between random and systematic variation? This is a familiar problem in hypothesis testing: we suppose first that variation is random, and then test whether the introduction of a systematic effect into the model improves the model, in a well-defined way. If the improvement is substantial, the systematic effect is retained and can be interpreted: if it is only minor, the systematic effect is not retained and is not interpreted. How large an improvement is substantial is partly a subjective matter, but well-established rules in hypothesis testing provide a good guide. The importance of hypothesis testing results from a general philosophical approach to scientific inference by statisticians and other scientists, based on the principle of parsimony, known historically as Occam's Razor (from William of Occam or William Ockham, 1280–1349): 'entia non sunt multiplicanda praeter necessitatem' – entities should not be needlessly multiplied. Systematic effects should be included in a model only if there is convincing evidence of the need for them: we should not spend time and effort interpreting effects which could just as well be random variation. The principle of parsimony provides an important guide to model simplification: though we may use very complex models for complex sample survey designs, our aim is always to simplify the model as much as possible, while remaining consistent with the observed data.

An important distinction needs to be made between the use of models for *representation* of an existing population, as we have just described, and their use for *prediction* of values of variables for future observations. These uses of models are often confused. The use of models for prediction will be considered in Chapter 3.

It is assumed in subsequent discussions that the data being modelled consist of a random sample of some kind from a population. In some cases, as noted earlier, the data available are a complete enumeration of a population. In such cases the simplification of models using hypothesis testing seems quite artificial, because no sampling is involved, and we already know whether or not the hypothesis is true.

There are two possible approaches to the statistical modelling of complete populations. One is to regard the population as a sample from a random process generating this population, and to model the population as though it were a sample. Such an approach is called *super-population* modelling (Smith, 1976) although no concept of sampling the observed population from a hypothetical superpopulation of populations is actually necessary.

The other approach, which we shall follow, is to regard models simply as smooth approximations to the rough, irregular complete population. The irregular variations about the smooth structure are treated as non-systematic variation to be ignored. The degree to which the population can be smoothed is now entirely subjective, since the formal rules of hypothesis testing are not appropriate, but smoothing to the extent appropriate if we had a sample, and not a population, will often be reasonable unless the population size is small.

2.2 Types of variables

The data for which models can be constructed in R consist of values of a *response variable* which we will denote by y, and a set of *explanatory* or *predictor* variables $x_1, x_2, \ldots, x_p$, which are assumed to be measured without error and recorded without missing values. These important restrictions are discussed in Chapter 3 and other chapters, and are removed to some extent there. The terms 'dependent' and 'independent' variables which are often used are confusing (since the 'independent' variables are not statistically independent) and will not be used in this book. The response and explanatory variables can take a wide range of forms, set out in Table 2.1.

Discussions of measurement in the social sciences frequently use Stevens's classification of nominal, ordinal, interval and ratio scales (Stevens, 1946). Nominal scales correspond to the first two variable types above and ordinal to the third. The distinction between interval and ratio scales (the latter has a fixed zero, the former does not) is not very useful, since it is often not relevant to the kind of analysis which is appropriate.

Quotation marks are used around 'continuous' above because all such variables are in practice measured to a finite precision, so they are actually discrete variables with a large number of numerical values. For example, height may be measured to the nearest half-inch or centimetre, survival time to the nearest day, week or month, and Stanford–Binet IQ may be given as an integer though it is defined as the ratio of mental age to chronological age multiplied by 100. Models for 'continuous' data usually ignore this measurement precision, but this leads to unnecessary difficulties

Table 2.1. Types of variables

Variable type	Examples
Categorical, two categories (binary)	Male/female, alive/dead, presence/absence of a disease, inoculated/not inoculated
Categorical, more than two unordered categories	Blood group, cause of death, type of cancer, political party vote, religious affiliation
Categorical with ordered categories	Severity of symptoms or illness, strength of agreement or disagreement, class of University degree
Discrete count	Number of children in a family, number of accidents at an intersection, number of ship collisions in a year
'Continuous'	Height, weight, response time, survival time, Stanford–Binet IQ

with statistical theory. Our formulation of models explicitly recognizes this aspect of 'continuous' data.

2.3 Population models

In this book, we use probability models as simplifying approximations to real or conceptual populations of response variables. In the experimental or survey study of the population, we measure or observe a sample of individuals, the value of the quantity or property measured for the i-th individual being denoted by the variable value $y_i, i = 1, \ldots, n$. In the population we have the set of values Y_J^*, with $J = 1, \ldots, N$, where N is the population size. These are not in general all observed, though we may be able to take a census of the whole population in some cases.

Because of the limited precision of any physical measurement, or because of natural discreteness in counting measure, the Y_J^* are generally not all distinct, and we represent them alternatively in terms of the D *distinct* values Y_I which occur with multiplicity N_I, for $I = 1, \ldots, D$, with $\sum_{I=1}^{D} N_I = N$. We include all possible distinct values Y even if they are not observed; these values are given multiplicity zero.

When the quantity Y is measured on an *ordered* scale, we take the Y_I to be in *increasing order* with I.

The values $P_I = N_I/N$ are the *relative frequencies* or *proportions* of the distinct values Y_I. A graph of the P_I against the Y_I is a *proportion graph*, and one of the N_I against the Y_I, is a *frequency graph*. These graphs differ only in the vertical scale.

More common in introductory courses is the *histogram*, which gives a solid bar plot of the frequencies N_I against the Y_I; these are almost invariably *grouped* into class intervals of Y, either by default or by the user specification.

For example, Fig. 2.1 shows the histogram of the counts of birthweights of 648 girls (part of the StatLab population of Hodges *et al.*, 1975) against birthweight, using the function defaults. Weight is measured to the nearest 0.1 pound (one pound = 0.4536 kg), so that a recorded value of 6.4 pounds corresponds to an actual value between 6.35 and 6.45 pounds.

The R commands are

```
> girls <- subset(statlab, sex == "girl")
> library(lattice)
> print(histogram(~c.b.wgt, data = girls,
+     ylab = "Freq/0.6 units of birthweight"))
```

The default grouping gives 10 class intervals. Smoothing using class intervals results in a loss of information or 'resolution'. Minimal smoothing results from choosing the interval width equal to the measurement precision of 0.1 pound. Inspection of the data, for example, using

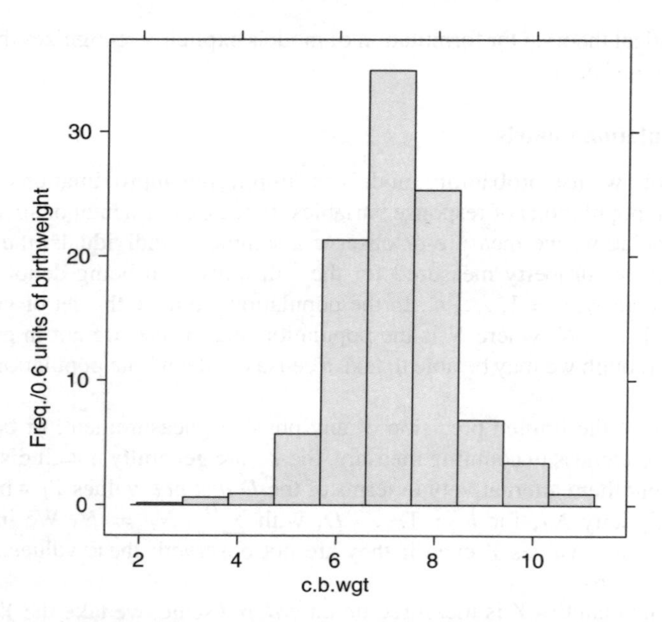

Fig. 2.1. The `lattice` package default birthweight histogram, girls

```
> summary(girls$c.b.wgt)

  Min.  1st Qu.  Median    Mean 3rd Qu.    Max.
 2.3000  6.6000  7.2000  7.2231  7.9000 10.9000
```

shows that the range of girl birthweights is from 2.3 to 10.9 pounds. To construct the histogram over a slightly larger range with defined class widths the `break` argument needs to be defined.

```
> print(histogram(~c.b.wgt, data = girls, breaks = seq(2,
+     12, by = 0.1), xlab = "Birth weight in pounds",
+     main = "Histogram of birthweight of girls"))
```

The histogram shown in Fig. 2.2 is now very irregular, though the two histograms suggest a generally symmetrical population model like the normal. The direct modelling of the frequencies to identify a suitable population model was proposed by Lindsey (1974) and Lindsey and Mersch (1992), and a simple example was given in Aitkin (1995). The modelling approach we follow in this book is to construct the *cumulative proportion* graph, in which we plot the cumulative proportions

$$C_I = \sum_{I' \leq I} P_{I'} = \sum_{I' \leq I} N_{I'}/N$$

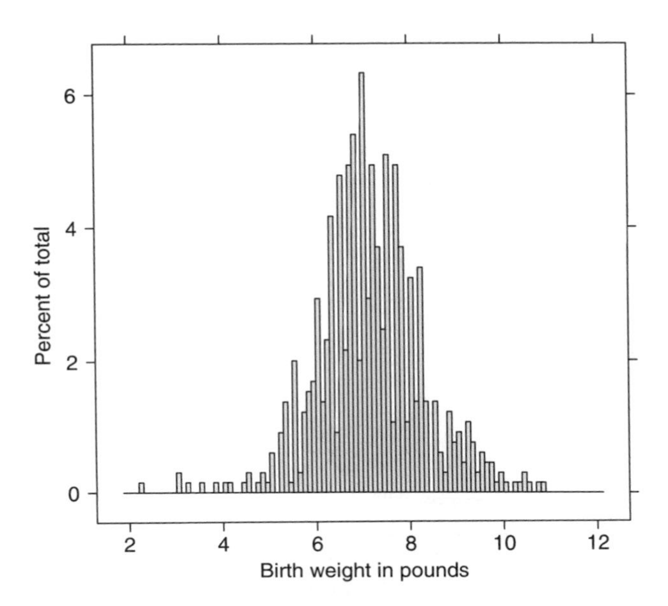

Fig. 2.2. Histogram of the birth weights of girls in the STATLAB population

against the Y_I. We will in fact change slightly this definition, and define

$$C_I = \sum_{I' \leq I} N_{I'}/(N+1).$$

The division by $N+1$ instead of N is to prevent the last value being 1.0, which causes difficulty in transformations of the probability scale. The difference in the definitions is at most $1/(N+1)$, which for large N is negligible. (Other definitions are commonly used also, like

$$C_I = \left(\sum_{I' \leq I} N_{I'} - 1/2 \right) \Big/ N.$$

These give very similar results.)

Figure 2.3 shows the cumulative proportions plotted against birthweight. We first save the counts using the `table` function.

```
> n <- with(girls, table(c.b.wgt))
> y <- as.numeric(names(n))
> cn <- cumsum(n)
> cn <- cn/(sum(n) + 1)
> print(xyplot(cn ~ y, type = "p", xlab = "birthweight",
+      ylab = "cumulative proportion"))
```

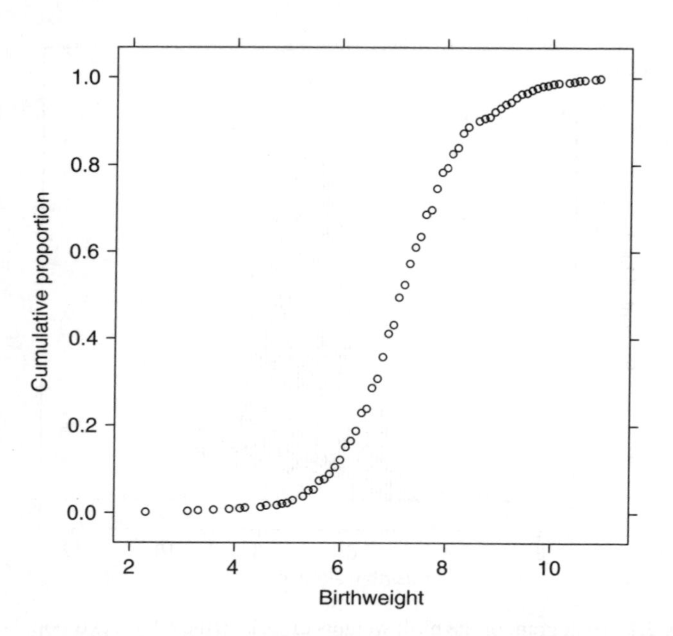

Fig. 2.3. Birthweight cumulative proportion, girls

The roughness of the maximum resolution histogram is greatly reduced.

The theoretical or experimental questions we have about the population values Y_I are generally expressed in terms of *parameters*: means or medians, or variances or standard deviations, rather than in terms of the individual proportions P_I themselves. In sample surveys of consumer purchasing, for example, we may be interested in inference about the average family income, which is

$$\mu = \sum_{I=1}^{D} P_I Y_I$$

for the population including all Y_I, not just those with $N_I > 0$.

The role of *statistical models* is to aid this process of inference about parameters like μ by making a *simplifying assumption* about the proportions P_I. Specifically, we assume that the cumulative proportions C_I at Y_I can be represented approximately by a *smooth function* $F(Y_I)$, the *cumulative distribution function* (*cdf*) of a random variable Y:

$$C_I \doteq F(Y_I).$$

A question of immediate interest is whether this approximating process is *necessary* – can inferences about parameters like μ be made *without* modelling or approximating the C_I? Remarkably, likelihood inferences (defined in Section 2.5) *can* be made without such approximations, through *empirical likelihood* (Owen, 1988, 2001); this is described briefly in Section 2.10.2. We concentrate in this book on explicit models for the C_I or P_I.

In general, the approximating *cdf* F will depend on one or more parameters $\boldsymbol{\theta}$ which specify the location and variation of the random variable Y. If the quantity y could in theory be measured on a continuous scale but is recorded only to a precision of δ, then we can represent the distribution of Y equivalently by its *probability density function*, which we write for a general argument y as $f(y)$:

$$f(y) = \frac{\mathrm{d}F(y)}{\mathrm{d}y} .$$

The proportions P_I can then be represented approximately by

$$P_I \doteq F(Y_I + \delta/2) - F(Y_I - \delta/2) \doteq \delta f(Y_I).$$

If y is measured on an inherently discrete scale (e.g. the non-negative integers) then the distribution of Y can be represented equivalently by its *probability mass function*

$$f(y) = F(y) - F(y - \delta) = F(y + \delta/2) - F(y - \delta/2),$$

where δ is the discrete measurement scale unit (1 if y is defined on the integers). The proportions P_I are then represented approximately by $P_I \doteq f(Y_I)$. (The second equality above may seem unusual – it is adopted for consistency of notation with the continuous case.)

The choice of the approximating model is based on the closeness of agreement between the population proportions C_I and the approximating model proportions $F(Y_I)$. The appearance of the histogram for the girl birthweights, and the general symmetry of the cumulative proportion graph, suggest a normal distribution model for the C_I. Figure 2.4 shows the normal *cdf* $F(y)$ for the normal distribution with mean 7.22 and standard deviation 1.14, the values for this variable in the population (continuous curve), superimposed on the proportions C_I. The fitted normal *cdf* is defined by the R commands

```
> curve(pnorm(x, 7.22, 1.14), from = 2.3, to = 10.9,
+     xlab = "birthweight", ylab = "cumulative proportion",
+     las = 1)
> points(y, cn)
```

Because of the non-linearity and the upper and lower bounds of the *cdf*, it is difficult to assess the agreement between C_I and $F(Y_I)$. It is therefore convenient to transform the *cdf* graph by adopting standard probability plotting procedures familiar from residual analysis in normal regression. For a continuous

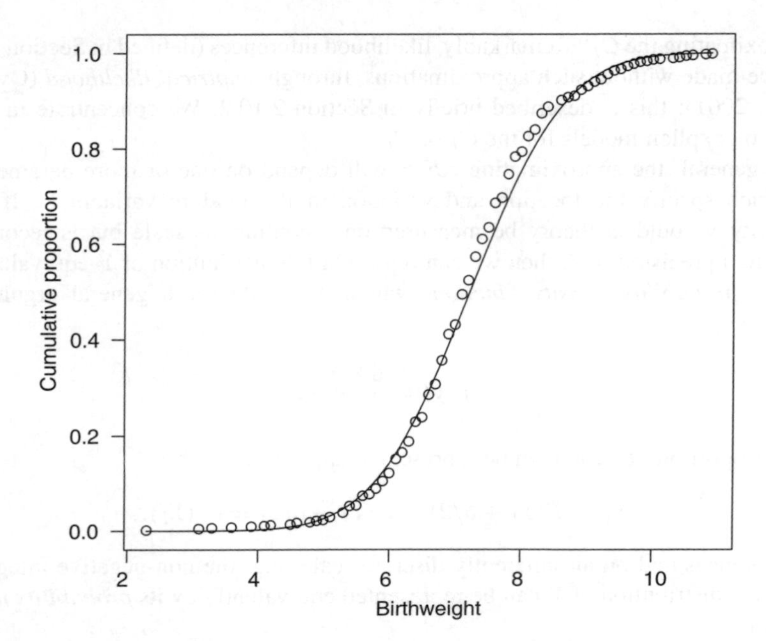

Fig. 2.4. Birthweight cumulative proportion, girls

approximating Model $f(y)$ we look for scales for F and Y on which the plot is approximately linear. We try to transform the scales so that $g(F)$ is linear in $h(Y) : g(F)$ is generally F^{-1}, so that $g(F)$ are the *quantiles* of the approximating distribution F. For a discrete model $f(y)$ such procedures are less satisfactory, but for both continuous and discrete models the goodness-of-fit of the model to the population can be assessed by further modelling.

We try the *inverse normal cdf* scale, or the *equivalent normal deviate* scale, for F, and leave Y unchanged. Figure 2.5 shows the cumulative proportion graph of girl birthweights using the inverse normal *cdf* scale for the proportion, together with the straight line corresponding to the mean 7.22 and standard deviation 1.14 of this variable. The R commands are

```
> curve(qnorm(pnorm(x, 7.22, 1.14)), from = 2.3, to = 10.9,
+     xlab = "birthweight", ylab = "normal deviate",
+     lwd = 2)
> points(y, qnorm(cn), lwd = 2)
```

The plot is generally linear, supporting a normal distribution model for birthweight, but there are notable departures from the line for both small and large babies. An acute difficulty in the inspection of such plots is how to decide

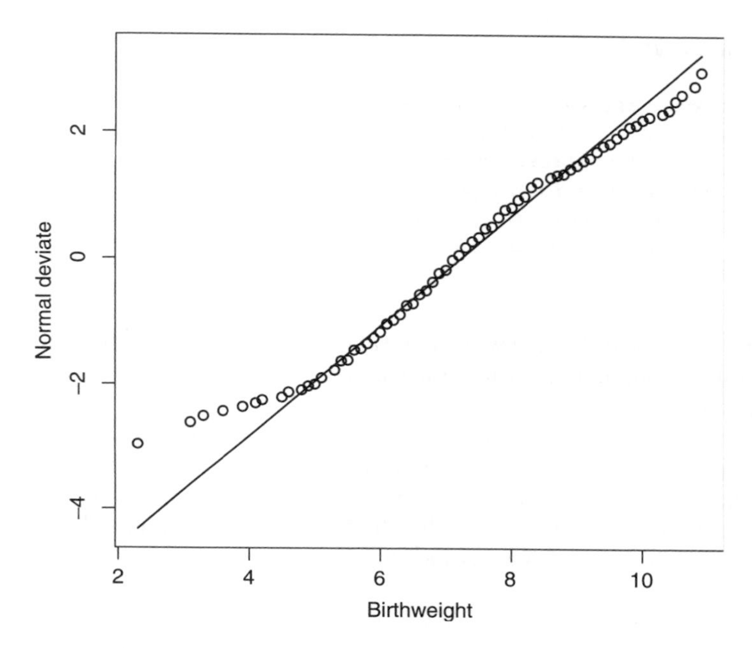

Fig. 2.5. Normal deviate of cumulative proportion, girls

whether the variation in the plot is too far from a straight line. We are concerned in this section with models for complete populations, but we follow the approach of treating the population as though it were a sample; this may be formalized as a superpopulation model. We will be applying the same approach to samples, and it is especially easy with small samples to over-interpret sampling variation as real model failure. Simulation of samples from normal population models gives some feeling for the behaviour of such plots in small samples, and *simulation envelopes* (Atkinson, 1985) formalize this variation.

It is useful to have a visual statistical bound on the reasonable sampling variation of the plot. This can be obtained from the *binomial distribution*, and provides a simultaneous confidence region or band for the population proportions (Owen, 1995). The band construction is discussed briefly in Section 2.7.5.

The band is constructed in the function `NPL.bands`, which requires the vector of data values as an argument. The function computes upper and lower endpoints of the 95% simultaneous confidence band for each Y; these vectors can then be graphed to give the confidence band, together with the fitted normal *cdf*. The simultaneous band is wider than that based on *pointwise* 95% confidence intervals at each Y.

```
> library(SMIR)

> print(xyplot(lower + upper ~ x, data = NPL.bands(girls$c.b.wgt),
+      pch = 20, xlab = "birthweight",
+      ylab = "cumulative proportion",
+      panel = function(x, y, ...) {
+          panel.xyplot(x, y, ...)
+          panel.curve(pnorm(x, 7.22, 1.14), lwd = 1.5)
+      }))
```

Figure 2.6 repeats Fig. 2.4 with the 95% confidence band instead of the observed proportion, using dot characters for the band, with the fitted normal `cdf` as a continuous curve.

The band provides a clear picture of the sampling variability in the *cdf* to be expected for the given sample size. Systematic crossings of the band by the model *cdf* indicate failure of the model to represent adequately the data: the cumulative proportions from the fitted model for the population proportions should all fall inside this band for a satisfactory model.

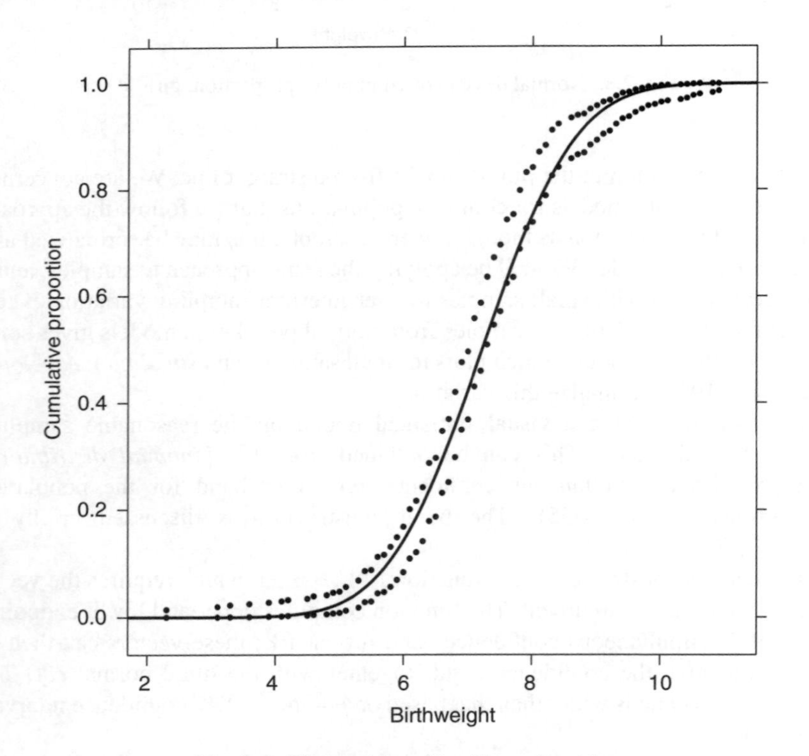

Fig. 2.6. Birthweight *cdf* and bounds, girls

It is difficult to read from the graph whether points fall outside the band. Since we are generally working on a transformed scale (the inverse normal scale in this example), it is convenient to transform the band endpoints in the same way to corresponding band endpoints on the transformed scale. At the extreme values of the *cdf* and with large N the confidence band may reach the limits 0 and 1.

```
> print(xyplot(qnorm(lower) + qnorm(upper) ~ x,
+     data = NPL.bands(girls\$c.b.wgt), pch = 20,
+     xlab = "Birth weight (pounds)", ylab = "normal deviate",
+     panel = function(x, y, ...) {
+         panel.xyplot(x, y, ...)
+         panel.curve(qnorm(pnorm(x, 7.22, 1.14)),
+             lwd = 1.5)
+     }))
```

Figure 2.7 repeats Fig. 2.6 with the corresponding transformed band.

It is now clear that the fluctuations in the *cdf* are not consistent with a normal model – the apparent departures from linearity for small babies are outside the band range, and those for babies around 8 pounds nearly outside it. The normal

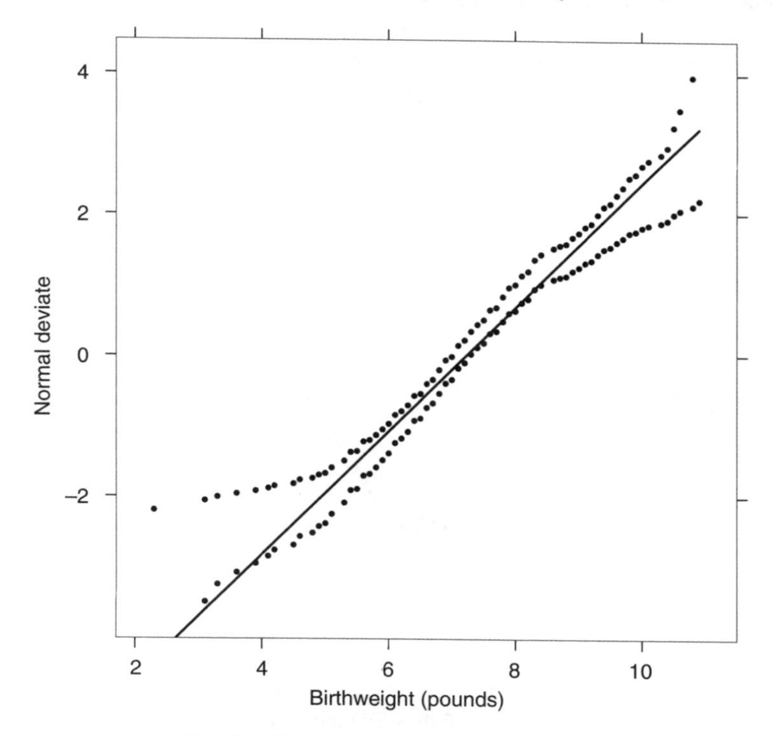

Fig. 2.7. Normal deviate and bounds, girls

population model is an unsatisfactory fit: there are more large and small babies than are consistent with a normal population model.

What alternative model or models should we consider? A model with heavier 'tails' is needed; the central part of the confidence band appears consistent with a normal model, but the tails are not consistent with the same model, though they appear to be consistent with other normal models. In Chapter 7 we discuss *mixtures* of distributions; here we consider the *mixture of normals* distribution as a possible model.

Figure 2.8 shows a plot on the inverse normal *cdf* scale of the *cdf* of a three-component normal mixture model, with means 3.89, 7.08, and 9.13 pounds and common standard deviation 0.865 pounds, in corresponding proportions 0.017, 0.886, and 0.097, together with the 95% confidence band. (The mixture distributions described here are fitted by maximum likelihood. See Chapter 7 for a full discussion.)

The mixture *cdf* follows the population cumulative proportions quite closely, the fitted mixture *cdf* lying fully inside the band, so the normal mixture distribution provides a good model, with the appealing interpretation of a 'normal' sub-population of about 89% with mean 7.1 pounds, a 'high birthweight' sub-population of about 10% with mean 9.1 pounds, and a 'low birthweight' sub-population of about 2% with mean 3.9 pounds. This population representation does not take account of possible explanatory factors like the mother's age and

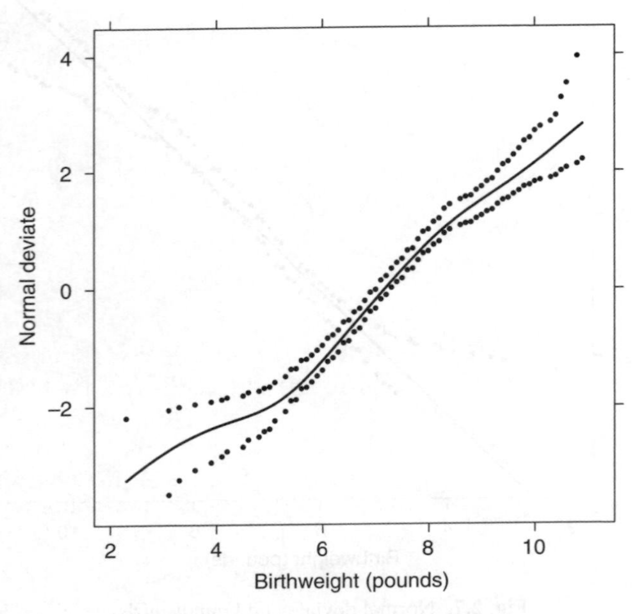

Fig. 2.8. Mixture cdf and bounds, girls

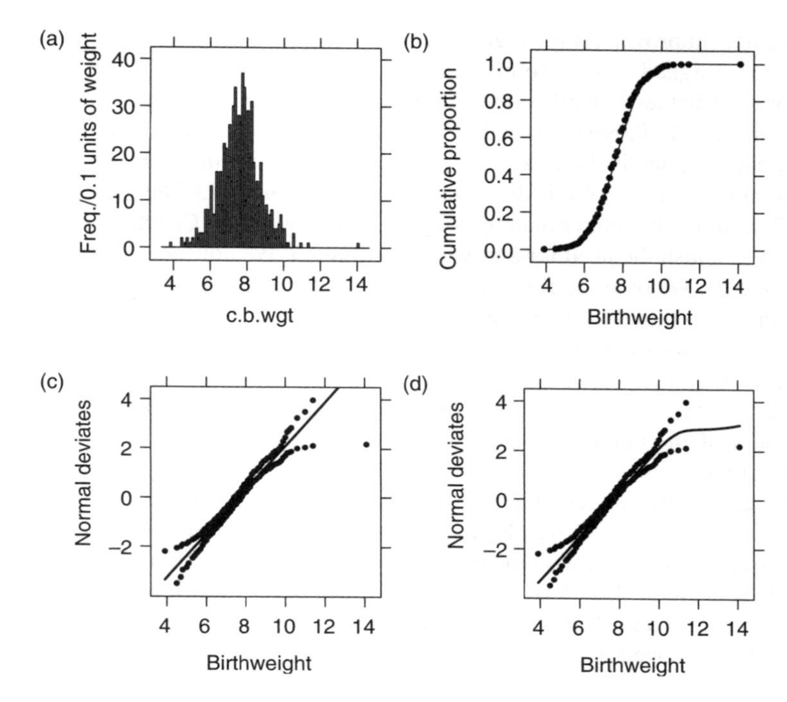

Fig. 2.9. (a) Maximum resolution birthweight histogram, boys; (b) birthweight cumulative proportion, boys; (c) birthweight cumulative proportion, boys; (d) mixture cdf and bounds, boys

weight, which when modelled might explain the low and high birthweight groups. Such investigations are the subject of regression analysis or linear modelling; in Chapter 7 we examine these variables.

We now examine the birthweights of the 648 boys in the STATLAB population. We do not repeat the R commands. Figure 2.9(a) shows the maximum resolution birthweight histogram; the range of boy birthweights is 3.9–14.1 pounds. The cumulative proportion graph is shown in Fig. 2.9(b) with the superimposed normal *cdf* with mean 7.65 and standard deviation 1.12, the values for birthweight in the boy population. The fitted normal model is shown in Fig. 2.9(c) on the inverse normal *cdf* scale with the 95% confidence band. Figure 2.9(c) shows some departures from the straight line for large babies, with one extreme outlier having weight 14.1 pounds, though all points except the extreme one fall inside the band.

Figure 2.9(d) shows a three-component normal mixture distribution with means 14.0, 7.7, and 6.9 pounds and common standard deviation 1.08 pounds in corresponding proportions 0.002, 0.939, and 0.059, on the inverse normal *cdf* scale, together with the 95% band.

The fit is improved for the high birthweight babies, mainly because the extreme outlier is assigned its own component; note that $1/648 = 0.0015$. The fit for low birthweight babies is hardly changed, because the two lower components differ in mean by only 0.69 standard deviations, and so are not really distinguishable: the model is effectively a two-component mixture with one component for the single outlier. (This point is discussed further in Chapter 7.) The normal mixture distribution is little better than the single normal model. In Chapter 7 we discuss how to establish the need for a mixture distribution. We now consider models for *skewed* data. We omit the R commands which parallel those above.

Figure 2.10 shows the maximum resolution histogram of weights of the mothers (at the diagnosis of pregnancy) of the 648 boys in the STATLAB population.

The histogram is skewed to the right. The cumulative proportion graph is shown in Fig. 2.11(a). The roughness of the histogram is again greatly reduced, and skewness of the cumulative proportions is visible as asymmetry, with several large values.

In Fig. 2.11(b) we graph the 95% simultaneous confidence band on the inverse normal *cdf* scale against weight, together with the straight line corresponding to the normal distribution with mean 143.2 and standard deviation 27.8, the values for this population. The fit is very bad, with much of the fitted model lying outside the confidence band. The consistent *single curvature* of the *cdf* points strongly to a positively skewed population model. A log transformation of the weight scale

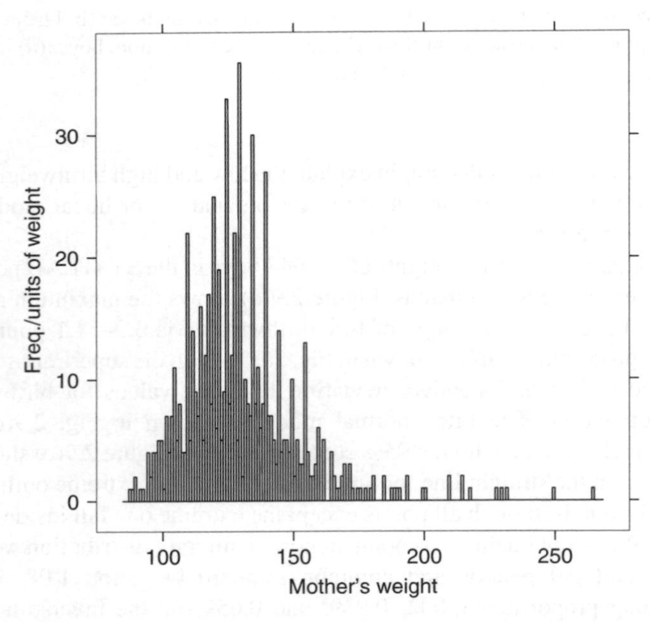

Fig. 2.10. Histogram of mother's weight at diagnosis of pregnancy, boys

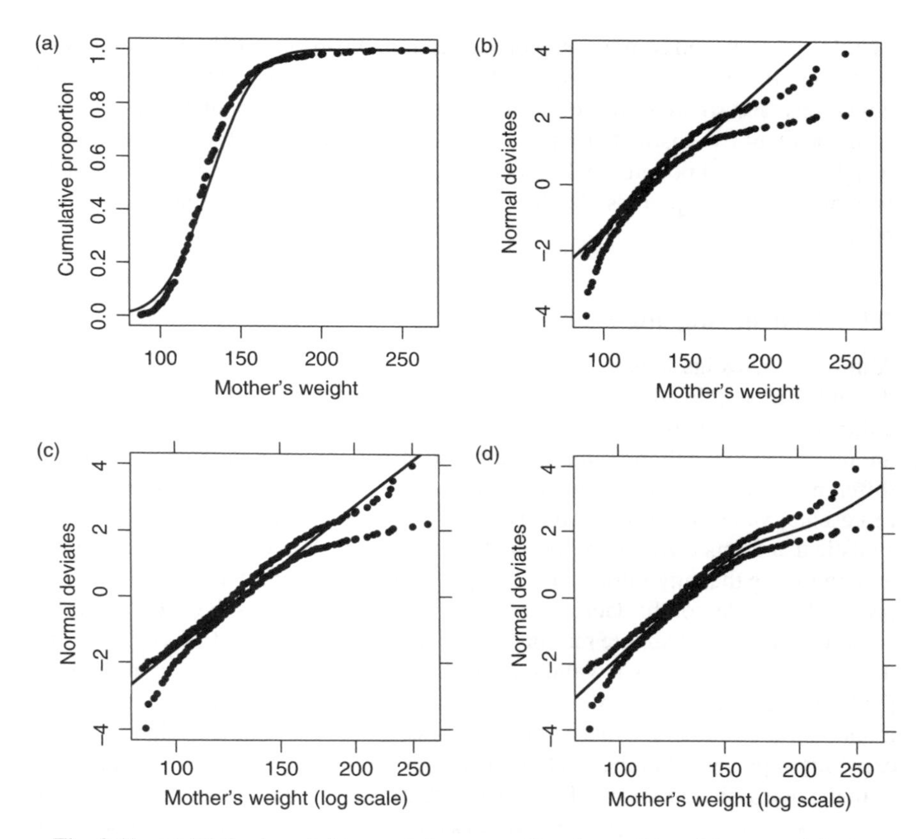

Fig. 2.11. (a) Mother's weight cumulative proportion, boys; (b) weight normal deviate and bounds; (c) log weight normal deviate and bounds; (d) normal mixture and bounds, log weight

before fitting the normal model is a common method for reducing the skew; this is equivalent to using a *lognormal* population model.

In Fig. 2.11(c) we graph the 95% confidence band on the inverse normal *cdf* scale against log weight with the fitted (log)normal model. The fit is improved but the lognormal model still falls outside the band at the upper end. The central region of the fitted model is nearly straight but the upper tail is not: the lognormal distribution is not a suitable model. The change in slope of the upper end of the graph suggests a mixture distribution.

In Fig. 2.11(d) we graph the *cdf* of a two-component mixture of lognormals on the inverse normal scale, with the 95% confidence band.

The fitted distribution lies near the centre of the band over the whole range, and provides a good model. The two components have means (on the log weight scale) of 4.84 and 5.29, and common standard deviation 0.136, and are in the proportions 0.962 and 0.038. The two components are separated by more than two standard deviations, and so are nearly distinct. The mothers' weight population appears to be made up of two distinct groups, a 'normal' subpopulation of 96% with median weight 128 ($e^{4.85}$) pounds, and a 'high weight' subpopulation of 4% with median weight 198 ($e^{5.29}$) pounds. Each of these groups has a lognormal distribution of weights.

2.4 Random sampling

Almost all investigations in which statistical modelling plays a fundamental role depend on a sampling design of some kind to obtain data from the target population, either by a formal process of random selection from a population list, or by an act of randomization in the assignment of patients or experimental units to different treatments. The purpose of the random selection or assignment is to ensure representativeness of the population by the sample as far as possible. In developing statistical inference about the population from the observed sample, a critical assumption is that any informative sample design is explicitly represented in the probability model for the data, and hence in the likelihood. Thus if observations on some sample individuals are missing, the analysis of the observed data by ignoring the missing observations is valid only if the missing data are missing at random [MAR in the notation of Little and Rubin (1987)] and the 'missingness process' is ignorable. If not, then the missingness process, that is the sample design, must be explicitly represented in the likelihood. This is discussed further in Chapter 3.

In the discussion in this book, simple or stratified random sampling will almost always be assumed, except in the discussion of variance component models in Chapter 9, where multi-stage sampling is assumed. In the remainder of this chapter we consider only simple random sampling; stratified sampling is discussed in Section 3.3.

2.5 The likelihood function

We now adopt a consistent notation for the rest of this book. The variable for which a probability model is to be adopted will be denoted by Y, with generic observed values y. These are measured to a finite measurement precision δ, so every possible observed y has a probability modelled by

$$\begin{aligned} p(y) &= \Pr(y - \delta/2 < Y < y + \delta/2) \\ &= F(y + \delta/2 \mid \boldsymbol{\theta}) - F(y - \delta/2 \mid \boldsymbol{\theta}), \end{aligned}$$

where $F(y \mid \boldsymbol{\theta})$ is the approximating *cdf* corresponding to the density or mass function $f(y \mid \boldsymbol{\theta})$ depending on the parameter $\boldsymbol{\theta}$. If Y is inherently discrete (e.g. defined on the non-negative integers) then

$$p(y) = f(y \mid \boldsymbol{\theta}),$$

while if Y is theoretically continuous then if the measurement precision is high (δ small compared with the effective range of Y),

$$p(y) \doteq \delta f(y \mid \boldsymbol{\theta}).$$

If the measurement precision is *not* high, for example when Y is heavily grouped, the previous *cdf* definition must be retained.

Observed sample values will be denoted by $y_1, \ldots, y_n$, and when necessary, unobserved random variables corresponding to sample values will be denoted by $Y_1, \ldots, Y_n$. We assume unless otherwise stated that the population size N is large compared to the sample size n. (The consequences of a small population size are discussed later in this chapter.) We may then treat the sample $y_1, \ldots, y_n$ as having been drawn randomly with replacement. The likelihood function is now defined as the probability of obtaining the actual sample values $\mathbf{y} = y_1, \ldots, y_n$, under the approximating model F:

$$\begin{aligned} L(\boldsymbol{\theta} \mid \mathbf{y}) &= \Pr(y_1, \ldots, y_n \mid \boldsymbol{\theta}) \\ &= \prod_{i=1}^{n} p(y_i) \\ &= \prod f(y_i \mid \boldsymbol{\theta}), \quad Y \text{ discrete} \\ &= \prod [F(y_i + \delta/2 \mid \boldsymbol{\theta}) - F(y_i - \delta/2 \mid \boldsymbol{\theta})], \quad Y \text{continuous}, \quad \delta \text{ large} \\ &\doteq \delta^n \prod f(y_i \mid \boldsymbol{\theta}), \quad Y \text{ continuous}, \quad \delta \text{ small}. \end{aligned}$$

In definitions of the likelihood function for continuous variables, the constant δ^n (or the equivalent 'differential element' $dy_1, \ldots, dy_n$) is usually dropped, and the likelihood function is defined as any function proportional to $\prod f(y_i \mid \boldsymbol{\theta})$. When we are comparing different continuous probability models for the same data, the constant δ^n always cancels and so can be omitted, but other constants in each model do not cancel in general, and must be retained in the likelihood. We will generally drop the $|\mathbf{y}$ in the likelihood $L(\boldsymbol{\theta} \mid \mathbf{y})$ for $\boldsymbol{\theta}$, but it should be remembered that in this book the likelihood is always defined for *observed* values: the y_i are not regarded as random variables but as observed numerical values. When we want to consider the repeated sampling properties of terms in the likelihood, the algebraic values y_i (or other functions of the data) will be replaced by capital letters Y_i to represent the unobserved random variables.

2.6 Inference for single parameter models

The likelihood function $L(\theta)$ is of fundamental importance in statistical inference, whether from a Neyman–Pearson, Bayes or pure likelihood perspective. Although there is a lively controversy over the role of models, and of likelihood, in survey sampling theory, we take the use of and reliance on models as axiomatic. We make some comments on non-model based methods in the discussion of empirical likelihood in Section 2.10.2.

We will be principally concerned with models in the *exponential family* and mixtures of them. The exponential family includes the binomial, Poisson, normal, gamma and inverse Gaussian distributions; the last of these is more specialized than the others and we do not discuss it in this book.

We consider the general properties of this family in Section 2.9, but discuss first the problems of inference in the simplest models. In our approach we follow closely the 'direct' or 'pure' likelihood theory of Barnard (1949), Birnbaum (1962), and Barnard *et al.* (1962), described at book length in Edwards (1972), Clayton and Hills (1994), Lindsey (1996), Royall (1997) and Sprott (2000). The Likelihood Principle book of Berger and Wolpert (1986) discusses many of the same issues from a Bayesian viewpoint. We give the very close connections between likelihood theory and conventional repeated-sampling (Neyman–Pearson) theory in many models. We give a short discussion of Bayes theory; this provides a unification of inferential approaches through the posterior distribution of the likelihood ratio between models (Dempster, 1974, 1997; Aitkin, 1997; Chadwick, 2002; Aitkin *et al.*, 2005), which gives a simple calibration for nested model comparisons, and for many of the non-nested and common-family model comparisons needed in later chapters.

We consider in this section single parameter models, and begin with the simplest normal model.

2.6.0.1 *Normal mean model*

A simple random sample $y_1, \ldots, y_n$ of size $n = 25$ is drawn from a population in which Y has a normal model $N(\mu, \sigma^2)$ with $\sigma^2 = 1$ known, and yields the value $\bar{y} = 0.4$. We will consider three inferential questions:

(i) Two different substantive hypotheses H_1 and H_2 specify values for μ, namely $H_1 : \mu = \mu_1 = 0$ and $H_2 : \mu = \mu_2 = 1$. What evidence does the sample provide about hypothesis H_1 relative to hypothesis H_2?

(ii) In the absence of any substantive hypothesis about the value of μ, what information does the sample provide about it?

(iii) A substantive null hypothesis H_0 specifies $\mu = \mu_0$; the alternative is the general H_1 that $\mu \neq \mu_0$. What evidence does the sample provide about the hypothesis H_0 relative to the hypothesis H_1?

These three questions are simple examples of 'hypothesis testing' and 'estimation' problems.

The likelihood function is

$$
\begin{aligned}
L(\mu) &= \prod_{i=1}^{n} \Big[F(y_i + \delta/2 \mid \mu) - F(y_i - \delta/2 \mid \mu) \Big] \\
&\doteq \prod_{i=1}^{n} \delta f(y_i \mid \mu) \\
&= \prod_{i=1}^{n} \delta \cdot \frac{1}{\sqrt{2\pi}\sigma} \exp\left\{ -\frac{1}{2\sigma^2} (y_i - \mu)^2 \right\} \\
&= \frac{\delta^n}{(\sqrt{2\pi})^n \sigma^n} \exp\left\{ -\frac{1}{2\sigma^2} \Big[\sum (y_i - \bar{y})^2 + n(\bar{y} - \mu)^2 \Big] \right\}.
\end{aligned}
$$

In this expression, δ, σ, and $\sqrt{2\pi}$ are known constants (though the value of δ was not specified in the question), and $\bar{y}$ and $\sum (y_i - \bar{y})^2$ are known once the data are observed.

In likelihood theory we interpret the likelihood $L(\theta)$ directly as the evidence (function), given the model and the data y, for different values of the parameter θ.

2.6.1 *Comparing two simple hypotheses*

To apply the theory to question (i), we assess the relative evidence for the hypothesis H_1 compared with the hypothesis H_2 by computing the *likelihood ratio* (LR) $= L(\mu_1)/L(\mu_2)$. If this ratio is small, substantially smaller than 1, then the data provide strong evidence against H_1 in favour of H_2. A *large* value of this ratio correspondingly provides strong evidence in *favour* of H_1 and against H_2. A value of 1 for the likelihood ratio means that both hypotheses are equally well supported by the data. An important point in this analysis is that we do not have to specify a 'null' hypothesis: the two hypotheses are treated symmetrically – if we change the labels of H_1 and H_2, the conclusions about the parameter values are unaffected.

In computing the likelihood ratio between two hypotheses, we can omit from the likelihood any constants which take the same value under both hypotheses, since these constants cancel in the ratio. So we may as well begin with the *kernel* of the likelihood, in which these constants are omitted. Note that it is not necessary to know the measurement precision δ to compute the likelihood ratio (though it *is* necessary to know that the measurement precision is sufficient for the likelihood representation by the normal density, as previously discussed).

For our normal model, the kernel of the likelihood is

$$K(\mu) = \exp\left\{-\frac{1}{2\sigma^2}\left[n(\bar{y}-\mu)^2\right]\right\}.$$

Evaluating at $\bar{y} = 0.4$, $\sigma = 1$, we have LR $= L(0)/L(1) = K(0)/K(1) = 12.18$. An equivalent comparison, of which we will make much use in later chapters, is through the value of the *likelihood ratio test statistic* (LRTS)

$$\begin{aligned}-2\log \mathrm{LR} = -2\log\left[L(\mu_1)/L(\mu_2)\right] &= \frac{n(\bar{y}-\mu_1)^2}{\sigma^2} - \frac{n(\bar{y}-\mu_2)^2}{\sigma^2} \\ &= Z_1^2 - Z_2^2\end{aligned}$$

where Z_j $(j = 1, 2)$ is the usual Z-test (normal test) statistic for the hypothesis H_j. Here $Z_1 = \sqrt{25} * 0.4 = 2,\ Z_2 = 3$ and $-2\log[L(\mu_1)/L(\mu_2)] = -5$.

We have sample evidence in favour of H_1 and against H_2, since $12.18 >> 1$. This *sample* evidence has to be interpreted in the light of other information about the hypotheses; Bayes theory does this formally through the *prior probabilities* π_1 and $\pi_2 = 1 - \pi_1$ of H_1 and H_2. The *prior odds* $p = \Pr(H_1)/\Pr(H_2) = \pi_1/\pi_2$ – the 'weight of evidence' in favour of H_1 compared to H_2 before observing the data – is 'updated' to the *posterior odds* using Bayes' theorem:

$$\begin{aligned}\frac{\Pr(H_1 \mid \mathbf{y})}{\Pr(H_2 \mid \mathbf{y})} &= \frac{\Pr(\mathbf{y} \mid H_1)\Pr(H_1)}{\Pr(\mathbf{y} \mid H_2)\Pr(H_2)} \\ &= \frac{L(\mu_1)}{L(\mu_2)} \cdot \frac{\pi_1}{\pi_2},\end{aligned}$$

by multiplying the prior odds p by the LR to give the posterior odds $p * \mathrm{LR}$.

The posterior probability of H_1 is $p * \mathrm{LR}/(p * \mathrm{LR} + 1)$. If the prior probabilities of the two hypotheses are equal, the posterior probability of H_1 is LR / (LR + 1).

How do we calibrate likelihood ratios as measures of evidence? What value of a likelihood ratio constitutes 'persuasive' sample evidence for H_1 compared to H_2? There is no unique answer to this question, just as there is no unique answer to the question of how small a P-value under the null hypothesis H_2, or how large a posterior probability of H_1, constitute 'persuasive' evidence in the Neyman–Pearson or Bayes theories. Different observers may be persuaded by different strengths of evidence. It is nevertheless convenient to have some scale reference values, like the 0.05 and 0.01 P-values. In Bayes theory, a posterior probability of 0.9 for H_1 would generally be regarded as quite strong evidence in favour of H_1; if the hypotheses have equal prior probability, the corresponding likelihood ratio would be 9. So the slightly larger rounded value 10 – an *order of magnitude* – for the likelihood ratio would be quite strong evidence, or a value of $-2\log\mathrm{LR}$

of 4.6, in the absence of prior information favouring one of the hypotheses over the other.

We adopt the scale below, adapted from one suggested by Jeffreys (1961) for Bayes factors, in terms of orders of magnitude.

We regard the sample evidence against H_1 and in favour of H_2, writing LR $= L(\mu_1)/L(\mu_2)$, as

- suggestive if LR $< 1/3\,(-2\log \text{LR} > 2.20)$,
- quite strong if LR $< 1/10\,(-2\log \text{LR} > 4.60)$,
- strong if LR $< 1/30\,(-2\log \text{LR} > 6.80)$, and
- very strong if LR $< 1/100\,(-2\log \text{LR} > 9.21)$.

If $1/3 <$ LR < 3, then we do not have even suggestive sample evidence in support of either model: the evidence is inconclusive.

The Neyman–Pearson theory does not treat the hypotheses symmetrically. We have to specify one of the hypotheses as the 'null', and the other as the 'alternative'. The repeated-sampling distribution of the likelihood ratio for the null to the alternative then has to be found under the null hypothesis, and this hypothesis is rejected if the likelihood ratio is too small, compared with a pre-specified percentage point of its sampling distribution. In practice the tail-area probability – the 'P-value' associated with the observed likelihood ratio – is found instead, and referred to standard critical values. A small P-value is interpreted as evidence against the corresponding null hypothesis.

The P-value depends on which hypothesis is specified as the null. Thus if the null hypothesis is $H_1 : \mu = 0$, the Z_1 test statistic is 2, and its P-value is 0.023, which would lead to rejection of this hypothesis at the usual 5% level. But if the null hypothesis is $H_2 : \mu = 1$, the Z_2 test statistic is -3, and its P-value is 0.0013, which would lead to rejection of *this* hypothesis at any conventional level. Given that both the P-values are small, it is unclear what we should conclude – do the data point to some other value of μ?

The Neyman–Pearson conclusions therefore depend on the prior specification of which hypothesis should be regarded as the null: the P-value for a particular specification of the null hypothesis cannot be used as a measure of the strength of evidence against this hypothesis, without also considering the other possible null hypothesis.

We now turn to question (ii).

2.6.2 *Information about a single parameter*

Given a single-parameter likelihood $L(\theta)$, the parameter value $\hat{\theta}$ with the highest likelihood is called the *maximum likelihood estimate* (MLE) of θ, and is a natural reference value for other values of θ. For many (but not all) of the statistical models considered in this book, the likelihood function has a unique maximum interior to the parameter space, and no minima, and in these cases the maximum can be found for a continuous parameter by (in general, partial) differentiation of the log-likelihood function $\ell(\theta) = \log L(\theta)$. In our example,

$$\frac{\partial \ell}{\partial \mu} = \sum (y_i - \mu)/\sigma^2$$

$$\frac{\partial^2 \ell}{\partial \mu^2} = -n/\sigma^2$$

giving the unique maximum at $\hat{\mu} = \bar{y}$. Values of θ near the MLE $\hat{\theta}$ have higher likelihoods than those remote from $\hat{\theta}$, so the *maximum* $L(\hat{\theta})$ of the likelihood is correspondingly a natural reference value for other values of $L(\theta)$. The *relative likelihood* function $R(\theta) = L(\theta)/L(\hat{\theta})$ expresses this directly, giving a measure of 'support' or strength of evidence on a 0–1 scale for each value of θ relative to the 'best supported' value $\hat{\theta}$. *Likelihood intervals* or *regions* of 'well-supported' values of θ are constructed by solving the inequality $R(\theta) > c$ for a suitable c.

The relative likelihood function in our example is easily seen to be

$$R(\mu) = \exp\left\{-\frac{n(\bar{y} - \mu)^2}{\sigma^2}\right\}.$$

This has the same form as a normal density function, because of the symmetrical appearance of $\bar{y}$ and μ in the likelihood. For this reason this likelihood is called a *normal* likelihood, or equivalently the log-likelihood is *quadratic* (in the parameter μ).

The importance of this name convention follows from the fact that in samples from regular parametric models, with increasing sample size the likelihood in the parameters approaches a normal likelihood. This is easily shown generally, assuming that the likelihood $L(\theta)$ has an internal maximum. We give the result for a single parameter; the general case follows directly. Expanding the log-likelihood function about the MLE $\hat{\theta}$ gives

$$\ell(\theta) = \ell(\hat{\theta}) + (\theta - \hat{\theta})\ell'(\hat{\theta}) + \frac{1}{2!}(\theta - \hat{\theta})^2\ell''(\hat{\theta}) + \frac{1}{3!}(\theta - \hat{\theta})^3\ell^{(3)}(\hat{\theta}) + \frac{1}{4!}(\theta - \hat{\theta})^4\ell^{(4)}(\hat{\theta}) + \cdots$$

Since the log-likelihood is a sum of n terms, write $\ell''(\hat{\theta})$ as $n\bar{\ell}''$ and similarly for the higher derivatives. Then $\bar{\ell}''$ and the higher derivatives are O(1) as n increases, and since $\ell'(\hat{\theta}) = 0$,

$$\ell(\theta) = \ell(\hat{\theta}) + \frac{n}{2!}(\theta - \hat{\theta})^2\bar{\ell}'' + \frac{n}{3!}(\theta - \hat{\theta})^3\bar{\ell}^{(3)} + \frac{n}{4!}(\theta - \hat{\theta})^4\bar{\ell}^{(4)} + \cdots$$

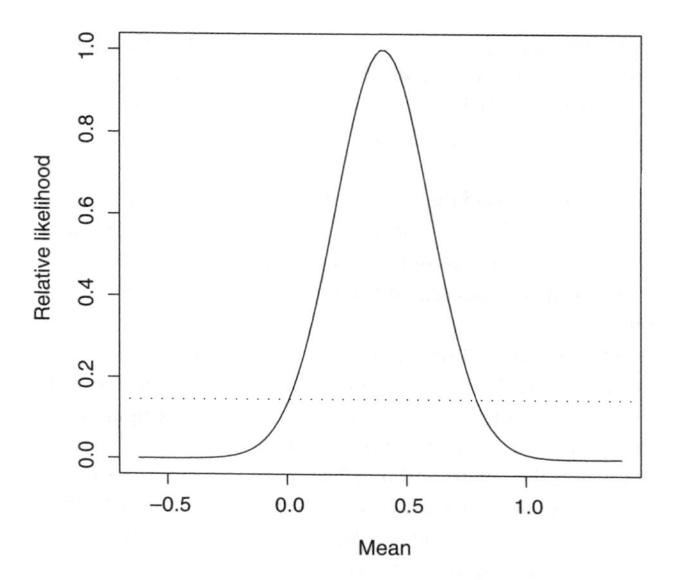

Fig. 2.12. Relative likelihood function

Write $-\bar{\ell}'' = 1/\sigma^2$ and $z = \sqrt{n}(\theta - \hat{\theta})/\sigma$. Then

$$\ell(\theta) = \ell(\hat{\theta}) - \frac{1}{2}z^2 + \sum_{j=3}^{\infty} \frac{n^{1-j/2}}{j!} z^j \bar{\ell}^{(j)} \sigma^j .$$

As n increases the cubic and higher terms in z tend to zero, giving

$$\ell(\theta) \to \ell(\hat{\theta}) - \tfrac{1}{2}z^2,$$

whence

$$L(\theta) \to L(\hat{\theta}) \exp\left\{-\tfrac{1}{2}z^2\right\}.$$

The relative likelihood is shown in Fig. 2.12, together with the line $R(\mu) = 0.146$. It is just a rescaling of the likelihood function itself, since only known constants have been omitted. The R commands are

```
> c <- 0.146
> curve(exp(-12.5 * (0.4 - x)$^2$), from = -0.62, to = (-0.6 +
+     0.02 * 100), xlab = "mean", ylab = "relative likelihood")
> abline(h = c, lty = "dotted")
```

As for the likelihood ratio for the two specified hypotheses, it is not necessary to know the measurement precision δ to compute the relative likelihood.

The likelihood interval for μ based on a relative likelihood of c is easily seen to be $\mu \in \bar{y} \pm \lambda\sigma/\sqrt{n}$ where $\lambda = \sqrt{-2\log c}$. This will be recognized as the usual Neyman–Pearson confidence interval for μ (but with the random variable $\bar{Y}$ replaced by the observed $\bar{y}$), with the confidence coefficient determined by the choice of c. Thus if $c = 0.146$, $\lambda = 1.96$, giving a 95% confidence interval for μ – a *likelihood-based confidence interval*. If $c = 0.0368$, $\lambda = 2.57$, giving a 99% confidence interval.

It is a feature of likelihood theory that the likelihood interval is determined *only* from the observed likelihood – we have not used the sampling distribution of $\bar{Y}$ to derive it. However, the confidence coefficient of the interval for a particular c depends on the form of repeated sampling assumed for $\bar{Y}$. We discuss this point further below.

The reader may notice that the value of c of 0.146 for a 95% confidence interval is *larger* than the value $1/10 = 0.1$ which we proposed above as the value of a likelihood ratio which would provide quite strong evidence against the 'numerator' hypothesis, when we are comparing two simple hypotheses. This reflects one of the differences between the likelihood and Neyman–Pearson theories – in likelihood theory (as in Bayes theory) we do not regard P-values of 0.05 as persuasive evidence against the null hypothesis. We discuss this point further below (Aitkin, 1991, 1997, gave a detailed discussion).

In the Neyman–Pearson theory, in the absence of any specific hypothesis about the parameter $\boldsymbol{\theta}$ (here taken to be p-dimensional), the sample information about it is conventionally expressed through the repeated sampling distribution of the MLE, the *score statistic*, or the *likelihood ratio statistic*. We define the log-likelihood function $\ell(\boldsymbol{\theta})$, score vector $\mathbf{s}(\boldsymbol{\theta})$, Hessian matrix $H(\boldsymbol{\theta})$, information matrix $I(\boldsymbol{\theta})$ and expected information matrix $\mathcal{I}(\boldsymbol{\theta})$ by

$$\begin{aligned}
\ell(\boldsymbol{\theta}) &= \log L(\boldsymbol{\theta}) \\
\mathbf{s}(\boldsymbol{\theta}) &= \frac{\partial \ell}{\partial \boldsymbol{\theta}} \\
H(\boldsymbol{\theta}) &= \frac{\partial \mathbf{s}}{\partial \boldsymbol{\theta}'} = \frac{\partial^2 \ell}{\partial \boldsymbol{\theta} \partial \boldsymbol{\theta}'} \\
I(\boldsymbol{\theta}) &= -H(\boldsymbol{\theta}) \\
\mathcal{I}(\boldsymbol{\theta}) &= \mathrm{E}\left[I(\boldsymbol{\theta})\right],
\end{aligned}$$

where the expectation is over the distribution of the random variables $Y_1, \ldots, Y_n$. Then under standard regularity conditions (that $\boldsymbol{\theta}$ is not an origin parameter of the support of Y, that $\hat{\boldsymbol{\theta}}$ is an interior point of the parameter space, and that the dimension of $\boldsymbol{\theta}$ does not increase with n), the sampling distribution of the MLE $\hat{\boldsymbol{\theta}}$ (regarded as a random variable in repeated sampling) is asymptotically

$$\hat{\boldsymbol{\theta}} \rightarrow N_p(\boldsymbol{\theta}, I^{-1}(\hat{\boldsymbol{\theta}})).$$

Alternatively, the asymptotic distribution of the score statistic is

$$\mathbf{S}(\boldsymbol{\theta}) \to N_p(0, \mathcal{I}(\boldsymbol{\theta})),$$

while the asymptotic distribution of the likelihood ratio statistic is

$$-2\log\{L(\boldsymbol{\theta})/L(\hat{\boldsymbol{\theta}})\} \to \chi_p^2.$$

(Strictly speaking the first two distributions are degenerate asymptotically since the variances tend to zero or infinity, so the MLE and score statistics should be standardized to give non-degenerate limiting distributions, but we give the usual results in the sense of 'for large n'.)

A 'point' estimate of $\boldsymbol{\theta}$ is $\hat{\boldsymbol{\theta}}$ (and if we are referring to the sampling distribution, it is of the estimat*or* $\hat{\boldsymbol{\theta}}$), and an 'interval' estimate is obtained by including in the interval all those values of $\boldsymbol{\theta}$ which would not be rejected by a formal test, using one of the three criteria above.

For the normal model example,

$$\begin{aligned}
L(\mu) &= \frac{\delta^n}{(\sqrt{2\pi})^n\sigma^n}\exp\left\{-\frac{1}{2\sigma^2}\left[\sum(y_i-\bar{y})^2 + n(\bar{y}-\mu)^2\right]\right\} \\
\ell(\mu) &= c - n(\bar{y}-\mu)^2/2\sigma^2 \\
\hat{\mu} &= \bar{y} \\
s(\mu) &= n(\bar{y}-\mu)/\sigma^2 \\
H(\mu) &= -n/\sigma^2 \\
I(\mu) &= I(\hat{\mu}) = \mathcal{I}(\mu) = \mathcal{I}(\hat{\mu}) = n/\sigma^2.
\end{aligned}$$

For this example all the asymptotic results are exact and equivalent:

$$\begin{aligned}
\hat{\mu} = \bar{Y} &\sim N(\mu, \sigma^2/n) \\
S(\mu) = n(\bar{Y}-\mu)/\sigma^2 &\sim N(0, n/\sigma^2) \\
-2\log\{L(\mu)/L(\bar{Y})\} = n(\bar{Y}-\mu)^2/\sigma^2 &\sim \chi_1^2.
\end{aligned}$$

Thus the point estimate of μ is $\bar{y}$, and the $100(1-\alpha)\%$ confidence interval for μ in repeated sampling (of fixed size n) is $\bar{Y} \pm \lambda_{1-\alpha/2}\sigma/\sqrt{n}$; this random interval covers the true value of μ with probability $1-\alpha$. For the observed interval in the given sample $\bar{y} \pm \lambda_{1-\alpha/2}\sigma/\sqrt{n}$ no such statement can be made: it either contains μ or does not.

As we noted above, the confidence interval for μ is formally identical to the likelihood interval through the relation between c and the percentage point of the normal distribution. However, their *meanings* are different: the likelihood interval contains those parameter values giving a relatively high likelihood to the observed sample, but the properties of the confidence interval do not refer to the observed sample: they are defined only with respect to the *family* or *reference set* of intervals produced by hypothetical replications of the sample.

Bayes' inference about $\boldsymbol{\theta}$ requires a *prior distribution* $\pi(\boldsymbol{\theta})$ for $\boldsymbol{\theta}$, representing the information about $\boldsymbol{\theta}$ before the current data are observed. On observing the data y, the prior distribution is updated to the *posterior distribution* $\pi(\boldsymbol{\theta} \mid y)$ using Bayes' theorem:

$$\pi(\boldsymbol{\theta} \mid y) = L(\boldsymbol{\theta})\pi(\boldsymbol{\theta}) / \int L(\boldsymbol{\theta})\pi(\boldsymbol{\theta})\, \mathrm{d}\boldsymbol{\theta},$$

where the integral is over the parameter space of $\boldsymbol{\theta}$.

There is a vigorous debate amongst Bayesians over the choice of priors, which we will not enter here; where necessary we will generally use *noninformative* or *flat* priors to represent prior information which is uninformative relative to the sample information. Such priors are useful in providing in many cases a *reference* analysis which can be modified by informative priors if needed.

For the normal mean example, a flat prior $\pi(\mu) = 1/2C$ on the interval $-C < \mu < C$, where C is large compared to $\sigma/\sqrt{n}$, gives (as $C \to \infty$) the *normal posterior distribution* $\mu \sim N(\bar{y}, \sigma^2/n)$, and a $100(1-\alpha)\%$ *highest posterior density* (HPD) or *credible* interval for μ is $\bar{y} \pm \lambda_{1-\alpha/2}\sigma/\sqrt{n}$. This is identical to the $100(1-\alpha)\%$ confidence interval, but as for the likelihood interval the (posterior) probability statement about μ refers to the observed data on which the distribution of μ is conditioned, and not to hypothetical replications of the sample.

If the prior distribution of μ is *not* flat, the posterior distribution will not have this simple normal form, and the HPD interval will not be the same as the confidence interval or the likelihood interval: this is to be expected from the additional information in the prior.

We now turn to question (iii).

2.6.3 *Comparing a simple null hypothesis and a composite alternative*

This problem is a central one in data analysis and model simplification.

In the Neyman–Pearson theory, the null hypothesis is tested using the previous sampling distribution results for the MLE, score and likelihood ratio statistic. The previous confidence interval constructions are formalized through the corresponding tests: the *Wald*, *score* and *likelihood ratio* tests. The LRTS for a null hypothesis $H_0 : \boldsymbol{\theta} = \boldsymbol{\theta}_0$ against a general alternative hypothesis $H_1 : \boldsymbol{\theta} \neq \boldsymbol{\theta}_0$ is

$$\text{LRTS} = -2\log\{L(\boldsymbol{\theta}_0)/L(\hat{\boldsymbol{\theta}})\}.$$

As alternatives to the LRT, the Wald and score or Lagrange multiplier tests are often used when the LRT is difficult or computationally expensive.

The Wald test statistic is

$$W = (\hat{\boldsymbol{\theta}} - \boldsymbol{\theta}_0)' I(\hat{\boldsymbol{\theta}})(\hat{\boldsymbol{\theta}} - \boldsymbol{\theta}_0),$$

and the score test statistic is

$$S = \mathbf{s}(\boldsymbol{\theta}_0)' \mathcal{I}^{-1}(\boldsymbol{\theta}_0)\mathbf{s}(\boldsymbol{\theta}_0).$$

Under the null hypothesis H_0, LRTS, W, and S are all asymptotically distributed as χ^2_p under the same regularity conditions (these conditions are in fact weaker for the score test, for which boundary values of $\hat{\boldsymbol{\theta}}$ may not invalidate the asymptotic distribution).

For the simple normal model,

$$\begin{aligned} s(\mu) &= n(\bar{y} - \mu)/\sigma^2 \\ I(\mu) &= n/\sigma^2 \\ \text{LRTS} &= n(\bar{y} - \mu_0)^2/\sigma^2 \\ W &= n(\bar{y} - \mu_0)^2/\sigma^2 \\ S &= \frac{n^2(\bar{y} - \mu_0)^2}{\sigma^4} / \frac{n}{\sigma^2} = W. \end{aligned}$$

The three test statistics are identical, and have the exact sampling distribution χ^2_1. For the example, the test statistics have the value 4.0, for which the corresponding P-value from the χ^2_1 distribution is 0.0455. This would lead to rejection of H_0 using the common critical P-value of 0.05. The three test procedures lead to the same test in this normal model (and the same set of confidence intervals for μ), and do so in large samples in many other models, but in small samples they may give markedly different results (as in the next binomial example). The score and LRT's are invariant to monotone transformation of $\boldsymbol{\theta}$, but the Wald test is not, as we will see shortly. There is one general likelihood approach corresponding closely to the Neyman–Pearson: the *maximized* likelihood approach. This uses the same LRTS, but interprets it directly as evidence in the same way as the likelihood ratio for simple hypotheses. In our example, the likelihood in maximized at $\mu = \hat{\mu} = \bar{y}$, and the MLR is therefore

$$L(\mu_0)/L(\bar{y}) = \exp\left\{-\frac{1}{2\sigma^2}\left[n(\bar{y} - \mu_0)^2)\right]\right\}$$

which has the value $\mathrm{e}^{-2} = 0.135$. Thus the relative likelihood at $\mu = \mu_0$ is quite high (relative to that at $\bar{y}$), and the LR 0.135 would not be considered convincing, or even quite strong, evidence against H_0, if $\bar{y}$ were a *specified* alternative value of μ. But $\bar{y}$ has been determined from the data, and any other *specified* alternative

$\mu_1 \neq \bar{y}$ will have lower likelihood than $L(\bar{y})$, and so an even greater likelihood ratio $L(\mu_0)/L(\mu_1)$. So the maximized likelihood ratio clearly *overstates* the evidence against H_0 – we are acting as though the unknown parameter under H_1 is *known* to be equal to its MLE – and cannot be interpreted directly as a measure of the strength of the sample evidence against H_0 without some further calibration.

This difficulty has limited the use of maximized likelihood ratios as direct measures of evidence. Considerable theoretical attention has been paid (Lindsey, 1996, gave details) to various forms of *adjustment* of the maximized likelihood, to better reflect the sample information about the 'nuisance' or unspecified parameters over which we have maximized. These adjusted likelihoods have generally been interpreted in a repeated sampling framework. The number of parameters p in the null hypothesis has to be allowed for as well, in any direct interpretation of the MLR; Aitkin (1991) and Lindsey (1999) gave discussions of possible approaches to this allowance.

In Bayes theory, the hypothesis comparison problem has a straightforward solution. We need the probability of the data under the alternative hypothesis, and so we integrate the likelihood (the conditional probability of the data, given the parameters) over the marginal (prior) distribution $\pi(\boldsymbol{\phi})$ of the unspecified parameters $\boldsymbol{\phi}$, and compare the specified likelihood under the null hypothesis with the integrated likelihood under the alternative; the ratio is called the *Bayes factor* for the two hypotheses, and is interpreted in the same way as a likelihood ratio for two simple hypotheses; the prior odds on the two hypotheses multiplies the Bayes factor to give the posterior odds as in the comparison of simple hypotheses.

However, this approach encounters serious difficulties, because the integration of the likelihood over the prior is in general very sensitive to the prior specification. For the normal example above, with the flat prior on μ over $(-C, C)$ and likelihood kernel

$$K(\mu) = \exp\left\{-\frac{1}{2\sigma^2}[n(\bar{y}-\mu)^2]\right\},$$

the integrated likelihood kernel is

$$\begin{aligned}
\bar{K} &= \int_{-\infty}^{\infty} K(\mu)\pi(\mu)\,\mathrm{d}\mu \\
&= \frac{1}{2C}\int_{-C}^{C} \exp\left\{-\frac{n}{2\sigma^2}(\bar{y}-\mu)^2\right\}\mathrm{d}\mu \\
&= \frac{1}{2C}\cdot\frac{\sigma}{\sqrt{n}}\int_{\frac{\sqrt{n}(\bar{y}-C)}{\sigma}}^{\frac{\sqrt{n}(\bar{y}+C)}{\sigma}} \exp\left\{-\frac{1}{2}Z^2\right\}\mathrm{d}Z \\
&= \frac{1}{2C}\frac{\sigma}{\sqrt{n}}\sqrt{2\pi}\left[\Phi\left(\frac{\sqrt{n}(\bar{y}+C)}{\sigma}\right) - \Phi\left(\frac{\sqrt{n}(\bar{y}-C)}{\sigma}\right)\right],
\end{aligned}$$

where Φ is the standard normal *cdf*. For C large relative to $\sigma/\sqrt{n}$, the term in brackets rapidly approaches 1, so for large C the integrated likelihood kernel is

$$\frac{1}{2C}\frac{\sigma}{\sqrt{n}}\sqrt{2\pi}.$$

This tends to 0 as $C \to \infty$, and it is clear that the integrated likelihood kernel can be made arbitrarily small by choice of C. The Bayes factor is

$$\begin{aligned} BF &= \frac{K(\mu_1)}{\bar{K}} = 2C \cdot \frac{\sqrt{n}}{\sqrt{2\pi}\sigma} \exp\left\{-\frac{n}{2\sigma^2}(\bar{y} - \mu_0)^2\right\} \\ &= 2C \cdot \frac{\sqrt{n}}{\sigma}\phi(Z_0) \end{aligned}$$

and this can be made arbitrarily large by choice of C. As $C \to \infty$ the Bayes factor $\to \infty$, whatever the (fixed) value of Z_0, and so the evidence in favour of H_0, whatever it is, becomes overwhelming if the prior is sufficiently diffuse!

This result is generally known as Lindley's paradox (Lindley, 1957; Bartlett, 1957), though the 'paradox' usually refers to the fact that, as $n \to \infty$ for *fixed* C, BF $\to \infty$ also. This is the same phenomenon: whether the likelihood becomes concentrated with n while the prior stays fixed, or the prior becomes diffuse with C while the likelihood stays fixed, the end result is the same: the data apparently overwhelmingly support the null hypothesis H_0, whatever it is. For the example above we have $Z_0 = 2, n = 25, \sigma = 1$, and the Bayes factor takes the values 0.27, 0.54, 5.4, and 54 for $C = 0.5$, 1, 10, and 100, respectively.

Thus direct interpretation of Bayes factors is unsatisfactory, and considerable efforts have been devoted to calibrating them. These are reviewed in Aitkin (1991) and Kass and Raftery (1995).

Aitkin (1997) gave an alternative Bayesian solution to this problem which does not suffer from Lindley's paradox, based on earlier results of Dempster (1974, 1997), which allows the general use of maximized likelihood ratios and profile likelihoods, though detailed applications are currently few.

We use the posterior distribution of θ under the alternative hypothesis to find the posterior distribution of the *true likelihood ratio* (TLR) $= L(\theta_0)/L(\theta)$, evaluated at the true value of θ. If the posterior probability of the event TLR < 0.1 is high, we have strong evidence against the null hypothesis.

For the above example, we take the flat prior on μ; the posterior distribution of μ is then $N(\bar{y}, \sigma^2/n)$, which is $N(0.4, 0.04)$. The posterior probability that $\mu > 0$ is $1 - \Phi(-2) = 0.9773 = 1 - P/2$. We work with the posterior distribution of $-2\log \text{TLR} = Z_0^2 - Z^2$. Then the posterior probability that the TLR is less than

0.1, a value which would be quite strong evidence against H_0, is

$$\begin{aligned}\Pr[\text{TLR} < 0.1 \mid y] &= \Pr[-2\log\text{TLR} > 4.6 \mid y]\\ &= \Pr[Z^2 < Z_0^2 - 4.6 \mid y]\\ &= \Pr[Z^2 < -0.6 \mid y]\\ &= 0\end{aligned}$$

since Z^2 must be positive! The TLR cannot be as small as 0.1 given the data. In fact, the TLR cannot be less than $\exp(-\frac{1}{2}Z_0^2) = 0.135$, since this is the value of the likelihood ratio resulting from maximizing the denominator.

For the much weaker criterion that the TLR is less than 1 – that is, that the null hypothesis is less well supported than the alternative – the posterior probability is

$$\begin{aligned}\Pr[\text{TLR} < 1 \mid y] &= \Pr[-2\log\text{TLR} > 0 \mid y]\\ &= \Pr[Z^2 < Z_0^2 \mid y]\\ &= \Pr[Z^2 < 4 \mid y]\\ &= 0.9546 = 1 - P\end{aligned}$$

a remarkable result pointed out by Dempster (1974): with normal likelihoods and flat priors, the P-value is equal to *the posterior probability that the* TLR *is greater than 1*, or equivalently, $1 - P$ is the posterior probability that the TLR is less than 1.

So the P-value *is* a measure of strength of evidence against the null hypothesis, but it is not a persuasive one, since it refers only to the null hypothesis being better or worse supported than the alternative, and not that the ratio of the likelihoods is small.

We show the value of this approach in the next section.

2.7 Inference with nuisance parameters

The discussion of inference in Section 2.6 was limited to single-parameter models. In practice, almost all inferential problems in statistics involve models with several unknown parameters. Research questions are usually about a subset of these parameters, called the *parameters of interest*, while the remaining parameters – the *nuisance parameters* – are not of interest in themselves, but affect the inference about the parameters of interest.

We begin again with the simplest normal mean model.

2.7.0.1 *Normal mean model*

A simple random sample $y_1, \ldots, y_n$ of size $n = 25$ is drawn from a population in which Y has a normal model $N(\mu, \sigma^2)$ with σ^2 now unknown, and yields the values $\bar{y} = 0.4$ and $s^2 = 1.0$, where $s^2 = \sum(y_i - \bar{y})^2/(n-1)$. We consider again the three inferential questions above, though we now change their order:

(i) In the absence of any substantive hypothesis about the value of μ, what information does the sample provide about it?
(ii) A substantive null hypothesis H_0 specifies $\mu = \mu_0$; the alternative is the general H_1 that $\mu \neq \mu_0$. What evidence does the sample provide about the hypothesis H_0 relative to the hypothesis H_1?
(iii) Two different substantive hypotheses H_1 and H_2 specify values for μ, namely $H_1 : \mu = \mu_1 = 0$ and $H_2 : \mu = \mu_2 = 1$. What evidence does the sample provide about hypothesis H_1 relative to hypothesis H_2?

The likelihood function is as before

$$
\begin{aligned}
L(\mu, \sigma) &= \prod_{i=1}^{n} [F(y_i + \delta/2 \mid \mu, \sigma) - F(y_i - \delta/2 \mid \mu, \sigma)] \\
&\doteq \prod_{i=1}^{n} \delta f(y_i \mid \mu, \sigma) \\
&= \frac{\delta^n}{(\sqrt{2\pi})^n \sigma^n} \exp\left\{-\frac{1}{2\sigma^2}\left[\sum (y_i - \bar{y})^2 + n(\bar{y} - \mu)^2\right]\right\}.
\end{aligned}
$$

Now σ is unknown, so the likelihood depends explicitly on the two parameters, though our inferential interest is only in μ. To make an inference about μ, we need to 'eliminate' the parameter σ from the likelihood in some way. We will give a general Neyman–Pearson approach based on *maximized* or *profile* likelihoods, calibrated through the sampling distribution of the LRT statistic, while referring in several applications to other non-general approaches based on *marginal* or *conditional* likelihoods. We calibrate the profile likelihood from a Bayes/likelihood theory point of view using the results of Aitkin (1997), described below; the same approach applies to the general problem with nuisance parameters as to the earlier case of a simple null hypothesis.

2.7.1 *Profile likelihoods*

Consider the general case of a likelihood $L(\boldsymbol{\theta}, \boldsymbol{\phi})$ depending on a p_1-dimensional parameter of interest $\boldsymbol{\theta}$ and a p_2-dimensional nuisance parameter $\boldsymbol{\phi}$. If for any outcome data y the likelihood can be written in the form

$$L(\boldsymbol{\theta}, \boldsymbol{\phi}) = L_1(\boldsymbol{\theta}) L_2(\boldsymbol{\phi}),$$

where L_1 is a function of $\boldsymbol{\theta}$ only and L_2 a function of $\boldsymbol{\phi}$ only, and the parameter spaces for $\boldsymbol{\theta}$ and $\boldsymbol{\phi}$ are unrelated, then we call the likelihood $L(\boldsymbol{\theta}, \boldsymbol{\phi})$ *separable*, and say the parameters $\boldsymbol{\theta}$ and $\boldsymbol{\phi}$ are *orthogonal*. In this case inference about $\boldsymbol{\theta}$ may be based on $L_1(\boldsymbol{\theta})$, since for any specific value $\boldsymbol{\phi}_0$ of $\boldsymbol{\phi}$,

$$L(\boldsymbol{\theta}_0, \boldsymbol{\phi}_0)/L(\boldsymbol{\theta}_1, \boldsymbol{\phi}_0) = L_1(\boldsymbol{\theta}_0)/L_1(\boldsymbol{\theta}_1).$$

Separable likelihoods are rare. If the likelihood is not separable, we define the *profile* likelihood $P(\boldsymbol{\theta})$ for $\boldsymbol{\theta}$ by

$$P(\boldsymbol{\theta}) = L(\boldsymbol{\theta}, \hat{\boldsymbol{\phi}}(\boldsymbol{\theta})),$$

where the notation $\hat{\boldsymbol{\phi}}(\boldsymbol{\theta})$ means the MLE of $\boldsymbol{\phi}$ as a function of the specified value $\boldsymbol{\theta}$. Thus the profile likelihood is the value of the likelihood as we follow a curved path through the parameter space, defined by $\boldsymbol{\phi} = \hat{\boldsymbol{\phi}}(\boldsymbol{\theta})$.

We now answer the first two questions above generally. For a general likelihood function $L(\boldsymbol{\theta}, \boldsymbol{\phi})$, the information provided by the sample about $\boldsymbol{\theta}$ is expressed through the profile likelihood $P(\boldsymbol{\theta})$, interpreted as though it were a likelihood function from a parametric model $f(z \mid \boldsymbol{\theta})$. The evidence provided by the sample about a null hypothesis $H_0 : \boldsymbol{\theta} = \boldsymbol{\theta}_0$ against a general alternative $H_1 : \boldsymbol{\theta} \neq \boldsymbol{\theta}_0$ is the profile likelihood ratio (profile relative likelihood)

$$P(\boldsymbol{\theta}_0)/P(\hat{\boldsymbol{\theta}}) = L(\boldsymbol{\theta}_0, \hat{\boldsymbol{\phi}}(\boldsymbol{\theta}_0))/L(\hat{\boldsymbol{\theta}}, \hat{\boldsymbol{\phi}}),$$

where $\hat{\boldsymbol{\phi}} = \boldsymbol{\phi}(\hat{\boldsymbol{\theta}})$ is the MLE of ϕ. The profile likelihood ratio is interpreted as though it were a likelihood ratio from the parametric model $f(z \mid \boldsymbol{\theta})$.

The specification of the same interpretation for a profile likelihood as for a 'real' likelihood clearly requires justification, and in some examples this specification fails to provide a satisfactory calibration of the profile likelihood. We discuss this below, but first illustrate with the normal model example.

For fixed μ, the MLE of σ is given by

$$\hat{\sigma}^2(\mu) = \left[\sum (y_i - \bar{y})^2 + n(\bar{y} - \mu)^2\right] \Big/ n.$$

Substituting this value into the likelihood gives the profile likelihood

$$P(\mu) = \frac{\delta^n}{(\sqrt{2\pi})^n \hat{\sigma}(\mu)^n} \exp\left\{-\frac{n}{2}\right\},$$

and the profile likelihood ratio is

$$P(\mu)/P(\hat{\mu}) = [\hat{\sigma}/\hat{\sigma}(\mu)]^n,$$

where

$$\hat{\sigma}^2 = \hat{\sigma}^2(\bar{y}) = \sum (y_i - \bar{y})^2/n = (n-1)s^2/n.$$

Straightforward algebra gives

$$P(\mu)/P(\hat{\mu}) = \left[1 + \frac{n(\bar{y}-\mu)^2}{(n-1)s^2}\right]^{-n/2}$$
$$= \left[1 + \frac{t^2}{(n-1)}\right]^{-n/2},$$

where t is the usual one-sample t-statistic: $t = \sqrt{n}(\bar{y} - \mu)/s$. Remarkably, the profile relative likelihood for μ is identical to the relative likelihood for μ based on just the observation of the t-statistic, which has a t_{n-1} distribution, with density

$$f(t) = \text{const} \cdot \left[1 + \frac{t^2}{(n-1)}\right]^{-n/2},$$

where the normalizing constant disappears in the relative likelihood. Thus a profile likelihood interval for μ based on a profile relative likelihood of c is identical to a confidence interval for μ based on the t distribution, using the critical value

$$t_c = [(n-1)(c^{-2/n} - 1)]^{1/2}.$$

For our example, with $n = 25$ and the observed $t = 2$, a 14.6% relative likelihood interval for μ uses $c = 0.146$, for which $t_c = 1.998$; the interval is $(0.4 \pm 1.998 \cdot 1.0/\sqrt{25})$, which is $(0.00, 0.80)$. The 95% confidence interval for μ from the t_{24}-distribution uses the critical value 2.064, slightly larger; this interval is $(-0.01, 0.81)$. The difference arises because the sampling distributions of the t and normal likelihood ratios, as random variables, are different.

We conclude that the observed t-value of 2 is not very strong evidence against H_0: the P-value just exceeds 0.05.

For smaller df the differences are larger: for $n = 11$ the 14.6% relative likelihood uses $t = 2.047$, while the 95% confidence interval uses $t = 2.230$. For $n = 5$ the values are 2.153 and 2.777, respectively. This illustrates the difficulty of direct interpretation of maximized likelihood ratios – for consistent confidence coverage we have to consider the different sampling distributions involved.

Figure 2.13 shows the t likelihood (solid curve) for the example, and the normal likelihood (dashed curve) assuming $\sigma^2 = \hat{\sigma}^2$ for comparison. The t likelihood is slightly more diffuse but very similar in shape. Extending the previous graph of the normal relative likelihood, the R commands are

```
> library(lattice)
> print(xyplot(0:1 ~ -0.5:1.5, ylab = "relative likelihoods",
```

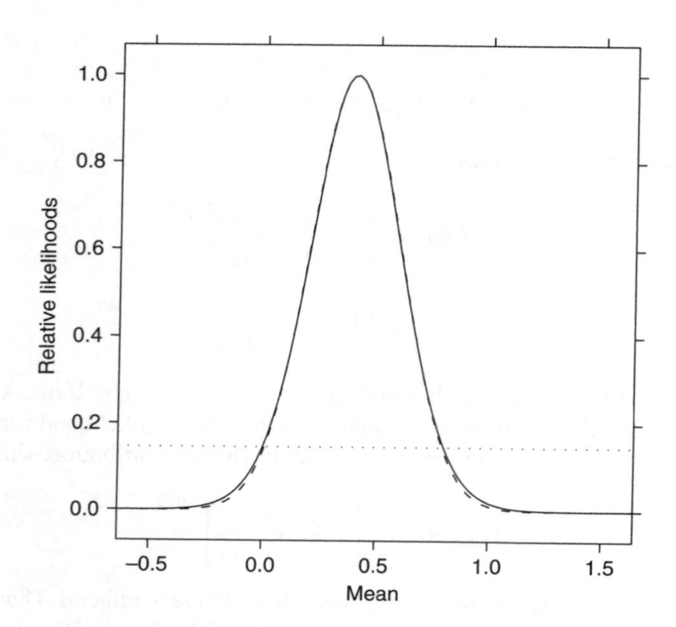

Fig. 2.13. Relative likelihood functions

```
+       xlab = "mean", type = "n", panel = function(...) {
+           panel.xyplot(...)
+           panel.curve((1 + (5 * (0.4 - x))^2/24)^{(-12.5)})
+           panel.curve(exp(-12.5 * (0.4 - x)^2), lty = 2)
+           panel.abline(h = c, lty = 3)
+       }))
```

The profile likelihood ratio is calibrated in the Neyman–Pearson theory through the LRTS, defined by

$$\mathrm{LRTS} = -2\log\{P(\boldsymbol{\theta})/P(\hat{\boldsymbol{\theta}})\}.$$

The asymptotic repeated-sampling distribution of this statistic (in samples of fixed size n) is $\chi^2_{p_1}$ under the regularity conditions of Section 2.6.2.

In the normal model above, this gives

$$n\log[1 + t^2/(n-1)] \to \chi^2_1;$$

for small t compared with n the LHS is approximately $nt^2/(n-1)$. The asymptotic distribution is not needed in this model because of the known exact distribution of t. This normal mean example is one of the few in which the LRTS has an exact small-sample distribution.

The profile likelihood method is readily extended to functions of the model parameters. For example, suppose that a variable Y has a lognormal distribution, with $Z = \log Y \sim N(\mu, \sigma^2)$. We consider an example of family income in Section 2.9. The standard analysis of such data would be to log transform them, and fit the normal model to Z. Suppose however we wish to draw an inference about the mean of Y, rather than of Z. We have $\mathrm{E}[Y] = \exp(\mu + \sigma^2/2)$, an awkward function. However, it is quite straightforward to compute the profile likelihood in the parameter $\theta = \mu + \sigma^2/2$, and then plot this likelihood against $\exp(\theta)$.

The likelihood in μ and σ is

$$L(\mu, \sigma) = \frac{\delta^n}{(\sqrt{2\pi})^n \sigma^n \prod y_i} \exp\left\{-\frac{1}{2\sigma^2}\left[\sum(z_i - \bar{z})^2 + n(\bar{z} - \mu)^2\right]\right\},$$

where $z_i = \log y_i$. We define $T = \sum(z_i - \bar{z})^2$, $\phi = \sigma^2$, $\mu = \theta - \phi/2$, and reparametrize the likelihood in terms of θ and ϕ. The log-likelihood is

$$\begin{aligned}
\ell(\theta, \phi) &= c - \frac{n}{2}\log\phi - \frac{1}{2\phi}\left[T + n(\bar{z} - \theta + \phi/2)^2\right] \\
&= c - \frac{n}{2}\log\phi - \frac{1}{2}\left[\frac{T + n(\bar{z} - \theta)^2}{\phi} + n(\bar{z} - \theta) + n\phi/4\right] \\
&= c - \frac{n}{2}\log\phi - \frac{n}{2}\left[\frac{\hat{\sigma}^2 + (\bar{z} - \theta)^2}{\phi} + \bar{z} - \theta + \phi/4\right] \\
\frac{\partial\ell}{\partial\phi} &= -\frac{n}{2\phi} - \frac{n}{2}\left[-\frac{\hat{\sigma}^2 + (\bar{z} - \theta)^2}{\phi^2} + \frac{1}{4}\right].
\end{aligned}$$

Solving $\partial\ell/\partial\phi = 0$ for ϕ gives

$$\hat{\phi}(\theta) = \widehat{\sigma}^2(\theta) = 2[\sqrt{1 + \hat{\sigma}^2 + (\bar{z} - \theta)^2} - 1].$$

Substituting for σ in the likelihood gives the profile likelihood in θ.

2.7.2 *Marginal likelihood for the variance*

We now consider the same normal model for Y, but suppose σ is the parameter of interest and μ the nuisance parameter. We have immediately $\hat{\mu}(\sigma) = \bar{y}$ for all σ, and so

$$P(\sigma) = \frac{\delta^n}{(\sqrt{2\pi})^n \sigma^n} \exp\left\{-\frac{\sum(y_i - \bar{y})^2}{2\sigma^2}\right\}.$$

The profile likelihood in this case is just the *section* of the likelihood at $\mu = \bar{y}$. The profile likelihood ratio, or profile relative (PR) likelihood, is

$$\begin{aligned} \mathrm{PR}(\sigma) = P(\sigma)/P(\hat{\sigma}) &= \left(\frac{\hat{\sigma}}{\sigma}\right)^n \exp\left\{\frac{n}{2} - \frac{\sum(y_i - \bar{y})^2}{2\sigma^2}\right\} \\ &= \left(\frac{\hat{\sigma}}{\sigma}\right)^n \exp\left\{\frac{n}{2}\left(1 - \frac{\hat{\sigma}^2}{\sigma^2}\right)\right\}. \end{aligned}$$

An unsatisfying feature of the profile likelihood is that it is the same as the likelihood which would be obtained if we *knew* that $\mu = \bar{y}$. This is a feature of *all* profile likelihoods, but the fact that $\hat{\mu}(\sigma)$ does not depend on σ in this case accentuates the overstatement of precision in the profile likelihood – surely we know *less* about σ than the profile likelihood suggests.

A re-expression of the likelihood provides an alternative approach. Write

$$L(\mu, \sigma) = L_1(\mu, \sigma \mid \bar{y}) M(\sigma \mid s),$$

where

$$\begin{aligned} L_1(\mu, \sigma \mid \bar{y}) &= \frac{\sqrt{n}}{\sqrt{2\pi}\sigma} \exp\left\{-\frac{n(\bar{y} - \mu)^2}{2\sigma^2}\right\}, \\ M(\sigma \mid s) &= c\left(\frac{s}{\sigma}\right)^{n-1} \exp\left\{-\frac{(n-1)s^2}{2\sigma^2}\right\}, \end{aligned}$$

and c is a constant not involving μ or σ. Here we denote explicitly the dependence of L_1 and M on the sufficient statistics. L_1 and M are true likelihoods, arising from the normal distribution $N(\mu, \sigma^2/n)$ for $\bar{Y}$ and the independent $\sigma^2\chi^2_{n-1}$ distribution for $(n-1)s^2$. M depends only on σ, since the distribution of the sum-of-squares statistic $T = (n-1)s^2 = \sum(y_i - \bar{y})^2$ does not depend on μ. The distribution of $\bar{Y}$ however depends on both μ and σ.

If we are willing to ignore the information about σ in L_1, then inference about σ can be based on M, which is called a *marginal* or *restricted* likelihood for σ (Kalbfleisch and Sprott, 1970; Patterson and Thompson, 1971). It is frequently argued, following Kalbfleisch and Sprott, that L_1 does not provide 'available' information about σ, since μ is also unknown. This argument is difficult to make precise (though it has been used to define weaker forms of sufficiency by Basu, 1977), and it is simpler to note that the additional information about σ in $\bar{y}$ is worth *at most* one 'degree of freedom' in addition to the $n-1$ in s^2, so by using the marginal likelihood we are ignoring a proportion of at most $1/n$ of the information about σ. As $n \to \infty$ this proportion goes to zero, and the marginal and profile likelihoods are equivalent.

The marginal likelihood is maximized at the MML or REML estimate

$$\tilde{\sigma}^2 = s^2 = \sum (y_i - \bar{y})^2/(n-1),$$

and the marginal likelihood ratio, or marginal relative likelihood, is

$$\mathrm{MR}(\sigma) = M(\sigma)/M(\tilde{\sigma}) = \left(\frac{s}{\sigma}\right)^{\nu} \exp\left\{\frac{\nu}{2}\left(1 - \frac{s^2}{\sigma^2}\right)\right\},$$

where $\nu = n - 1$.

The profile and marginal relative likelihoods for σ are shown in Fig. 2.14 for our example. The profile likelihood (solid curve) is offset slightly towards smaller values of σ, and is slightly more concentrated, than the marginal likelihood. As n increases the two likelihoods agree more closely. The R commands are

```
> print(xyplot(0:1 ~ 0.6:2, type = "n", xlab = expression(sigma),
+     ylab = "relative likelihood", panel = function(...) {
+         panel.curve((0.98/x)$^25$ * exp(12.5 * (1 -
+             (0.96/x$^2$))))
+         panel.curve((1/x)$^24$ * exp(12 * (1 - (1/x$^2$))),
+             lty = "dotted")
+         panel.abline(h = 0.146, lty = "dotted")
+     }))
```

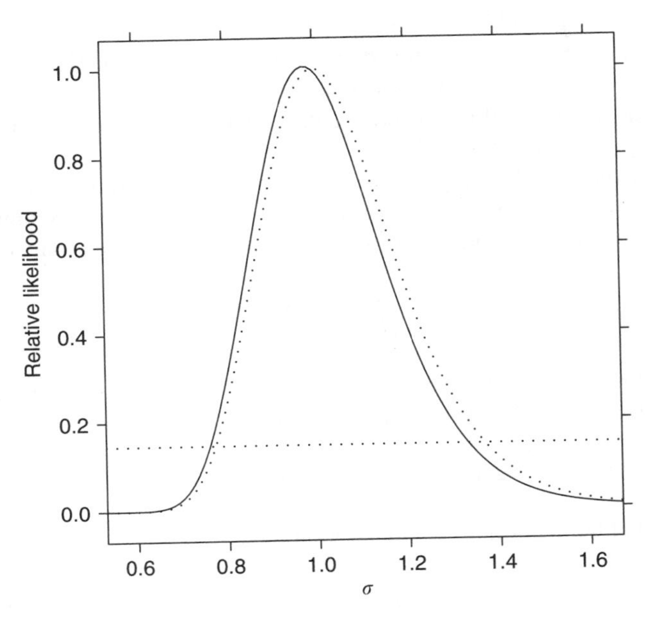

Fig. 2.14. Relative likelihood functions

Profile or marginal likelihood intervals for σ are somewhat tedious to compute by interpolation, Newton methods or grid search, as the relative likelihood functions do not have explicit inverses. Very accurate approximate marginal likelihood intervals for σ may be obtained easily by using the *likelihood normalizing transformation* of σ, which we now describe.

2.7.3 *Likelihood normalizing transformations*

We find the transformation $\phi = \phi(\theta)$ of θ which makes the likelihood $L(\phi)$ as 'normal' as possible. This is achieved (Anscombe, 1964; Sprott, 1973) by making the third derivative of the log-likelihood $\ell(\phi)$ zero at the MLE $\hat{\phi}$. From Section 2.6.2, departures from normality at $\hat{\phi}$ are then confined to the fourth and higher derivatives of the log-likelihood in ϕ. Then provided $\hat{\phi}$ is an internal point (not on the boundary of the parameter space), ϕ has an approximately normal likelihood:

$$L(\phi) \doteq L(\hat{\phi}) \exp\left\{-\tfrac{1}{2}(\phi - \hat{\phi})^2 I_{\hat{\phi}}\right\},$$

where $I_{\hat{\phi}}$ is the observed information on the ϕ scale, and the likelihood interval for ϕ such that $L(\phi)/L(\hat{\phi}) \geq c$ is given approximately by

$$\phi \in \hat{\phi} \pm \lambda_c \mathrm{SE}(\hat{\phi}),$$

where $\mathrm{SE}(\hat{\phi}) = I_{\hat{\phi}}^{-1}$ and $\lambda_c = \sqrt{-2 \log c}$. The corresponding likelihood interval for θ can then be found by reverse-transforming the interval for ϕ. The approximation on the ϕ scale is closer than on the θ scale because the term in $n^{-1/2}$ in the Taylor series expansion has coefficient zero, so the convergence to the normal likelihood is at the rate n^{-1} instead of $n^{-1/2}$.

For the exponential family, the necessary transformation does not depend on the data, only on the form of the variance/mean relationship Anscombe (1964). Writing μ for the mean and $V(\mu)$ for the variance, the required transformation is

$$\phi = \int^{\mu} [V(t)]^{-2/3}\, \mathrm{d}t.$$

For likelihood inference about $\theta = \sigma^2$ from the marginal likelihood based on T, which has the $\sigma^2 \chi_\nu^2$ distribution (a gamma distribution), we have

$$\mathrm{E}[T] = \mu = \nu\sigma^2, \quad \mathrm{Var}[T] = V(\mu) = 2\nu\sigma^4 = 2\mu^2/\nu.$$

Then the likelihood normalizing transformation is

$$\phi = \int^{\mu} [V(t)]^{-2/3}\, \mathrm{d}t = \int^{\mu} t^{-4/3}\, \mathrm{d}t = \mu^{-1/3} = \sigma^{-2/3} = \theta^{-1/3},$$

omitting irrelevant constants. The likelihood for ϕ is very accurately normal even in very small samples, and is expressed in terms of the REML estimate $\tilde{\phi} = s^{-2/3}$ with standard error

$$\text{SE}(\tilde{\phi}) = \frac{\sqrt{2}}{3}\frac{\tilde{\phi}}{\sqrt{\nu}}.$$

So a 14.6% likelihood interval for ϕ is $\tilde{\phi} \pm 1.96\sqrt{2}\tilde{\phi}/(3\sqrt{\nu})$, giving the corresponding interval for σ of

$$s(1 \pm 0.924/\sqrt{\nu})^{-3/2}.$$

This is particularly simple to calculate. Figure 2.15 shows the exact marginal likelihood (solid curve) and the normal approximation (dashed curve) on the σ scale, for our example. The R commands are

```
> print(xyplot(0:1 ~ 0.6:2, type = "n", xlab = expression(sigma),
+     ylab = "relative likelihood", panel = function(...) {
+         panel.xyplot(...)
+         panel.curve((1/x)$^24$ * exp(12 * (1 - (1/x$^2$))))
+         panel.curve(exp(-0.5 * (((x$^(-2/3)$) - 1) *
+             (3 * sqrt(12)))$^2$), lty = "dotted", lwd = 3)
+         panel.abline(h = 0.146, lty = "dotted")
+     }))
```

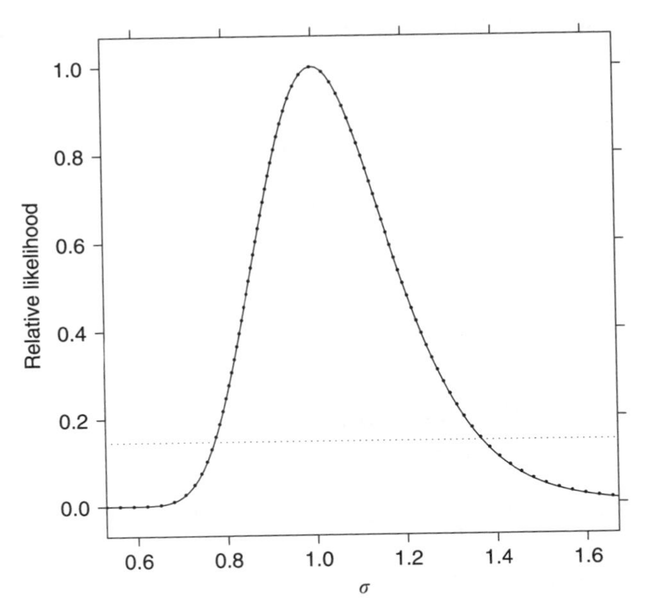

Fig. 2.15. Exact and approximate (dotted) marginal likelihoods

The accuracy of the approximation is remarkable, and is the basis of the very accurate Wilson–Hilferty (1931) normal approximation to the distribution of χ^2. Note that the interval endpoints of s^2/σ^2 for a $100(1-\alpha)\%$ confidence interval are in general $(1 \pm 0.471\lambda_{1-\alpha/2}/\sqrt{\nu})^3$, where λ is the normal percentage point.

The usual Neyman–Pearson approach is to use the χ^2_ν sampling distribution of T/σ^2 to construct equal-tailed confidence intervals for σ^2. It has long been known that these are unsatisfactory (Tate and Klett, 1959): they are not likelihood intervals, are offset relative to the likelihood intervals with the same confidence coverage, and give biased tests for a specific σ.

For our example with $n = 25$ and $s = 1$, the 95% likelihood-based confidence limits for s^2/σ^2 are (0.535, 1.679), while those based on the usual equal-tailed χ^2_{24} distribution are (0.517, 1.640). For small ν the differences are greater: for $\nu = 10$ the corresponding limits are (0.355, 2.156) and (0.325, 2.048), respectively.

It should be noted that for very small ν and α the lower endpoint of the approximate likelihood interval may become negative. The approximation fails in this case, but we rarely ask for very high confidence with very small samples.

2.7.4 *Alternative test procedures*

In the Neyman–Pearson theory, alternatives to the LRT are widely used. The Wald and score tests of Section 2.6 can be readily generalized: let $\ell(\boldsymbol{\theta}, \boldsymbol{\phi})$ be the log-likelihood, $(\mathbf{s}_{\boldsymbol{\theta}}, \mathbf{s}_{\boldsymbol{\phi}})$ be the partitioned score vector, and $I_{\boldsymbol{\theta}\boldsymbol{\theta}}, I_{\boldsymbol{\theta}\boldsymbol{\phi}}, I_{\boldsymbol{\phi}\boldsymbol{\phi}}$ be the component submatrices of the observed information matrix, defined for example by

$$I_{\boldsymbol{\theta}\boldsymbol{\phi}} = -\frac{\partial^2 \ell}{\partial \boldsymbol{\theta} \partial \boldsymbol{\phi}'},$$

and $\mathcal{I}_{\boldsymbol{\theta}\boldsymbol{\phi}}$ be the corresponding expected information components. Then under the same regularity conditions, the sampling distribution of the MLE $\hat{\boldsymbol{\theta}}$ is asymptotically

$$\hat{\boldsymbol{\theta}} \to N_{p_1}(\boldsymbol{\theta}, V_{\hat{\boldsymbol{\theta}}}),$$

where $V_{\hat{\boldsymbol{\theta}}} = I^{-1}(\hat{\boldsymbol{\theta}} \mid \hat{\boldsymbol{\phi}}) = (I_{\hat{\boldsymbol{\theta}}\hat{\boldsymbol{\theta}}} - I_{\hat{\boldsymbol{\theta}}\hat{\boldsymbol{\phi}}} I^{-1}_{\hat{\boldsymbol{\phi}}\hat{\boldsymbol{\phi}}} I_{\hat{\boldsymbol{\phi}}\hat{\boldsymbol{\theta}}})^{-1}$. The corresponding Wald test statistic for a hypothetical $\boldsymbol{\theta} = \boldsymbol{\theta}_0$ is

$$W = (\hat{\boldsymbol{\theta}} - \boldsymbol{\theta}_0)' V^{-1}_{\hat{\boldsymbol{\theta}}} (\hat{\boldsymbol{\theta}} - \boldsymbol{\theta}_0)' = (\hat{\boldsymbol{\theta}} - \boldsymbol{\theta}_0) I(\hat{\boldsymbol{\theta}} \mid \hat{\boldsymbol{\phi}})(\hat{\boldsymbol{\theta}} - \boldsymbol{\theta}_0).$$

For the score test, we evaluate the score components and expected information at the MLE of the nuisance parameter under the null hypothesis as $\left[\mathbf{s}(\boldsymbol{\theta}_0), \mathbf{s}(\hat{\boldsymbol{\phi}}(\boldsymbol{\theta}_0))\right]$

and $\mathcal{I}_{\boldsymbol{\theta}_0\hat{\boldsymbol{\phi}}(\boldsymbol{\theta}_0)}$. The score test statistic is then

$$S = \left[\mathbf{s}(\boldsymbol{\theta}_0), \mathbf{s}(\hat{\boldsymbol{\phi}}(\boldsymbol{\theta}_0))\right]' \mathcal{I}^{-1}_{\boldsymbol{\theta}_0\hat{\boldsymbol{\phi}}(\boldsymbol{\theta}_0)} \left[\mathbf{s}(\boldsymbol{\theta}_0), \mathbf{s}(\hat{\boldsymbol{\phi}}(\boldsymbol{\theta}_0))\right].$$

Both W and S, like the LRTS, are asymptotically distributed as $\chi^2_{p_1}$ under the null hpothesis. However they may give quite different results in small samples.

The score test has the computational advantage that the information has to be evaluated only under the null hypothesis: the full alternative hypothesis model does not have to be fitted. This is very convenient in many standard models where some additional complexity is introduced by an additional parameter whose ML estimation is difficult. Score tests for overdispersion (Chapter 8) or variance component structure (Chapter 9) are examples.

The Wald test requires the fitting of only the alternative hypothesis model; the LRT requires the fitting of both models.

We illustrate the Wald and score tests on the normal model, where the null hypothesis is $H_0 : \mu = \mu_0$, and σ is the nuisance parameter. The likelihood is

$$L(\mu, \sigma) = \frac{\delta^n}{(\sqrt{2\pi})^n \sigma^n} \exp\left\{-\frac{1}{2\sigma^2}\left[\sum(y_i - \bar{y})^2 + n(\bar{y} - \mu)^2\right]\right\}.$$

The score components are

$$\frac{\partial \ell}{\partial \mu} = n(\bar{y} - \mu)/\sigma^2$$
$$\frac{\partial \ell}{\partial \sigma} = -n/\sigma + [T + n(\bar{y} - \mu)^2]/\sigma^3$$

and the elements of the Hessian matrix are

$$\frac{\partial^2 \ell}{\partial \mu^2} = -n/\sigma^2$$
$$\frac{\partial^2 \ell}{\partial \mu \partial \sigma} = -2n(\bar{y} - \mu)/\sigma^3$$
$$\frac{\partial^2 \ell}{\partial \sigma^2} = n/\sigma^2 - 3[T + n(\bar{y} - \mu)^2]/\sigma^4.$$

The observed and expected information are both diagonal. For the Wald test, the MLE of σ^2 is $\hat{\sigma}^2 = T/n$, and hence

$$W = \frac{n(\bar{y} - \mu_0)^2}{\hat{\sigma}^2} = \frac{nt^2}{n - 1}.$$

For the score test, the MLE of σ^2 given $\mu = \mu_0$ is

$$\hat{\sigma}^2(\mu_0) = [T + n(\bar{y} - \mu_0)^2]/n,$$

and the σ-component of the score is zero when evaluated at $\mu_0, \hat{\sigma}(\mu_0)$. The expected information for the μ-component of the score, evaluated at $\sigma(\mu_0)$, is $n/\hat{\sigma}^2(\mu_0)$ and so the score test statistic is

$$\left[\frac{n(\bar{y} - \mu_0)}{\hat{\sigma}^2(\mu_0)}\right]^2 \Big/ \frac{n}{\hat{\sigma}^2(\mu_0)} = \frac{n(\bar{y} - \mu_0)^2}{\hat{\sigma}^2(\mu_0)} = \frac{nt^2}{n - 1 + t^2}.$$

For the numerical example with $t = 2$ and $n = 25$, the LRTS is 3.85, the Wald test statistic is 4.17 and the score test statistic is 3.57. The score test statistic will be far from the Wald or squared t-statistic if the sample mean $\bar{y}$ is far from the null hypothesis value μ_0. This is a general feature of the score test: since it does not evaluate the nuisance parameter estimate under the alternative hypothesis, the curvature of the likelihood in this parameter at the null hypothesis may be far from that at the MLE, giving poor agreement with the LRT. When the null hypothesis is $\sigma = \sigma_0$ with μ the nuisance parameter, the Wald and score tests can diverge substantially from the LRT in small samples. The reader can verify that the score test statistic is

$$S = n\left(\frac{\hat{\sigma}^2}{\sigma_0^2} - 1\right)^2 \Big/ 2,$$

while the Wald test statistic is

$$W = 2n\left(\frac{\sigma_0}{\hat{\sigma}} - 1\right)^2$$

for this parametrization, but if σ^2 is taken as the parameter of interest and the null hypothesis is expressed as $H_0 : \sigma^2 = \sigma_0^2$, the Wald test statistic is

$$W^* = n\left(\frac{\sigma_0^2}{\hat{\sigma}^2} - 1\right)^2 \Big/ 2.$$

While all the test statistics depend only on the ratio $\hat{\sigma}^2/\sigma_0^2$, and all have the same asymptotic χ_1^2 distribution under H_0, their values can be very different. For example if $n = 10$, $\hat{\sigma} = 1$ and $\sigma_0 = 2$, we have

$$S = 2.813, \quad W = 20, \quad W^* = 5,$$

while the LRTS has the value 6.36. The lack of invariance of the Wald test is a serious drawback. The substantial difference between the score and LRTS is due to the substantial difference in the curvature $2n/\sigma^2$ evaluated at $\sigma_0 = 2$ (5) and at $\hat{\sigma} = 1$ (20).

2.7.5 *Bayes inference*

The Bayes analysis of question (i) is again direct. The *joint* prior distribution $\pi(\boldsymbol{\theta}, \boldsymbol{\phi})$ combines with the likelihood $L(\boldsymbol{\theta}, \boldsymbol{\phi})$ to give the *joint posterior distribution* $\pi(\boldsymbol{\theta}, \boldsymbol{\phi} \mid y)$. The *marginal posterior distribution* of $\boldsymbol{\theta}$ is then obtained by integrating over $\boldsymbol{\phi}$:

$$\begin{aligned}\pi(\boldsymbol{\theta} \mid y) &= \int \pi(\boldsymbol{\theta}, \boldsymbol{\phi} \mid y)\, \mathrm{d}\boldsymbol{\phi} \\ &= \frac{\int L(\boldsymbol{\theta}, \boldsymbol{\phi})\pi(\boldsymbol{\theta}, \boldsymbol{\phi})\, \mathrm{d}\boldsymbol{\phi}}{\int\int L(\boldsymbol{\theta}, \boldsymbol{\phi})\pi(\boldsymbol{\theta}, \boldsymbol{\phi})\, \mathrm{d}\boldsymbol{\phi}\, \mathrm{d}\boldsymbol{\theta}}.\end{aligned}$$

For the normal mean–variance model, if we take the usual flat priors on μ and $\log \sigma$, it is easily shown that the joint posterior distribution of μ and σ can be expressed as the conditional normal distribution $N(\bar{y}, \sigma^2/n)$ of $\mu \mid \sigma$, and the marginal χ^2_{n-1} distribution of T/σ^2. The marginal distribution of μ, integrating over σ, is such that $t = \sqrt{n}(\bar{y} - \mu)/s$ has a t_{n-1} distribution. This choice of priors gives the analogue of the repeated-sampling distributions; other choices of priors do not. The $100(1 - \alpha)\%$ HPD interval for μ is identical to the $100(1 - \alpha)\%$ confidence interval.

Hypothesis testing using Bayes factors suffers the same difficulty as in the simpler case above. The posterior distribution of the TLR was extended to the general case by Aitkin (1997). For a null hypothesis $H_0 : \boldsymbol{\theta} = \boldsymbol{\theta}_0$ with a nuisance parameter $\boldsymbol{\phi}$, the posterior distribution $\pi(\boldsymbol{\theta}, \boldsymbol{\phi} \mid y)$ of all the parameters is used to find the posterior distribution of the TLR $= L(\boldsymbol{\theta}_0, \boldsymbol{\phi})/L(\boldsymbol{\theta}, \boldsymbol{\phi})$ (evaluated at the true values of the nuisance parameters), or of $-2 \log \mathrm{TLR}$, and hence the posterior probability that the TLR is less than k, for any specified k. This posterior probability can be expressed (in large samples, and with diffuse priors) in terms of the maximized likelihood ratio and hence in terms of the P-value.

This allows the simple recalibration of maximized likelihood ratios, or P-values, to provide direct measures of strength of evidence; a particularly striking result [extended from Dempster (1974)] is that, with normal likelihoods and flat priors, the P-value is equal to *the posterior probability that the TLR is greater than 1*, or equivalently, $1 - P$ is the posterior probability that the TLR is less than 1. So a P-value of 0.05 means that the posterior probability is 0.95 that the TLR is less than 1. Since we need a likelihood ratio of 1/10 or less for quite strong evidence, this P-value does not provide that evidence.

In general the recalibration requires us to interpret P-values more conservatively; to achieve a posterior probability of $\pi_{k,p}$ that the TLR for the null hypothesis to the alternative is less than k, the necessary P-value (for normal

Table 2.2. P-value required for Pr[TLR < 0.1] = 0.84

p	1	2	3	4	5	6	7	8	9	10
PV	0.010	0.016	0.021	0.025	0.028	0.031	0.034	0.037	0.039	0.042
p	11	12	13	14	15	16	17	18	19	20
PV	0.044	0.046	0.048	0.049	0.051	0.053	0.054	0.055	0.057	0.058

likelihoods, and flat priors) is

$$P = 1 - F_p(F_p^{-1}(\pi_{k,p}) - 2\log k),$$

where $F_p(x)$ is the *cdf* of the χ_p^2 distribution at x. For reasonably strong evidence against the null hypothesis, we will require $k = 0.1$ and $\pi_{kp} = 0.84$, giving the P-values (PV) shown in Table 2.2 (adapted from Aitkin, 1997).

This P-value table is for large samples; small-sample results remain to be developed for specific models. A detailed discussion is given in Chadwick (2002), and some examples in Aitkin *et al.* (2005).

We will use this table as a guideline for model interpretation in subsequent chapters: a single-parameter hypothesis requires a P-value of 0.01 or less (or an equivalent critical Z-value of 2.57, or a relative likelihood of $\exp\{-\frac{1}{2}\cdot 2.57^2\} = 0.037$) for quite strong evidence against the hypothesis, but as the number p of specified parameters in the null hypothesis increases, so does the P-value required, but at a quite slow rate for large p.

For the normal mean–variance model, we have as before

$$\begin{aligned}
\text{TLR} &= L(\mu_0, \sigma)/L(\mu, \sigma) \\
-2\log\text{TLR} &= \frac{n(\bar{y}-\mu_0)^2}{\sigma^2} - \frac{n(\bar{y}-\mu)^2}{\sigma^2} \\
&= \frac{n(\bar{y}-\mu_0)^2}{s^2}\cdot\frac{s^2}{\sigma^2} - \frac{n(\bar{y}-\mu)^2}{\sigma^2} \\
&= t_0^2 \cdot \frac{T}{(n-1)\sigma^2} - Z^2,
\end{aligned}$$

where $Z \sim N(0,1)$ conditional on σ, and $T/\sigma^2 \sim \chi_{n-1}^2$ marginally. The posterior distribution of $-2\log\text{TLR}$ does not have a closed form in this case, though we can evaluate immediately

$$\begin{aligned}
\Pr[\text{TLR} < 1] &= \Pr[-2\log\text{TLR} > 0] \\
&= \Pr\left[\frac{Z^2}{T/\sigma^2} < t_0^2\right] \\
&= \Pr[F_{1,n-1} < t_0^2] \\
&= 1 - P,
\end{aligned}$$

where P is the P-value of the observed t_0. So as in the simpler cases above, the P-value is again the posterior probability that the TLR is greater than 1. For other values of k, and for other models without closed-form posteriors, *simulation* of the posterior distribution provides a general solution. Further discussion is given in Chadwick (2002) and Aitkin *et al.* (2005).

We discuss finally the third inferential question, assessing the evidence for $H_1 : \mu = \mu_1$ against $H_2 : \mu = \mu_2$. This is the most difficult for the Neyman–Pearson theory, as for the simpler case of two specified null hypotheses, but it has a surprisingly simple answer from the posterior distribution of the TLR. Proceeding as above, we have

$$\begin{aligned}
\text{TLR} &= L(\mu_1, \sigma)/L(\mu_2, \sigma) \\
-2\log \text{TLR} &= \frac{n(\bar{y}-\mu_1)^2}{\sigma^2} - \frac{n(\bar{y}-\mu_2)^2}{\sigma^2} \\
&= \left[\frac{n(\bar{y}-\mu_1)^2}{s^2} - \frac{n(\bar{y}-\mu_2)^2}{s^2}\right] \cdot \frac{s^2}{\sigma^2} \\
&= \left[t_1^2 - t_2^2\right] \cdot \frac{T}{(n-1)\sigma^2}.
\end{aligned}$$

Thus $-2\log \text{TLR} \sim [t_1^2 - t_2^2]\chi^2_{n-1}/(n-1)$. Here $[t_1^2 - t_2^2]/(n-1) = -5/24 = -0.208$, and taking $k = 10$

$$\begin{aligned}
\Pr[\text{TLR} > 10] &= \Pr[-2\log \text{TLR} < -2\log 10] \\
&= \Pr\left[\chi^2_{n-1} < \frac{n-1}{t_1^2 - t_2^2} \cdot (-2\log 10)\right] \\
&= \Pr[\chi^2_{24} < -4.8 \cdot -4.60 = 22.08] \\
&= 0.426.
\end{aligned}$$

The posterior probability that H_1 is quite strongly supported (TLR > 10) relative to H_2 is less than 0.5, but it is certainly better supported.

As n increases, $\chi^2_{n-1}/(n-1)$ rapidly approaches $N(1, 2/(n-1))$ which converges to 1 in probability. For large n the disparity or deviance difference between the models is effectively compared with $-2\log k$ as in the case of two simple hypotheses. This result (for large n) applies generally to the comparison of 'common-family' models with the same number of parameters which are both nested in a larger family.

We consider now a second example.

2.7.6 *Binomial model*

A random sample of $n = 10$ is drawn from the STATLAB population and the number of 'smoking mothers' – mothers smoking at the time of birth of the baby – is found to be $r = 1$. What information do we have about θ, the proportion of smoking mothers in the STATLAB population? There is no specific hypothesis about its value. The probability model here is a Bernoulli trial model, in which 'trials' are draws of a mother from the population, and θ is the probability of a 'success' – the drawing of a smoking mother from the population. The population is large compared to the sample and so sampling with and without replacement are equivalent.

Using the standard device of a dummy or indicator variable y_i which takes the value 1 if the i-th trial results in success and 0 if failure, the probability of the sequence $y_1, \ldots, y_n$ of successes and failures is

$$L(\theta) = \prod_{i=1}^{n} \theta^{y_i}(1-\theta)^{1-y_i} = \theta^r(1-\theta)^{n-r},$$

where $r = \sum y_i$ is the number of successes, and this does not depend on the particular sequence of zeroes and ones, only on the total number of successes. Thus r is a sufficient statistic for θ. The likelihood function is shown in Fig. 2.16(a) on a fine grid. It is remarkably skewed. R functions are

```
> theta <- seq(from = 0, to = 1, by = 0.001)
> bl <- theta * (1 - theta)$^9$
> plot(theta, bl, xlab = expression(theta), ylab = "likelihood",
+      type = "l")
```

The MLE is $\hat{\theta} = 0.1$, and a likelihood interval for θ is obtained from $R(\theta) < c$. The equation $R(\theta) = c$ cannot be inverted to give θ analytically, so numerical computation is necessary. The simplest approach is to solve it iteratively by the Newton–Raphson method. We do not give details here. For a relative likelihood of 0.146, the likelihood interval endpoints are 0.006 and 0.372 (to 3 dp). The somewhat tedious computation of the interval endpoints can be avoided in single-parameter models in the exponential family by using the (approximate) likelihood normalizing transformation, as in the previous discussion. For the binomial or Bernoulli model

$$f(y \mid \theta) = \theta^y(1-\theta)^{1-y},$$

we have

$$\mathrm{E}[Y] = \mu = \theta, \quad \mathrm{Var}[Y] = V(\mu) = \mu(1-\mu),$$

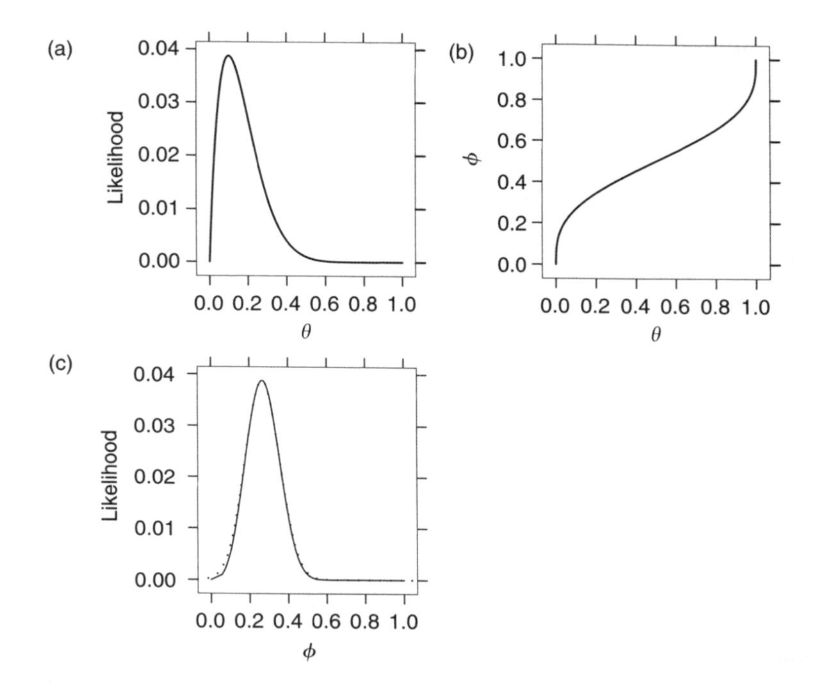

Fig. 2.16. (a) Likelihood for one success in 10 trials; (b) likelihood normalizing transformation of θ; and (c) exact likelihood and normal approximation, ϕ scale

and the required transformation is

$$\phi = \int_0^\theta t^{-2/3}(1-t)^{-2/3}\,\mathrm{d}t,$$

an incomplete Beta integral. This was tabulated by Anscombe (1964) in the form $\frac{1}{3}B_\theta(\frac{1}{3},\frac{1}{3})$. We adopt a slightly different form in which ϕ has the range (0, 1) like θ: define

$$\phi = \int_0^\theta t^{-2/3}(1-t)^{-2/3}\,\mathrm{d}t \Big/ \int_0^1 t^{-2/3}(1-t)^{-2/3}\,\mathrm{d}t$$

so that ϕ is the *cdf* at θ of the Beta $\left(\frac{1}{3},\frac{1}{3}\right)$ distribution. A graph of ϕ against θ is shown in Fig. 2.16(c). The transformation is similar in shape to the logit transformation (Chapter 4) but has finite limits 0 and 1, like the arc-sine transformation. The transformation is easily calculated in R by

```
> phi <- pbeta(theta, 1/3, 1/3)
> plot(theta, phi, xlab = expression(theta),
+      ylab = expression(phi), type = "l")
```

The information for $L(\phi)$ is easily obtained by chain differentiation, giving

$$I(\hat{\phi}) = I(\hat{\theta})/[\phi'(\hat{\theta})]^2$$

and so

$$\text{SE}(\hat{\phi}) = \text{SE}(\hat{\theta})\phi'(\hat{\theta}).$$

In our binomial example

$$I(\hat{\theta}) = \frac{n}{\hat{\theta}(1-\hat{\theta})}$$

$$\phi'(\hat{\theta}) = \hat{\theta}^{-2/3}(1-\hat{\theta})^{-2/3}/B\left(\frac{1}{3}, \frac{1}{3}\right)$$

giving

$$\text{SE}(\hat{\phi}) = \hat{\theta}^{-1/6}(1-\hat{\theta})^{-1/6}/B\left(\frac{1}{3}, \frac{1}{3}\right)\sqrt{n}.$$

To evaluate this, we need $B\left(\frac{1}{3}, \frac{1}{3}\right) = \Gamma^2\left(\frac{1}{3}\right)/\Gamma\left(\frac{1}{3}\right) = 5.300$.

With $\hat{\theta} = 0.1$ and $n = 10$, the value of $\hat{\phi}$ is 0.267 and $\text{SE}(\hat{\phi}) = 0.0891$. The approximate 0.146 likelihood interval for ϕ is then $\hat{\phi} \pm 1.96\text{SE}(\hat{\phi})$, that is (0.092, 0.442), which transforms to $\theta \in (0.004, 0.380)$. This agrees quite well with the exact interval (0.006, 0.372) found earlier, though it is slightly longer. The relative likelihoods at the endpoints are 0.100 and 0.133, not quite equal to 0.146. Figure 2.16(c) shows the relative likelihood on the ϕ scale (solid line) and the normal approximation (dotted line). R functions are

```
> theta <- seq(from = 0, to = 1, by = 0.001)
> bl <- theta * (1 - theta)$^9$
> plot.binomial.a <- xyplot(bl ~ theta, xlab = expression(theta),
+     ylab = "likelihood", type = "l", main = "(a)")
> phi <- pbeta(theta, 1/3, 1/3)
> plot.binomial.b <- xyplot(phi ~ theta, xlab = expression(theta),
+     ylab = expression(phi), type = "l", main = "(b)")
> s <- 0.1$^(-1/6)$ * 0.9$^(-1/6)$/(5.3 * sqrt(10))
> nl <- 0.1 * 0.9$^9$ * exp(-0.5 * ((phi - 0.267)/s)$^2$)
> plot.binomial.c <- xyplot(bl ~ phi, type = "l",
+     xlab = expression(phi), ylab = "likelihood",
+     main = "(c)", panel = function(x, y, ...) {
+         panel.xyplot(x, y, ...)
+         s <- 0.1$^(-1/6)$ * 0.9$^(-1/6)$/(5.3 * sqrt(10))
+         panel.curve(0.1 * 0.9$^9$ * exp(-0.5 * ((x -
+             0.267)/s)$^2$), lty = "dotted", cex = 1.5)
+     })
```

```
> print(plot.binomial.a, position = c(0, 0.5, 0.5,
+     1), more = TRUE)
> print(plot.binomial.b, position = c(0.5, 0.5, 1,
+     1), more = TRUE)
> print(plot.binomial.c, position = c(0, 0, 0.5, 0.5))
```

The visual agreement is close even for this small sample and extreme result except in the tails, where the normal approximation gives intervals which are slightly too long. The normal approximation can be further improved using a t-approximation to the log-likelihood with the degrees of freedom determined by the fourth derivative of the log-likelihood (Sprott, 1980). We do not pursue this as even in small samples the estimated degrees of freedom is usually very large and the normal approximation is recovered.

When $r = 0$ or n the normalizing transformation fails, as the maximum is on the boundary of the parameter space. However, relative likelihood intervals can be easily calculated analytically in these cases: for example when $r = 0$, the likelihood is

$$L(\theta) = (1 - \theta)^n,$$

and a relative likelihood of c gives the interval for θ of $(0, 1 - c^{1/n})$. If $r = n$ the interval is $(c^{1/n}, 1)$. For $n = 10$ and $c = 0.146$, $c^{1/n} = 0.825$.

For the event of $r = 1$ success in $n = 10$ trials, the (approximate) 95% likelihood-based confidence interval is obtained (to 3 dp) as (0.004, 0.380).

```
> theta <- seq(0, 1, length = 1000)
> L <- dbinom(1, size = 10, prob = theta)
> maxL <- max(L)
> L <- L/maxL
> plot(theta, L, type = "l", lwd = 2,
+     ylab = "Relative likelihood, binomial(1,10)",
+     xlab = expression(theta))
> abline(h = 0.146, lty = 2, lwd = 2)
> a <- theta[L > 0.146]
> limits <- range(a)
> round(limits, 3)
[1] 0.004 0.380
```

The confidence interval construction above does not guarantee a minimum 95% coverage over all θ. For this we need the classical interval construction, in which we consider other possible values of r, besides the observed value 1, which might have been generated by the the given value $n = 10$. The $100(1 - \alpha)\%$ confidence limits for θ are the solutions of

$$\Pr(R \geq 1 \mid \theta_{\mathrm{L}}) = 1 - (1 - \theta)^{10} = \alpha/2$$

$$\Pr(R \leq 1 \mid \theta_{\mathrm{U}}) = (1 - \theta)^{10} + 10\theta(1 - \theta)^9 = \alpha/2,$$

giving for $\alpha = 0.05$, $\theta_L = 0.0025$, $\theta_U = 0.445$. These values are easily obtained in R from the cumulative binomial distribution function *pbinom(r,n,θ)* and the function *uniroot* which finds the root of a function with respect to its first argument.

```
> uniroot(function(theta) 0.025 - (1 - pbinom(0, 10,
+      theta)), c(0, 1))$root

[1] 0.0025315970

> uniroot(function(theta) 0.025 - (pbinom(1, 10, theta)),
+      c(0, 1))$root
[1] 0.44501716
```

They are substantially wider than the likelihood-based intervals. The construction of the classical confidence interval from the exact binomial distribution of R is frequently avoided by relying on the large-sample normality of the distribution of $\hat{\Theta}$ or of the score. The log-likelihood, score and information are

$$\begin{aligned}
\ell(\theta) &= r\log\theta + (n-r)\log(1-\theta)\\
s(\theta) &= \frac{r}{\theta} - \frac{n-r}{1-\theta}\\
&= \frac{r-n\theta}{\theta(1-\theta)}\\
I(\theta) &= \frac{r}{\theta^2} + \frac{n-r}{(1-\theta)^2}\\
I(\hat{\theta}) &= \frac{n}{\hat{\theta}(1-\hat{\theta})}\\
\mathcal{I}(\theta) &= \frac{n}{\theta(1-\theta)}.
\end{aligned}$$

The confidence intervals based on the asymptotic normal distribution of the MLE $\hat{\Theta} \sim N(\theta, \hat{\theta}(1-\hat{\theta})/n)$ (equivalent to inverting the Wald test) are $\theta \in \hat{\theta} \pm \lambda_{\alpha/2}\sqrt{\hat{\theta}(1-\hat{\theta})/n}$. In our example with $\hat{\theta} = 0.1$ and $n = 10$, the approximate 95% confidence interval using $\lambda = 1.96$ is $(-0.086, 0.286)$ which includes impossible negative values of θ. Truncating the interval to $(0, 0.286)$ does not help matters because we know θ must be *positive*, since one success has been observed. If $r = 0$ successes are observed, this interval construction collapses with a zero standard error.

Proper intervals may be obtained by not substituting $\hat{\theta}$ for θ in the information (equivalent to inverting the score test rather than the Wald test). The confidence interval is

$$\theta \in \hat{\theta} \pm \lambda_{\alpha/2}\sqrt{\theta(1-\theta)/n},$$

which is equivalent to

$$\theta \in (\theta_L, \theta_U),$$

where $\theta_{\mathrm{L}} < \theta_{\mathrm{U}}$ are the roots of the quadratic equation (writing $\lambda = \lambda_{\alpha/2}$)

$$(\theta - \hat{\theta})^2 = \lambda^2\theta(1-\theta)/n.$$

The roots are

$$\frac{\hat{\theta} + \lambda^2/2n \pm \lambda\sqrt{\hat{\theta}(1-\hat{\theta})/n + \lambda^2/4n^2}}{1 + \lambda^2/n}.$$

This interval always lies in $[0, 1]$. If $r = 0$ the interval endpoints are 0 and $(\lambda^2/n)/(1 + \lambda^2/n) = 0.278$ for $n = 10$. In our example with $\lambda = 1.96, \theta_{\mathrm{L}} = 0.018, \theta_{\mathrm{U}} = 0.404$. An R function to calculate this interval is

```
> bin.conf.int <- function(r, n, alpha = 0.05) {
+       theta <- r/n
+       lambda <- qnorm(1 - alpha/2)
+       ci2 <- lambda * sqrt(theta * (1 - theta)/n +
+           lambda$^2$4/n/n)
+       lower <- (theta + lambda$^2$/2/n - ci2)/(1 + lambda$^2/n$)
+       upper <- (theta + lambda$^2$/2/n + ci2)/(1 + lambda$^2/n$)
+       list(lower = lower, upper = upper)
+ }
```

Bayes inference is straightforward. The uniform prior on (0, 1) gives a Beta posterior density:

$$\pi(\theta \mid y) = \theta^r(1-\theta)^{n-r}/B(r+1, n-r+1),$$

which is the likelihood scaled to integrate to one. Tail area probabilities or critical values are easily found from the Beta distribution using `pbeta` or `qbeta`, but the HPD interval is more difficult.

For the example with $r = 1$ and $n = 10$, a 95% equal-tailed credible interval for θ from the Beta(2, 10) distribution is (0.023, 0.413). This is shifted upwards substantially relative to the likelihood interval.

If $r = 0$, the (one-tailed) 95% credible interval is (0, 0.238).

In R, HPD intervals can be found when $r > 0$ using the `betaHPD` function from the `pscl` package. The HPD interval for the $r = 1$ and $n = 10$ example using this function is (0.006, 0.368).

2.7.7 *Hypergeometric sampling from finite populations*

We noted in Section 2.5 that the construction of the binomial likelihood depended on the assumption that the population size N was large compared to the sample size n, and so sampling with and without replacement were equivalent. For general response variables Y it is difficult to construct the likelihood for sampling without replacement, since the approximating population model has to change with the sampled population, but for categorical response variables the likelihood construction is straightforward. We illustrate with the binary response model for the smoking mothers example.

Let R be the number of smoking mothers in the population, with $\theta = R/N$. We draw a random sample of size n without replacement, and find r smoking mothers. The likelihood function (in the parameter $R = N\theta$) is given by the hypergeometric sampling model

$$L(R) = \binom{R}{r}\binom{N-r}{n-r}\Big/\binom{N}{n}, \qquad R = r, r+1, \ldots, N-(n-r).$$

This can readily be computed as a function of R using `lgamma`, the log gamma function, to evaluate the factorials. Expanding the binomial coefficients, we have

$$\begin{aligned}
L(R) &= c \times \prod_{i=1}^{r}(R-i+1)\prod_{j=1}^{n-r}(N-R-j+1) \\
&= c' \times \prod_{i=1}^{r}\left(\frac{R}{N} - \frac{i-1}{N}\right)\prod_{j=1}^{n-r}\left(\frac{N-R}{N} - \frac{j-1}{N}\right) \\
&= c'\theta^r(1-\theta)^{n-r}\prod_{i=1}^{r}\left(1 - \frac{i-1}{R}\right)\prod_{j=1}^{n-r}\left(1 - \frac{j-1}{N-R}\right) \\
&= c'\theta^r(1-\theta)^{n-r} \\
&\quad \times \exp\left[-\frac{r(r-1)}{2R} - \frac{(n-r)(n-r-1)}{2(N-R)} + O\left(\frac{1}{R^2}\right) + O\left(\frac{1}{(N-R)^2}\right)\right].
\end{aligned}$$

If r and $n-r$ are small compared to R and $N-R$, the exponential term will be close to 1, and the likelihood will be equivalent to the previous binomial likelihood. It is easily shown by considering the ratio $L(R+1)/L(R)$ that the likelihood is maximized when

$$R = \hat{R} = [(N+1)r/n + 1] = [N\tilde{\theta} + 1 + \tilde{\theta}],$$

where $\tilde{\theta} = r/n$ is the binomial model MLE and $[x]$ is the integral part of x. For our example

$$\hat{\theta} = \hat{R}/N = 130/1296 = 0.1003,$$

and the binomial and hypergeometric likelihoods are indistinguishable since n is very small compared to N.

2.8 The effect of the sample design on inference

In our second example we have assumed that a simple random sample of fixed size n is taken, and r 'successes' are observed. But other sample designs might have led to the same data. For example, individuals might have been sampled randomly until the r-th 'success' – smoking mother – was found, this occurring at the n-th observation. The likelihood function would, as a function of θ, be the same as before:

$$L(\theta) = \theta^{r-1}(1-\theta)^{n-r} \cdot \theta,$$

but now r is fixed by the sample design, and the sample size N is the random variable. It is a striking feature of the difference between the Bayes and likelihood theories, and the Neyman–Pearson theory, that this different sample design affects the inference about θ in the latter theory, but not in the former.

For the Bayes and likelihood theories, this is obvious: the likelihood interval depends only on the likelihood from the observed data, so for the same likelihood and the same prior the same Bayes and likelihood intervals will be obtained.

To see the effect on the Neyman–Pearson theory, note that the sufficient statistic N has a negative binomial distribution with parameters r and θ:

$$\Pr(N = n \mid r, \theta) = \binom{n-1}{r-1}\theta^r(1-\theta)^{n-r}$$

for $n = r, r+1, \ldots$ since the last trial must result in a success, and the preceding $r-1$ successes can occur anywhere in the preceding $n-1$ trials. Note that $\mathrm{E}[N] = r/\theta$. To construct a Neyman–Pearson 95% confidence interval for θ, we now have to consider other possible values of n, besides the observed value 10, which might have generated the given value $r = 1$. Now the confidence limits for θ are the solutions of

$$\Pr(N \geq 10 \mid \theta_{\mathrm{U}}) = \alpha/2,$$
$$\Pr(N \leq 10 \mid \theta_{\mathrm{L}}) = \alpha/2,$$

which require the *cdf* of the negative binomial distribution, available in R as `pnbinom`. The solutions are for $\alpha = 0.05, \theta_{\mathrm{L}} = 0.0025, \theta_{\mathrm{U}} = 0.336$. The lower endpoint is the same, but the upper endpoint is quite different from the endpoint 0.445 for binomial sampling.

This dependence of the family of intervals on the sampling design used to obtain the data, commonly called the *stopping rule*, is a major complication of conventional confidence interval construction. A striking example of the difficulty of Neyman–Pearson inference in such cases is the ECMO trial data, discussed in Begg (1990), in which eight different P-values are given for the same 2×2 contingency table. In this example the precise design or stopping rule was not clearly specified, or if specified, was not followed, and so strictly speaking *no* classical confidence interval statement can be made! The different P-values are based on different hypothetical replications of the study. Chapter 4 gives more details.

In the Neyman–Pearson theory, we have to be aware of the possibility of different forms of hypothetical replication, each of which may lead to a different statement of precision of the likelihood interval. Defining the *reference set* of hypothetical replications with which our sample is to be compared becomes an important issue in this theory. This is another aspect of the stopping rule issue discussed above: if we do not know exactly how our sample was obtained, how can we know what reference set should be used for its evaluation?

In the next section we discuss the exponential family of distributions.

2.9 The exponential family

The exponential family of distributions is central to data analysis in most scientific fields. Members of this family can represent variation in proportions, counts and time durations as well as the usual 'continuous' measurements for which the normal distribution was traditionally used. Regression modelling using the binomial, Poisson and gamma distributions has largely replaced the older approximate methods involving transformations of the response to more nearly approximate a normal distribution.

The probability function $f(y)$ for this family has the general form

$$\log f(y) = [y\psi - b(\psi)]/\phi + c(y, \phi).$$

The parameter ψ is the 'natural' or 'canonical' parameter of the distribution, and ϕ is a 'scale' parameter. The binomial and Poisson distributions do not have a scale parameter (formally we can set it to 1.0), while in the normal distribution ϕ is related to σ^2 and in the gamma distribution to the shape parameter r. We now consider some general properties of the exponential family.

2.9.1 *Mean and variance*

Differentiating with respect to ψ, we have

$$\frac{\partial \log f}{\partial \psi} = [y - b'(\psi)]/\phi,$$
$$\frac{\partial^2 \log f}{\partial \psi^2} = -b''(\psi)/\phi.$$

Since $\int f(y)\,\mathrm{d}y = 1$, differentiating with respect to ψ gives

$$\begin{aligned} 0 &= \int \frac{\partial f}{\partial \psi}\,\mathrm{d}y = \int \frac{\partial \log f}{\partial \psi} \times f\,\mathrm{d}y \\ &= \mathrm{E}\left[\frac{\partial \log f}{\partial \psi}\right] \\ &= \{\mathrm{E}[Y] - b'(\psi)\}/\phi \\ 0 &= \int \frac{\partial^2 f}{\partial \psi^2}\,\mathrm{d}y = \int \left\{\frac{\partial^2 \log f}{\partial \psi^2}\cdot f + \left(\frac{\partial \log f}{\partial \psi}\right)^2 \cdot f\right\}\mathrm{d}y \\ &= \mathrm{E}\left[\frac{\partial^2 \log f}{\partial \psi^2}\right] + \mathrm{E}\left[\frac{\partial \log f}{\partial \psi}\right]^2. \end{aligned}$$

Thus

$$\begin{aligned} \mu = \mathrm{E}[Y] &= b'(\psi) \\ \mathrm{Var}[Y] &= \mathrm{E}[Y - b'(\psi)]^2 \\ &= \phi^2 \mathrm{E}\left[\frac{\partial \log f}{\partial \psi}\right]^2 \\ &= -\phi^2 \mathrm{E}\left[\frac{\partial^2 \log f}{\partial \psi^2}\right] \\ &= \phi b''(\psi). \end{aligned}$$

The mean involves only ψ while in the variance ϕ appears as a scale multiplier (whence its name).

2.9.2 *Generalized linear models*

A generalized linear model is defined by the following three components:

(1) an exponential family distribution for the response variable Y;
(2) a linear regression function or *linear predictor* η in the explanatory variables

$x_1, \ldots, x_p$:

$$\eta = \beta' x = \beta_0 x_0 + \beta_1 x_1 + \ldots + \beta_p x_p,$$

where x_0 is identically 1;

(3) a *parameter transformation* or *link function* $g(\mu)$ which relates the linear predictor η to the mean μ:

$$\eta = g(\mu).$$

We will use the standard abbreviation of GLM for generalized linear model, though this causes frequent confusion with the older meaning of GLM – *general* linear model, which is the generalized linear model in the normal distribution.

It is an important restriction of GLMs that the explanatory variables *do not affect the scale parameter*. This restriction is relaxed in Chapter 3 on the normal distribution and in Chapter 6 on the gamma distribution, where we consider *double modelling* of both parameters by regression models.

R provides general methods for fitting GLMs, for a wide range of standard link functions for each distribution, and also for non-standard user-defined link functions. We have to specify the response variable Y to be modelled, the probability ('error') distribution for Y, the link function and the regresssion model to be fitted. The `glm` functions are specified in symbolic form

$$y \sim \text{model}.$$

1. The response variable is placed on the left of the $\sim$ operator while the model is specified on the right.
2. The error distribution is specified by the argument `family`, which has distributional values `normal`, `poisson`, `binomial`, `Gamma`, `inverse.gaussian`, `quasi`, `quasibinomial` and `quasipoisson`. See `help(family)` for more details.
3. The link function can be explicitly specified as an argument within the function defining the error distribution.
4. The regression model is specified as a series of variable and factor names separated by the `+` operator. Interactions between variables and factors are defined by placing the `:` operator between the terms.

Much more general models (e.g. non-linear in the parameters) can be fitted by generalizations of the computational algorithm used in R. We shall concentrate on the standard models in R, with extensions to other distributions (like the exponential, Weibull and extreme value, and mixtures of the exponential family) which are fitted by extensions of the standard models.

2.9.3 *Maximum likelihood fitting of the GLM*

We consider the GLM with $\eta_i = g(\mu_i) = \boldsymbol{\beta}' \mathbf{x}_i$. The log-likelihood function is

$$\ell(\boldsymbol{\beta}, \phi) = \sum_i [y_i \psi_i - b(\psi_i)]/\phi + \sum_i c(y_i, \phi)$$

and its derivatives, the score function components, are

$$\mathbf{s}(\boldsymbol{\beta}) = \frac{\partial \ell}{\partial \boldsymbol{\beta}} = \sum_i [y_i - b'(\psi_i)] \cdot \frac{\partial \psi_i}{\partial \boldsymbol{\beta}} \Big/ \phi$$

$$s(\phi) = \frac{\partial \ell}{\partial \phi} = -\sum [y_i \psi_i - b(\psi_i)]/\phi^2 + \sum_i c'(y_i, \phi),$$

where c' is the derivative of c with respect to ϕ. Substitution of

$$\frac{\partial \psi_i}{\partial \boldsymbol{\beta}} = \frac{d\psi_i}{d\mu_i} \cdot \frac{d\mu_i}{d\eta_i} \cdot \frac{\partial \eta_i}{\partial \boldsymbol{\beta}} = \frac{1}{b''(\psi_i)} \cdot \frac{1}{g'(\mu_i)} \cdot \mathbf{x}_i$$

and

$$\phi b''(\psi_i) = \mathrm{Var}[Y_i] = V_i, \quad b'(\psi_i) = \mu_i, \quad g'(\mu_i) = g'_i$$

gives

$$\mathbf{s}(\boldsymbol{\beta}) = \frac{\partial \ell}{\partial \boldsymbol{\beta}} = \sum_i (y_i - \mu_i)\mathbf{x}_i / V_i g'_i.$$

For the second derivatives, we have

$$\frac{\partial^2 \ell}{\partial \boldsymbol{\beta} \partial \boldsymbol{\beta}'} = \sum_i [-b''(\psi_i)] \frac{\partial \psi_i}{\partial \boldsymbol{\beta}} \frac{\partial \psi_i}{\partial \boldsymbol{\beta}'} \Big/ \phi + \sum_i [y_i - b'(\psi_i)] \frac{\partial^2 \psi_i}{\partial \boldsymbol{\beta} \partial \boldsymbol{\beta}'} \Big/ \phi$$

$$\frac{\partial^2 \ell}{\partial \boldsymbol{\beta} \partial \phi} = -\sum_i [y_i - b'(\psi_i)] \cdot \frac{\partial \psi_i}{\partial \boldsymbol{\beta}} \Big/ \phi^2 = -\mathbf{s}(\boldsymbol{\beta})/\phi$$

$$c\frac{\partial^2 \ell}{\partial \phi^2} = 2\sum_i [y_i \psi_i - b(\psi_i)]/\phi^3 + \sum_i c''(y_i, \phi).$$

Straightforward algebra gives

$$\frac{\partial^2 \ell}{\partial \boldsymbol{\beta} \partial \boldsymbol{\beta}'} = -\sum_i \mathbf{x}_i \mathbf{x}'_i / V_i g'^2_i - \sum_i (y_i - \mu_i) \mathbf{x}_i \mathbf{x}'_i (V_i g''_i + V'_i g'_i)/V_i^2 g'^3_i = -X' W^* X,$$

where W^* is a diagonal weight matrix with elements

$$w_i^* = w_i + (y_i - \mu_i)(V_i g''_i + V'_i g'_i)/V_i^2 g'^3_i$$

and

$$w_i = (V_i g_i'^2)^{-1}, \quad V_i' = \frac{dV_i}{d\mu_i}, \quad g_i'' = \frac{d^2 g(\mu_i)}{d\mu_i^2},$$

while X is the $n \times (p+1)$ design matrix of the explanatory variables $\mathbf{x}$. If the likelihood has an internal maximum over $\boldsymbol{\beta}$ at $\hat{\boldsymbol{\beta}}$, then $\mathbf{s}(\hat{\boldsymbol{\beta}}) = \mathbf{0}$. Also the expected score in $\boldsymbol{\beta}$ is zero, and hence both the observed and expected information matrices have a block-diagonal structure, with the off-diagonal block of cross-derivatives betweeen $\boldsymbol{\beta}$ and ϕ equal to zero. Further, since ϕ appears in the score equation for $\boldsymbol{\beta}$ only as a scale constant in V_i, the MLE of $\boldsymbol{\beta}$ can be obtained independently of ϕ. The MLE of ϕ can then be found by solving the single score equation in ϕ evaluated at the MLE $\hat{\boldsymbol{\beta}}$. This process is familiar from the normal distribution.

We now give the form of the Newton–Raphson (NR) and Fisher scoring (FS) algorithms for the GLM. We solve iteratively for the zero of the score function $\mathbf{s}(\boldsymbol{\beta})$. Let the estimate at the r-th iteration be $\boldsymbol{\beta}_r$. Then the estimate at the $(r+1)$-th iteration of the NR algorithm is given by solving

$$\mathbf{0} = \mathbf{s}(\boldsymbol{\beta}_r) + \frac{\partial \mathbf{s}(\boldsymbol{\beta})}{\partial \boldsymbol{\beta}'}(\boldsymbol{\beta}_{r+1} - \boldsymbol{\beta}_r)$$

giving

$$\begin{aligned}\boldsymbol{\beta}_{r+1} &= \boldsymbol{\beta}_r - H^{-1}(\boldsymbol{\beta}_r)\mathbf{s}(\boldsymbol{\beta}_r) \\ &= \boldsymbol{\beta}_r + (X'W_r^*X)^{-1}\mathbf{s}(\boldsymbol{\beta}_r),\end{aligned}$$

where H is the Hessian matrix and W_r^* is W^* evaluated at the estimate $\boldsymbol{\beta}_r$. Now

$$\begin{aligned}\mathbf{s}(\boldsymbol{\beta}) &= \sum_i (y_i - \mu_i) g'(\mu_i) w_i \mathbf{x}_i \\ &= X'W\mathbf{u},\end{aligned}$$

where

$$u_i = (y_i - \mu_i) g'(\mu_i).$$

Then each iteration of the NR algorithm can be expressed as

$$\begin{aligned}\boldsymbol{\beta}_{r+1} &= \boldsymbol{\beta}_r + (X'W_r^*X)^{-1}X'W_r\mathbf{u}_r \\ &= (X'W_r^*X)^{-1}(X'W_r^*X\boldsymbol{\beta}_r + X'W_r\mathbf{u}_r) \\ &= (X'W_r^*X)^{-1}(X'W_r^*\boldsymbol{\eta}_r + X'W_r\mathbf{u}_r) \\ &= (X'W_r^*X)^{-1}X'W_r^*\mathbf{z}_r^*,\end{aligned}$$

where $\boldsymbol{\eta}_r$ is the vector of 'fitted values' of the linear predictor $\boldsymbol{\eta}$ at the r-th iteration, and

$$\mathbf{z}^* = \boldsymbol{\eta} + W^{*-1}W\mathbf{u}.$$

Thus each iteration of the NR algorithm can be expressed as a weighted least squares regression of an 'adjusted dependent variate' $\mathbf{z}^*$ on the explanatory variables $\mathbf{x}$ with weight vector (diagonal weight matrix) W^*.

R uses the FS algorithm rather than the NR algorithm. This uses the expected rather than the observed information. Since $\mathrm{E}[Y_i - b'(\psi_i)] = 0$ for all i, the second term in the Hessian matrix is identically zero, and the weights w_i and w_i^* are identical. Thus each iteration of the FS algorithm can be expressed as

$$\begin{aligned}\boldsymbol{\beta}_{r+1} &= \boldsymbol{\beta}_r + (X'W_rX)^{-1}X'W_r\mathbf{u}_r \\ &= (X'W_rX)^{-1}(X'W_rX\boldsymbol{\beta}_r + X'W_r\mathbf{u}_r) \\ &= (X'W_rX)^{-1}X'W_r\mathbf{z}_r,\end{aligned}$$

where

$$\mathbf{z} = \boldsymbol{\eta} + \mathbf{u}.$$

The adjusted dependent variate in this case is much simpler. The FS algorithm is generally more stable than the NR, and converges at about the same rate. In rare cases it may not converge but oscillate in successive iterations (Ridout, 1990).

The NR algorithm gives the correct observed information for $\boldsymbol{\beta}$, after scaling by the MLE of ϕ (used in the weight matrix W^*), though as noted the parameter estimates do not depend on ϕ. The FS algorithm gives the (estimated) expected information matrix, when correspondingly scaled. This scaling is not needed for the Poisson and binomial distributions, where $\phi = 1$. In R the value of ϕ may be set as the `dispersion` argument in the `summary` method for the `glm` class.

If it is *not* set by the user, it is automatically calculated by R for the normal and gamma distributions. Details are given in the chapters on these distributions.

The observed information represents the curvature of the observed log-likelihood at the MLE. The expected information represents the *average* curvature over outcomes Y with the same explanatory variable values. In models with canonical links (when $\eta = \psi$) the informations are identical, but for other links they are different.

For individual parameter estimates $\hat{\beta}_j$, the standard error is s_j, where s_j^2 is the j-th diagonal element of the inverse of the (observed or expected) information matrix.

2.9.4 *Model comparisons through maximized likelihoods*

The comparison of competing regression models for the data is based on the LRTS.

Consider two candidate models for the data: a 'full' model $\eta_f = \boldsymbol{\beta}'\mathbf{x}$ and a 'reduced' model $\eta_r = \boldsymbol{\beta}_r'\mathbf{x}_r$, where $\mathbf{x}_r$ is a vector of length $p_1 + 1$, made up of the constant 1 plus p_1 of the explanatory variables in $\mathbf{x}$, and $\boldsymbol{\beta}_r$ is the corresponding vector of regression coefficients. We say that the reduced model is 'nested' in the full model. Write $\mathbf{x}_d$ for the vector of explanatory variables deleted from $\mathbf{x}$ in defining $\mathbf{x}_r$, and $\boldsymbol{\beta}_d$ for the corresponding vector of regression coefficients, so that

$$\boldsymbol{\beta}'\mathbf{x} = \boldsymbol{\beta}_r'\mathbf{x}_r + \boldsymbol{\beta}_d'\mathbf{x}_d.$$

Then if $\boldsymbol{\beta}_d = \mathbf{0}$, the full model η_f can be simplified to the reduced model η_r, and so the use of the reduced model is justified (through considerations of parsimony discussed in Section 2.1) if the evidence against $\boldsymbol{\beta}_d = \mathbf{0}$ is not strong. This evidence is provided in the Neyman–Pearson theory by the LRT of the hypothesis $\boldsymbol{\beta}_d = \mathbf{0}$, which compares the maximized likelihood functions under the two models. The LRTS is

$$\lambda = -2\left\{\ell(\hat{\boldsymbol{\beta}}_r, \hat{\phi}_r) - \ell(\hat{\boldsymbol{\beta}}, \hat{\phi})\right\},$$

where $\hat{\phi}_r$ is the MLE of ϕ in the reduced model. If the parameter estimates under both models are not on the boundary of the parameter space, and if the number of variables p in the full model does not increase with the sample size n, then in large samples the distribution of the LRTS under the null hypothesis is $\chi^2_{p-p_1}$.

We will adopt the name *disparity* for the expression $-2\log L(\hat{\boldsymbol{\theta}})$ for a specific model, when the likelihood $L(\theta)$ includes all the constants (apart from the measurement precision term δ^n) from the probability model $f(y \mid \theta)$. As we noted earlier, this allows the comparison of maximized likelihoods across different ('non-nested') models; considerable confusion has arisen in the past from the omission of such constants, making cross-family comparisons very difficult.

The macros that we use in this book have been standardized to include all constants in the density or mass function, and so they allow a direct comparison of disparities across different families.

The disparity for a model is closely related to the *deviance* for a model as defined by Nelder and Wedderburn (1972) and as computed in R: the deviance is the disparity for the model relative to the disparity for a 'saturated' model with a parameter for each observation. *Differences* in disparities between nested models for the single-parameter Poisson and binomial distributions are thus identical to differences in deviance, and the disparity does not have to be calculated explicitly for nested model comparisons in these distributions. For the two-parameter normal and gamma distributions, mixture distributions, and more generally when different distributions are being compared for the same data, disparities have to be calculated explicitly, retaining all the constants in the density or mass function.

We conclude this chapter with a discussion of likelihood methods without a specific model. This section is not central to the book and may be skipped on first reading.

2.10 Likelihood inference without models

The title of this section may seem paradoxical. If we do not have an explicit model for the population probabilities p_I of a variable Y, how can we write down a likelihood function? Two different approaches provide answers to this question, by constructing likelihoods for population parameters even though the population model itself is not given. In the first case the object of inference is a *percentile* of the population; in the second, it is the conventional mean of the population.

2.10.1 *Likelihoods for percentiles*

We begin for simplicity with the median. For the population cumulative proportions C_I, the median ψ is defined as the value of y_I such that

$$C_I \leq 0.5 \quad \text{for } y_I \leq \psi,$$
$$C_I > 0.5 \quad \text{for } y_I > \psi.$$

Suppose now that we have a simple random sample of n observations from the population, giving values $y_1, \ldots, y_n$ of Y. For example, we give in Table 2.3 the values of family income for a random sample of 40 families from the STATLAB population. For subsequent convenience the 40 values have been ordered from smallest to largest. Given the population median ψ, if the random variable Y is assumed *continuous*, that is if the measurement precision is very high, then it will be approximately true that

$$\Pr[Y < \psi] = \Pr[Y > \psi] = 0.5$$

and

$$\Pr[Y = \psi] = 0.$$

Then the probability that of the n sample values, r are less than the median and $(n - r)$ greater than the median is the symmetric binomial probability

$$b(r; n, 0.5) = \binom{n}{r} \cdot \frac{1}{2^n}.$$

Table 2.3. Family incomes for random sample of 40 families (hundreds of dollars)

26	35	38	39	42	46	47	47	47	52	53	55	55	56
58	60	60	60	60	60	65	65	67	67	69	70	71	72
75	77	80	81	85	93	96	104	104	107	119	120		

This provides immediately a likelihood function for the median:

$$L(\psi) = \binom{n}{r} \cdot \frac{1}{2^n}$$

for $L(\psi) = \binom{n}{r} \cdot \frac{1}{2^n}$ for $y_{(1)} \leq, \ldots, \leq y_{(r)} \leq \psi \leq y_{(r+1)} \leq, \ldots, \leq y_{(n)}, \quad r = 0, 1, \ldots, n$. This function is *piecewise constant* between successive distinct ordered observations and can take only the values of the symmetric binomial probabilities $b(r; n, 0.5)$. Figure 2.17(a) shows the likelihood for the median for the above sample of 40 family incomes. R functions are

```
> inc <- c(26, 35, 38, 39, 42, 46, 47, 47, 47, 52,
+        53, 55, 55, 56, 58, 60, 60, 60, 60, 60, 65, 65,
+        67, 67, 69, 70, 71, 72, 75, 77, 80, 81, 85, 93,
+        96, 104, 104, 107, 119, 120)
> psi <- seq(40, 90)
> like <- NULL
> for (i in seq(along = psi)) {
+       like[i] <- dbinom(sum(psi[i] > inc), length(inc),
+         0.5)
+ }
> print(xyplot(like ~ psi, type = "s", xlab = "median",
+       ylab = "likelihood"))
```

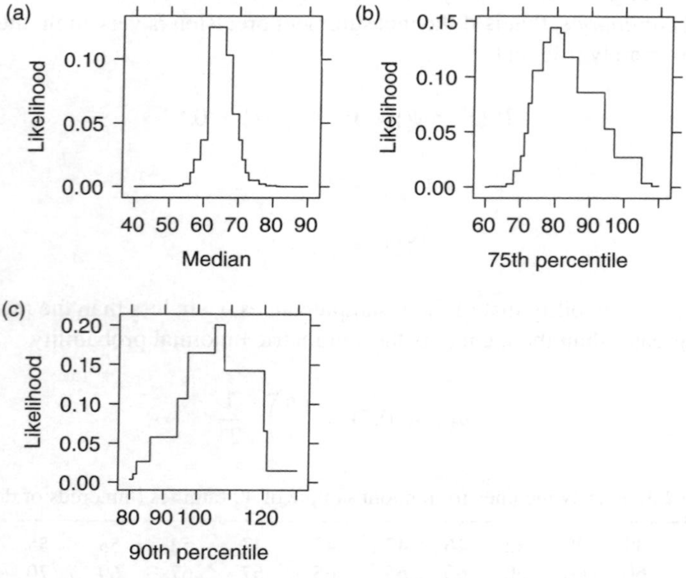

Fig. 2.17. (a) Likelihood for median; (b) likelihood for 75th percentile; and (c) likelihood for 90th percentile

Note that the likelihood does not quite go to zero as $\psi \rightarrow \pm\infty$, because there is a non-zero, though extremely small, probability that all 40 observations exceed the median, whatever its value. A likelihood interval for the median is constructed as before, with the unusual feature that only certain values of the relative likelihood are possible, corresponding to the probabilities of the discrete binomial distribution. The probabilities $b = b(r; 40, 0.5)$ are shown in Table 2.4 for $r = 11, \ldots, 29$. A probability (likelihood) ratio of 0.168 occurs for $r = 14$ or 26 relative to $r = 20$, and one of 0.087 for $r = 13$ or 27 relative to $r = 20$. So a relative likelihood of 0.168 is achieved for the interval (y_{14}, y_{26}), that is, (56, 70), and one of 0.087 for the interval (y_{13}, y_{27}), that is, (55, 71).

The likelihood approach to interval construction here is quite different from inverting the usual Neyman–Pearson hypothesis testing approach. If we want to test the null hypothesis H_0 that $\psi = \psi_0$ against a general alternative H_1, we would proceed in the Neyman–Pearson approach by rejecting H_0 if the number of observations exceeding ψ_0 were too large or too small (the median test). As for the binomial confidence interval procedures described in Section 2.7.5, we then have to find the values of ψ for which the hypothesis would just be rejected at the 5% level. The resulting interval for ψ will in general be different from the likelihood interval.

The above approach to the median generalizes directly to the case of an arbitrary percentile. Let ψ_p be the $100p$-th percentile of the distribution of Y, so that

$$\Pr[Y \leq \psi_p] = p, \; \Pr[Y > \psi_p] = 1 - p,$$

with $\psi_{0.5} = \psi$, and the variable Y again assumed continuous. Then for the sample $y_1, \ldots, y_n$ ordered as before, the likelihood function for ψ_p is

$$L(\psi_p) = \binom{n}{r} p^r (1 - p)^{n-r}$$

for $y_{(1)} \leq, \ldots, \leq y_{(r)} \leq \psi_p \leq y_{(r+1)} \leq, \ldots, \leq y_{(n)}, r = 0, 1, \ldots, n$. The likelihood is again piecewise constant but asymmetric in its values. For p near 0 or 1 the corresponding tail of the likelihood is relatively flat in small samples, reflecting the obvious fact that a small sample can provide little information about extreme percentiles, in the absence of a specific distributional model.

Table 2.4. Binomial probabilities for $n = 40$, $p = 0.5$

r	11	12	13	14	15	16	17	18	19	20
b	0.0021	0.0051	0.0109	0.0211	0.0366	0.0572	0.0807	0.1031	0.1194	0.1254
r	21	22	23	24	25	26	27	28	29	
b	0.1194	0.1031	0.0807	0.0572	0.0366	0.0211	0.0109	0.0051	0.0021	

Figure 2.17(b) shows the likelihood for ψ_p for $p = 0.75$, and Fig. 2.17(c) that for $p = 0.9$ for the family income sample. The R commands are

```
> psi <- seq(60, 110)
> like <- NULL
> for (i in seq(along = psi)) {
+     like[i] <- dbinom(sum(psi[i] > inc), length(inc),
+         0.75)
+ }
> print(xyplot(like ~ psi, type = "s", xlab = "75-th percentile",
+     ylab = "likelihood"))

> psi <- seq(80, 130)
> like <- NULL
> for (i in seq(along = psi)) {
+     like[i] <- dbinom(sum(psi[i] > inc), length(inc),
+         0.9)
+ }
> print(xyplot(like ~ psi, type = "s", xlab = "90-th percentile",
+     ylab = "likelihood"))
```

For $p = 0.9$ the likelihood is constant for all values of $\psi_p > 120$; it is possible, though very unlikely, that *all* the sample observations are less than the 90th percentile. Thus as noted above precise information about extreme percentiles requires large samples.

2.10.2 *Empirical likelihood*

This term was introduced by Owen (1988, 2001) to describe a likelihood function for a population parameter, a function of the population proportions p_I, when no specific model is assumed for the p_I. The original use of the empirical likelihood was by Lindsey (1974), though his purpose was different, namely to model the multinomial probabilities.

Consider the family income sample above. As in Section 2.3, let Y_I be the distinct values of Y occurring in the population. We now allow Y values with multiplicity zero to be included in this set, since as we will see this makes no difference to the conclusions. Without any loss of generality, we may assume that the Y_I form a grid, or mesh, at intervals given by the measurement precision δ. Let D be the number of distinct values of Y, with non-zero or zero probabilities. We want to draw inferences about the *mean* family income $\mu = \sum_I P_I Y_I$ without assuming a model for the P_I. The likelihood function for the P_I is just the multinomial likelihood

$$L(P_1, \ldots, P_D) = \prod_{I=1}^{D} P_I^{n_I}$$

with

$$\sum_{I=1}^{D} P_I = 1.$$

We base inference about μ on the profile likelihood in μ, which is

$$P(\mu) = L(\hat{P}_1(\mu), \ldots, \hat{P}_D(\mu)),$$

where the likelihood is maximized over the P_I subject to the constraints $P_I \geq 0$, $\sum P_I = 1$, and $\sum P_I y_I = \mu$. Note that μ is constrained to lie in the interval $(y_{(1)} < \mu < y_{(D)})$. The maximization is easily accomplished using Lagrange multipliers: let

$$G(P_1, \ldots, P_D) = \log L(P_1, \ldots, P_D) - \phi\left(\sum P_I - 1\right) - n\lambda\left(\sum P_I y_I - \mu\right).$$

Then

$$\frac{\partial G}{\partial P_I} = \frac{n_I}{P_I} - \phi - n\lambda y_I = 0$$
$$\frac{\partial G}{\partial \lambda} = \sum P_I y_I - \mu = 0$$
$$\frac{\partial G}{\partial \phi} = \sum P_I - 1 = 0$$

for a maximum of the constrained likelihood. Multiplying the first equation by P_I and summing over I, we have

$$n = \hat{\phi}(\mu) + n\mu\hat{\lambda}(\mu)$$

giving $\hat{\phi}(\mu) = n(1 - \mu\hat{\lambda}(\mu))$ and

$$\frac{n_I}{\hat{P}_I(\mu)} = n[1 + \hat{\lambda}(\mu)(y_I - \mu)]$$

so that

$$\hat{P}_I(\mu) = \frac{n_I/n}{1 + \hat{\lambda}(\mu)(y_I - \mu)} = \frac{\tilde{P}_I}{1 + \hat{\lambda}(\mu)(y_I - \mu)},$$

where $\tilde{P}_I = n_I/n$. Summing over I again gives $\hat{\lambda}(\mu)$ as the implicit solution of

$$f(\lambda) = 1 - \sum_{I=1}^{D} \frac{\tilde{P}_I}{1 + \lambda(y_I - \mu)} = 0.$$

Note that $\lambda = 0$ always satisfies this equation: it gives the *unrestricted* maximum of the likelihood since the constraint $\sum_I P_I y_I = \mu$ is not active. Since the $\hat{P}_I(\mu)$ must be non-negative, we require $1 + \hat{\lambda}(\mu)(y_I - \mu) > 0$ for all I, which implies

$$-(y_{(D)} - \mu)^{-1} < \hat{\lambda}(\mu) < (\mu - y_{(1)})^{-1}.$$

Values of Y_I which do not occur in the sample give values of $\tilde{P}_I$ and therefore $\hat{P}_I(\mu)$ of zero, and so the implicit equation for $\hat{\lambda}(\mu)$ involves only the Y_I with $n_I > 0$.

Since this section is not central to the book, we do not give details of the computational method to compute the empirical likelihood. Owen (2001) gave full details.

Figure 2.18 shows the empirical profile relative likelihood for the example. It is shifted considerably to the right compared to that for the median ψ, reflecting the skewness of the income distribution, with the mean income larger than the median. Despite this skewness, the profile likelihood in μ is only slightly skewed.

Figure 2.19 repeats the empirical profile relative likelihood together with the profile relative likelihoods for μ, assuming lognormal (dashed curve) and gamma (dotted curve) distributions for income.

The gamma and empirical likelihoods are very close; the lognormal likelihood is less close on the right to the empirical likelihood. Computation of the lognormal

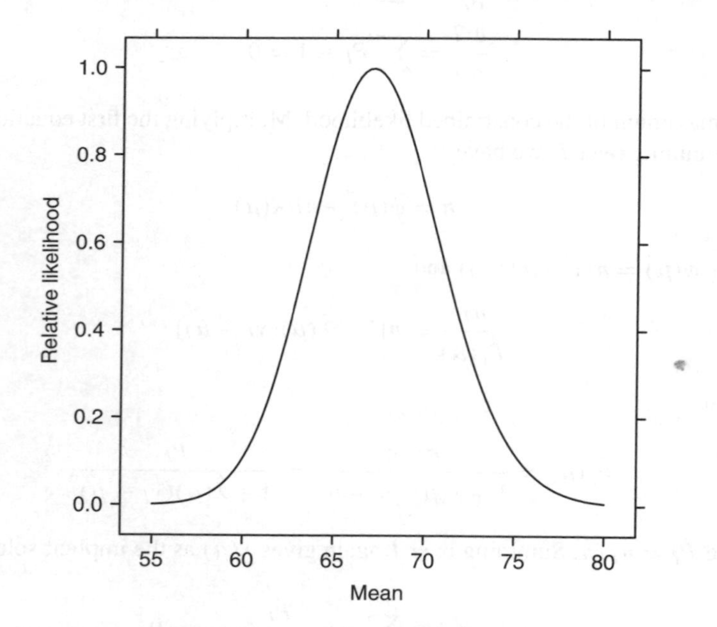

Fig. 2.18. Empirical likelihood

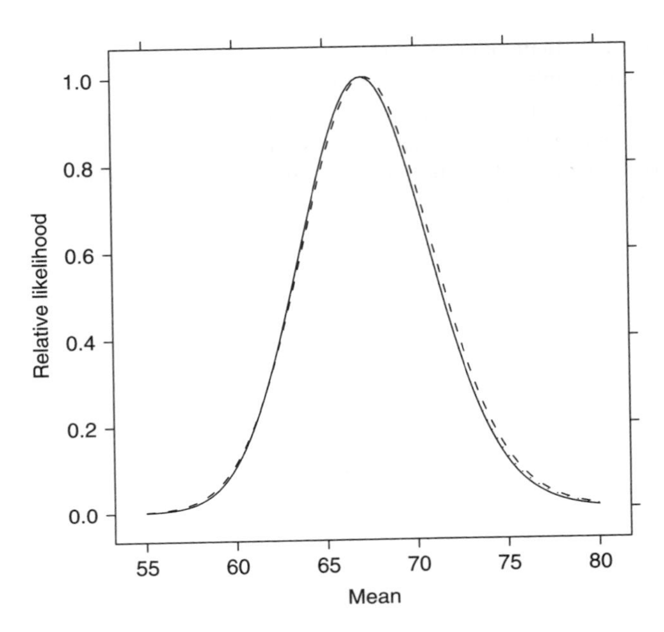

Fig. 2.19. Empirical and parametric likelihoods

and gamma profile likelihoods is discussed in Section 2.7.1 and Chapter 6, respectively.

In Owen's derivation of the empirical likelihood, the asymptotic repeated sampling result $-2\log\{L(\mu)/L(\hat{\mu})\} \sim \chi_1^2$ is used to set confidence intervals for μ. Remarkably, this asymptotic result holds despite the apparent violation of the condition that the number of nuisance parameters P_I does not increase with n, since the number of distinct possible values D certainly increases (though slowly) with n. This is a special case of a very general result by Murphy and van der Vaart (2000) to which we refer in Chapter 8.

Owen and others (see Owen, 2001, for full details) have extended the empirical likelihood approach to other parameters, including variances and regression coefficients. *Joint* empirical likelihood regions in several parameters can be constructed in the same way. However, the small-sample coverage of empirical likelihood intervals and regions may be more affected by the large number of nuisance parameters than profile likelihood intervals and regions in parametric models.

This approach to likelihood construction in the absence of a model provides an important theoretical response to criticisms of model-based approaches, as being based on models which may be incorrect and which may thefore give misleading inferences. Since no approximating population model is assumed for the P_I, the empirical likelihood provides a completely model-robust likelihood inference about μ. In theory this approach can be extended very widely. In practice, computational complexity has so far limited its generality.

Intervals based on the empirical likelihood have also been regarded as an alternative to bootstrap confidence intervals (Hall and La Scala, 1990), which also do not make any population model assumption but perform physical resampling from the observed sample.

3 Regression and analysis of variance

3.1 An example

We consider a practical example from a psychological study. A sample of twenty-four children was randomly drawn from the population of fifth-grade children attending a state primary school in a Sydney suburb. Each child was assigned to one of two experimental groups, and given instructions by the experimenter on how to construct, from nine differently coloured blocks, one of the 3×3 square designs in the Block Design subtest of the Wechsler Intelligence Scale for Children (WISC, see Wechsler, 1949). Children in the first group were told to construct the design by starting with a row of three blocks (row group), and those in the second group were told to start with a corner of three blocks (corner group). The total time in seconds to construct four different designs was then measured for each child.

Before the experiment began, the extent of each child's 'field dependence' was tested by the embedded figures test (EFT), which measures the extent to which subjects can abstract the essential logical structure of a problem from its context (high scores corresponding to high field dependence and low ability).

The data are given in Table 3.1, and are held in the dataset `solv`. The file also contains a `group` factor taking values 1 and 2 for row and corner groups.

The experimenter was interested in whether the different instructions produced any change in the average time required to construct the designs, and whether this time was affected by field dependence.

In practical data analysis, we usually begin with a graph of the data. The *`lattice`* package in R provides a versatile function *`xyplot`*, which we use here to plot the response variable `time` against the covariate `eft` for each level of the factor `group` on a separate graph and add a simple regression line for each group.

```
> data(solv, package = "SMIR")

> library(lattice)
> print(xyplot(time ~ eft | group, data = solv,
+       type = c("p", "r")))
```

The resulting figure is shown in Fig. 3.1. There appears to be a general increase in time required as `eft` increases, with much individual variation.

Table 3.1. Block design completion times

Row group						
`time`	317	464	525	298	491	196
`eft`	59	33	49	69	65	26
`time`	268	372	370	739	430	410
`eft`	29	62	31	139	74	31
Corner group						
`time`	342	222	219	513	295	285
`eft`	48	23	9	128	44	49
`time`	408	543	298	494	317	407
`eft`	87	43	55	58	113	7

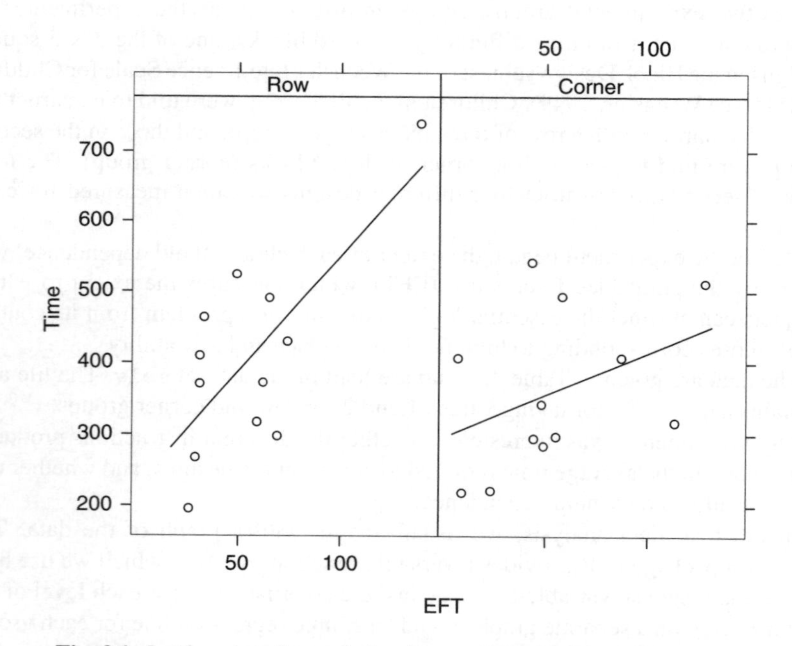

Fig. 3.1. Lattice plot of completion time vs EFT conditioned on group

The usual approach to the analysis of the data above is by multiple regression, or analysis of covariance. We formulate the model, expressed in a traditional form,

$$Y_i = \beta_0 + \beta_1 x_{1i} + \beta_2 x_{2i} + \epsilon_i, \quad i = 1, \dots, 24,$$

where Y_i is the time taken by the i-th child, x_{1i} is the `eft` score of the i-th child, and x_{2i} is a dummy variable, with $x_{2i} = 0$ for the row group children and $x_{2i} = 1$ for the corner group children. The 'error' variables ϵ_i are assumed to be independent and normally distributed with mean zero and common variance σ^2, written $\epsilon_i \sim N(0, \sigma^2)$. If not included in the dataset factors can be defined directly by

```
> solv$group <- gl(2, 12, labels = c("row", "corner"))
```

The `gl` function offers the user one way to enter data for factors. The number of levels and the number of replications can be defined to produce the pattern required (as defined in Chapter 1).

Dummy variables for categorical (here binary) explanatory variables are automatically defined by R if the variables are declared to be factors. The dummy variable used by R for x_2 in the regression is denoted by `groupcorner` because the `row` level has been set as the first level of the factor `group`. For a factor `A` with k levels ($k \geq 2$), R defines $(k-1)$ dummy variables, pasting the level name to the factor name, with $A(j) = 1$, $j > 1$ for the j-th level of `A`, and zero otherwise.

In the formal model structure of Chapter 2, the *systematic part* of the model is the regression function or linear model $\beta_0 + \beta_1 x_1 + \beta_2 x_2$ for the mean time taken, and the *random part* is the normal error variation about the regression with variance σ^2. If we write μ_i for the mean time taken for the i-th child, then

$$\mu_i = \beta_0 + \beta_1 x_{1i} + \beta_2 x_{2i}, \quad Y_i = \mu_i + \epsilon_i, \quad \epsilon_i \sim N(0, \sigma^2),$$

which we will call model `solv.glm2`; in all we will consider five models for these data. In the remainder of this discussion we will suppress the specification of the random part of the model as it is the same for all the regression models for μ_i.

Since x_2 is a dummy variable, we may write this regression function equivalently as

$$\begin{aligned} \mu_i &= \beta_0 + \beta_1 x_{1i}, & x_{2i} &= 0: \text{Row group} \\ \mu_i &= \beta_0 + \beta_2 + \beta_1 x_{1i}, & x_{2i} &= 1: \text{Corner group} \end{aligned}$$

The two regression lines for the row group and the corner group for the five models discussed are shown in Fig. 3.2.

```
> solv$group <- relevel(solv$group, "row")
> solv.glm5 <- glm(time ~ 1, data = solv)
> solv.glm4 <- update(solv.glm5, . ~ group)
> solv.glm3 <- update(solv.glm5, . ~ eft)
> solv.glm2 <- update(solv.glm4, . ~ . + eft)
> solv.glm1 <- update(solv.glm2, . ~ . + eft:group)
```

```
> all <- rbind(
+     cbind(solv, model = "Model 5", fitted = fitted(solv.glm5)),
+     cbind(solv, model = "Model 4", fitted = fitted(solv.glm4)),
+     cbind(solv, model = "Model 3", fitted = fitted(solv.glm3)),
+     cbind(solv, model = "Model 2", fitted = fitted(solv.glm2)),
```

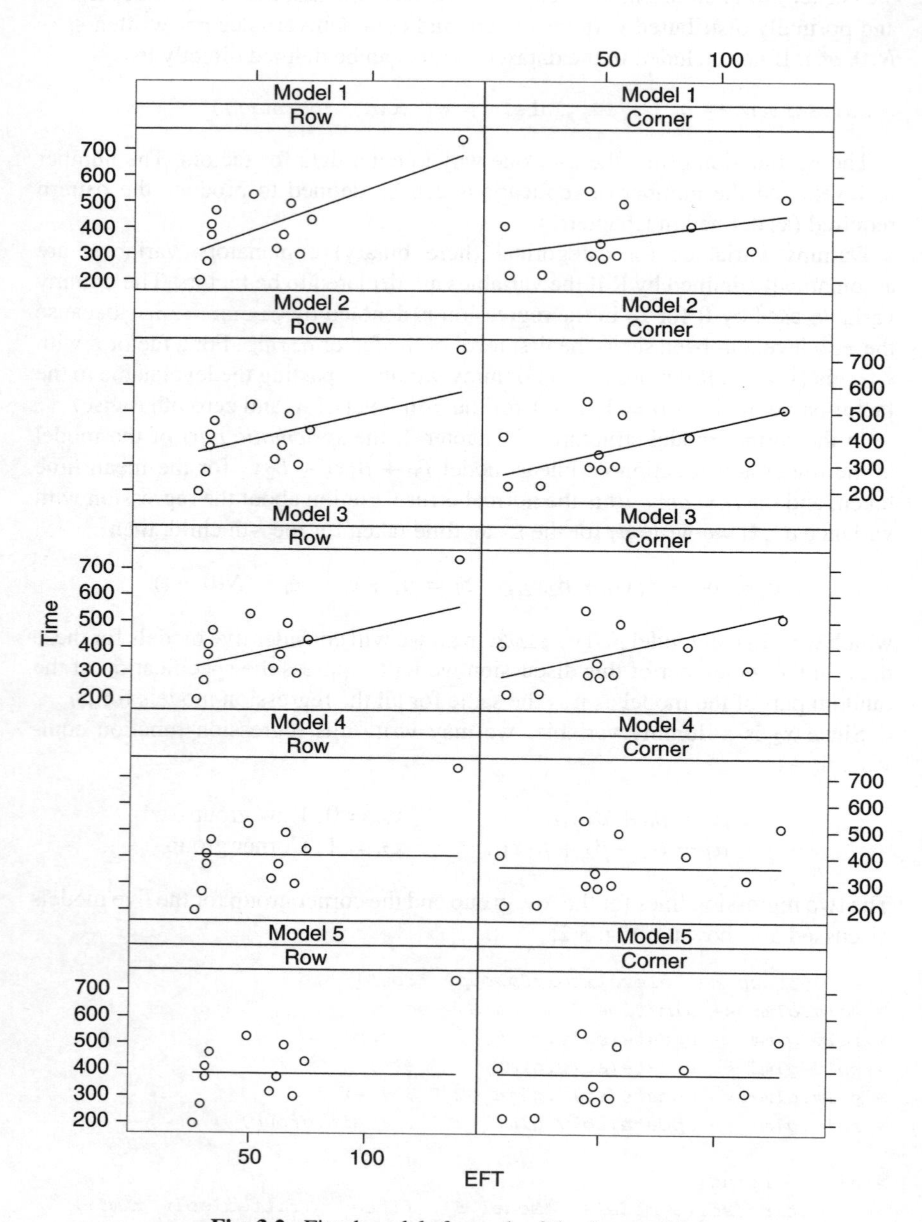

Fig. 3.2. Fitted models for each of the five models

```
+       cbind(solv, model = "Model 1", fitted = fitted(solv.glm1)))
> print(xyplot(time ~ eft | group * model, data = all,
+       subscripts = TRUE, panel = function(x, y, subscripts,
+           ...) {
+           panel.xyplot(x, y, ...)
+           llines(x, all$fitted[subscripts], ...)
+       }))
```

The regression lines for model `solv.glm2` are parallel, with slope β_1 representing the increase in mean time taken for a one point increase in `eft` score, β_2 representing the difference in mean time taken between the corner and row groups with the same value of `eft`, and β_0 the (hypothetical) mean time taken for a child with `eft` score zero in the row group.

If the instructions have no differential effect on completion time, then $\beta_2 = 0$, and model `solv.glm2` reduces to the model `solv.glm3`:

$$\mu_i = \beta_0 + \beta_1 x_{1i},$$

identical regressions for the two instruction groups. If the instructions have a different effect, but `eft` is unrelated to completion time, then $\beta_1 = 0$, and model `solv.glm2` reduces to the model `solv.glm4`:

$$\mu_i = \beta_0 + \beta_2 x_{2i}.$$

It is also possible that neither `eft` nor instructions has any effect, in which case the *null* model `solv.glm5` results:

$$\mu_i = \beta_0.$$

A further possibility is that the regressions of mean time on `eft` score for the two instruction groups are linear, but not parallel. This can be represented by the *interaction term* in model `solv.glm1`:

$$\mu_i = \beta_0 + \beta_1 x_{1i} + \beta_2 x_{2i} + \beta_3 x_{1i} x_{2i},$$

which is equivalent to

$$\begin{aligned} \mu_i &= \beta_0 + \beta_1 x_{1i} && x_{2i} = 0\text{: Row group} \\ \mu_i &= (\beta_0 + \beta_2) + (\beta_1 + \beta_3) x_{1i} && x_{2i} = 1\text{: Corner group.} \end{aligned}$$

For model `solv.glm1`, the difference between mean completion times in the two experimental groups, for the same `eft` score x, is

$$(\beta_0 + \beta_2) + (\beta_1 + \beta_3)x - (\beta_0 + \beta_1 x) = \beta_2 + \beta_3 x,$$

a linear function of x. Thus the effect of different instructions on mean completion time depends on `eft` score, a much more complicated interpretation than that of model *solv.glm2*, where the difference is a constant β_2 because β_3 is zero.

The interpretation of the experiment depends, therefore, on which of the five (or possibly other) models is the 'best' representation of the data, in the sense of 'simplest consistent with the data'. In Chapter 2 we developed the general theory for this question; we now apply it.

It is worth stressing first though that *none* of the models is a true representation of the population. If we could take a complete census of the population of fifth-grade children in the school, and administer the EFT and WISC to all of them, we would find that the mean completion time for children with each `eft` score in each experimental group did not lie on a straight line: the means would be quite irregular about a general trend. Our model ignores the variation, and uses only a linear trend: it thus represents an idealized or smooth version of the population. This is most obvious when we are in fact modelling complete populations: we still smooth the complete population using the model so that the gross systematic features of the population are retained, but the minor irregularities or fine structure are lost, being represented as random variation.

To fit the models using the *glm* function in R, we declare:

(1) the *response variable* Y as *time* by placing the name of the response variable on the left-hand side of the '$\sim$' operator in the formula defining the regression model.
(2) the *probability distribution* for Y is normal $N(\mu, \sigma^2)$ with mean μ and variance σ^2:
The distribution is defined by the family argument, *family=gaussian* (this is the default distribution and can be omitted);
(3) the *link function* is the identity $\eta = g(\mu) = \mu$:
The 'gaussian' family accepts the links 'identity' (this is the default link for the normal distribution and can be omitted), as well as the 'log', and 'inverse';
(4) the *linear predictor* is $\eta = \beta_0 + \beta_1 x_1 + \beta_2 x_2$, where x_1 is `eft` and x_2 `group`, or one of the other models 1–5:

```
> glm(time ~ eft + group, data = solv)
```

R provides minimal default output: the print method for a *glm* object gives the model call, the estimated coefficients, degrees of freedom, null deviance and residual deviance and the model AIC.

For linear Gaussian models, R provides a more convenient modelling function, *lm*. The summary function for *lm* objects includes summary statistics such as the parameter estimates, marginal t-tests and R^2 statistic.

For model comparisons by the likelihood ratio test, it is easily shown (following Section 2.9.4 – we do not give the derivation) that for a regression model $\boldsymbol{\beta}'\mathbf{x}$ the

MLEs of the model parameters are

$$\hat{\boldsymbol{\beta}} = \left(\sum \mathbf{x}_i\mathbf{x}_i'\right)^{-1} \sum \mathbf{x}_i y_i = (\boldsymbol{X}'\boldsymbol{X})^{-1}\boldsymbol{X}'\mathbf{y}$$

$$\hat{\sigma}^2 = \sum (y_i - \hat{\boldsymbol{\beta}}'\mathbf{x}_i)^2/n = \text{RSS}/n,$$

where $\boldsymbol{X}$ is the 'design matrix': $\boldsymbol{X} = (\mathbf{x}_1, \ldots, \mathbf{x}_n)'$, $\mathbf{y} = (y_1, \ldots, y_n)'$ and RSS is the residual sum of squares from the fitted model. Note that R uses the unbiased variance estimate $s^2 = \text{RSS}/(n-p-1)$ instead of $\hat{\sigma}^2$ to calculate standard errors. The value of the disparity as we define it is

$$-2\ell(\hat{\boldsymbol{\beta}}, \hat{\sigma}) = c + n\log(\text{RSS}),$$

where

$$c = n[1 + \log(2\pi) - \log n],$$

and $\ell(\beta, \sigma)$ is the log-likelihood for the normal model. A formal comparison of two models is expressed as in Section 2.9.4 through the LRT of the hypothesis that the variables $\mathbf{x}_d$ omitted from $\mathbf{x}$ in the partition $\mathbf{x}' = (\mathbf{x}_r, \mathbf{x}_d)'$ have zero regression coefficients and do not contribute to the regression function, so that $\boldsymbol{\beta}_d = \mathbf{0}$. The LRTS for this hypothesis is

$$\begin{aligned} \lambda &= -2\{\ell(\hat{\boldsymbol{\beta}}_r, \hat{\sigma}_r) - \ell(\hat{\boldsymbol{\beta}}, \hat{\sigma})\} \\ &= n\{\log \text{RSS}_r - \log \text{RSS}_f\} \\ &= n\log(\text{RSS}_r/\text{RSS}_f). \end{aligned}$$

(Note that the constant c disappears from the LRTS. For this reason, numerical constants which are not functions of the parameters can be omitted in the comparison of nested models, as we noted in Section 2.6 and elsewhere.)

If the hypothesis is true, then the residual sum of squares for the reduced model will be not much larger than that for the full model, and so λ will be small. If the hypothesis is false, then λ will be large.

For the hypothesis that a single component β_j of $\boldsymbol{\beta}$ is zero, the Wald test compares $\hat{\beta}_j$ with its estimated asymptotic standard error $\sqrt{\hat{v}_{jj}}$, and treats $W = t_j^2 = \hat{\beta}_j^2/\hat{v}_{jj}$ as approximately χ_1^2 under the hypothesis. For large samples in which the likelihood function is normal in the parameters, this test will be closely equivalent to the likelihood ratio test, with $W \approx \lambda$. For small samples and for regression models in which the response variable is not normally distributed, the likelihood may be far from normal and the value of W may be substantially different from λ. Critical values for λ are based on its asymptotic distribution when the null hypothesis is true, which is $\chi^2_{p-p_1}$ if both X and X_r are of full rank p and p_1, respectively.

For the normal distribution this asymptotic result is not used, since exact results are available. If we define the *hypothesis sum of squares* (HSS) by

$$\text{HSS} = \text{RSS}_r - \text{RSS}_f$$

then

$$\begin{aligned}\lambda &= n\log(1 + \text{HSS}/\text{RSS}_f)\\ &= n\log\left\{1 + \frac{p-p_1}{n-p-1}\left[\frac{\text{HSS}/(p-p_1)}{\text{RSS}_f/(n-p-1)}\right]\right\}\\ &= n\log(1 + (p-p_1)F/(n-p-1)),\end{aligned}$$

where

$$F = \left[\frac{\text{HSS}/(p-p_1)}{\text{RSS}_f/(n-p-1)}\right]$$

is the usual F-statistic (with $p - p_1$ and $n - p - 1$ df) for the test of the hypothesis $\boldsymbol{\beta}_d = 0$ in multiple regression or the analysis of variance. The exact distribution of λ could therefore be obtained from that of F (or from the beta distribution of $\text{RSS}_f/\text{RSS}_r$), but it is simpler to use the F-distribution itself which is well tabulated.

Under the hypothesis that a single component β_j of $\boldsymbol{\beta}$ is zero, the distribution of $t_j = \hat{\beta}_j/\sqrt{\hat{v}_{jj}}$ is exactly t_{n-p-1}. Here the t-test is identical to the F-test, as $t_j^2 = F$, and the distribution of t^2_{n-p-1} is $F_{1,n-p-1}$.

In practice, the single F-test described above is insufficient to identify parsimonious models, because we need to examine the importance of individual terms in the model. In our example we have five candidate models, so *multiple* tests will be necessary for this identification. These tests are constructed by *partitioning* the hypothesis sum of squares into individual sums of squares for the model terms, by fitting a hierarchical sequence of models of increasing complexity, from the simplest model to the full model. The table of sums of squares for individual terms is called an *analysis of variance* (ANOVA) table. For example, for the sequence of models 5, 4, 2, 1, listing the models inside the `anova` function gives the RSSs for each models in the order specified and differences their RSSs, giving the ANOVA table (Table 3.2)

```
> (anova2 <- anova(solv.glm5, solv.glm4, solv.glm2,
+      solv.glm1, test = "F"))
```

Table 3.2. ANOVA table for the `solv` dataset

```
Analysis of Deviance Table

Model 1: time ~ 1
Model 2: time ~ group
Model 3: time ~ group + eft
Model 4: time ~ group + eft + group:eft
  Resid. Df Resid. Dev Df Deviance       F   Pr(>F)
1        23     367537
2        22     355522  1    12015  1.1019 0.306360
3        21     245531  1   109991 10.0873 0.004749
4        20     218076  1    27455  2.5179 0.128247
```

The use of the *update* function allows modification of the previous model formula without defining all remaining terms. The RSSs are 367537, 355522, 245531, and 218076.

The usual approach to model comparisons in analysis of variance tables is to calculate mean squares by dividing each sum of squares by its degrees of freedom and then to assess the importance of each effect by testing its mean square against the residual mean square using the F-test with 1 and 20 df as reported in Table 3.2. R will generate this more familiar method of reporting the ANOVA table if the largest model (here *solv.glm1*) is used as the argument to the *anova* function. However, this procedure is ambiguous in the analysis of survey and much experimental data because the sum of squares for each effect generally depends on its order of fitting in the model. Thus, if we fit the models in the order 5, 3, 2, and 1, and difference the RSSs, we obtain the different ANOVA table (3.3).

```
> solv.glm3 <- update(solv.glm5, . ~ . + eft)
> (anova3 <- anova(solv.glm5, solv.glm3, solv.glm2,
+       solv.glm1, test = "F"))
```

The analysis of variance tables differ because the variables `eft` and `group` are slightly correlated, this correlation being due to the small difference in `eft` means between the row and corner groups. The difference is only small because of the random allocation of children to groups. Only for *orthogonal* experimental designs is the sum of squares breakdown unique, independent of the fitting order. In observational studies such differences can be large; we discuss this important issue further below (Section 3.16).

Qualitatively, we can see that the `eft` mean square is large (more than 10 times the residual mean square in either model) while the `group` mean square is small (about the same size as the residual mean square). The interaction mean square is the same in both tables, since it is always the last term fitted, and its F-value is 2.52(27455/10903.8). This value is nowhere near the 5% level

Table 3.3. ANOVA table

```
Analysis of Deviance Table

Model 1: time ~ 1
Model 2: time ~ eft
Model 3: time ~ group + eft
Model 4: time ~ group + eft + group:eft
  Resid. Df Resid. Dev Df Deviance       F   Pr(>F)
1        23     367537
2        22     257274  1   110263 10.1123 0.004706
3        21     245531  1    11743  1.0770 0.311760
4        20     218076  1    27455  2.5179 0.128247
```

for the $F_{1,20}$ distribution, and so we can clearly reduce model `solv.glm1` to model `solv.glm2`. Comparing model `solv.glm2` with models `solv.glm3` and `solv.glm4`, we see that model `solv.glm2` can be reduced to model `solv.glm3`, since the SS for `group` is of the same order as the residual mean square. However, model `solv.glm2` cannot be reduced to model `solv.glm4`, because the `eft` variable cannot be omitted as its F-value is more than 10. We conclude that model `solv.glm3` is the most appropriate and parsimonious of the models considered to represent the data.

The fitted values of the response variable can be extracted from the `glm` object using the `fitted` function, and can be plotted as a line with the observed times as follows (see Fig. 3.2):

```
> plot(time ~ eft, data = solv)
> lines(solv$eft, fitted(solv.glm3))
```

The fitted regression obtained is

$$271.1 + 2.04\,\texttt{eft},$$

a one-point increase in `eft` score being associated with a two-second increase in mean time. Thus field dependence, as measured by the EFT, is associated with increased completion time, but the type of instruction given is not. Completion times vary randomly about the mean with variance estimated by $s^2 = 257274/22 = 11694$, or standard deviation $s = 108$ sec, slightly larger than $104 = \sqrt{10903.8}$ from the full model.

To readers used to classical analysis of variance and covariance, this conclusion may seem unsatisfying. Where is the estimate of the treatment effect and its standard error? In simplifying the model, we have set the treatment effect (the coefficient β_2 of `group`) equal to zero because there is no strong evidence from the data against this value. If we are particularly interested in certain parameters of the model because these represent experimentally controlled variables, then non-parsimonious models containing these variables may be presented as the final summary of the data. The parameter estimates and standard errors for the fitted model *solv.glm2* are generated using the *summary* and *coef* functions.

```
> solv.glm2$call

glm(formula = time ~ group + eft, data = solv)

> coef(summary(solv.glm2))

              Estimate Std. Error   t value      Pr(>|t|)
(Intercept) 293.387    48.359      6.0671    5.078345e-06
groupcorner -44.240    44.144     -1.0022    3.276618e-01
eft           2.038     0.664      3.0671    5.848880e-03
```

The residual standard deviation estimate is $s = 108$ sec.

A 95% confidence interval for β_2 is

$$-44.2 \pm t_{0.025,21}(44.1), \qquad \text{i.e. } (-136, 48).$$

The `confint` function in R assumes the normal distribution (with known variance equal to the MLE) and so corresponds to the profile likelihood intervals which assume the asympototic χ_1^2 distribution as explained in Chapter 2. This interval is noticeably shorter as it does not allow for the variability in the sample variance.

```
> confint(solv.glm2)

                   2.5 %       97.5 %
(Intercept)    198.610      388.165
groupcorner  -130.761       42.280
eft              0.736        3.340
```

The corner group has an *estimated* mean 44 sec below that for the row group (for the same value of `eft`), but the population value could be any value in the interval $(-136, 48)$, including zero.

3.2 Strategies for model simplification

The simplification of the model in the above sequence – interaction, then main effects – is the most common and useful strategy for model simplification. High-order interactions are usually small, and their retention in a model makes the interpretation of the model much more difficult. Successively removing interactions from the model as far as possible, starting with the most complex, is therefore appealing.

This is, however, not the only possible strategy. Another possible approach is to examine the lower-order interactions or main effects and to simplify them, while retaining the high-order interactions in a simplified form. The example of Section 3.1 provides a good example. The parameter estimates and their standard errors (in parentheses) from the full interaction model *solv.glm1* are

```
> round(coef(summary(solv.glm1)), 4)

                 Estimate Std. Error t value Pr(>|t|)
(Intercept)       226.052     63.099   3.583   0.0019
groupcorner        70.449     83.913   0.840   0.4111
eft                 3.249      0.997   3.258   0.0039
groupcorner:eft    -2.067      1.303  -1.587   0.1282
```

As we have already seen, the interaction term can be omitted from the model. However, examination of the regression coefficients for `eft` in the two instruction groups reveals that in the row group, the slope is large: 3.25, while in the corner group it is small: $3.25 - 2.07 = 1.18$. This suggests that a different simplification of the model is possible: we could set the slope of the `eft` regression to zero in the corner group, but keep it non-zero in the row group. Since the slope is $(\beta_1 + \beta_3)$ in the corner group, this constraint is achieved by setting $\beta_1 + \beta_3 = 0$, or $\beta_3 = -\beta_1$. The interaction model (model *solv.glm5*) then becomes model *solv.glm6*:

$$\begin{aligned}\mu &= \beta_0 + \beta_1 x_1 + \beta_2 x_2 - \beta_1 x_1 x_2 \\ &= \beta_0 + \beta_1 x_1 (1 - x_2) + \beta_2 x_2.\end{aligned}$$

This model specifies a linear regression

$$\mu = \beta_0 + \beta_1 x_1$$

for the row group, and a constant mean

$$\mu = \beta_0 + \beta_2$$

for the corner group. Model *solv.glm6* is equivalent to model *solv.glm1* for the row group and model *solv.glm5* for the corner group. It is important to note that β_2 again does not represent a constant difference in means between the groups for children with the same `eft` score x_1: this difference is now $\beta_2 - \beta_1 x_1$.

The model can be fitted by constructing a new variable *eft.row* and fitting it with `group`.

```
> solv$eft.row <- solv$eft * (solv$group == "row")
> solv.glm6 <- glm(time ~ eft.row + group, data = solv)
> coef(summary(solv.glm6))

            Estimate Std. Error  t value     Pr(>|t|)
(Intercept) 226.052      64.568    3.501  0.002127401
```

```
eft.row        3.249      1.021    3.184     0.004465092
groupcorner  135.865     71.558    1.899     0.071434816

> anova(solv.glm6, solv.glm1, test = "F")

Analysis of Deviance Table

Model 1: time ~ eft.row + group
Model 2: time ~ group + eft + group:eft
  Resid. Df Resid. Dev Df Deviance      F Pr(>F)
1        21     239766
2        20     218076  1    21689 1.9891 0.1738
```

The RSS is 239766, giving an $F_{1,20}$-value of 1.99 for the comparison of the two models, or an equivalent t-statistic of 1.41. The parameter estimates and standard errors are

```
226.1 + 3.25 eft.row + 135.9 groupcorner
(64.6) (1.02)          (71.6)
```

The slope of the regression on `eft.row` for the row group is the same as in the interaction model, though its standard error is slightly different since the model is different.

If we omit `eft.row` from this model we obtain model *solv.glm4* which we know is not adequate. The only other possibility is to omit `group`. Omission of this variable is equivalent to fitting the model

$$\mu = \beta_0 + \beta_1 x_1$$

for the row group, and

$$\mu = \beta_0$$

for the corner group. This implies that the group regression lines intersect at $x_1 = 0$, so that the mean completion time for a corner group child is the same as that for a row group child with an EFT score of zero. Since there are no such children, this model has a strong property which is not verifiable from the observable data. In the absence of compelling experimental reasons for this strong mathematical property, we do not proceed further with this model, but return to model *solv.glm6*.

The interpretation of model *solv.glm6* is quite different from that of model *solv.glm3*. Mean completion time is affected by field dependence under row instructions, but not under corner instructions. The mean time taken under row instructions is less than that under corner instructions for low field dependence, but greater for high field dependence.

These conclusions are quite complicated. The interpretation of model 3 was much simpler: no difference in mean time between instructions, and a common effect of field dependence for both instruction groups. But both models are consistent with the data, and we have no statistical criterion for choosing between them other than parsimony: model `solv.glm3` has only one variable, model `solv.glm6` has two.

A problem occurring frequently with models with non-parallel regressions is that of determining the values of the explanatory variable for which a significant difference between means can be asserted. In model `solv.glm6` the difference between the mean completion times between the corner and row groups, for a given EFT score x, is $\beta_2 - \beta_1 x$. For what values of x can we assert that this is non-zero? The fitted regression lines cross when $x = \hat{\beta}_2/\hat{\beta}_1 = 41.8$, so around this value the population means are indistinguishable. For any x the variable $\hat{\beta}_2 - \hat{\beta}_1 x$ is normally distributed with mean $\beta_2 - \beta_1 x$ and variance $(v_{22} - 2v_{12}x + v_{11}x^2)$, where $V = (v_{jk})$ is the covariance matrix of the parameter estimates, and is estimated by $\hat{V} = s^2(X'X)^{-1}$. The hypothesis $\beta_2 - \beta_1 x = 0$ can be rejected by a simultaneous test of level γ valid for all x if

$$t^2(x) = \frac{(\hat{\beta}_2 - \hat{\beta}_1 x)^2}{\hat{v}_{22} - 2\hat{v}_{12}x + \hat{v}_{11}x^2} > 2F_{\gamma,2,n-p-1}.$$

This procedure was first proposed by Johnson and Neyman (1936) and is often called the Johnson–Neyman 'technique'; the simultaneous test formulation was given by Potthoff (1964). The above inequality is equivalent to

$$Q(x) = (\hat{\beta}_1^2 - c\hat{v}_{11})x^2 - 2(\hat{\beta}_1\hat{\beta}_2 - c\hat{v}_{12})x + (\hat{\beta}_2^2 - c\hat{v}_{22}) > 0,$$

where $c = 2F_{\gamma,2,n-p-1}$. Thus if $Q(x) = 0$ has real roots $x_L < x_U$, the rejection region is $x > x_U, x < x_L$ if $\hat{\beta}_1^2 - c\hat{v}_{11}$ is positive, or $x_L < x < x_U$ if $\hat{\beta}_1^2 - c\hat{v}_{11}$ is negative. If there are no real roots, the rejection region in x is empty.

In the above example, we will take $\gamma = 0.05$ for illustration, so that $c = 2F_{0.05,2,21} = 6.94$. The parameter estimates are given earlier in this section: we reproduce them here with the variances and covariances of the parameter estimates, obtained using:

```
> summary(solv.glm6)$sigma^2 * summary(solv.glm6)$cov.unscaled
```

$$\begin{array}{llll} \hat{\beta}_1 = 3.249 & \hat{v}_{11} = 1.041 & \\ \hat{\beta}_2 = 135.9 & \hat{v}_{22} = 5120 & \hat{v}_{12} = 57.89 = \hat{v}_{21}. \end{array}$$

The quadratic is

$$Q(x) = 3.3315x^2 - 2(39.7825)x - 17064$$

and the roots of $Q(x) = 0$ are -60.6 and 84.5. These can be calculated using R's `uniroot` function and restricting the solution to be non-negative.

```
> Qx <- function(x) 3.3315 * x^2 - 2 * 39.7825 * x -
+      17064
> uniroot(Qx, lower = 0, upper = 200)$root

[1] 84.49897
```

For values of `eft` 85 or greater, the mean completion time under row instructions is significantly greater than that under corner instructions. For values of `eft` less than 85, the mean completion times do not differ significantly.

We should note that only three observations have `eft`$>$ 85, one in the row group with `time` 739, and two in the corner group with `time` 513 and 317. These three observations are thus *influential* in this model; we discuss influential observations in Section 3.4.3.

Given the quite different conclusions from models `solv.glm3` and `solv.glm6`, what can we confidently say? The answer is that we are tantalizingly short of data. The conclusions from both models are strongly influenced by two of the observations, as we will see in the next sections. With small experimental or observational studies it will frequently happen that several different models are equally well supported by the data. In such cases all the competing models and their implications should be presented, and firm judgements suspended. This example is considered further later in the chapter.

3.3 Stratified, weighted and clustered samples

In most sample surveys the sample design is not simple random sampling, but some form of stratified design with known but unequal sampling fractions. Multi-stage *cluster* sampling is often used, and this form of sampling induces correlations between the observations in a cluster, invalidating the assumption of independence used in constructing the likelihood function in Chapter 2. Standard regression methods, as applied above to the data of Section 3.1, are invalid when applied to the observations from clustered samples. Regression models can be adapted for the cluster design, leading to *variance component* or *mixed* models; these are considered in Chapter 9.

Stratified and weighted sample designs which are not clustered can be analysed by the standard methods for GLMs, with slight modifications.

For the psychological example of Section 3.1, we stated that the sample was randomly drawn from the population of fifth-grade children in one primary school. In fact, the sample design was stratified: the population was divided into sexes, and twelve children of each sex were randomly drawn from the sex sub-populations. The twelve children of each sex were then assigned to each experimental group by restricted randomization so that each group had six children of each sex. Thus sex and experimental group are orthogonal in the analysis.

The first six children in each experimental group in Section 3.1 are girls, the last six are boys. Sex is a stratifying factor in the design, and it is modelled as an explanatory variable like experimental group. The sex identification is not given in the data listing, but we can generate it as follows:

```
> solv$sex <- gl(2, 6, labels = c("girl", "boy"))
```

Why is it necessary to model the stratifying factor? There are two reasons: First, there may be substantial parameter differences between different strata: this is often the reason for stratification in the first place. Omission of the stratum variable from the model may then substantially bias the estimates of the model parameters, and will certainly increase their standard errors.

Second, the estimates of parameters based on aggregating over the strata assume that the population has the same structure over strata as the sample. Unequal sampling fractions (i.e. unequal proportions of the population from different strata included in the sample) combined with large stratum differences may give quite misleading population estimates if the stratum variable is ignored. On the other hand, if there are no stratum differences in parameters then unequal sampling fractions make no difference and the model can be collapsed over strata.

One point which often causes confusion is the use of 'sample weights' in regression. Survey studies sometimes substantially over-sample small strata or sub-populations to provide sample sizes similar to those from (under-sampled) large sub-populations. A 'sample weight' is often provided for each observation in the sample data set to allow the re-aggregation of the final model to provide population predictions. The sample weight is the reciprocal of the probability of inclusion in the sample of an observation from each sub-population. The sample weight will be high for the large sub-populations, and low for the small sub-populations.

These weights can be used formally to define a *weighted* or *pseudo* likelihood: for the sample weight w_i for y_i, the weighted likelihood is

$$\mathrm{WL} = \prod_{i=1}^{n} f(y_i \mid \theta)^{w_i}.$$

Then the weighted MLEs $\tilde{\theta}$ from the score equation satisfy

$$\sum_i w_i \frac{\partial \log f(y_i)}{\partial \theta} = 0.$$

If θ is the population mean and the model for Y is $N(\mu, \sigma^2)$, the weighted MLE is $\tilde{\mu} = \sum w_i y_i / \sum w_i$. This correctly weights for the disproportionate sampling.

However, it is an important point that these sample weights should *not* be used as formal weights in a regression analysis: the observations should be equally weighted (i.e. unweighted) in this analysis, and the model should always include the stratifying factor, together with its interactions with the other variables in the model. Model reduction then proceeds as in Chapter 2.

Weights are used only at the *predictive* stage, when population predictions are required taking the sample design into account. If the model reduction has allowed the elimination of the stratum interactions and main effect, then no weighting adjustment is needed, as the strata are homogeneous with respect to the residuals from the regression of the outcome variable on the other explanatory variables. If the stratum interactions or main effect *are* needed in the model, then the predicted values for the final model are aggregated across the strata using the sampling weights.

These weights are inappropriate as formal weights in regression for two reasons: they change the parameter estimates by giving higher weight to the samples from the larger populations, though each observation in fact represents a single individual, not an aggregate (like a mean) of several individuals, and the weights increase the standard errors of the estimated coefficients. Pfefferman (1993) noted 'Clearly, the use of (1.1) [weighted regression coefficients] cannot be justified in general based on optimality considerations'(p. 318).

The reason generally given for weighting is that it accounts for non-random sampling, if the sample inclusion probabilities depend on the response Y. In this case the likelihood has to represent explicitly the sample design. This kind of *biased sampling mechanism* is discussed in Section 3.17 on missing data.

Returning to the example, the number of possible models now increases considerably since it is possible that regressions of time on `eft` are different for each sex. A sequence of models of increasing complexity could be fitted as before. For reasons which will be made clear in the next section, we will consider only the most complex model, in which the regression of time on `eft` is different for each sex/instruction group:

```
> round(coef(summary(glm(time ~ sex * group * eft,
+       data = solv))), 4)

                        Estimate Std. Error t value Pr(>|t|)
(Intercept)             303.9024   126.5748  2.4010   0.0289
sexboy                  -84.9280   145.3573 -0.5843   0.5672
groupcorner            -120.4862   141.8091 -0.8496   0.4081
eft                       1.5534     2.4040  0.6462   0.5273
sexboy:groupcorner      373.4782   177.6511  2.1023   0.0517
sexboy:eft                1.9306     2.5991  0.7428   0.4684
groupcorner:eft           1.0230     2.6110  0.3918   0.7004
sexboy:groupcorner:eft   -5.5120     3.0204 -1.8249   0.0867
```

The six observations for each group and the fitted regressions are shown in Fig. 3.3.

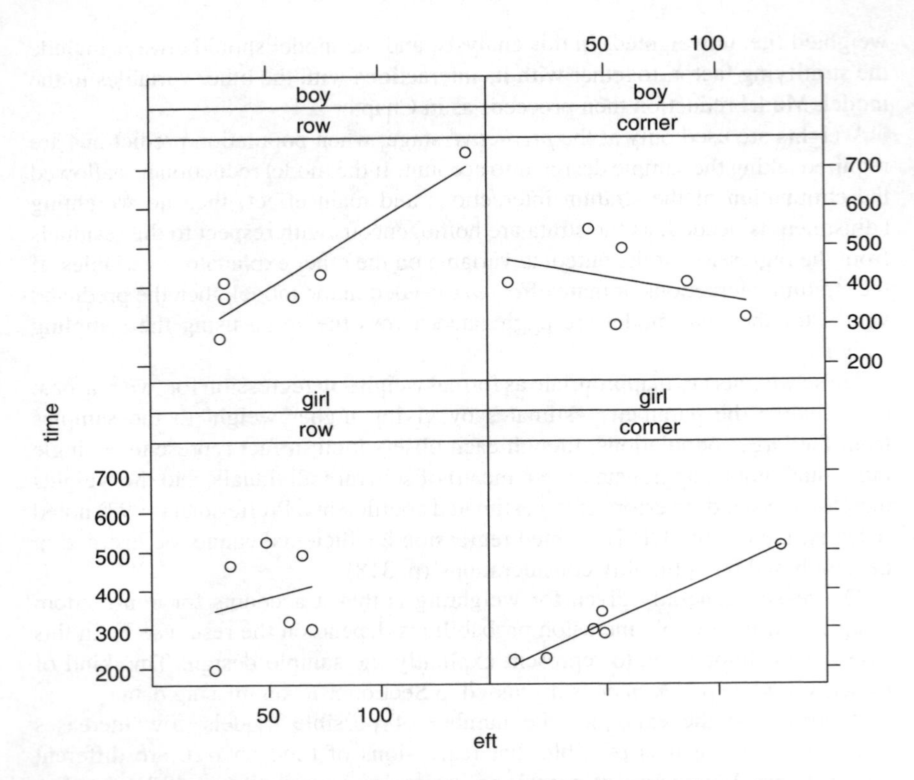

Fig. 3.3. Lattice plot for time versus eft conditioning on group and sex

```
> print(xyplot(time ~ eft | group * sex, data = solv,
+       type = c("p", "r")))
```

It is immediately striking that two observations (one in the girl/corner group and one in the boy/row group) are remote from the other observations, and without these two observations there would be little evidence of any regression on `eft` at all. We have already noticed these observations in model *solv.glm6*.

We now turn to an examination of such features of a model.

3.4 Model criticism

The conclusions drawn from a statistical model depend on the validity of the model. We pointed out in Chapter 2 that models are not exact representations of the population. We require only that they reproduce the main features of the population without major distortion.

A careful examination of the correspondence between data and model should be part of any statistical modelling of data. Examination of the data for failure of the model has been called *model criticism* by Box (1980, 1983), a term which we shall adopt, though the systematic examination of the model through *residuals* has a long history. Draper and Smith (1998) gave a discussion of residuals in normal models, and the field is extensively developed [see, e.g. Atkinson (1982); and the books by Atkinson (1985); Belsley *et al.* (1980); Barnett and Lewis (1994), and Cook and Weisberg (1982)].

There are four important areas in which failures of the model may occur:

(1) mis-specification of the probability distribution for y, leading to an inappropriate likelihood function and inappropriate MLEs and standard errors for the parameters;
(2) mis-specification of the link function;
(3) the occurrence of aberrant observations, distorting either the probability distribution or the parameter estimates from the model;
(4) mis-specification of the systematic part of the model, leading to incorrect interpretations.

Before considering these failures, we quote some standard results for the normal model.

The (raw or 'Pearson') residual e_i for the i-th observation from the model is

$$\begin{aligned} e_i &= y_i - \hat{\mu}_i \\ &= y_i - \mathbf{x}_i'\hat{\boldsymbol{\beta}}. \end{aligned}$$

The R method `residuals` (with alias `resid`) extracts model residuals from model objects, specifically 'working', 'response', 'deviance', 'Pearson', and 'partial' from both `lm` and `glm` model objects.

In matrix terms

$$\begin{aligned} \mathbf{e} &= \mathbf{y} - X\hat{\boldsymbol{\beta}} \\ &= (I - X(X'X)^{-1}X')\mathbf{y} \\ &= (I - H)\mathbf{y}, \end{aligned}$$

where H is the projection or 'hat' matrix

$$H = X(X'X)^{-1}X',$$

so-called since $\hat{\boldsymbol{\mu}} = H\mathbf{y}$: the fitted values are a projection of the original data $\mathbf{y}$ into the space spanned by X. The distribution of $\mathbf{e}$ as a random variable in repeated sampling is

$$\mathbf{E} \sim N_n(\mathbf{0}, \sigma^2(I - H))$$

which is a singular n-dimensional multivariate normal distribution with rank $n-p-1$, since there are only $n-p-1$ linearly independent residuals e_i. Denote the diagonal elements of H by h_i; then

$$h_i = \mathbf{x}_i'(X'X)^{-1}\mathbf{x}_i,$$
$$\text{var } E_i = \sigma^2(1-h_i), \quad \text{cov}(E_i, E_j) = -\sigma^2 h_{ij}, \ \ i \neq j,$$
$$\text{var } \hat{\mu}_i = \sigma^2 h_i.$$

The values of h_i are between zero and unity with $\Sigma h_i = p+1$. A value of zero for h_i means the fitted value of μ_i must be fixed, independently of the y_i, while a value of unity for h_i means that the residual is identically zero, and the model must exactly reproduce the i-th observation y_i. Thus large values of h_i are an indication that the corresponding observations may be influential in determining the position of the fitted model; for this reason the h_i are often called 'leverage' values. They are accessible by the function `hatvalues`.

The residuals have different variances, but can be converted to *standardized residuals* f_i:

$$f_i = e_i/s(1-h_i)^{\frac{1}{2}}.$$

These are obtained from the model object using the function `stdres` in the `MASS` package (Venables and Ripley, 2002). Most of the assessment of model failure is based on the standardized residuals. Other forms of residual are sometimes useful: the *jack-knife*, *studentized* or *cross-validatory* residuals are

$$j_i = f_i/((n-p-1-f_i^2)/(n-p-2))^{\frac{1}{2}}.$$

These are obtained from the model object using the `studres` function in `MASS`. The individual j_i have t_{n-p-2} distributions, though they are not independent; the distribution of the f_i, though it is not normal or t, does not depend on σ. Other statistics, functions of the residuals and/or the influence values h_i, are also used by several authors. R provides *Cook's distances* (Cook, 1977), defined as

$$cd_i = \frac{h_i}{p(1-h_i)} f_i^2.$$

These are available after extraction of the model object using the function `cooks.distance`.

We now consider the four areas of model failure introduced above.

3.4.1 *Mis-specification of the probability distribution*

Since the probability distribution $f(y)$ is fundamental to the likelihood function and therefore to the parameter estimates and standard errors in the model, an incorrect specification of the distribution can have serious consequences for interpretation of the data.

Examination of the data for failure of this assumption is usually through a *probability plot* (actually a *quantile* or *Q–Q* plot) of the standardized (or sometimes even the raw) residuals. The residuals f_i are ordered, and the ordered values $f_{(i)}$ plotted against the normal quantiles

$$z_i(a) = \Phi^{-1}\{(i - a)/(n + 1 - 2a)\},$$

where a is a suitably chosen constant ($0 \le a < 1$), and $\Phi(x)$ is the standard normal cumulative distribution function. The usual choice of a is either zero or $\frac{1}{2}$. For discussions of such choices, see Barnett (1975), Filliben (1975), and Draper and Smith (1998). If the probability distribution is correctly specified, the plot should be roughly a straight line. Systematic curvature, or individual observations far from the straight line, indicate failures of the probability distribution specification. The function `qqnorm` produces a normal Q–Q plot and `qqline` adds a line to a normal Q–Q plot.

As we noted in Chapter 2, an acute difficulty in the inspection of such plots is how to decide whether the variation in the plot is too far from a straight line. In Chapter 2 we used the simultaneous confidence band from the *cdf* of the observed data to check the probability model specification; the same approach may be used for residuals though it is now less theoretically sound since the residuals are not independent and do not have identical normal distributions. This difficulty can be resolved by a *simulation envelope* constructed by simulating from the fitted model (Atkinson, 1981; 1982; 1983).

We illustrate with the example of Section 3.1. We will first ignore sex and fit the `eft*group` model; we need to sort the residuals before constructing the simultaneous band. Residual examination is usually done with the full or most complex model, since simplifications of this model may be affected by failure of the normal distribution assumption.

```
> library(SMIR)

> print(xyplot(lower + upper ~ x,
+      data = NPL.bands(resid(solv.glm1)),
+      pch = 20, xlab = "residual", ylab = "cumulative proportion",
+      panel = function(x, y, ...) {
+          panel.xyplot(x, y, ...)
+          panel.curve(pnorm(x, 0,
+      sqrt(summary(solv.glm1)$dispersion)))}))
```

It is immediately clear from Fig. 3.4 that the residual distribution is poorly defined in a sample of 24 (with 20 df): the confidence band is so wide that only severe non-normality could be identified. There is no need to transform to the normal deviate scale.

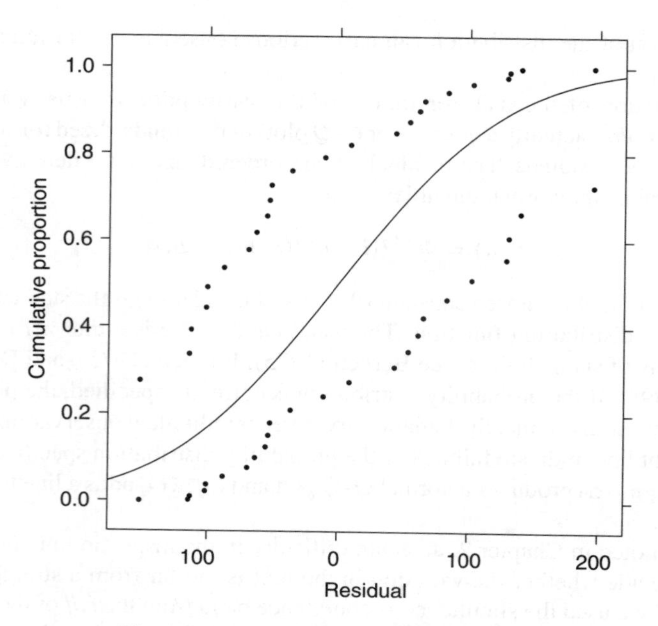

Fig. 3.4. Cumulative proportion, residuals

What should we do if the graph shows a major discrepancy from normality? There are two possibilities, as we discussed in Chapter 2: try a transformation of the variable to produce a closer agreement – usually the log transformation, or less commonly the reciprocal transformation – or use a different type of continuous probability distribution. These first of these possibilities is examined in detail below, and the second is examined in Chapter 6.

The standardized and jack-knife residuals can be explicitly calculated into named variables and used subsequently, provided that `hatvalue`, the leverage value, is first extracted.

For example, to obtain the standardized residuals (given here to 3 dp), we use the `stdres` function provided by the MASS package:

```
> library(MASS)
> round(stdres(solv.glm1), 3)

     1      2      3      4      5      6      7      8      9
-1.009  1.342  1.401 -1.537  0.540 -1.199 -0.542 -0.556  0.446
    10     11     12     13     14     15     16     17     18
 1.105 -0.372  0.859 -0.113 -1.057 -0.957  0.822 -0.538 -0.695
    19     20     21     22     23     24
 0.090  1.968 -0.635  1.290 -1.292  1.118
```

A second and less common form of failure of the probability model assumption is that the nuisance parameters, assumed constant for all observations, are in fact varying randomly, or systematically with the explanatory variables. A simple example of this would be if the variances were different for the two instruction groups in the example of Section 3.1. This could be assessed by plotting the residuals separately for each group against the fitted values. In general, it is necessary to plot the residuals against the fitted values and each explanatory variable: if different variability is evident then the model can be modified appropriately.

Heterogeneity of variance is examined in a factorial design in Section 3.15, and the 'double modelling' of both mean *and* variance is discussed in detail in Section 3.17.

3.4.2 *Mis-specification of the link function*

The choice of parameter scale on which the systematic effects are modelled is an important part of statistical modelling. When the response variable is normal it is almost universal practice to work on the scale of the normal mean μ, using the identity link: this has therefore been taken as the default option in R for the link function with a normal error specification.

In non-normal distributions the choice of scale is much more open, as we shall see in later chapters. However, even for normal distribution models, the possibility of working on the log scale particularly should be borne in mind. This possibility is investigated in Section 3.7, where both the lognormal distribution for the response variable and a log link with the normal distribution are examined.

3.4.3 *The occurrence of aberrant and influential observations*

Good data analysis depends heavily on the correct recording of variable values. Misrecording of data values can result in substantial changes to the fitted model. If an observation happens to lie on the boundary of the space of the explanatory variables, misrecording of its response variable may have a powerful effect on the position of the fitted regression, but may not produce a large residual. The same effect can occur by misrecording an explanatory variable, thus moving the observation out of the cluster of correctly recorded observations. The influence values h_i are particularly useful for diagnosing observations of this type.

Even when no misrecording occurs it is valuable to know whether a small number of observations is substantially influencing the position of the fitted regression. These observations can then be carefully examined and referred back to the original survey or experimental scientists for comment. Observations with large residuals or influence values are not automatically rejected from the model: they always require evaluation.

Table 3.4. Influence values

1	21	22	3	5	8	13	17	18	20	4	11	2
0.08	0.08	0.08	0.09	0.09	0.09	0.09	0.09	0.09	0.09	0.10	0.11	0.13

9	12	7	14	19	6	15	24	23	16	10
0.14	0.14	0.15	0.15	0.15	0.16	0.22	0.23	0.30	0.42	0.72

We illustrate the use of influence values with the example from Section 3.1. For the `group*eft` model, we can list the influence values against observation number:

```
> sort(round(hatvalues(solv.glm1), 2))
```

The influence values are shown in Table 3.4.

How should we interpret the influence values? The model specifies a linear regression in each experimental group, with different slopes for each group. For a linear regression model, it is easily shown (see, e.g. Hoaglin and Welsch, 1978) that

$$\mu_i = \beta_0 + \beta_1 x_i$$

gives

$$h_i = \frac{1}{n} + \frac{(x_i - \bar{x})^2}{\sum_{j=1}^{n}(x_j - \bar{x})^2}.$$

Thus h_i takes its minimum value of $1/n$ for observations at the explanatory variable mean, and its maximum value for the observation furthest from the mean: extreme values of x are the most influential. If all x_i are equidistant from the mean, $h_i = 2/n$; if all but one observation have identical values of x_i, these will have $h_i = 1/(n-1)$, and the remaining observation will have $h_i = 1$. In this case the position of the regression is determined completely by the single observation, and we could have no confidence at all in the regression.

For the general regression model $\beta_0 + \boldsymbol{\beta}'\mathbf{x}_i$ with p explanatory variables $\mathbf{x}$, the corresponding result is

$$h_i = \frac{1}{n} + (\mathbf{x}_i - \bar{\mathbf{x}})' S^{-1} (\mathbf{x}_i - \bar{\mathbf{x}}),$$

where $\bar{\mathbf{x}}$ is the vector of means and S is the SSP matrix of the explanatory variables. In general, $(p+1)/n$ is the average value of all the h_i and also corresponds to equally influential observations. Hoaglin and Welsch proposed regarding $2(p+1)/n$ as a value of h indicating important influence.

For the model `group*eft` with $p = 3$ and $n = 24$, $(p+1)/n$ is 0.167. Since the model `group*eft` corresponds to unrelated simple linear regressions in each instruction group of 12 children, observations near the `eft` mean in each group will have influence values near $1/12 = 0.0833$.

One value stands out: 0.72 for the 10th observation in the row group. This observation has the largest `eft` value of 139, and also the largest completion time of 739 sec. It has a powerful effect on the fitted regression for any subgroup in which it appears. For the `eft + group` model, its influence value is reduced to 0.35, because the slope of the regression is determined by both row and column groups, and the effect of this observation is diluted by the doubled sample size.

The influence value of 0.42 for the fourth observation in the corner group (sixteenth overall) is also high (in the *group*eft* model). This observation has the largest `eft` value of 128 in the corner group, though its completion time value of 513 sec is not exceptionally large. Its influence value is reduced to 0.28 in the main effect model.

It is notable that neither of these observations has a large residual: in both cases the fitted regression passes close to the observed value, partly because of the influence of the observation on the position of the fitted line.

In model *solv.glm6*, only the first of these observations is influential: the 10th observation in the row group has influence 0.72 as before, because the regression on `eft` is determined by only the row group. The column group observations all have the same influence 0.083, as a constant mean is being fitted for this group.

In the full *sex*group*eft* model, the two influential observations above (again given to 2 dp) have influence values of more than 0.8.

```
> sort(round(hatvalues(glm(time ~ sex * group * eft,
+       data = solv)), 2))

   3    8   13   17   18   21   22   11   20    1   14    9   12
0.17 0.17 0.17 0.17 0.17 0.17 0.17 0.19 0.21 0.22 0.25 0.27 0.27
  19    7    5    2   15    4    6   23   24   10   16
0.27 0.28 0.31 0.36 0.37 0.40 0.55 0.58 0.60 0.84 0.88
```

The reasons are clear from Fig. 3.3: the more we subdivide or stratify the observations by explanatory variables, the greater the effect these observations have on the regressions within subgroup. The regressions in the corner group for girls and the row group for boys are being very largely determined by the single largest observation in each case. We did not pursue the simplification of the `sex*group*eft` model in Section 3.3 because the conclusions from this model depend so heavily on these two observations. This is a frequent problem when samples are extensively cross-classified by many explanatory factors, so that many cells have very small numbers of observations.

To examine the effect of restricting the analysis to those observations with influence values greater than $2(p+1)/n$ we remove those values with influence values $< 2(p+1)/n$,

```
> model <- glm(time ~ sex * group * eft, data = solv)
> summary(update(model, subset = (hatvalues(model) <
+       (2 * model$rank/length(solv$time)))))

Coefficients:
                       Estimate Std. Error t value Pr(>|t|)
(Intercept)             303.902    131.522   2.311   0.0366
sexboy                   -6.606    174.081  -0.038   0.9703
groupcorner            -120.953    168.111  -0.719   0.4837
eft                       1.553      2.498   0.622   0.5440
sexboy:groupcorner      295.622    219.352   1.348   0.1992
sexboy:eft                0.048      3.409   0.014   0.9890
groupcorner:eft           1.038      3.715   0.279   0.7841
sexboy:groupcorner:eft   -3.644      4.541  -0.803   0.4357

(Dispersion parameter for gaussian family taken to be 9564.76)

    Null deviance: 214531  on 21  degrees of freedom
Residual deviance: 133907  on 14  degrees of freedom
AIC: 272.14

Number of Fisher Scoring iterations: 2
```

The three-way interaction parameter estimate is now smaller than its standard error, and successive elimination of terms shows that all the `eft` terms can be omitted from the model. The two-way interaction `group.sex` now becomes quite large (as is its standard error). This interaction can be identified as due to one discrepant cell: girls in the corner group have a mean completion time of 272.6 sec, while boys in both groups and girls in the row group have similar means, with a common mean completion time of 388.7 sec.

These conclusions are markedly different from those obtained by retaining the two influential observations: in the latter case the `sex*group*eft` model can be reduced to the `eft` only model *solv.glm3* of Section 3.1, with no group difference and a common regression on `eft`.

What action, if any, should we take? The original data should be checked with the experimenter: are the `eft` and time values correct for these two observations? If they are not, the correct values can be entered. If they are, then the experiment can be interpreted as before. The difficulty, as previously mentioned, is that because the sample sizes are small several different conflicting interpretations of the data are possible.

It may not be possible to check the correctness of the original data. In this case an analysis of the data might be carried out as above by including and excluding the two highly influential observations, and comparing the results. It is not, however, good statistical practice to set aside routinely or exclude observations with large residuals and/or influence simply because these values are large.

3.4.4 *Mis-specification of the systematic part of the model*

This form of model failure is different from the others because it is rectified by further modelling, though diagnostic plotting of the standardized residuals against other variables may be useful.

When the explanatory variables are continuous, one form of model failure is the reliance on linear regressions when the relationships are non-linear. Inclusion of quadratic or higher-order polynomial terms in the regression is a common solution; however, this may lead to unreasonable features in the model, for example, the fitted values may reach a maximum and then decrease while the observed values approach an asymptote. Transformations of the variables or of the scale of μ (a different link function) may be preferable.

Interactions between variables may also be necessary. In fitting such terms in complex models a careful eye should be kept on the influence values, because in small samples the possibility of single observations producing apparent interactions is considerable, as in the above example.

The need for non-linear terms or interactions is usually assessed graphically by plotting the raw or standardized residuals against the explanatory variables and the fitted values. Non-random scatter patterns suggest the need for model changes. We do not discuss this further here as the examples presented in later chapters are analysed in considerable detail.

3.5 The Box–Cox transformation family

In this section we extend the usefulness of the normal distribution by embedding it in a larger family of distributions: the Box–Cox transformation family (Box and Cox, 1964).

The motivation for this general family is the frequent occurrence of skewed data in which a log transformation, or less commonly a reciprocal transformation, of the response variable produces a nearly normal distribution. This should be carefully distinguished from the use of a log or reciprocal link function: in the first case the observed data have a lognormal or reciprocal normal distribution, in the second they have a normal distribution. We will illustrate both cases with an example in Section 3.7.

The general form of the transformation can be represented by

$$\begin{aligned} y^{(\lambda)} &= (y^{\lambda} - 1)/\lambda, \quad \lambda \neq 0 \\ &= \log y, \quad \lambda = 0, \end{aligned}$$

where λ is the transformation parameter and y the response variable. Here y must be *positive*. For $\lambda = 1$, $y^{(1)} = y - 1$, and as $\lambda \to 0$, $y^{(\lambda)} \to \log y$, so that $y^{(\lambda)}$ is a continuous function of λ. For $\lambda = -1$, $y^{(-1)} = 1 - y^{-1}$.

Given data $(y_i, \mathbf{x}_i)$, we assume that there is some value of λ for which $Y_i^{(\lambda)}$ has a normal distribution with mean $\boldsymbol{\beta}'\mathbf{x}_i$ and variance σ^2. We want to estimate λ, and decide the appropriate scale of y on which to fit the regression model. The choice of the scale for y, that is, of the value of λ, is determined both by the plausible values of λ from the fitted model and by the interpretability of the scale. Because the transformation is defined by a parameter, we can proceed by maximum likelihood. For each fixed value of λ, we estimate $\boldsymbol{\beta}$ and σ by maximum likelihood, and substitute these values in the likelihood. The resulting function of λ only is the *profile likelihood in* λ; its log is the *profile-log likelihood* (these are discussed in Chapter 2). It is used to construct likelihood-based confidence intervals for λ.

The probability density function of $Y^{(\lambda)}$ is normal (μ, σ^2) and hence that of Y is for $\lambda \neq 0$,

$$f(y|\lambda, \mu, \sigma) = \frac{1}{\sigma\sqrt{2\pi}}|\lambda|y^{\lambda-1}\exp\{-(y^{(\lambda)} - \mu)^2/2\sigma^2\},$$

and for $\lambda = 0$,

$$f(y|0, \mu, \sigma) = \frac{1}{\sigma\sqrt{2\pi}}y^{-1}\exp\{-(\log y - \mu)^2/2\sigma^2\}.$$

We assume that, whatever the value of λ, $\mu = \boldsymbol{\beta}'\mathbf{x}$. For the given observations $(y_i, \mathbf{x}_i)$, the log-likelihood function is

$$\ell(\lambda, \boldsymbol{\beta}, \sigma) = -n\log\sqrt{2\pi} - n\log\sigma + n\log|\lambda| + (\lambda - 1)\sum\log y_i - \sum(y_i^{(\lambda)} - \boldsymbol{\beta}'\mathbf{x}_i)^2/2\sigma^2,$$

where for $\lambda = 0$, $\log|\lambda|$ is defined to be zero. For fixed λ, the partial derivatives with respect to $\boldsymbol{\beta}$ and σ are then

$$\frac{\partial\ell}{\partial\boldsymbol{\beta}} = \Sigma\mathbf{x}_i(y_i^{(\lambda)} - \mathbf{x}_i'\boldsymbol{\beta})/\sigma^2$$

$$\frac{\partial\ell}{\partial\sigma} = -n/\sigma + \Sigma(y_i^{(\lambda)} - \mathbf{x}_i'\boldsymbol{\beta})^2/\sigma^3.$$

Denote the solutions of the equations $\partial\ell/\partial\boldsymbol{\beta} = 0$, $\partial\ell/\partial\sigma = 0$ by $\hat{\boldsymbol{\beta}}(\lambda)$, $\hat{\sigma}(\lambda)$. These are the same as in Section 3.1, with $y_i^{(\lambda)}$ replacing y_i:

$$\hat{\boldsymbol{\beta}}(\lambda) = (X'X)^{-1}X'\mathbf{y}^{(\lambda)}$$

$$\hat{\sigma}^2(\lambda) = \Sigma\{y_i^{(\lambda)} - \mathbf{x}_i'\hat{\boldsymbol{\beta}}(\lambda)\}^2/n = \text{RSS}(\lambda)/n,$$

where $\text{RSS}(\lambda)$ is the residual sum of squares for the given λ.

Substituting into the log-likelihood function, we obtain the profile log-likelihood function

$$p\ell(\lambda) = -n \log \sqrt{2\pi} - \frac{n}{2} \log n - \frac{n}{2} - \tfrac{1}{2} n \log \mathrm{RSS}(\lambda)$$
$$+ n \log |\lambda| + (\lambda - 1) \sum \log y_i .$$

An (approximate) $100(1-\alpha)\%$ confidence interval for λ consists of those values of λ for which $p\ell(\lambda)$ is within $\frac{1}{2}\chi^2_{\alpha,1}$ units of its maximum: the interval is most simply found by tabulation of $p\ell(\lambda)$ over a grid of values of λ. This interval is used to identify the values of λ giving an interpretable scale.

The function `boxcox` from the MASS package (Venables and Ripley, 2002) allows for the construction and plotting of the profile log-likelihood. Illustration of its use is postponed to the next section.

An important feature of response-variable transformations is that on the transformed scale the model represents variation in the *mean* of the (normal) transformed variable, but on the original scale the variation is in the *median* of the variable.

This is most simply seen for the log transformation. Suppose that $\log Y \sim N(\mu, \sigma^2)$. Then Y has a lognormal distribution, and

$$\begin{aligned} \text{median}\,[Y] &= \exp(\mu) \\ \mathrm{E}[Y] &= \exp(\mu + \tfrac{1}{2}\sigma^2) \\ \mathrm{Var}[Y] &= [\exp(\sigma^2) - 1] \exp(2\mu + \sigma^2). \end{aligned}$$

Thus the (additive) regression model for the mean of log Y is a multiplicative model for the median of Y, and also for the mean of Y, though the intercept is changed by $\frac{1}{2}\sigma^2$, and the variance of Y is not constant.

Though the model is fitted on the log scale, we usually want to interpret the model on the original scale: if fitted values from the model are transformed by `exp(fitted(model.object))` to the original scale, these are fitted values for the median response, not for the mean: for the mean the fitted values are `exp(fitted(model.object))+summary(model.object)$dispersion/2`, where here `summary(model.object)$dispersion` is the unbiased estimate s^2 of σ^2.

For transformations Y^λ with $\lambda \neq 0$, if μ is the mean of Y^λ then

$$\begin{aligned} \text{median}\,[Y] &= \mu^{1/\lambda} \\ \mathrm{E}[Y] &\approx \mu^{1/\lambda}\{1 + \sigma^2(1-\lambda)/(2\lambda^2\mu^2)\} \\ \mathrm{Var}[Y] &\approx \mu^{2/\lambda}\sigma^2/(\lambda^2\mu^2). \end{aligned}$$

The apparent discontinuity between $\lambda = 0$ and $\lambda \neq 0$ is caused by the use here of y^λ rather than $(y^\lambda - 1)/\lambda$.

3.6 Modelling and background information

The most useful models are those which use background information or theory from the field of application. However, theory may be incorrect, or speculative, or there may be no adequate theory and the modelling may be meant to assist the development of an adequate theory.

We illustrate with a set of data on tree volumes taken from the Minitab Handbook, (Ryan *et al.*, 1976) and discussed at length by Atkinson (1982) and other authors. The volume of usable wood `v` in cubic feet (1 foot = 30.48 cm) is given for each of a sample of 31 black cherry trees, and the height `h` in feet and the diameter `d` in inches (1 inch = 2.54 cm) at a height 4.5 feet above the ground. We want to develop a model which will predict the usable wood volume from the easily measured height and diameter.

The data are in the file `trees`, and are listed in Table 3.5. We begin by graphing `v` against `d` and `h` displayed here as a scatterplot matrix (Fig. 3.5).

```
> data(trees, package = "SMIR")
```

A basic matrix of scatterplots can be generated by

```
> plot(~., data = trees)
```

the '.' referring to all variables in the dataframe, but an enhanced version can be generated using the *scatterplot.matrix* function from the *car* package. This package provides a Gaussian kernel density estimate for each of the variables; kernel density estimation is discussed in Section 7.9. A *rugplot* (Chambers and Hastie, 1992) is added to each density plot. From the height rug plot it is clear that the trees were selected over a nearly equally spaced grid of heights.

Table 3.5. Tree data

d	h	v	d	h	v	d	h	v	d	h	v
8.3	70	10.3	8.6	65	10.3	8.8	63	10.2	10.5	72	16.4
10.7	81	18.8	10.8	83	19.7	11.0	66	15.6	11.0	75	18.2
11.1	80	22.6	11.2	75	19.9	11.3	79	24.2	11.4	76	21.0
11.4	76	21.4	11.7	69	21.3	12.0	75	19.1	12.9	74	22.2
12.9	85	33.8	13.3	86	27.4	13.7	71	25.7	13.8	64	24.9
14.0	78	34.5	14.2	80	31.7	14.5	74	36.3	16.0	72	38.3
16.3	77	42.6	17.3	81	55.4	17.5	82	55.7	17.9	80	58.3
18.0	80	51.5	18.0	80	51.0	20.6	87	77.0			

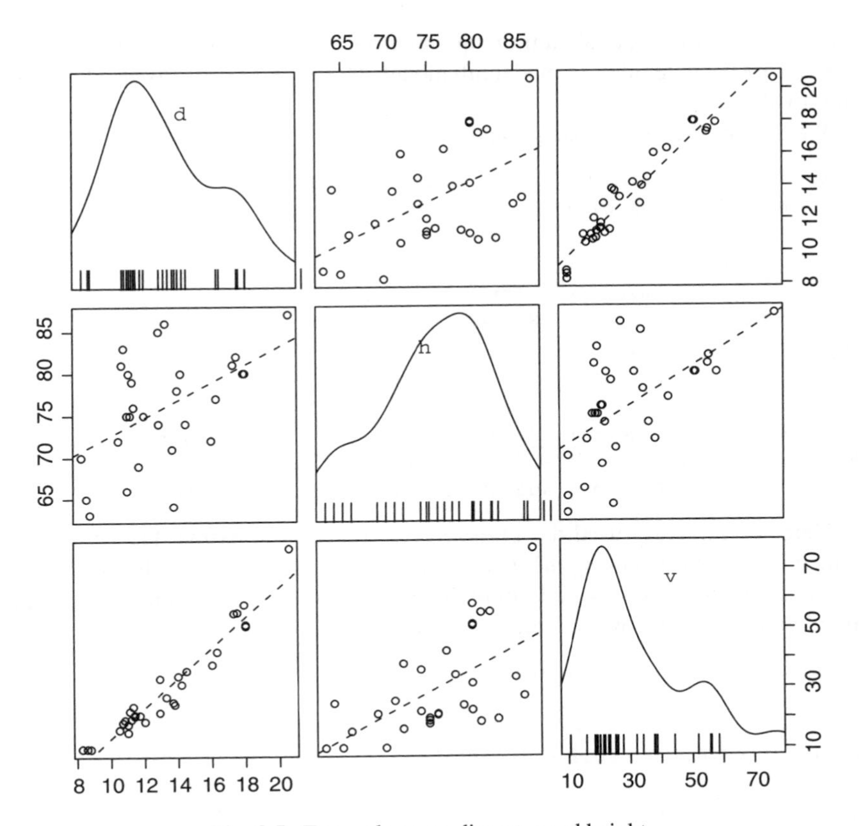

Fig. 3.5. Tree volume vs diameter and height

```
> library(car)
> scatterplot.matrix(~d + h + v, data = trees, smooth = FALSE,
+       col = c(1, 1, 1))
```

Clearly accurate prediction of `v` can be made from `d`: the variation about a smooth curve is quite small. Height is less useful, though obviously related to volume. Curvature in the `v,d` graph is evident: the extreme points at both ends are well above the nearly linear ellipse in the centre. The `(v, h)` graph shows increasing variability of `v` with increasing `h`. A linear relationship with constant variance will not be adequate.

We will use the multiple correlation R (or more usually its square, R^2) as a measure of the predictability of the response from the explanatory variables: R is just the correlation between the response and the fitted values from the model. Equivalently,

$$R^2 = 1 - \mathrm{RSS}/\mathrm{TSS},$$

where RSS is the residual sum of squares from the given model and TSS is the RSSs from the indexidxnull modelnull model: TSS $= \sum(y_i - \bar{y})^2$ is usually called the 'total sum of squares'.

We first try the model using h and d.

```
> trees.glm1 <- glm(v ~ h + d, data = trees)
> summary.lm(trees.glm1)

Coefficients:
            Estimate Std. Error t value Pr(>|t|)
(Intercept) -57.9877     8.6382  -6.713 2.75e-07
h             0.3393     0.1302   2.607   0.0145
d             4.7082     0.2643  17.816  < 2e-16

Residual standard error: 3.882 on 28 degrees of freedom
Multiple R-Squared: 0.948,      Adjusted R-squared: 0.9442
F-statistic:   255 on 2 and 28 DF,  p-value: < 2.2e-16
```

Here we explicitly used the *summary.lm* function to extract the R^2 statistic which is not returned using the default summary method for *glm* objects. For this model R^2 is very high, implying close agreement between observed and fitted volumes. We now examine the residuals and influence values.

```
> round(hatvalues(trees.glm1), 3)

    1     2     3     4     5     6     7     8     9    10
0.116 0.147 0.177 0.059 0.121 0.156 0.115 0.051 0.092 0.048
   11    12    13    14    15    16    17    18    19    20
0.074 0.048 0.048 0.073 0.038 0.036 0.131 0.143 0.067 0.211
   21    22    23    24    25    26    27    28    29    30
0.036 0.045 0.050 0.111 0.069 0.088 0.096 0.106 0.110 0.110
   31
0.227
```

The 20th and the last observations have influence values of 0.21 and 0.23, respectively, just larger than 6/31 = 0.194. The last tree is the largest, the 20th has a large diameter for its height.

```
> scatter.smooth(trees$d, residuals(trees.glm1, type = "working"),
+      xlab = "d", ylab = expression(paste("residual from model   ",
+           v, " ~ ", h + d)))
```

The graph of residuals against h (not shown) looks fairly random but that against d (Fig. 3.6) shows a marked dip in the middle, with large positive residuals at each end and smaller and negative residuals in the centre. This suggests that the model is mis-specified, and the curvature we saw in the v versus d graph suggests that we may be working on the wrong scale. To investigate this we construct a Box–Cox plot.

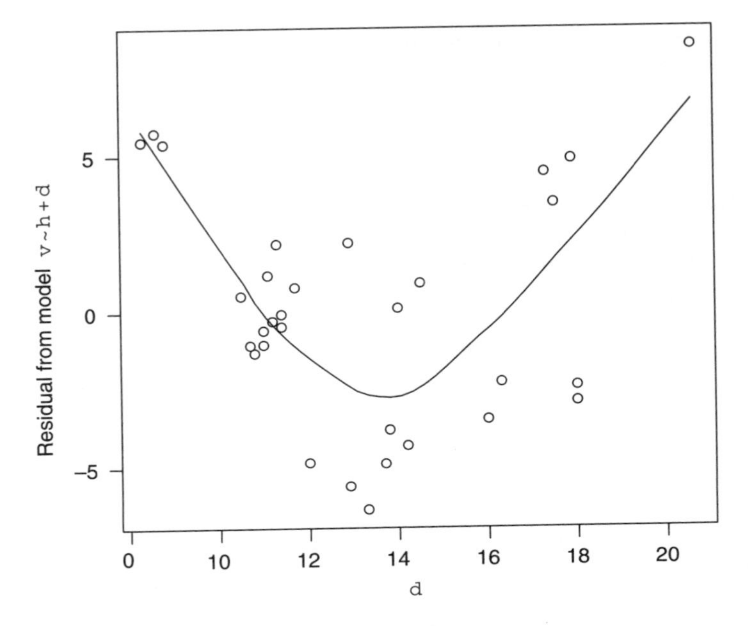

Fig. 3.6. Residual vs diameter

We first load the `MASS` package (Venables and Ripley, 2002) which contains the *boxcox* function:

```
> require(MASS)
```

The *boxcox* function allows the user to specify the range of values of λ over which the values of the disparity are to be plotted. It is seldom necessary to choose the range wider than -2 to $+2$ which is the default, with steps of 0.1 usually sufficient. We do not reproduce the plot, but give the maximized log-likelihoods produced by the Box–Cox function.

```
> tree.bc <- boxcox(trees.glm1, plotit = FALSE, lambda = seq(-2,
+     2, by = 0.5))
> names(tree.bc$y) <- tree.bc$x
> print(as.vector(tree.bc$y), digits = 3)

[1] -134.3 -122.4 -109.8  -95.7  -80.7  -78.2  -93.7 -110.0
[9] -124.8
```

The maximum of the log-likelihood occurs near $\lambda = 0.5$. A finer tabulation in steps of 0.1 over the range 0 to 1 is required.

```
> tree.boxcox <- boxcox(trees.glm1, plotit = FALSE,
+      lambda = seq(from = 0, to = 1, by = 0.1))
> tree.boxcox$x[which.max(tree.boxcox$y)]

[1] 0.3

> lambda.max <- tree.boxcox$x
> logl <- tree.boxcox$y
> names(logl) <- tree.boxcox$x
> print(logl, digits = 3)

    0   0.1   0.2   0.3   0.4   0.5   0.6   0.7   0.8   0.9
-80.7 -78.3 -76.7 -76.1 -76.6 -78.2 -80.6 -83.6 -86.8 -90.3
    1
-93.7

> rbind(lambda.max, logl)

                   0         0.1         0.2         0.3         0.4
lambda.max   0.00000     0.10000     0.20000     0.30000     0.40000
logl       -80.70206   -78.34921   -76.71938   -76.08257   -76.58266
                 0.5         0.6         0.7        0.8         0.9
lambda.max   0.50000     0.60000     0.70000     0.8000     0.90000
logl       -78.15815   -80.57769   -83.55375   -86.8375   -90.25369
                   1
lambda.max    1.0000
logl        -93.6947

> logl <- logl - max(logl)
> a <- seq(from = 0, to = 1, by = 0.1)[logl > log(0.15)]
```

Computing $\log L$ over a fine mesh, we find the maximum of -76.1 occurs at $\lambda = 0.31$, and the approximate 95% confidence interval, based on a value of -76.1 ± 1.92, is (0.12, 0.49), which is quite narrow. How should we interpret this result?

The nature of the problem throws light on this question. Volume is measured in cubic feet, but height and diameter in feet (or inches). We are attempting to predict a volumetric measurement from two linear measurements. This suggests that the 'side of the equivalent cube' might be more appropriate as a response, the cube root of volume. The profile likelihood in λ points very closely to this value. To see why, we calculate $\mathrm{v}^{1/3}$ and graph it against d and h (Fig. 3.7):

```
> trees$vth <- trees$v^(1/3)
> par(mfrow = c(1, 2))
> plot(vth ~ d + h, data = trees, ask = FALSE)
```

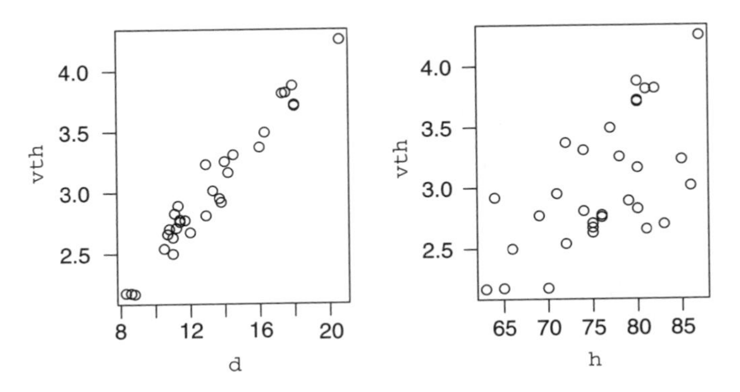

Fig. 3.7. Cube root volume vs diameter and height

The graph against d is now very closely linear. The curvature has been removed by the cube root transformation: the extreme points at each end now line up with the central ellipse. The graph against h also shows homogeneity of variability with increasing h. We now refit the model:

```
> trees.vth.glm1 <- glm(vth ~ h + d, data = trees)
> summary.lm(trees.vth.glm1)

Coefficients:
              Estimate Std. Error t value Pr(>|t|)
(Intercept) -0.085388   0.184315  -0.463    0.647
h            0.014472   0.002777   5.211 1.56e-05
d            0.151516   0.005639  26.871  < 2e-16

Residual standard error: 0.08283 on 28 degrees of freedom
Multiple R-Squared: 0.9777,       Adjusted R-squared: 0.9761
F-statistic: 612.5 on 2 and 28 DF,  p-value: < 2.2e-16
```

The value of R^2 is a substantial increase over that for $\lambda = 1$. Residual graphs against d and h (not shown) show a random scatter, though the residuals seem to increase noticeably in magnitude with h, and a quantile plot gives a closely linear fit.

The fitted median values of v from the model are $\hat{\mu}^3$, and the fitted mean values of v are approximately $\hat{\mu}^3(1+3s^2/\hat{\mu}^2)$. These can be compared with the observed values, for example, using

```
> round(rbind(fitted = fitted(trees.vth.glm1)^3,
+       observed = trees$v), 2)

             1     2     3     4     5     6     7     8     9
fitted   10.43 10.05 10.07 16.53 19.86 20.85 16.32 18.96 20.89
observed 10.30 10.30 10.20 16.40 18.80 19.70 15.60 18.20 22.60
```

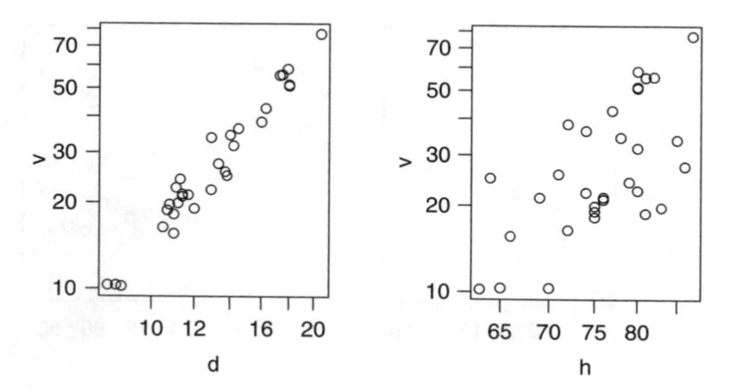

Fig. 3.8. Log volume vs log diameter and log height

```
          10    11    12    13    14    15    16    17    18
fitted   19.62 21.25 20.61 20.61 19.38 22.38 25.41 29.77 31.99
observed 19.90 24.20 21.00 21.40 21.30 19.10 22.20 33.80 27.40
          19    20    21    22    23    24    25    26    27
fitted   27.49 25.2 31.69 33.51 32.23 38.64 42.83 50.98 52.85
observed 25.70 24.9 34.50 31.70 36.30 38.30 42.60 55.40 55.70
          28    29    30    31
fitted   54.2 54.86 54.86 79.22
observed 58.3 51.50 51.00 77.00
```

Fitted median and mean values both agree closely with the observed values. We have a closely fitting model to which we are led by physical considerations of dimensionality, and the Box–Cox plot. Other physical considerations, however, lead to a different model. The curvature which we noted in the `v` against `d` graph can also be removed by a double log transformation (Fig. 3.8), which improves variability heterogeneity as well:

```
> trees$lv <- log(trees$v)
> trees$ld <- log(trees$d)
> trees$lh <- log(trees$h)
> par(mfrow = c(1, 2))
> plot(v ~ d + h, data = trees, ask = FALSE, log = "xy")
```

This suggests a regression of log `v` on the logged explanatory variables.

```
> trees.lv.glm1 <- glm(lv ~ lh + ld, data = trees)
> summary.lm(trees.lv.glm1)

Coefficients:
            Estimate Std. Error t value Pr(>|t|)
(Intercept) -6.63162    0.79979  -8.292 5.06e-09
lh           1.11712    0.20444   5.464 7.81e-06
```

```
ld            1.98265    0.07501   26.432  < 2e-16

Residual standard error: 0.08139 on 28 degrees of freedom
Multiple R-Squared: 0.9777, Adjusted R-squared: 0.9761
F-statistic: 613.2 on 2 and 28 DF,  p-value: < 2.2e-16
```

The value of R^2 is exactly the same (to four decimal places) as that for the $\mathrm{v}^{1/3}$ model. Residual plots (not shown) show little unusual apart from three rather large negative residuals, though the same pattern of residuals increasing with h is visible as with the cube root transformation. The quantile plot of the standardized residuals shows some variation. The influence values are much the same as before, though that for the 20th observation is larger (0.24) and that for the last is smaller (0.18).

Should we be working on the log scale of v? What is the effect of the log transformations of h and d on the Box–Cox transformation of v?

We specify a new model:

```
> library(MASS)
> boxcox(v ~ lh + ld, data = trees, plotit = FALSE,
+      lambda = seq(from = -2, to = 2, by = 0.5))

$x
[1] -2.0 -1.5 -1.0 -0.5  0.0  0.5  1.0  1.5  2.0

$y
[1] -129.73506 -116.45269 -101.57192  -84.50386  -75.33877
[6]  -88.63894 -104.42525 -118.30619 -131.27850
```

Using the grid $-2(0.5)2$ for λ shows a likelihood maximum near $\lambda = 0$. A finer tabulation in steps of 0.01 around zero shows a maximum of -75.0 at $\lambda = -0.07$, and a 95% confidence interval of $(-0.24, 0.11)$, which includes zero but excludes $\lambda = 1/3$. Thus if we log transform h and d, the cube root transformation of v is not appropriate, but the log transformation is. The value of the disparity at $\lambda = 0$ of 132.20 is slightly smaller than for $\mathrm{v}^{1/3}$, h, and d, which was 133.76.

How can we compare these models? It would appear that we can use the likelihood ratio test, since both models can be expressed in the form

$$\mathrm{v}^* \sim \mathrm{N}(\beta_0 + \beta_1 \mathrm{h}^* + \beta_2 \mathrm{d}^*, \sigma^2)$$

with

$$\mathrm{v}^* = (\mathrm{v}^{\lambda_\mathrm{v}} - 1)/\lambda_\mathrm{v}$$
$$\mathrm{h}^* = (\mathrm{h}^{\lambda_\mathrm{h}} - 1)/\lambda_\mathrm{h}$$
$$\mathrm{d}^* = (\mathrm{d}^{\lambda_\mathrm{d}} - 1)/\lambda_\mathrm{d}.$$

The comparison is of the set of values $\lambda_{\mathrm{v}} = 1/3, \lambda_{\mathrm{h}} = \lambda_{\mathrm{d}} = 1$ with the set $\lambda_{\mathrm{v}} = \lambda_{\mathrm{h}} = \lambda_{\mathrm{d}} = 0$, the other parameters being nuisance parameters in each case, the likelihood being maximized over them. The disparity difference is 1.56, but the models are not nested and have the same number of parameters, so this difference cannot be distributed as χ^2 with any number of degrees of freedom.

Model comparisons of this kind can be treated as discussed in Section 2.7.4 through the posterior distribution of the TLR for the two models, evaluated at the true values of the nuisance parameters $\boldsymbol{\beta}$ and σ. For large samples the disparity difference, or the ratio of maximized likelihoods, is interpreted directly as the weight of evidence for one model relative to the other.

The disparity difference of 1.56 is equivalent to a maximized likelihood ratio of $e^{0.78} = 2.18$; the sample evidence gives only a very weak preference for the log model over the cube root model.

In the fitted model for log `v`, the coefficient of `lh` is close to 1, and that of `ld` close to 2. Again the nature of the problem throws light on this result. The tree may be idealized (modelled) as a regular solid figure, like a cylinder or a cone. The volume `v` of a cylinder of height `h` and diameter `d` is $\pi\mathrm{d}^2\mathrm{h}/4$, and of that of a cone of height `h` and base diameter d is $\pi\mathrm{d}^2\mathrm{h}/12$. In either case

$$\log \mathrm{v} = c + \log \mathrm{h} + 2 \log \mathrm{d},$$

where c is log $\pi/4$ for the cylinder and log $\pi/12$ for the cone. These volume formulae assume that `h` and `d` are measured in the same units. In our example, `h` is in feet but `d` is in inches. We first convert `d` to feet:

```
> trees <- transform(trees, df = d/12)
> trees <- transform(trees, ldf = log(df))
```

We redefine the y-variate and refit the model.

```
> (trees.log.glm1 <- glm(lv ~ ldf + lh, data = trees))

Coefficients:
(Intercept)          ldf           lh
     -1.705        1.983        1.117

Degrees of Freedom: 30 Total (i.e. Null);  28 Residual
Null Deviance:          8.309
Residual Deviance: 0.1855        AIC: -62.71
```

The intercept is now -1.705. Now we fit the model in which the coefficients are fixed at 1 and 2. The model is then

$$\mu = \beta_0 + x_1 + 2x_2$$

in which only β_0 is to be estimated. Here $(x_1 + 2x_2)$ is treated as a single variable whose coefficient is to be fixed at 1.0. This is achieved by declaring this variable to be an *offset* and fitting the null model in which only β_0 is estimated.

```
> trees$z <- trees$lh + 2 * trees$ldf
> trees.log.glm2 <- glm(lv ~ offset(z), data = trees)
> summary.lm(trees.log.glm2)

Coefficients:
            Estimate Std. Error t value Pr(>|t|)
(Intercept) -1.19935    0.01421  -84.42   <2e-16

Residual standard error: 0.0791 on 30 degrees of freedom
```

R does not report an R^2 value for this model because it incorrectly calculates the R^2 using the sums of squares corrected for the mean and the offset. We need to be more specific.

```
> 1 - (sum(resid(trees.log.glm2)^2))/(sum((trees$lv -
+      mean(trees$lv))^2))

[1] 0.977411
```

The value of R^2 is a trivial change from the full model. Does the value of β_0 correspond to a cylinder or a cone or neither?

```
> log(c(pi/4, pi/12))

[1] -0.2415645 -1.3401768
```

The tree is closer to a cone than a cylinder, but it has a greater volume than the cone, which seems reasonable.

The fitted median values of v from the model are $\exp(\hat{\mu})$, which can be compared with the observed values

```
> fv <- exp(fitted(trees.log.glm2))
> round(rbind("f median" = fv, observed = trees$v),
+      1)

            1    2    3    4    5    6    7    8    9   10   11
f median 10.1 10.1 10.2 16.6 19.4 20.3 16.7 19.0 20.6 19.7 21.1
observed 10.3 10.3 10.2 16.4 18.8 19.7 15.6 18.2 22.6 19.9 24.2
           12   13   14   15   16   17   18   19   20   21   22
f median 20.7 20.7 19.8 22.6 25.8 29.6 31.8 27.9 25.5 32.0 33.8
observed 21.0 21.4 21.3 19.1 22.2 33.8 27.4 25.7 24.9 34.5 31.7
           23   24   25   26   27   28   29   30   31
f median 32.6 38.6 42.8 50.7 52.6 53.6 54.2 54.2 77.3
observed 36.3 38.3 42.6 55.4 55.7 58.3 51.5 51.0 77.0
```

The fitted mean values of v are $\exp(\hat{\mu} + \frac{1}{2}s^2)$, which are only 0.3% greater than the median values.

The two fitted models agree closely with the observed values (a maximum error of 4.59 cubic feet for the $\mathtt{v}^{1/3}$ model and 4.66 cubic feet for the log v model), with the error standard deviation of 0.079 in the log model corresponding to a percentage error of 8% in the fitted volume. Here the data are insufficient to discriminate between models based on different physical arguments.

3.7 Link functions and transformations

We noted above that even in normal models the choice of the link function relating the parameter of the probability distribution to the linear predictor is not always obvious. The preceding example provides a good illustration. For the first model relating tree volume to diameter and height, a cube root transformation of v linearized the graph of v against d. It is not, however, necessary to assume that $\mathtt{v}^{1/3}$ has a normal distribution with mean θ equal to the linear predictor $\beta_0 + \beta_1\mathtt{h} + \beta_2\mathtt{d}$, in order to fit such a model. We can assume instead that V is normal, but the link function relating V to the linear predictor is the cube root, that is, if μ is the mean tree volume,

$$\eta = g(\mu) = \mu^{1/3}, \quad \text{or} \quad \mu = \eta^3 = (\beta_0 + \beta_1\mathtt{h} + \beta_2\mathtt{d})^3,$$

while $V \sim N(\mu, \sigma^2)$.

The fitting of this model by maximum likelihood is equivalent to fitting by non-linear least squares, since the model for μ is now non-linear in the parameters. A simple change in the link function to the Exponent link is all that is needed; the offset also needs to be removed:

```
> (trees.glm1 <- glm(v ~ h + d, data = trees,
+       family = quasi(power(lambda = 1/3),
+       variance = "constant")))

Coefficients:
(Intercept)           h           d
   -0.05132     0.01429     0.15033

Degrees of Freedom: 30 Total (i.e. Null);  28 Residual
Null Deviance:              8106
Residual Deviance: 184.2        AIC: NA
```

R does not report an R^2 value for *glm* objects but we can entice it by using the *summary.lm* function.

The value of R^2 is very slightly less than for the $\mathtt{v}^{1/3}$ model. The fitted model is almost the same as in Section 3.6. The RSSs for this model is 184.16, and converting this to the disparity $n(\log \text{RSS} + \log(2\pi) + 1 - \log n)$ gives a value of

143.21. Residual plots show the same feature as for the cube root transformation of v, the residuals again increasing in absolute size with h, though not with d.

```
> qqnorm(resid(trees.glm1))
> qqline(resid(trees.glm1))
```

The quantile plot (not shown) looks linear; this model also seems satisfactory.

We can repeat this procedure using the log link instead of the lognormal distribution for v:

```
> trees.glm2 <- glm(v ~ lh + ld, data = trees, quasi(link = "log",
+      variance = "constant"))
> summary.glm(trees.glm2)

Coefficients:
            Estimate Std. Error t value Pr(>|t|)
(Intercept) -6.53700    0.94352  -6.928 1.57e-07
lh           1.08765    0.24216   4.491 0.000111
ld           1.99692    0.08208  24.330  < 2e-16

(Dispersion parameter for quasi family taken to be 6.41642)

    Null deviance: 8106.08  on 30  degrees of freedom
Residual deviance:  179.66  on 28  degrees of freedom
AIC: NA

Number of Fisher Scoring iterations: 4
```

The fitted model is again very similar to that in Section 3.6, and the residual sum of squares gives a slightly better fit, with a disparity of 142.44. The residual graph against h shows the same feature of rapidly increasing residuals.

Now we have four possible models. Can we choose between response variable and link transformations? Extending the argument of Section 3.6, we can write the competing models in terms of *four* λ parameters:

$$\mathrm{v}^* \sim N(\eta^*, \sigma^2)$$

with

$$\eta^* = (\eta^{\lambda_\ell} - 1)/\lambda_\ell$$

and

$$\eta = \beta_0 + \beta_1 \mathrm{h}^* + \beta_2 \mathrm{d}^*$$

as in Section 3.6. Here λ_ℓ is the link function parameter. The two models of Section 3.6 had $\lambda_\ell = 1$ with $\lambda_\mathrm{v} = 1/3$ and $\lambda_\mathrm{v} = 0$; the two models above have $\lambda_\ell = 1/3$ and $\lambda_\ell = 0$ with $\lambda_\mathrm{v} = 1$.

Compared with the disparity value of 132.20 for the lognormal model, both the link function transformation models are considerably worse (a disparity difference of 10 is equivalent to a ratio of maximized likelihoods of 0.0067, with the same number of parameters). We therefore conclude that the log or cube root variable transformation models are better, and we prefer the log model because of its 'solid body' interpretation. Thus the maximized likelihoods enable us to choose between transforming the distribution and transforming the link function.

A detailed discussion of transformations of the explanatory variables was given by Box and Tidwell (1962). Scallan *et al.* (1984) extended the idea of the Box–Cox transformation to 'Box–Coxing the link function' by parametrizing the link function as $\eta = g(\mu) = (\mu^\lambda - 1)/\lambda$ and estimating λ.

3.8 Regression models for prediction

In Chapter 2 we noted two uses of regression models: for simplified descriptions of populations, and for prediction of values of the response variable for new observations. We now consider prediction.

The tree volume example provides a good illustration. The trees were felled and the volume of usable wood measured so that the volume of usable wood could be estimated for stands or forests of similar trees. The real purpose of the model is not for simple description – though this is an essential part of the modelling – but for the prediction of wood volume of similar trees.

This prediction is easily obtained from the fitted model. We are given $Y_i \sim \mathbf{N}(\boldsymbol{\beta}'\mathbf{x}_i, \sigma^2)$ independently, $i = 1, \ldots, n$, and a new observation $\mathbf{x}_{n+1}$ on $\mathbf{x}$, and we want to predict the corresponding value Y_{n+1} of Y. We assume that the same distributional model applies for Y_{n+1}. For simplicity of notation, we write

$$Y^* = Y_{n+1}, \mathbf{x} = \mathbf{x}_{n+1}.$$

Then

$$Y^* \sim N(\boldsymbol{\beta}'\mathbf{x}, \sigma^2)$$

and

$$\hat{\boldsymbol{\beta}}'\mathbf{x} \sim \mathbf{N}(\boldsymbol{\beta}'\mathbf{x}, \sigma^2\mathbf{x}'(X'X)^{-1}\mathbf{x}),$$
$$\mathrm{RSS} \sim \sigma^2\chi^2_{n-p-1}$$

independently of Y^*. It follows immediately that,

$$Y^* - \hat{\boldsymbol{\beta}}'\mathbf{x} \sim N(0, \sigma^2(1+h))$$
$$(Y^* - \hat{\boldsymbol{\beta}}'\mathbf{x})/s(1+h)^{\frac{1}{2}} \sim t_{n-p-1},$$

where $s^2 = \text{RSS}/(n-p-1)$ and $h = \mathbf{x}'(X'X)^{-1}\mathbf{x}$ is the variance of the estimated linear predictor $\hat{\boldsymbol{\beta}}'\mathbf{x}$, divided by σ^2.

Thus probability statements may be made about the random variable Y^* from the t-distribution, for example,

$$\Pr(|Y^* - \hat{\boldsymbol{\beta}}'\mathbf{x}| < t_{\alpha/2,n-p-1}s(1+h)^{\frac{1}{2}}) = 1-\alpha$$

is a $100(1-\alpha)\%$ *prediction* interval for y. It is not a *confidence* interval since Y^* is a random variable, not a parameter. Confidence interval statements can be made about the mean of Y^*, $\boldsymbol{\beta}'\mathbf{x}$, as this is a parametric function of $\boldsymbol{\beta}$. Since

$$\hat{\boldsymbol{\beta}}'\mathbf{x} \sim N(\boldsymbol{\beta}'\mathbf{x}, \sigma^2 h),$$

a $100(1-\alpha)\%$ confidence interval for $\boldsymbol{\beta}'x$ is

$$|\boldsymbol{\beta}'\mathbf{x} - \hat{\boldsymbol{\beta}}'\mathbf{x}| < t_{\alpha/2,n-p-1}sh^{\frac{1}{2}}.$$

This is shorter than that for Y^* as the additional variation in the random variable does not have to be allowed for.

The same results for Y^* can be obtained by treating Y^* as a formal unknown parameter in the model for $y_1, \ldots, y_n, Y^*$ and constructing a profile likelihood for Y^* by maximizing the likelihood of $y_1, \ldots, y_n, Y^*$ over $\boldsymbol{\beta}$ and σ for a fixed value of Y^*. This is known as a *predictive likelihood*; see Butler (1986) for details and references. This device is very useful in non-normal models, and is used in subsequent chapters.

Construction of the prediction interval depends on the predicted value $\hat{Y}^* = \hat{\boldsymbol{\beta}}'\mathbf{x}$ and the variance $s^2 h$ of the linear predictor $\hat{\boldsymbol{\beta}}'\mathbf{x}$. These are easily obtained in R with the `predict` function and setting the `interval` argument to '`prediction`'.

We illustrate with the trees data. Suppose in addition to the 31 observations we have a new tree with height `h` = 80 feet and diameter `d` = 15 inches. What predictive statement can be made about this tree's usable wood volume?

We first fit the required model – we will use the log model for illustration.

```
> trees.log.glm3 <- glm(lv ~ lh + ld, data = trees)
> new <- data.frame(lh = log(80), ld = log(15))
> predict.lm(trees.log.glm3, new, interval = "prediction")

          fit      lwr      upr
[1,] 3.632763 3.461916 3.803610
```

The fitted value is 3.633, and a 95% prediction interval for the tree log-volume is then (3.462, 3.804). The corresponding interval for the predicted tree volume is (31.9, 44.9).

3.9 Model choice and mean square prediction error

We noted in Chapter 2 that regression models are used both for smooth representation and for prediction. The model selection procedure described in Section 3.1 is intended to provide the model which gives the simplest representation of the population consistent with the data. It does not necessarily follow that this model is the best one for prediction.

Since the outcome of the prediction process is a predicted value (and an interval for the true value), it seems reasonable to choose the model which gives the closest prediction. This can be defined in several ways; we will adopt the usual convention that the *squared prediction error*

$$\sum_{i \in N} (y_i - \hat{y}_i)^2$$

is to be minimized, where the minimization is over the set N of future observations. Since this is usually unknown, we are forced to characterize the set of future observations by reference to the calibration set of observations already available: the new observations $\mathbf{x}$ are assumed to be similar to the existing observations $\mathbf{x}_i$. (If they were not, then we should have reservations about extrapolating the fitted regression into regions of the explanatory variable space not represented in the data.)

This similarity is expressed in one of two ways: the existing $\mathbf{x}_i$ are thought of as fixed, and the new observations $\mathbf{x}$ have a uniform distribution over the $\mathbf{x}_i$; or the existing $\mathbf{x}_i$ are thought of as values of a random variable $\mathbf{X}$ (apart from the constant 1), and the new $\mathbf{x}$ are independent values of this random variable. Minimization of the *expected* or *mean* square prediction error (MSPE) is then taken to be the aim of the model selection procedure, where the expectation is over the uniform distribution of the $\mathbf{x}_i$, or the common multivariate distribution of $\mathbf{X}$. Because of the difficulty of specifying the latter distribution (a multivariate normal distribution is a common, but unrealistic, specification) we consider only the first case.

For the normal model $Y \sim N(\boldsymbol{\beta}'\mathbf{x}, \sigma^2)$, the mean square prediction error is (e.g. Aitkin, 1974)

$$E_f = (n + p + 1)\sigma^2/n.$$

For the subset model $\boldsymbol{\beta}'_r\mathbf{x}_r$, omitting the component $\boldsymbol{\beta}'_d\mathbf{x}_d$ from $\boldsymbol{\beta}'\mathbf{x}$, the MSPE is (using the notation of Section 3.1)

$$E_r = (n + p_1 + 1)\sigma^2/n + \boldsymbol{\beta}'_d S_{r.d} \boldsymbol{\beta}_d/n,$$

where

$$S_{d.r} = X'_d X_d - X'_d X_r (X'_r X_r)^{-1} X'_r X_d.$$

Thus prediction is improved (MSPE decreases) if

$$E_f - E_r > 0$$

that is,

$$p_2\sigma^2 > \boldsymbol{\beta}_d' S_{d.r} \boldsymbol{\beta}_d.$$

Thus if $\boldsymbol{\beta}_d$ is small, prediction will be improved by omitting **x**_d. It is possible to test formally whether this inequality is violated for any subset **x**_r, thus giving a simultaneous test for the non-reduction of the MSPE (Aitkin, 1974) for all possible subsets.

However, this test does not identify the model with the *smallest* MSPE, only those models with MSPEs *not significantly larger* than that of the full model. Replacing parameter values in E_f and E_r by sample estimates from each model allows a comparison of the *estimated* MSPEs, though a formal test is not available.

The above inequality becomes, on substituting sample estimates,

$$\frac{\hat{\boldsymbol{\beta}}_d' S_{d.r} \hat{\boldsymbol{\beta}}_d}{p_2 s^2} < 1$$

which is equivalent to $F_{d.r} < 1$ where $F_{d.r}$ is the F-statistic for the significance of the omitted variables **x**_d.

The use of this criterion is considered in an example in Section 3.11, but we first examine the use of *cross-validation*.

3.10 Model selection through cross-validation

Considerable experience with the use of regression models for prediction has shown that when applied to new observations or samples, the model predicts much less well than in the calibration sample. This phenomenon is called 'shrinkage on cross-validation', the shrinkage referred to being that of R^2. Stone (1974) gave some historical examples. Copas (1983) gave a detailed discussion.

In regression studies with large samples it is possible to divide the sample (randomly) into two halves. The model is fitted on the first half, and used for the prediction of the values of y in the second half. This *cross-validation* of the model on the second independent sample gives a more realistic assessment of its predictive value, and of the value of reduced models. An example with small samples was given by Copas (1983).

With small samples cross-validation can be achieved by omitting each observation in turn from the data, fitting the model (or models) to the remaining observations, predicting the value of y for the omitted observation, and comparing the prediction with the observed value. Let $\hat{y}_{(i)}$ be the predicted value of y_i when the i-th observation is omitted from the data. Then

$$\text{CVE} = \sum_{i=1}^{n} (y_i - \hat{y}_{(i)})^2/n$$

is the *cross-validation estimate* (CVE) of the MSPE; it is an unbiased estimate of σ^2. The sum $n\text{CVE} = \sum (y_i - \hat{y}_{(i)})^2$ was called PRESS (prediction error sum of squares) by Allen (1971) who gave a more general definition. The computation of PRESS for any normal linear model is easily achieved, without omitting observations, from any model object.

Let $X_{(-i)}$ be the matrix of explanatory variables with $\mathbf{x}_i$ deleted and $\mathbf{y}_{(-i)}$ the vector of response values with y_i deleted, and let $\hat{\boldsymbol{\beta}}_{(-i)}$ be the corresponding estimate of $\boldsymbol{\beta}$. Then

$$\hat{y}_{(i)} = \mathbf{x}_i' \hat{\boldsymbol{\beta}}_{(-i)}$$

and

$$\begin{aligned} X'X &= X'_{(-i)} X_{(-i)} + \mathbf{x}_i \mathbf{x}_i' \\ X'\mathbf{y} &= X'_{(-i)} \mathbf{y}_{(-i)} + \mathbf{x}_i y_i. \end{aligned}$$

Then

$$\begin{aligned} \left(X'_{(-i)} X_{(-i)}\right)^{-1} &= (X'X - \mathbf{x}_i \mathbf{x}_i')^{-1} \\ &= (X'X)^{-1} + (X'X)^{-1} \mathbf{x}_i (1 - \mathbf{x}_i'(X'X)^{-1}\mathbf{x}_i)^{-1} \mathbf{x}_i'(X'X)^{-1} \\ X'_{(-i)} \mathbf{y}_{(-i)} &= X'\mathbf{y} - \mathbf{x}_i y_i \\ \hat{\boldsymbol{\beta}}_{(-i)} &= (X'_{(-i)} X_{(-i)})^{-1} X'_{(-i)} \mathbf{y}_{(-i)} \end{aligned}$$

and

$$\begin{aligned} \hat{y}_{(i)} &= \mathbf{x}_i' \boldsymbol{\beta}_{(-i)} = (\hat{y}_i - h_i y_i)/(1 - h_i) \\ e_{(i)} &= y_i - \hat{y}_{(i)} = (y_i - \hat{y}_i)/(1 - h_i) = e_i/(1 - h_i), \end{aligned}$$

where

$$h_i = \mathbf{x}_i'(X'X)^{-1}\mathbf{x}_i$$

is the influence of the i-th observation in the complete sample. The PRESS criterion is thus

$$\text{PRESS} = \sum_{i=1}^{n} e_{(i)}^2 = \sum_{i=1}^{n} e_i^2/(1 - h_i)^2$$

which is easily calculated using a small function:

```
> PRESS <- function(x) sum((resid(x)^2/(1 - hatvalues(x))^2))
```

that employs the `resid` and `hatvalues` functions and extracting the degrees of freedom from the model object to obtain the CVE estimate and the RSS values directly.

```
> CVE <- function(x) sum((resid(x)^2/
+      (1 - hatvalues(x))^2))/length(fitted(x))
> RSS <- function(x) sum(resid(x)^2)
```

Again, we might choose for prediction the model with the smallest value of PRESS, or more realistically, one of the models with a small value of this criterion. We examine in the next section the use of these criteria for prediction model choice on a complex example. We conclude this section with a simple illustration on the trees data, using the log volume model. The coefficients of log height and log diameter have simple physical interpretations: should we use the model 2*`ld + lh` for prediction, or the model with estimated coefficients?

```
> data(trees, package = "SMIR")

> trees <- transform(trees, lv = log(v), lh = log(h),
+     ld = log(d))
> trees.glm <- glm(lv ~ ld + lh, data = trees)
> PRESS(trees.glm)

[1] 0.2186

> CVE(trees.glm)

[1] 0.00705
```

For the full model we have RSS $= 0.1855$, $s^2 = 0.006624$, PRESS $= 0.2186$ and CVE $= 0.00705$ which is 6% larger than s^2. The cross-validation R^2_{CV} is 0.9737, very little less than the value 0.9777 for R^2 itself.

We define the functions R2 and R2CV to help extract the information from the model objects as:

```
> R2 <- function(x) {
+     f <- fitted(x)
+     1 - (sum(resid(x)^2))/(sum((x$y - mean(f))^2))
+ }
> R2CV <- function(x) {
+     yj <- (fitted(x) - hatvalues(x) * x$y)/(1 - hatvalues(x))
+     cor(x$y, yj)^2
+ }
```

```
> trees$z <- 2 * trees$ld + trees$lh
> trees.glm1 <- glm(lv ~ offset(z), data = trees)
> RSS(trees.glm1)

[1] 0.1877

> summary(trees.glm1)$dispersion

[1] 0.006256

> PRESS(trees.glm1)

[1] 0.2004

> R2(trees.glm1)

[1] 0.9774

> CVE(trees.glm1)

[1] 0.006465

> R2CV(trees.glm1)

[1] 0.9759
```

For the reduced model RSS $= 0.1877$, $s^2 = 0.006256$, PRESS $= 0.2004$, $R^2 = 0.9774$ and CVE $= 0.006465$, which is now only 3% larger than s^2. The cross-validation R^2_{CV} is 0.9759, *greater* than R^2_{CV} for the full model. Both s^2 and PRESS decrease in the reduced model, by 6–8% of their values in the full model. This shows that prediction should be based on the reduced model.

3.11 Reduction of complex regression models

The value of modelling and model simplification becomes clear when we are dealing with complex data sets with many possible explanatory variables. A good example is given by Henderson and Velleman (1981), also discussed by Aitkin and Francis (1982). The data are shown below and are in the file `car`:

```
> data(car, package = "SMIR")

> row.names(car) <- abbreviate(car$model, minlength = 14)
> car$model <- NULL
> format(car)

               s c t g  disp  hp cb drat    wt   qmt  mpg
MAZDA RX-4     0 6 1 4 160.0 110  4 3.90 2.620 16.46 21.0
MAZDARX-4WAGON 0 6 1 4 160.0 110  4 3.90 2.875 17.02 21.0
```

DATSUN 710	1	4	1	4	108.0	93	1	3.85	2.320	18.61	22.8
HORNET 4 DRIVE	1	4	0	3	258.0	110	1	3.08	3.215	19.44	21.4
HORNETSPORTABO	0	8	0	3	360.0	175	2	3.15	3.440	17.02	18.7
VALIANT	1	6	0	3	225.0	105	1	2.76	3.460	20.22	18.1
DUSTER 360	0	8	0	3	360.0	245	4	3.21	3.570	15.84	14.3
MERCEDES 240D	1	4	0	4	146.7	62	2	3.69	3.190	20.00	24.4
MERCEDES 230	1	4	0	4	140.8	95	2	3.92	3.150	22.90	22.8
MERCEDES 280	1	6	0	4	167.6	123	4	3.92	3.440	18.30	19.2
MERCEDES 280C	1	6	0	4	167.6	123	4	3.92	3.440	18.90	17.8
MERCEDES 450SE	0	8	0	3	275.8	180	3	3.07	4.070	17.40	16.4
MERCEDES 450SL	0	8	0	3	275.8	180	3	3.07	3.730	17.60	17.3
MERCEDES450SLC	0	8	0	3	275.8	180	3	3.07	3.780	18.00	15.2
CADILLACFLEETW	0	8	0	3	472.0	205	4	2.93	5.250	17.98	10.4
LINCOLNCONTINE	0	8	0	3	460.0	215	4	3.00	5.425	17.82	10.4
IMPERIAL	0	8	0	3	440.0	230	4	3.23	5.345	17.42	14.7
FIAT 128	1	4	1	4	78.7	66	1	4.08	2.200	19.47	32.4
HONDA CIVIC	1	4	1	4	75.7	52	2	4.93	1.615	18.52	30.4
TOYOTA COROLLA	1	4	1	4	71.1	65	1	4.22	1.835	19.90	33.9
TOYOTA CORONA	1	4	0	3	120.1	97	1	3.70	2.465	20.01	21.5
DODGECHALLENGE	0	8	0	3	318.0	150	2	2.76	3.520	16.87	15.5
AMC JAVELIN	0	8	0	3	304.0	150	2	3.15	3.435	17.30	15.2
CHEVROLETCAMAZ	0	8	0	3	350.0	245	4	3.73	3.840	15.41	13.3
PONTIACFIREBIR	0	8	0	3	400.0	175	2	3.08	3.845	17.05	19.2
FIAT X1-9	1	4	1	4	79.0	66	1	4.08	1.935	18.90	27.3
PORSCHE 914-2	0	4	1	5	120.3	91	2	4.43	2.140	16.70	26.0
LOTUS EUROPA	1	4	1	5	95.1	113	2	3.77	1.513	16.90	30.4
FORD PANTERA L	0	8	1	5	351.0	264	4	4.22	3.170	14.50	15.8
FERRARIDINO197	0	6	1	5	145.0	175	6	3.62	2.770	15.50	19.7
MASERATI BORA	0	8	1	5	301.0	335	8	3.54	3.570	14.60	15.0
VOLVO 142E	1	4	1	4	121.0	109	2	4.11	2.780	18.60	21.4

The data are quarter-mile acceleration time in seconds (`qmt`) and fuel consumption in miles per (US) gallon (`mpg`) for 32 cars tested by the US *Motor Trend* magazine in 1974. Nine explanatory variables are given: shape of engine `s` (straight = 1, vee = 0), number of cylinders `c`, transmission type `t` (automatic = 0, manual = 1), number of gears `g`, engine displacement in cubic inches `disp`, horsepower `hp`, number of carburettor barrels `cb`, final drive ratio `drat`, and weight of the car in thousands of pounds `wt`. In the original analyses referenced above, quarter-mile time is taken as an explanatory variable for `mpg`, but it is not a basic design variable and is therefore omitted in this analysis of fuel consumption. A separate analysis of `qmt` as a response variable is given in Section 3.12.

Our object is to obtain a simple model relating `mpg` to the explanatory variables. The 32 cars are not a random sample of the car population to which this model can be generalized: the *Motor Trend* sample is heavily weighted to European and US high-performance sports cars and luxury cars.

We begin by graphing mpg against the explanatory variables. This can help identify outlying observations, suggest the important explanatory variables and give a general feel for the data. We show here only a few of these graphs.

```
> par(mfrow = c(3, 3), mar = c(4, 4, 2, 1.5))
> plot(mpg ~ s + c + t + g + disp + hp + cb + drat +
+      wt, data = car, ask = FALSE)
```

Curvature in the graph against disp is very noticeable (Fig. 3.9), suggesting that a scale change is required. We try the log scale.

```
> car <- transform(car, lmpg = log(mpg), ldis = log(disp))
> plot(lmpg ~ ldis, data = car, xlab = "log displacement",
+      ylab = "log mpg")
```

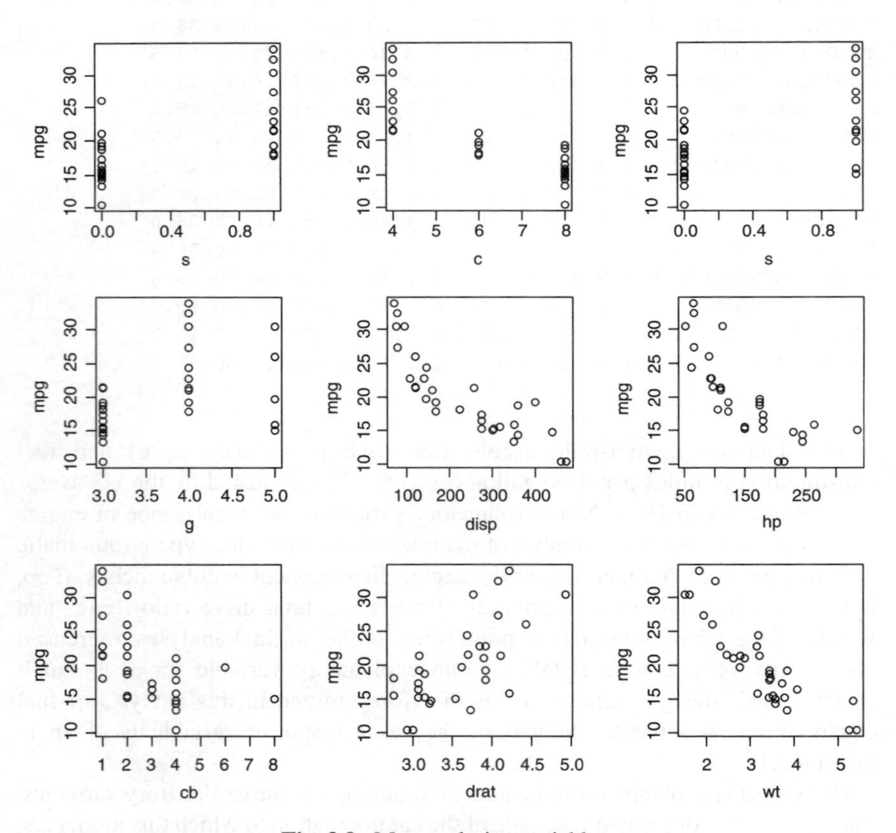

Fig. 3.9. Mpg vs design variables

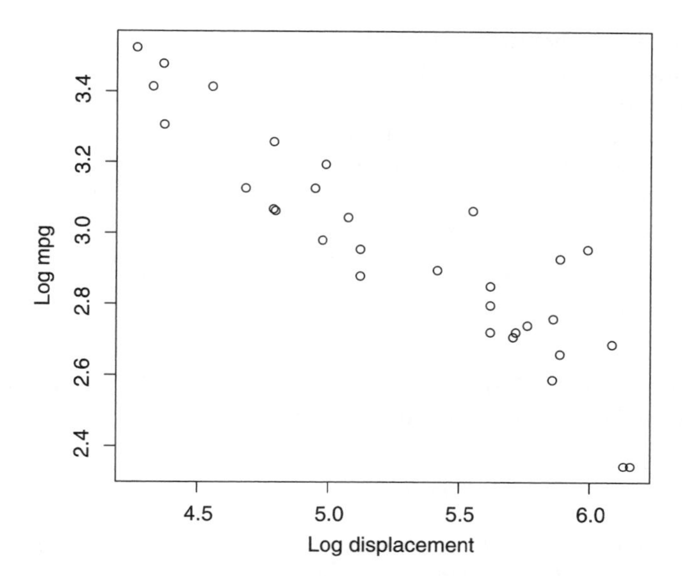

Fig. 3.10. Log mpg vs log displacement

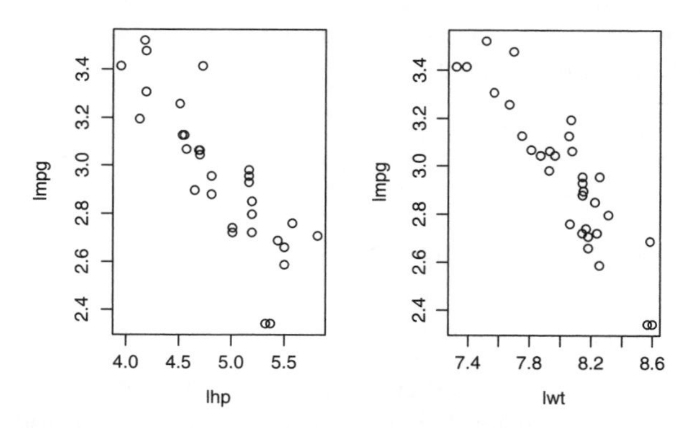

Fig. 3.11. Log mpg vs log hp and log wt

The graph is now much more nearly linear (Fig. 3.10). The graphs against `hp` and `wt` also show curvature and these are similarly linearized by a log transformation; we show below both these variables graphed on the log scale (Fig. 3.11).

```
> car <- transform(car, lhp = log(hp), lwt = log(wt *
+       1000))
> par(mfrow = c(1, 2), mar = c(4, 4, 2, 1.5))
> plot(lmpg ~ lhp + lwt, data = car, ask = FALSE)
```

The graphs do not show any marked outliers, and it is clear that hp, wt, and disp are important explanatory variables, as one might expect.

We now proceed to fit a model using all the explanatory variables, with hp, wt, and disp replaced by their log transformations. Should the model be fitted to mpg or log mpg, or perhaps gpm = 1/mpg – gallons per mile, the European standard for fuel consumption (litres per km) used by Henderson and Velleman (1981)? The Box–Cox family suggests a clear answer.

```
> car.bc <- boxcox(mpg ~ s + c + t + g + ldis + lhp +
+      cb + drat + lwt, data = car, plotit = FALSE,
+      lambda = seq(-2, 2, by = 0.5))
```

Specifying the range -2 to 2 in steps of 0.5 in response to the prompts, we obtain the values of the log-likelihood reproduced below:

```
   -2   -1.5    -1   -0.5     0    0.5     1    1.5     2
-97.8 -90.9 -84.2 -78.5 -74.6 -73.8 -76.2 -81.0 -87.0
```

The maximum of -73.67 occurs near $\lambda = 0.4$, and the approximate 95% confidence interval for λ includes $\lambda = 0$ but neither $\lambda = -1$ nor $\lambda = 1$. Thus the log scale for mpg is indicated.

Does the transformation of disp, hp and wt to their logs affect the transformation of mpg? To check, we can repeat the Box–Cox transformation without transforming these variables.

```
> car.bc2 <- boxcox(mpg ~ s + c + t + g + disp + hp +
+      cb + drat + wt, data = car, plotit = FALSE, lambda = seq(-2,
+      2, by = 0.5))
```

The values of the log-likelihood are now:

```
   -2   -1.5    -1   -0.5     0    0.5     1    1.5     2
-92.4 -86.3 -81.1 -77.5 -76.1 -77.4 -80.9 -85.7 -91.4
```

The maximum now occurs almost at $\lambda = 0$, and again neither $\lambda = -1$ nor $\lambda = 1$ is included in the approximate 95% confidence interval. The transformation of the explanatory variables has had very little effect on the estimation of λ, though the log-likelihood is increased by 1.5 at $\lambda = 0$ by the log transformations, showing a slightly better fit.

We proceed then with an examination of the first model above, using ldis, lhp and lwt. One point we need to consider first is the status of the Mercedes 240D. The 'D' stands for 'diesel': this car is the only diesel-engined one in the sample. Since the fuel consumption of diesel engines is generally much less than that of petrol engines of equivalent power, we need to treat this car separately from the others. We do this by defining a dummy variable d which takes the value 1 for this car and 0 for all the others, and then fit this variable in the model, in effect defining a two-group structure; the coefficient of d is the difference in mean log mpg between diesel and petrol engines in otherwise identical cars.

```
> car$d <- c(rep(0, 7), 1, rep(0, 24))
> summary(glm(lmpg ~ s + c + t + g + ldis + lhp + cb +
+      drat + lwt + d, data = car))
```

We suppress the coefficients, and report only that for d, which is 0.06647, with standard error 0.1564. The t-value for this parameter is only 0.425; there seems to be no real difference in fuel consumption between the diesel and the petrol engines; this is clear also from comparing the very similar Mercedes 230 and 240D: although the 240D has a higher mpg it is much less powerful, and would therefore be expected to be more economical. We therefore omit this dummy variable in the remaining analyses, and treat the diesel- and petrol-engined cars together. We now examine the residuals and influence values from the model.

```
> summary(car.glm <- glm(lmpg ~ s + c + t + g + ldis +
+      lhp + cb + drat + lwt, data = car))

Coefficients:
             Estimate Std. Error t value Pr(>|t|)
(Intercept)  7.35253    1.48080    4.97  5.7e-05
s           -0.02823    0.08830   -0.32    0.752
c           -0.00295    0.03604   -0.08    0.936
t           -0.04464    0.09116   -0.49    0.629
g            0.08111    0.06315    1.28    0.212
ldis        -0.11696    0.15435   -0.76    0.457
lhp         -0.16521    0.13888   -1.19    0.247
cb          -0.02928    0.02881   -1.02    0.320
drat        -0.02226    0.07097   -0.31    0.757
lwt         -0.38091    0.19702   -1.93    0.066

(Dispersion parameter for gaussian family taken to be 0.01300)

    Null deviance: 2.74874  on 31  degrees of freedom
Residual deviance: 0.28609  on 22  degrees of freedom
AIC: -38.14

Number of Fisher Scoring iterations: 2
```

The quantile plot for the model residuals is closely linear, apart from two rather large negative residuals. A graph of residuals against observation number shows that these large values are for the Cadillac and Continental, the luxury American cars. Which observations are influential?

```
> round(sort(hatvalues(car.glm)), 3)

MERCEDES 450SL MERCEDES450SLC PONTIACFIREBIR    AMC JAVELIN
         0.147          0.150          0.155          0.160
HORNETSPORTABO      FIAT X1-9 LINCOLNCONTINE MERCEDES 450SE
         0.163          0.163          0.180          0.185
CADILLACFLEETW TOYOTA COROLLA       IMPERIAL DODGECHALLENGE
         0.186          0.188          0.191          0.223
      FIAT 128     DATSUN 710     DUSTER 360 MAZDARX-4WAGON
         0.230          0.232          0.233          0.245
    MAZDA RX-4   MERCEDES 230   MERCEDES 280  MERCEDES 280C
         0.256          0.299          0.313          0.313
CHEVROLETCAMAZ        VALIANT     VOLVO 142E  TOYOTA CORONA
         0.326          0.331          0.354          0.375
FERRARIDINO197  MERCEDES 240D  MASERATI BORA   LOTUS EUROPA
         0.417          0.448          0.512          0.538
HORNET 4 DRIVE  PORSCHE 914-2    HONDA CIVIC FORD PANTERA L
         0.564          0.618          0.625          0.677
```

Here $p = 9, n = 32$ and $2(p + 1)/n = 0.625$. Only two values exceed this, 0.625 for the Honda Civic and 0.677 for the Ford Pantera. The observation for the Porsche 914 might also be considered influential with a value of 0.618. What is unusual about these cars? The Ford Pantera has the second largest `hp`, but has low gearing for its power – the third highest value of `drat`. Most of the powerful cars have low values of `drat`. The Honda Civic has the lowest `hp`, the highest `drat`, and the second lowest `disp` and `wt`. The Porsche 914 has low `disp` and `hp`, the second highest `drat` and low `wt`. These three cars all lie on the edge of the `hp/drat` scatter, as can be seen from a graph of `hp` against `drat`. The Cadillac and Continental, which had the largest negative residuals, by contrast have very little influence, with values of 0.186 and 0.180.

What happens if we exclude the three observations with high influence? We return to this point below, but for the moment assume that no action is needed. We now examine the fitted model.

At first glance the regression coefficients in the model output look disappointingly non-significant. Only the coefficient of `lwt` appears even near twice its standard error, and most of the others are around one SE or even less. The regression appears nearly useless for prediction of fuel consumption from the design variables. This impression is however quite false: as we will see below the value of R^2 for this model is very high. There is no doubt of the importance of at least some of the variables, but because of the high correlations among them, many are redundant given the others. This phenomenon occurs quite commonly in regressions with highly correlated explanatory variables. Our job is to determine which are the important variables. We proceed to simplify the full regression model.

A convenient model simplification procedure is *backward elimination*, in which at each step the least important variable is dropped from the current model. We prefer this approach to the common alternative of *forward selection*, in which variables are added to the model, starting from the null model. Forward selection requires a 'look ahead' feature in which each variable in turn is added to the model; this increases the number of models examined but does not allow the joint structure of all the explanatory variables to be considered. *Stepwise* procedures combine forward selection and backward elimination but also do not consider the full joint structure of the explanatory variables.

Elimination of variables continues as long as the t-statistics do not reach large values; how large is 'large' is expressed through the likelihood approach in Chapter 2, relating the P-values for the F-tests with different numbers of omitted variables through the table in Section 2.7.4.

```
> round(PRESS(car.glm), 4)

[1] 0.5469

> round(CVE(car.glm), 5)

[1] 0.01709
```

We find $R^2 = 0.896$, $R^2_{\mathrm{CV}} = 0.8081$. Here $R = 0.9465$ is the correlation between the observed `lmpg` values and the fitted values from the full regression model, which is very high: the full model gives a close reproduction of the actual `lmpg` values. The estimate of σ^2 from the full model is 0.0130 with $s = 0.114$.

Backward elimination now proceeds from the full model in the following sequence, using at each step:

```
model2 <- update(model1,.~.- [least important variable])
summary(model2)
```

Step	RSS	Rsq	RsqCV	s^2	CVE	Variable to be omitted	t	Omitted variables F	P-value P
0	0.2861	0.8959	0.8081	0.0130	0.0171	c	-0.082	0.007	0.934
1	0.2862	0.8959	0.8298	0.0124	0.0150	drat	-0.320	0.054	0.948
2	0.2875	0.8954	0.8337	0.0120	0.0146	s	-0.340	0.069	0.976
3	0.2888	0.8949	0.8422	0.0116	0.0137	t	-0.461	0.100	0.982
4	0.2913	0.8940	0.8476	0.0112	0.0132	ldis	-0.723	0.155	0.977
5	0.2971	0.8919	0.8529	0.0110	0.0127	cb	-1.070	0.363	0.895
6	0.3097	0.8873	0.8569	0.0111	0.0123	g	1.052	0.460	0.854
7	0.3220	0.8829	0.8549	0.0111	0.0125	lhp	-4.372	2.726	0.028
	0.5342	0.8057	0.7766	0.0178	0.0192				

At each step we list the model RSS, R^2, R^2_{CV} and s^2 values, the cross validation estimate CVE of σ^2, the variable with the smallest t-value in magnitude of those remaining in the model, and the cumulative F-statistic for the set of omitted variables with its P-value.

After step 6 elimination ceases, as the omission of `lhp` results in a sudden jump in all the statistics in the table at this point. The P-value for the F-test drops from 0.854 for seven omitted variables to 0.028 for eight, well below the value of 0.037 in the table in Section 2.7.4. The final model found by backward elimination uses `lhp` and `lwt`.

The fitted model for `lmpg`, with standard errors, is

```
8.72 - 0.255 lhp - 0.562 lwt
(0.53) (0.058)      (0.087)
```

while $s = 0.105$; the error standard deviation corresponds to a percentage error of 11% in the fitted `mpg`.

It is worth noting that both coefficients are substantially larger in this model than in the full model, as these two variables now subsume the effects of the other variables correlated with them. This model could also be used for prediction, if this were appropriate: its values of s^2 and CVE are very slightly greater than the minimum values. It is notable that the cross-validation R^2_{CV} is initially substantially less than R^2, showing shrinkage of R^2 from the full model on cross-validation, but as the model is progressively simplified R^2_{CV} *increases*, though its maximum value of 0.8569 is still well below R^2 for the full model.

We noted above that three observations had substantial influence on the model. If these are deleted by being given weight zero in a weighted fit, we find that backward elimination leads to the same model with very similar estimated coefficients:

```
8.98 - 0.268 lhp - 0.587 lwt
```

with slightly larger standard errors, reflecting the smaller sample size. Thus despite their high influence values these observations are not discrepant from the rest. We retain the preceding model as the final model.

It is tempting to interpret the final model causally: if weight were increased by 9% (an increase of 0.087 in `lhp`) keeping horsepower constant, then predicted `lmpg` would decrease by 0.049, giving a 4.8% decrease in `mpg`. The Mercedes 450 cars provide an example of this: all their design variables are identical apart from `wt`, which is 4070 (with `mpg` = 16.4) for the SE, 3780 (`mpg` = 15.2) for the SLC, and 3730 (`mpg` = 17.3) for the SL. The `mpg` (median) predictions from the model are 15.3, 15.9 and 16.0, respectively, with a maximum relative error of 8% for the SL.

Whether such an interpretation is possible or not, it is more realistic to think of the final model as a simple, parsimonious and accurate representation of the fuel consumption results for this set of cars. It may not generalize to other cars, or to the same cars tested in other years, because of the selective sample of cars on which the model has been based. Consequently, this model is inappropriate for prediction to other cars.

It is important to stress, as we noted above, that backward elimination is only one way of proceeding to simplify the model. Generating all possible regressions (2^p models with p explanatory variables) is feasible if p is not too large; since many of these models will not be adequate the amount of computation can be substantially reduced by careful bookkeeping. We have followed the practice in this book of not searching extensively for additional models; it sometimes happens, however, that several parsimonious and adequate models exist using quite different subsets of variables. Causal interpretations in such cases are extremely hazardous.

We conclude by noting that the model predicts median `mpg` by

$$\exp(\hat{\mu}) = 126.0\,\texttt{hp}^{-0.255}\,\texttt{wt}^{-0.562};$$

the prediction of median `gpm` is just the reciprocal of this.

3.12 Sensitivity of the Box–Cox transformation

Regression diagnostics allow us to identify influential observations which may substantially affect the position of the fitted model. The extension of the model to include the Box–Cox transformation parameter raises the possibility that the estimate of this parameter may also be strongly affected by influential observations. The identification of observations influencing the transformation parameter was discussed by Atkinson (1982); he described the use of an additional 'constructed variable' in the model. The need for caution in the use of the Box–Cox transformation is clearly shown in the next example, the modelling of acceleration time `qmt` in the `car` data.

```
> boxcox(qmt ~ s + c + t + g + disp + hp + cb + drat +
+       wt, data = car, plotit = TRUE)
```

We first use the explanatory variables `disp`, `hp` and `wt` without transformation. Tabulating over the usual range −2 to +2, we find that the disparity decreases monotonically with λ, and is still decreasing rapidly at $\lambda = -2$. Extending the grid, the minimum is found to occur at about $\lambda = -3.5$. (Note that it is possible for numerical instability to occur for large negative values of λ. This problem, however, is easily resolved by scaling, e.g. by dividing y by 10.)

How should we interpret this unusual value? A normal quantile plot on this scale shows a good straight-line fit, with no marked outliers, but the unusual estimate of λ makes us suspect an error in the data. Refitting the model on the log scale with

$\lambda = 0$ gives a quantile plot with a large negative residual for the Mercedes 230. Background knowledge of the scientific field is helpful, as in the tree example. The Mercedes has the slowest of all acceleration times, and is substantially slower than the diesel Mercedes 240D, which is much less powerful, has higher gearing, and is slightly heavier. The body shapes are identical, so this slow acceleration is very hard to understand. One possible explanation is that the acceleration times of these two cars have been reversed at some point in their publication, or in transcription from *Motor Trend* magazine.

We now examine the effect of excluding the Mercedes 230 from the analysis.

```
> boxcox(qmt ~ s + c + t + g + disp + hp + cb + drat +
+      wt, data = car[rownames(car) != "MERCEDES 230",
+      ])
```

Both the reciprocal and log transformations of `qmt` are appropriate but `qmt` itself is not. Thus the omission of a single observation has considerably changed the conclusion about the appropriate scale of y. The large outlier is accommodated in a normal model only by an extreme transformation of scale.

The reciprocal transformation corresponds to using average speed over the quarter-mile as the response variable. The interpretation of the model coefficients (not shown here) is hindered by their small size; if we note that `sp` $= 900/$`qmt` is the average speed in miles per hour (1 mph $=$ 1.6 kph) over the distance, the corresponding estimates and standard errors for the model for `sp` are just 900 times those for `qmt`, and are given below:

```
> car <- transform(car, sp = 900/qmt)
> summary.lm(glm(sp ~ s + c + t + g + disp + hp + cb +
+      drat + wt, data = car, subset = rownames(car) !=
+      "MERCEDES 230"))

Coefficients:
            Estimate Std. Error t value Pr(>|t|)
(Intercept) 44.93966    5.02680    8.94  1.3e-08
s           -3.91445    1.02079   -3.83  0.00096
c            0.16295    0.42003    0.39  0.70196
t           -0.71766    1.04367   -0.69  0.49921
g            1.62720    0.72948    2.23  0.03674
disp         0.01620    0.00796    2.04  0.05464
hp           0.03037    0.01083    2.80  0.01065
cb           0.56171    0.38135    1.47  0.15559
drat         0.67019    0.77867    0.86  0.39913
wt          -3.41496    0.77777   -4.39  0.00026

Residual standard error: 1.29 on 21 degrees of freedom
Multiple R-Squared: 0.947,    Adjusted R-squared: 0.924
F-statistic: 41.5 on 9 and 21 DF,  p-value: 2.22e-11
```

The full model gives $s^2 = 1.665$ and an R^2 of 0.9468. The number of gears appears to be important, but we may question whether treating it as a continuous variable is appropriate: this implies that average speed increases uniformly with each extra gear. We can check this by fitting a fully parametrized model for the number of gears. There are only three values for this variable (3, 4, and 5) so 2 df are required to saturate it. This may be done by declaring g to be a factor with five levels (though two are missing), or by defining a quadratic term in g and adding this to the model. These give the same disparity; for simplicity we use the second.

```
> car <- transform(car, g2 = g^2)
> summary.lm(glm(sp ~ s + c + t + g + disp + hp + cb +
+      drat + wt + g2, data = car, subset = rownames(car) !=
+      "MERCEDES 230"))

Coefficients:
              Estimate Std. Error t value Pr(>|t|)
(Intercept) 34.571760  16.483209    2.10  0.04887
s           -4.420059   1.286500   -3.44  0.00262
c            0.000747   0.491351  0.0015  0.99880
t           -1.172379   1.261709   -0.93  0.36387
g            8.154020   9.895942    0.82  0.41967
disp         0.016674   0.008102    2.06  0.05286
hp           0.035932   0.013834    2.60  0.01722
cb           0.448459   0.422787    1.06  0.30146
drat         0.336223   0.937008    0.36  0.72349
wt          -3.572275   0.823494   -4.34  0.00032
g2          -0.791797   1.197165   -0.66  0.51591

Residual standard error: 1.31 on 20 degrees of freedom
Multiple R-Squared: 0.948,        Adjusted R-squared: 0.922
F-statistic: 36.4 on 10 and 20 DF,  p-value: 1.21e-10
```

The coefficient of g2 is -0.792 with a standard error of 1.20, giving a t-value of only -0.661. The model linear in g is completely satisfactory, and we now reduce it by backward elimination as in Section 3.11, leading to the model below containing s, g, hp and wt, with $s^2 = 1.757$ and an R^2 of 0.9305.

```
46.9 - 4.82 s + 1.60 g + 0.0485 hp - 2.078 wt
(2.5) (0.70)    (0.45)   (0.0062)     (0.446)
```

Speed increases with the number of gears and horsepower, and decreases with weight and a straight rather than vee engine.

Now we examine the effect of log transforming hp, disp, and wt as in Section 3.11.

```
> car <- transform(car, lhp = log(hp), lwt = log(wt),
+      ldis = log(disp))
```

```
> boxcox(sp ~ s + c + t + g + ldis + lhp + cb + drat +
+       lwt, data = car[-9, ], plotit = FALSE)
```

The log-likelihood maximum of -53.56 now occurs at $\lambda = 0.25$, with values at $\lambda = -1, 0$, and 1 of -53.90, -53.60, and -54.54, respectively. All three values are now plausible for λ, though $\lambda = 0$ is better supported than $\lambda = -1$. The effect of log transforming the explanatory variables has been to increase the log-likelihood at $\lambda = -1, 0$, and 1 by 3.47, 2.10, and 0.21, respectively. Thus if log `qmt` or `qmt` itself are to be used, the explanatory variables should be transformed; if 1/`qmt` is to be used, the transformation of these variables has little effect.

Backward elimination using these logged design variables with log `qmt` allows the elimination of only `t, c` and `cb`, while with `qmt` the minimal model contains `s, g, lhp` and `lwt`:

```
28.1 + 1.66 s - 0.591 g - 2.34 lhp + 2.21 lwt
(1.6) (0.27)    (0.16)     (0.36)      (0.53)
```

with an s^2 of 0.226 and an R^2 of 0.9190, the latter somewhat less than for the `sp` response. This model has the same interpretation as that for speed, though the signs of the variable coefficients are naturally reversed.

3.13 The use of regression models for calibration

In some applications of regression, we want to estimate a new value x_{n+1} of a single explanatory variable, given data (y_i, x_i) and a new observed response value y_{n+1}. This *calibration* problem arises, for example, when x is a precise but expensive measurement, and y a cheap but imprecise measurement of the same quantity. The calibration sample (y_i, x_i) is chosen to cover the range of values of x of interest to the experimenter.

A likelihood treatment of the problem was given by Minder and Whitney (1975); Brown (1982) gave a more recent discussion and references. Confidence intervals for x_{n+1} are easily obtained. For simplicity of notation we write $Y = y_{n+1}$, $x = x_{n+1}$. Then

$$Y - \hat{\alpha} - \hat{\beta}x \sim N(0, \sigma^2 + v_{11} + 2v_{12}x + v_{22}x^2),$$

where

$$V = (v_{jk}) = \sigma^2(X'X)^{-1}.$$

Then

$$(Y - \hat{\alpha} - \hat{\beta}x)/(s^2 + \hat{v}_{11} + 2\hat{v}_{12}x + \hat{v}_{22}x^2)^{\frac{1}{2}} \sim t_{n-2}$$

and therefore with probability $1 - \alpha$,

$$|Y - \hat{\alpha} - \hat{\beta}x| < t_{1-\alpha/2,n-2}(s^2 + \hat{v}_{11} + 2\hat{v}_{12}x + \hat{v}_{22}x^2)^{\frac{1}{2}}.$$

This is equivalent to the quadratic inequality, for the observed y,

$$Q(x) = x^2(\hat{\beta}^2 - c\hat{v}_{22}) - 2x(\hat{\beta}(y - \hat{\alpha}) + c\hat{v}_{12}) + (y - \hat{\alpha})^2 - c(s^2 + \hat{v}_{11}) < 0$$

where

$$c = t^2_{1-\alpha/2,n-2},$$

and $\hat{V}$ is the estimated covariance matrix of the parameter estimates, available from R model objects as components of the *summary* methods. The confidence interval for x is defined by the roots of $Q(x) = 0$.

As a numerical illustration, for the data in Section 3.1 on completion times for the Block Design test, we obtained a final model using only EFT (model *solv.glm3*). Suppose we have an additional observation with completion time equal to 400 sec. What can we say about EFT for this observation?

```
> options(digits = 8)
> solv.glm <- glm(time ~ eft, data = solv)
> coef(summary(solv.glm))

               Estimate  Std. Error   t value     Pr(>|t|)
(Intercept) 271.1281926 42.95859263 6.3113844 2.3710893e-06
eft           2.0405134  0.66452557 3.0706319 5.5959738e-03

> summary(solv.glm)$dispersion

[1] 11694.28

> summary(solv.glm)$dispersion * summary(solv.glm)$cov.unscaled

             (Intercept)          eft
(Intercept)   1845.44068 -24.49007993
eft            -24.49008   0.44159423

> qt(0.975, solv.glm$df.residual)

[1] 2.0738731
```

The estimates are

$$\begin{aligned} \hat{\alpha} &= 271.128 & \hat{v}_{11} &= 1845.44 \\ \hat{\beta} &= 2.0405 & \hat{v}_{12} &= -24.490 \quad \hat{v}_{22} = 0.441594 \\ y &= 400 & s^2 &= 11694.3 \end{aligned}$$

and for a 95% confidence interval, $t_{0.975,22} = 2.074$, so $c = 4.301$. The quadratic in x is

$$Q(x) = 2.264x^2 - 2(157.6)x - 41626.4$$

and the equation $Q(x) = 0$ has roots -82.8 and 222.0. For $Q(x)$ to be negative, x must be between the roots so the 95% interval for x is $(-82.8, 222.0)$. Negative values are impossible from the nature of EFT, which is an unsatisfying feature of the confidence interval construction procedure. (This difficulty can be avoided by constructing a profile likelihood for x over the range of possible values of x. Minder and Whitney constructed a marginal likelihood for x.)

An R function to calculate these for the simple regression model is

```
> predictx <- function(model, y, alpha = 0.05) {
+     a <- coefficients(model)[1]
+     b <- coefficients(model)[2]
+     s2 <- if (class(model) == "lm") {
+         summary(model)$sigma^2
+     }
+     else {
+         if ((class(model)[1] == "glm") &
+             (family(model)$family == "gaussian")) {
+             summary(model)$dispersion
+         }
+         else {
+             stop("only available for normal models")
+         }
+     }
+     V <- summary(model)$cov.unscaled * s2
+     v11 <- V[1, 1]
+     v12 <- V[1, 2]
+     v22 <- V[2, 2]
+     df <- model$df.residual
+     c <- qt(1 - alpha/2, df)^2
+     z3 <- (b^2 - c * v22)
+     z2 <- (b * (y - a) + c * v12)
+     z1 <- ((y - a)^2 - c * (s2 + v11))
+     z2 <- z2/z3 * 2
+     z1 <- z1/z3
+     z <- -as.real(polyroot(c(z1, z2, 1)))
+     z <- c(z[1], ((y - a)/b), z[2])
+     names(z) <- c("ll", "x", "ul")
+     z
+ }
```

Applying it to this case gives

```
> predictx(lm(time ~ eft, data = solv), y = 400)

        ll          x         ul
-82.795664  63.156559 222.022500
```

The obvious (and ML) estimate for x is $(y - \hat{\alpha})/\hat{\beta}$ which is 63.2; the interval is extremely wide here because of the large residual variation about the regression line. For regressions which are not well determined, it is possible for $Q(x)$ to have no real roots, with $Q(x) < 0$ for all x. Then *all* values of x are consistent with the observed y. This may seem strange, but the special case $\beta = 0$ makes the result obvious: if β is actually zero, then y can provide no information about x, and a value of $\hat{\beta}$ near zero (relative to its standard error) will lead to the same result.

3.14 Measurement error in the explanatory variables

We noted in Chapter 2 that an important assumption of the model was that the explanatory variables are measured without error. In survey studies in the social sciences, this is frequently not the case: many variables contain measurement error, in the sense that repeated measurements of the same individual under the same circumstances give different values of the variable. In psychometric testing it is standard practice to quote the reliability of a test, which is the correlation of two parallel measurements on the same individual under identical conditions. For example, the quoted reliability for the Block Design Subtest of the WISC is 0.87 for 10-year-old (American) children (Wechsler, 1949).

How should measurement error be allowed for in the model and the analysis? It is currently difficult to allow for measurement error when modelling with R except when replicate or 'parallel' measurements on the unreliable explanatory variables are available: we give therefore only a short discussion of the problem. More detailed discussions are given in Schafer (2001) and Aitkin and Rocci (2002).

Consider the simple linear model with one explanatory variable. Let T_i be the true value of the explanatory variable for the i-th observation, assuming it could be measured without error, and suppose that the observed value X_i is related to T_i by

$$X_i = T_i + \xi_i,$$

where ξ_i is the measurement error, which we will assume to be normally distributed $N(0, \sigma_m^2)$, independently of the random component of the model for Y.

One possibility, perhaps the most commonly used, is to ignore measurement error. Suppose Y is a student achievement test at the end of an academic year, and X is a measure of initial ability, for example, Stanford–Binet IQ or Verbal

Reasoning Quotient. We might argue that the model

$$Y_i|X_i \sim N(\beta_0 + \beta_1 X_i, \sigma^2)$$

is still appropriate despite unreliability in X, because X is the only measure we have of the student's initial achievement, and the regression has to be based on observable data. For a descriptive use of the model, this may be reasonable, but if the observations come from an experimental design (as in the first example of Chapter 3) from which causal conclusions are to be drawn, the ignoring of measurement error may give quite misleading conclusions about experimental variables in the model.

To proceed further we need to make an assumption about the distribution of the T_i, the unobserved true values of the X_i. We will assume they are normally distributed $N(\mu, \sigma_t^2)$. Then the complete model is

$$\begin{aligned} Y_i|X_i, T_i &\sim N(\beta_0 + \beta_1 T_i, \sigma^2) \\ X_i|T_i &\sim N(T_i, \sigma_m^2) \\ T_i &\sim N(\mu, \sigma_t^2). \end{aligned}$$

Write

$$\begin{aligned} \sigma_y^2 &= \sigma^2 + \beta_1^2 \sigma_t^2, \\ \sigma_x^2 &= \sigma_m^2 + \sigma_t^2. \end{aligned}$$

Then the joint (marginal) distribution of (Y_i, X_i) is bivariate normal with means $(\beta_0 + \beta_1 \mu, \mu)$, variances (σ_y^2, σ_x^2) and covariance $\beta_1 \sigma_t^2$.

The conditional distribution of Y_i given the observable X_i is then

$$\begin{aligned} Y_i|X_i &\sim N(\beta_0 + \beta_1(\rho X_i + (1-\rho)\mu), \sigma^2 + \beta_1^2 \rho(1-\rho)\sigma_x^2) \\ &= N(\beta_0^* + \rho\beta_1 X_i, \sigma^2 + \beta_1^2 \rho(1-\rho)\sigma_x^2), \end{aligned}$$

where $\beta_0^* = \beta_0 + (1-\rho)\beta_1\mu$ and $\rho = \sigma_t^2/\sigma_x^2$.

Thus the slope of the regression of Y_i on X_i is not β_1 but $\rho\beta_1$, and the standard least squares estimate of β_1 systematically underestimates the true parameter value (i.e. is inconsistent). The estimates of any other parameters in a regression function will also be inconsistent. Further, the variance of the residuals, though constant, also depends on β_1, so that MLF of β_1 even if ρ is known is not equivalent to least squares estimation. A consistent estimate of the regression coefficient β_1 can be obtained if ρ is known by dividing the least squares estimate by ρ; the estimate obtained, however, will not be efficient since the additional information about β_1 in the variance is neglected; further, the other parameters in the model are not correctly estimated.

A particular difficulty with this model is that there are six model parameters but only five sufficient statistics for these parameters, so the model is unidentifiable without knowledge of one parameter. MLE in such models can be achieved in several cases: when reliabilities or error variances are known; when the ratio of the error variance to the residual variance in Y is known; when some ('gold standard') observations have both T and X measured, or when 'parallel' measurements of the unreliable explanatory variables are available (from parallel forms or split-half measures).

This discouraging unidentifiability result is, however, peculiar to the normal true-score/normal error-score model; if either T or $X|T$ is *not* normal, the model *is* identifiable (Reiersöl, 1950). This result allows ML estimation even in models without the additional information specified above, so long as the observed score distribution of X is *not* normal. Discussions of this case are given in Schafer (2001) and Aitkin and Rocci (2002).

Extensive discussions of consistent estimation in general regression models with measurement error in all the explanatory variables can be found in Fuller (1987) and Carroll *et al.* (1995).

Quite general models with measurement error in the explanatory variables can be fitted by maximum likelihood (given parallel measurements) using LISREL (Jöreskog and Sörbom, 1981). A very recent development in STATA, `gllamm` is described in Skrondal and Rabe-Hesketh (2003).

3.15 Factorial designs

We noted in Section 3.1 that the choice of an appropriate model is greatly simplified in orthogonal factorial designs because the analysis of variance table is unique. Orthogonal factorial designs have many other well-known advantages, which we will not discuss here. In many sample surveys with categorical explanatory variables, formally similar cross-classifications result, but the sample sizes in the cells of the table are generally unequal, and sometimes show marked associations with the explanatory variables. In such cases great care is necessary in model simplification and interpretation. We conclude this section with an orthogonal two-factor design discussed by Box and Cox (1964); unbalanced cross classifications are discussed in Section 3.16.

The file `poison` contains the survival times of rats in units of 10 h after poisoning and antidote treatment. The design is a completely randomized 3×4 factorial with three replicates, with three types of poisons and four treatments. The data are given in Table 3.6.

We input the data.

```
> data(poison, package = "SMIR")
```

```
> poison.glm1 <- glm(time ~ 1, data = poison)
```

Table 3.6. Survival times (unit, 10 h) of animals in a 3 × 4 factorial experiment (3 poison types by four treatments).

Type	Treat	Time			
I	A	0.31	0.45	0.46	0.43
I	B	0.82	1.10	0.88	0.72
I	C	0.43	0.45	0.63	0.76
I	D	0.45	0.71	0.66	0.62
II	A	0.36	0.29	0.40	0.23
II	B	0.92	0.61	0.49	1.24
II	C	0.44	0.35	0.31	0.40
II	D	0.56	1.02	0.71	0.38
III	A	0.22	0.21	0.18	0.23
III	B	0.30	0.37	0.38	0.29
III	C	0.23	0.25	0.24	0.22
III	D	0.30	0.36	0.31	0.33

```
> poison.glm2 <- update(poison.glm1, . ~ . + type)
> poison.glm3 <- update(poison.glm2, . ~ . + treat)
> poison.glm4 <- update(poison.glm3, . ~ . + type:treat)
> round(coef(summary(poison.glm4)), 4)

                 Estimate Std. Error t value Pr(>|t|)
(Intercept)        0.4125     0.0746  5.5318   0.0000
typeII            -0.0925     0.1055 -0.8771   0.3862
typeIII           -0.2025     0.1055 -1.9202   0.0628
treatB             0.4675     0.1055  4.4331   0.0001
treatC             0.1550     0.1055  1.4698   0.1503
treatD             0.1975     0.1055  1.8728   0.0692
typeII:treatB      0.0275     0.1491  0.1844   0.8547
typeIII:treatB    -0.3425     0.1491 -2.2965   0.0276
typeII:treatC     -0.1000     0.1491 -0.6705   0.5068
typeIII:treatC    -0.1300     0.1491 -0.8717   0.3892
typeII:treatD      0.1500     0.1491  1.0058   0.3212
typeIII:treatD    -0.0825     0.1491 -0.5532   0.5836
```

The analysis of variance table is given in Table 3.7.

Since the design is orthogonal, the sums of squares of the main effects are invariant under reordering, and each mean square can be compared directly with the residual mean square by the usual F-test.

The main effects are highly significant, the interaction not significant at the 10% level ($F_{0.10,6,36} = 1.945$), though one of the `type:treat` interaction parameters is rather large.

Inspection of the mean survival times is possible, but not convenient, from the parameter estimates for the interaction model. It is simpler to construct the table of means using *tapply*

```
> round(with(poison, tapply(time, list(type = type,
+       treatment = treat), mean)), 4)

      treatment
type       A     B      C      D
  I   0.4125 0.880 0.5675 0.6100
  II  0.3200 0.815 0.3750 0.6675
  III 0.2100 0.335 0.2350 0.3250
```

The means are arranged by rows for `type` and columns for `treat`.

Since the design is balanced, the marginal means can be inspected by simple one-way tabulations:

```
> round(with(poison, tapply(time, treat, mean)), 4)

     A      B      C      D
0.3142 0.6767 0.3925 0.5342
```

Tabulation of the variances is always useful in cross-classifications. We store them in the vector `tvar`:

```
> round(tvar <- with(poison, tapply(time, list(type = type,
+       treatment = treat), var)), 7)

      treatment
type          A         B         C         D
  I   0.0048250 0.0258667 0.0245583 0.0127333
  II  0.0056667 0.1131000 0.0032333 0.0734250
  III 0.0004667 0.0021667 0.0001667 0.0007000
```

There are remarkable differences in variance, with

$$s^2_{\max}/s^2_{\min} = 0.1131/0.000167 = 677.2.$$

Table 3.7. ANOVA table, time

	df	Deviance	Resid. df	Resid. dev.	F	Pr $(>F)$
NULL			47	3.01		
type	2	1.03	45	1.97	23.22	0.0000
treat	3	0.92	42	1.05	13.81	0.0000
type:treat	6	0.25	36	0.80	1.87	0.1123

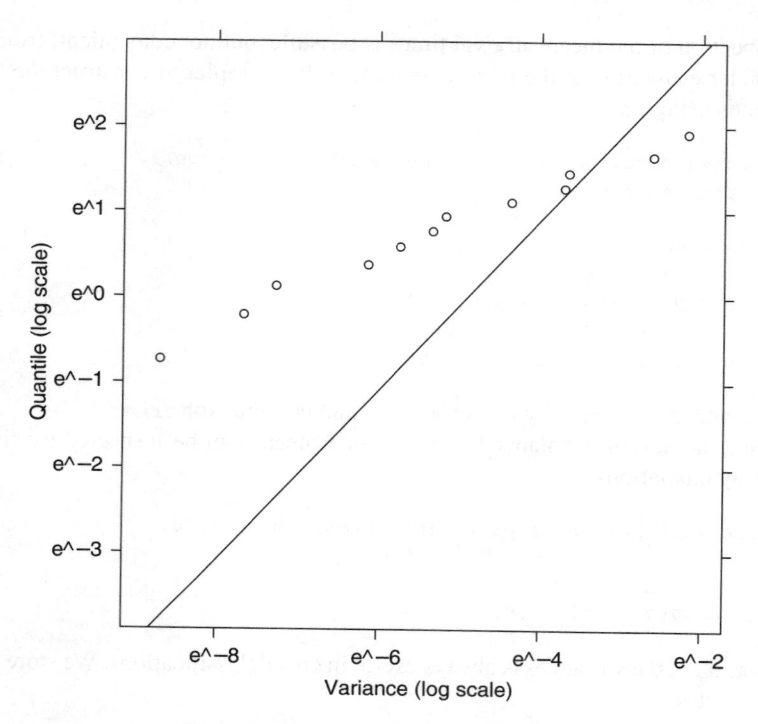

Fig. 3.12. Log quantile vs log variance

This is very large compared with $F_{\max,0.01,12,3} = 361$. [An ordinary $F_{3,3}$-test is not appropriate because this value is the largest of all the possible variance ratios; the $F_{\max}$ test due to Hartley (1950) must be used. Critical values can be found in the Biometrika Tables by Pearson and Hartley (1966).] Further, the small variances are associated with small means, as is easily seen by comparing the two tabulated outputs. Stronger evidence comes from a χ_3^2 quantile plot of the ordered variances shown in Fig. 3.12. The plotting points are $(q_i, s_{(i)}^2)$ where $s_{(i)}^2$ is the i-th smallest variance, and we take

$$q_i = F^{-1}(i/(n+1))$$

as for normal quantile plotting, but F is now the c.d.f. of the χ_3^2 distribution. Also plotted is the line through the origin $q = 3s^2/\tilde{\sigma}^2$, where $\tilde{\sigma}^2$ is the residual mean square from the full model. The points should fall close to the line since each variance should be distributed as $(\sigma^2\chi_3^2)/3$ if the model is correct. We use log scales for both axes.

```
> tvar <- as.vector(tvar)
> q <- qchisq(rank(tvar)/13, 3)
```

```
> print(xyplot(q ~ tvar, ylab = "quantile (log scale)",
+     xlab = "variance (log scale)", ylim = c(0.02, 20),
+     scales = list(y = list(log = "e"), x = list(log = "e")),
+     panel = function(...) {
+         panel.xyplot(...)
+         panel.curve(x + log(3) - log(0.0222))
+     }))
```

The assumption of constant variance is clearly violated, and the mean–variance association suggests the scale of the response is wrong. We look for an appropriate scale.

```
> poison.bc <- boxcox(poison.glm4, plotit = FALSE,
+     lambda = seq(-1.5, -0.2, by = 0.1))
> names(poison.bc$y) <- poison.bc$x
> round(poison.bc$y, 2)

 -1.5  -1.4  -1.3  -1.2  -1.1    -1  -0.9  -0.8  -0.7  -0.6
26.81 27.84 28.72 29.44 30.00 30.39 30.62 30.68 30.57 30.29
 -0.5  -0.4  -0.3  -0.2
29.83 29.21 28.42 27.47
```

The log-likelihood changes rapidly over the range -2 to $+2$, and a finer tabulation is necessary over the range -1.5 to -0.2. The maximum occurs at $\lambda = -0.82$, and the approximate 95% confidence interval is $(-1.29, -0.34)$ which includes $\lambda = -1$ but not $\lambda = 0$ or 1. The reciprocal scale is clearly indicated; the response variable on this scale can be interpreted as the *rate of dying*.

```
> poison.rate.glm1 <- glm(1/time ~ 1, data = poison)
> poison.rate.glm2 <- update(poison.rate.glm1, . ~
+     . + type)
> poison.rate.glm3 <- update(poison.rate.glm2, . ~
+     . + treat)
> poison.rate.glm4 <- update(poison.rate.glm3, . ~
+     . + type:treat)
```

The analysis of variance table is given in Table 3.8.

The F-values for the main effects are now much larger than they were before, but that for interaction is smaller: the interaction mean square is almost the same size as the residual mean square.

```
> round(with(poison, tapply(1/time, list(type = type,
+     treatment = treat), mean)), 3)

     treatment
type      A     B     C     D
  I   2.487 1.163 1.863 1.690
  II  3.268 1.393 2.714 1.702
```

Table 3.8. ANOVA table, rate

```
Analysis of Deviance Table

Model: gaussian, link: identity

Response: 1/time

Terms added sequentially (first to last)

           Df Deviance Resid. Df Resid. Dev
NULL                          47    65.5053
type        2  34.8771        45    30.6281
treat       3  20.4143        42    10.2139
type:treat  6   1.5708        36     8.6431
```

```
III 4.803 3.029 4.265 3.092

> round(tvar <- with(poison, tapply(1/time, list(type = type,
+     treatment = treat), var)), 5)

      treatment
type        A       B       C       D
  I   0.24667 0.03980 0.23949 0.13302
  II  0.67622 0.30602 0.17432 0.49267
  III 0.28051 0.17761 0.05514 0.05956
```

The variance heterogeneity is greatly reduced, with $s^2_{\max}/s^2_{\min} = 16.99$, much less than $F_{\max\,0.05,12,3} = 124$. More importantly, there is no association between cell means and variances. A χ^2_3 (log) quantile plot of the variances is shown in Fig. 3.13, with the theoretical straight line which now fits closely to the plotted points.

```
> tvar <- as.vector(tvar)
> q <- qchisq(rank(tvar)/13, 3)
> print(xyplot(log(q) ~ log(tvar), ylab = "quantile (log scale)",
+     xlab = "variance (log scale)", panel = function(...) {
+         panel.xyplot(...)
+         panel.curve(x + log(3) - log(0.24))
+     }))
```

The reciprocal transformation therefore seems satisfactory. (In Section 3.17 however we modify this conclusion.)

The rate means for `type` are 1.801, 2.269, and 3.797, and those for `treat` are 3.519, 1.862, 2.947, and 2.161. Transforming back the means, the fitted median

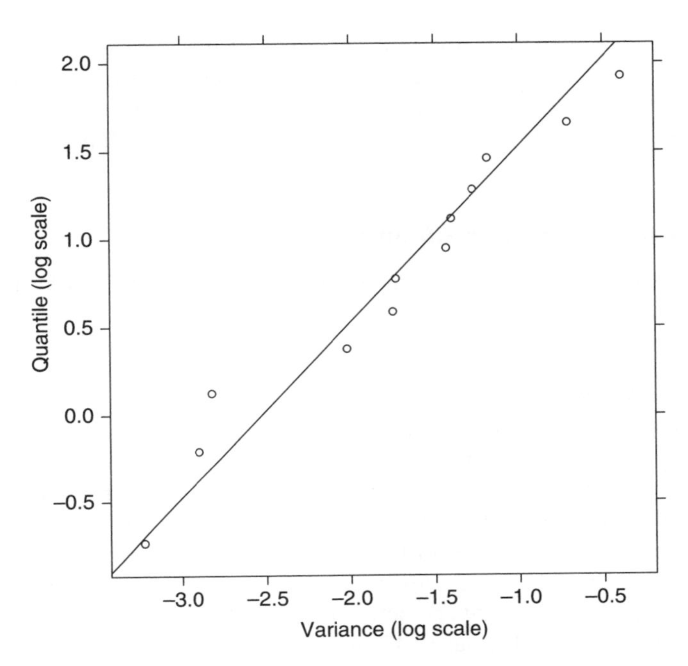

Fig. 3.13. Log quantile vs log variance

survival times are 0.555, 0.441, and 0.236 for the poison types, and 0.284, 0.537, 0.339, and 0.463 for the treatments.

The effect of the reciprocal transformation on the precision of the analysis in this example is important. The assumption of constant variance on the original scale was invalid, and the pooled variance estimate on that scale was inflated by the large cell variance estimates for the cells with large means. The reciprocal transformation stabilizes the variance estimates, increasing the precision of the analysis.

We saw earlier in this chapter that the estimate of λ in the Box–Cox transformation could be affected by a change in model. What happens if we fit the main effects model instead of the interaction model?

We suppress the output. The disparity minimum now occurs at -0.75, and the 95% confidence interval is $(-1.14, -0.36)$. The model change has had very little effect.

Box and Cox (1964) pointed out that the transformation in this case is trying simultaneously to achieve three ends: normality of the residuals, additivity of the means and homogeneity of variance on the transformed scale. By including the interaction, we require the transformation to attempt to satisfy only two of these requirements. We could relax the requirements further by fitting the interaction model with a different variance term for each cell, thus removing the requirement of homogeneity of variance. Box and Cox discussed this; we do not pursue it here because the *boxcox* function does not allow non-homogeneous variances. The requirement of homogeneity of variance is a strong one; without it, any value of λ from below -1 to about 2.5 is plausible (Box and Cox, 1964; p. 236).

3.16 Unbalanced cross-classifications

We illustrate the difficulties of the analysis of unbalanced cross-classifications with an example.

3.16.1 *The Bennett hostility data*

Bennett (1974) studied the emotional reactions of husbands to unsuccessful suicide attempts by their wives. He studied 25 husbands of wives who had been admitted to hospital after suicide attempts by taking drug overdoses. The husbands were interviewed in the hospital after the wife was out of danger, and responses to three specific questions were tape recorded, and subsequently analysed for emotional affect using the Gottschalk and Gleser (1969) scales of hostility and affection. The interview length was standardized between husbands.

Bennett used a 'control' group of 42 husbands of wives admitted to hospital with acute organic abdominal conditions, to allow for the effect of the hospital admission and environment on affect. The control husbands were interviewed in the same way. Bennett subsequently divided this control group into two, as some of the controls had experienced previous difficulties in the marriage requiring counselling, which might affect their responses in the stressful hospital environment. Controls without previous marriage difficulties are termed 'true controls', the others are termed 'false controls'; the group factor `g` is defined by `g` = 1 overdoses, `g` = 2 false controls, `g` = 3 true controls. Two other explanatory factors were recorded: the national origin `n` of the husband (`n` = 1 Australian, `n` = 2 British), and the previous occurrence `po` of a suicide attempt or acute abdominal hospitalization (`po` = 1 no previous occurrence, `po` = 2 a previous occurrence).

Bennett analysed the tape recordings for word frequency on five measures: inward hostility `i` (guilt), ambivalent hostility `a` (shown by others to self – paranoia), overt outward hostility `o` (shown by self to others – anger), covert outward hostility `c` (shown by others to others) and positivity `p` (affection). The word counts on these measures are normalized by a square root transformation and can be assumed normally distributed. The values of the five scale variables, and the factor values for `g`, `n` and `po` are in the file `hostility`. Parts of the data are listed in Table 3.9.

An initial issue is the non-orthogonality of the three-way classification, which can be assessed by tabulation:

Table 3.9. Hostility data

g	n	po	i	a	o	c	p
1	1	1	1.90	1.26	2.21	1.51	1.69
1	1	1	1.58	1.02	2.11	0.67	2.62
1	1	1	1.80	0.81	2.59	1.86	2.12
1	1	1	0.81	1.65	3.30	2.17	2.30
1	1	1	0.49	0.54	1.65	1.40	0.90
1	1	1	0.85	1.81	2.53	1.89	1.17
1	1	1	0.39	0.29	2.87	1.07	2.15
1	1	1	0.61	1.31	2.94	2.01	1.39
1	1	1	1.15	0.81	3.14	1.39	2.02
1	1	1	0.56	1.47	2.67	1.29	2.16
1	1	1	1.09	1.48	2.47	0.99	2.33
1	1	1	0.41	0.30	1.33	2.07	2.22
.	.	.					
3	2	2	0.48	0.38	0.68	1.56	3.16
3	2	2	0.99	0.30	1.81	0.91	1.39
3	2	2	1.47	1.20	1.83	0.60	3.23
3	2	2	0.83	1.31	1.57	0.70	2.56
3	2	2	1.49	0.36	1.53	0.85	2.31
3	2	2	1.07	0.24	1.37	0.44	2.84
3	2	2	1.15	0.47	1.69	2.10	2.61
3	2	2	0.38	0.38	2.41	0.38	2.89
3	2	2	0.31	0.43	1.61	1.35	1.68

```
> data(hostility, package = "SMIR")

> ftable(xtabs(~g + po + nation, data = hostility),
+      row.vars = c(1), col.vars = c(3, 2))

           nation Australian           British
           po           none previous     none previous
g
overdoses                 12        8        4        1
F controls                 4        5        3        1
T controls                13        6        1        9
```

For the Australian husbands, previous occurrences are nearly proportional over the three groups, with 40% having had a previous occurrence, but for British husbands they are nearly all in the true control group – 90% of true control husbands, but only 22% of overdose or false control husbands, had had a previous occurrence.

3.16.2 *ANOVA of the cross-classification*

We analyse the positivity response. Since the cross-classification is non-orthogonal, the sums of squares in the ANOVA tables will depend on the order of fitting the terms in the model. We give three tables constructed by permuting the orders of main effects and two-way interactions so that each term is fitted last in the set in one of the three tables, allowing its (conditional) contribution to be assessed, given the other terms in the model.

```
> host.glm1 <- glm(positivity ~ g * nation * po, data = hostility)
> host.glm2 <- glm(positivity ~ po * g * nation, data = hostility)
> host.glm3 <- glm(positivity ~ nation * po * g, data = hostility)
```

The ANOVA tables for main effects and interactions are constructed from the residual sums of squares (Table 3.10).

There is a considerable change in the sums of squares for the main effects depending on their order of fitting. The changes in the interaction sums of squares are much smaller.

The three-way interaction can be omitted, with an $F_{2,55}$ value of 0.169 ($P = 0.845$). From the first column we see that `nation:po` and `g.po` can be omitted similarly (giving a pooled $F_{5,55}$ value of 0.552, $P = 0.736$), while the F-value for `g:nation` of 2.00 has a P-value of 0.145. This is not significant at conventional levels, but the two df may be concealing one larger component. We examine the parameter estimates for the `g:nation` model:

```
> host.glm4 <- glm(positivity ~ (g + nation)^2 + po,
+       data = hostility)
> print(summary(host.glm4), digits = 3)

Coefficients:
                          Estimate Std. Error t value Pr(>|t|)
(Intercept)                  2.053      0.129   15.93   <2e-16
```

Table 3.10. ANOVA tables, positivity

Source	SS	df	MS	Source	SS	df	MS	Source	SS	df	MS
g	4.353	2	2.177	po	0.842	1	0.842	n	1.148	1	1.148
n	0.601	1	0.601	g	5.007	2	2.504	po	1.233	1	1.233
po	1.830	1	1.830	n	0.935	1	0.935	g	4.403	2	2.202
g:n	1.111	2	0.556	n:po	0.030	1	0.030	g:po	0.628	2	0.314
g:po	0.669	2	0.335	g:n	1.161	2	0.581	n:po	0.089	1	0.089
n:po	0.051	1	0.051	g:po	0.640	2	0.320	g:n	1.114	2	0.557
g:n:po	0.094	2	0.047	g:n:po	0.094	2	0.047	g:n:po	0.094	2	0.047
resid	15.316	55	0.278	resid	15.316	55	0.278	resid	15.316	55	0.278

```
gF controls                   0.324       0.209     1.55    0.1269
gT controls                   0.414       0.167     2.48    0.0158
nationBritish                -0.175       0.261    -0.67    0.5059
poprevious                   -0.403       0.141    -2.86    0.0058
gF controls:nationBritish     0.623       0.406     1.54    0.1297
gT controls:nationBritish     0.657       0.347     1.89    0.0631

(Dispersion parameter for gaussian family taken to be 0.26884179)

    Null deviance: 24.025  on 66  degrees of freedom
Residual deviance: 16.131  on 60  degrees of freedom
AIC: 110.7

Number of Fisher Scoring iterations: 2
```

The parameter estimates for the two components of the `g:nation` interaction are not individually large, but they are nearly equal, and we equate them before assessing whether they can be omitted.

```
> hostility$g23 <- factor(hostility$group)
> levels(hostility$g23) <- c("overdoses", "control",
+      "control")
> host.glm5 <- update(host.glm4, . ~ . - g:nation +
+      g23 + g23:nation)
> print(summary(host.glm5), digits = 4)

Coefficients: (1 not defined because of singularities)
                         Estimate Std. Error t value Pr(>|t|)
(Intercept)                2.0515     0.1266  16.207  < 2e-16
gF controls                0.3167     0.1905   1.663  0.10145
gT controls                0.4170     0.1604   2.600  0.01166
nationBritish             -0.1737     0.2585  -0.672  0.50406
poprevious                -0.3986     0.1323  -3.014  0.00375
g23control                     NA         NA      NA       NA
nationBritish:g23control   0.6451     0.3150   2.048  0.04488

(Dispersion parameter for gaussian family taken to be 0.26446773)

    Null deviance: 24.025  on 66  degrees of freedom
Residual deviance: 16.133  on 61  degrees of freedom
AIC: 108.74

Number of Fisher Scoring iterations: 2
```

The aliased term `g23` is included to obtain comparable parameter estimates. The change in deviance or disparity is almost zero, and we try replacing `g` in the same way by `g23`; that is, we collapse the group factor into overdoses versus controls:

```
> host.glm6 <- update(host.glm5, . ~ . - g)
> round(coef(summary(host.glm6)), 4)
```

```
                         Estimate Std. Error t value Pr(>|t|)
(Intercept)                2.0504     0.1259 16.2871   0.0000
nationBritish             -0.1732     0.2571 -0.6737   0.5030
poprevious                -0.3960     0.1315 -3.0122   0.0038
g23control                 0.3848     0.1498  2.5695   0.0126
nationBritish:g23control   0.6473     0.3133  2.0661   0.0430
```

We now observe that the `nationBritish` parameter estimate is less than its standard error, though `nation` is involved in an interaction with `g23`. The interpretation of this 'no main effect' in the presence of interaction is, however, quite clear. With the R factor coding, `nationBritish` represents the British/Australian difference for the first level of `g23`, that is, for overdoses. There is no significant mean difference in affection for British and Australian husbands for the overdose group, but there is for the two control groups. So we may set this non-significant difference to zero to simplify the model:

```
> hostility$nation.for.controls <- with(hostility,
+      (g23 == "control") & (nation == "British"))
> host.glm7 <- update(host.glm6, . ~ po + g23 +
+        nation.for.controls)
> round(coef(summary(host.glm7)), 4)
```

```
                        Estimate Std. Error t value Pr(>|t|)
(Intercept)               2.0125     0.1121 17.9480   0.0000
poprevious               -0.3870     0.1302 -2.9717   0.0042
g23control                0.4192     0.1402  2.9896   0.0040
nation.for.controlsTRUE   0.4712     0.1719  2.7413   0.0080
```

Model simplification could conclude at this point, with a pooled $F_{8,55}$ value of 0.460.

```
> anova(host.glm7, host.glm1, test = "F")
```

```
Analysis of Deviance Table

Model 1: positivity ~ po + g23 + nation.for.controls
Model 2: positivity ~ g * nation * po
  Resid. Df Resid. Dev Df Deviance       F  Pr(>F)
1        63   16.34136
2        55   15.31617  8  1.02519 0.46018 0.87861
```

However, we can smooth the fitted values a little more, and simplify the resulting interpretation, by noting that the three parameter estimates are very close to each other, and can be equated with little increase in disparity.

```
> hostility$comp <- with(hostility, (g23 == "control") +
+     ((nation == "British") * (g23 == "control")) -
+      (po == "previous"))
```

```
> host.glm8 <- update(host.glm7, . ~ comp, data = hostility)
> summary(host.glm8)

Coefficients:
            Estimate Std. Error t value  Pr(>|t|)
(Intercept) 2.034907   0.068352 29.7710 < 2.2e-16
comp        0.426585   0.077587  5.4982 6.911e-07

(Dispersion parameter for gaussian family taken to be 0.25228695)

    Null deviance: 24.0253  on 66  degrees of freedom
Residual deviance: 16.3987  on 65  degrees of freedom
AIC: 101.836

Number of Fisher Scoring iterations: 2
```

We finally tabulate the fitted means, and the cell means, on positivity:

```
> round(ftable(with(hostility, tapply(fitted(host.glm8),
+     list(g, po, nation), mean)), row.vars = c(1),
+     col.vars = c(3, 2)), 2)
> round(ftable(with(hostility, tapply(positivity, list(g,
+     po, nation), mean)), row.vars = c(1), col.vars = c(3,
+     2)), 2)

            Australian          British
                  none previous     none previous

overdoses         2.03     1.61     2.03     1.61
F controls        2.46     2.03     2.89     2.46
T controls        2.46     2.03     2.89     2.46

            Australian          British
                  none previous     none previous

overdoses         1.92     1.85     1.90     1.38
F controls        2.46     1.91     2.84     2.37
T controls        2.53     1.93     3.20     2.52
```

The fitted mean affection scores are on a four-point scale, with steps of 0.43:

(1) previous occurrence in an overdose: 1.61;
(2) no previous occurrence in an overdose, or previous occurrence in an Australian control: 2.03;
(3) no previous occurrence in an Australian control, or previous occurrence in a British control: 2.46;
(4) no previous occurrence in a British control: 2.89.

It will be seen that this final model is not easily expressed in terms of a classical ANOVA table with a sum of squares for each effect in the model. While such a table can be constructed from the successive steps of the model simplification, it is simpler to observe that the ANOVA table construction is only a formalization of regression model simplification, and that the whole process can be expressed as a backward elimination from the full model, which we now describe.

3.16.3 *Regression analysis of the cross-classification*

The process of reducing the full model can be simplified by following the usual backward elimination procedure for a regression model, though factors with multiple degrees of freedom need care. The description of the process takes some space, since new sets of parameter estimates have to be inspected at each step, but it is very important to follow this process, as we show in Section 3.16.4.

The parameter estimates for the full model are:

```
> round(coef(summary(host.glm1))[, 1:2], 4)

                                     Estimate Std. Error
(Intercept)                            1.9225     0.1523
gF controls                            0.5400     0.3047
gT controls                            0.6075     0.2113
nationBritish                         -0.0200     0.3047
poprevious                            -0.0762     0.2409
gF controls:nationBritish              0.4008     0.5052
gT controls:nationBritish              0.6900     0.6267
gF controls:poprevious                -0.4803     0.4282
gT controls:poprevious                -0.5271     0.3548
nationBritish:poprevious              -0.4463     0.6373
gF controls:nationBritish:poprevious   0.5294     0.9501
gT controls:nationBritish:poprevious   0.3685     0.8851
```

Inspection of the parameter estimates suggests that the three-way interactions are negligible, there is a large `gT controls` effect, and no other important effects. However, it is very important not to misinterpret these estimates; as with the general multiple regression model, the t-values for these model coefficients represent the importance of each variable if it *alone* is omitted from the model, all other variables being retained. But as we have seen in backward elimination procedures, the importance of each variable in general depends on which other variables are in the model. This issue is discussed extensively in Aitkin (1978) in terms of *effect coding*; the form of dummy variable coding used in the statistical package determines the parameter estimates and standard errors for these estimates. Whatever be this coding, to determine the importance of the variables requires a backward elimination in the non-orthogonal cross-classification, just as it does in the general multiple regression model.

Thus in inspecting the model estimates above, the only relevant estimates are those for the three-way interaction, since the estimates for the other parameters will change when these interaction terms are omitted. These terms are omitted *first* from the model because of their complexity in interpretation; as in Section 3.1 we generally simplify models according to the hierarchy of main effects, first-, second-order interactions, and so on.

Omitting the three-way interaction, we obtain new estimates and standard errors:

```
> host.glm2 <- update(host.glm1, . ~ . - nation:po:g)
> anova(host.glm2, host.glm1, test = "F")

Analysis of Deviance Table

Model 1: positivity ~ g + nation + po + g:nation + g:po + nation:po
Model 2: positivity ~ g * nation * po
  Resid. Df Resid. Dev Df Deviance        F  Pr(>F)
1        57   15.41051
2        55   15.31617  2  0.09433 0.16938 0.84463

> round(coef(summary(host.glm2))[, 1:3], 4)

                            Estimate Std. Error t value
(Intercept)                   1.9389     0.1473 13.1658
gF controls                   0.4896     0.2834  1.7277
gT controls                   0.5865     0.2021  2.9014
nationBritish                -0.0855     0.2768 -0.3089
poprevious                   -0.1172     0.2260 -0.5187
gF controls:nationBritish     0.5458     0.4207  1.2974
gT controls:nationBritish     0.8206     0.4158  1.9733
gF controls:poprevious       -0.3780     0.3753 -1.0072
gT controls:poprevious       -0.4714     0.3201 -1.4726
nationBritish:poprevious     -0.1596     0.3691 -0.4325
```

The two omitted terms give an F-value of 0.169 with 2 df. In addition to the large `gT controls` term, we now see a large `gT controls:nationBritish` interaction which was not present in the full interaction model.

The `nation:po` interaction is small and can be omitted; the `g:po` interaction terms are also small. We try omitting both of them:

```
> host.glm3 <- update(host.glm2, . ~ . - g:po - nation:po)
> anova(host.glm3, host.glm2, test = "F")

Analysis of Deviance Table

Model 1: positivity ~ g + nation + po + g:nation
Model 2: positivity ~ g + nation + po + g:nation + g:po
                      + nation:po
```

```
  Resid. Df Resid. Dev Df Deviance       F  Pr(>F)
1        60    16.1305
2        57    15.4105  3   0.7200 0.88771 0.45306
```

```
> round(coef(summary(host.glm3))[, 1:3], 4)

                           Estimate Std. Error t value
(Intercept)                  2.0530     0.1289 15.9296
gF controls                  0.3240     0.2093  1.5480
gT controls                  0.4136     0.1665  2.4835
nationBritish               -0.1745     0.2608 -0.6692
poprevious                  -0.4025     0.1407 -2.8607
gF controls:nationBritish    0.6232     0.4056  1.5364
gT controls:nationBritish    0.6572     0.3470  1.8939
```

The F-value for the 3 df omitted is less than 1. We now proceed as above for the further reduction of the model.

3.16.4 *Statistical package treatments of cross-classifications*

The backward elimination procedure above, and the multiple ANOVA tables resulting from the non-orthogonality of the cross-classification, complicate the analysis of such classifications. There have been many attempts to simplify the analysis by constructing a 'composite' ANOVA table with one unique entry for each effect. This is done in many packages by reporting the sum of squares for each effect, 'adjusted for' (after fitting) all other effects of the same order in the hierarchy (or less commonly, adjusted for all other effects of all orders, discussed below). In the Bennett data, this leads to a composite 'ANOVA' table (Table 3.11).

Table 3.11. Composite 'ANOVA' table, positivity

Source	SS	df	MS	F	P-value
g	4.403	2	2.202	7.92	0.0009
n	0.935	1	0.935	3.36	0.072
po	1.830	1	1.830	6.58	0.013
(Total)	7.168				
g.n	1.114	2	0.557	2.00	0.145
g.po	0.640	2	0.320	1.15	0.324
n.po	0.051	1	0.051	0.18	0.673
(Total)	1.805				
g.n.po	0.094	2	0.047	0.17	0.844
res	15.316	55	0.278		

Inspection of the table suggests that there are important main effects of g and po, but no important interactions and no significant main effect of n. Tabulation of the positivity means for g and po gives

```
> round(with(hostility, tapply(positivity, list(g = g,
+     po = po), mean, na.rm = TRUE)), 2)

            po
g             none previous
  overdoses  1.92     1.79
  F controls 2.63     1.98
  T controls 2.58     2.28
```

This analysis misses the *g:nation* interaction between nationality and the overdoses and controls (which requires careful examination of the parameter estimates to identify), but the main difficulty in the use of such composite 'ANOVA' tables is that they do not represent a breakdown of the 'hypothesis' sum of squares for all the full model effects into additive components from a hierarchical sequence of added or omitted terms. This is most clearly seen if we add the sums of squares for the three main effects in the composite table: the sum is 7.195 while the sum of the additive components in the three breakdowns is 6.784, whichever sequence is followed, since this represents the sum of squares difference between the null model and the model with all three main effects.

Thus the composite table overstates the importance of the main effects, as it does (slightly) for the two-way interactions. In other examples the composite ANOVA table can drastically *understate* the importance of effects, when these are positively correlated: the sums of squares in the composite table can add to much less than the real total in the hierarchical sequence: there may be a 'missing sum of squares' in the composite analysis. Aitkin (1978) gave examples.

If the composite 'ANOVA' table is produced by adjusting each effect for *all* other effects in the model, an even more serious problem occurs, in which the quoted sum of squares for each effect depends critically on the form of dummy variable (or contrast) coding used for the effects. Changes in the form of coding [e.g. from (0, 1) to (1, 0)] can completely change the composite table: such tables cannot be used to identify important terms in the model.

We emphasize that to obtain correct conclusions from unbalanced cross-classifications, it is *essential* to proceed by elimination of effects from the full interaction model, in a sequence of model fits.

This conclusion contradicts a large part of the literature on unbalanced cross-classifications, which attempts to produce a single composite ANOVA table (by the method above, or other methods) without explicit fitting of reduced models. Such tables are unfortunately produced by the default option in many statistical packages, and great care is needed in using these packages for unbalanced cross-classifications. The user should choose the option for 'full hierarchical partitioning' to avoid this problem. Whatever package is used, the backward elimination regression procedure from the full interaction model described above will identify the appropriate model (or models – there may be more than one with highly correlated factors); it is a serious and unfortunate consequence of non-orthogonality that no *single* model analysis can guarantee this.

Our conclusions may be thought controversial; the reader should consult Nelder (1977) and Aitkin (1978) for very detailed statements and examples (with contributed discussions).

3.17 Missing data

Throughout this chapter we have assumed that complete data on response and explanatory variables are available for every sample member. The reality of survey data collection, however, is that missing values always occur, through non-response, misrecording or accident. The data matrix of individuals by variables generally has a scatter of missing values, and this considerably complicates model fitting by maximum likelihood. In some cases the entire set of variables may be missing because the respondent could not be contacted – in the language of sample surveys, this is a missing 'unit-level' observation, rather than a missing 'item-level' observation on an individual variable.

The simplest case, which we consider first, is when the missing values are 'missing completely at random' (MCAR – this and other terms were introduced by Rubin (1976). We may imagine a completely random 'failure process', in which the value of any variable for each individual has a constant probability (possibly different for each variable) of failing, that is, of being missing, and failures occur independently. More realistically, the missingness probability for a variable may depend on other observed variables, but not on the missing variable – the value of the variable is 'missing at random' (MAR).

In both these cases fully efficient MLEs of the model parameters may be obtained from the observed data using the EM algorithm (Little and Rubin, 1987). We do not have to model explicitly the missingness process, since the likelihood factors into independent components for the response model and the missingness model. However, it is necessary to assume an explicit probability model for the variables for which values are missing. Little and Schluchter (1985) gave computational details for a normal regression model with normal or categorical explanatory variables. The EM algorithm does not provide standard errors, and computation of these from the information matrix using the result of Louis (1982) is particularly complicated.

These cases of values missing at random may not hold. The probability of a variable value being missing may depend on the value that *would have been* observed. This case of 'non-random missingness' or 'missing non-randomly' (MNR) is particularly difficult to deal with as the model for missingness is in general unidentifiable, since we are missing the data which would allow it to be estimated. In this case the most that can be done is to construct a model (usually logistic) relating the missingness probability to the explanatory variables, and assume a value for the parameter governing the dependence of the missingness on the value of the missing variable; this model is then identifiable, and a sensitivity analysis can be carried out to assess the sensitivity of parameter estimates and tests to the magnitude of the non-random dependence. Copas (1997) gave a detailed discussion.

As a simple example, consider a survey of monthly income of individuals classified by age, sex, educational level and industrial sector of employment. The object of analysis is to model average income by a regression on the other explanatory variables, but the value of income only is missing for some individuals. If these values are missing at random, then MLE of the regression model parameters is equivalent to omitting the individuals with missing values and analysing only the complete cases.

Suppose now that the probability of an income value being missing increases with age of the respondent. This can be established by a logistic regression (Section 4.2) of the missing data indicator on the explanatory variables. Then higher ages are under-represented in the complete cases, and if we intend to report the overall mean monthly income for the population the sample means will have to be weighted by the inverse of the sample inclusion probabilities to allow for the under-representation of older people (see Section 3.3 for a discussion of weighting).

The fitted regression of earnings on the explanatory variables is, however, not affected by the selection occurring in the sample, because the fitted model is conditional on the observed values of the explanatory variables. It is only marginal, or unconditional inferences which are affected. The variances of the regression coefficients are, however, increased by the restriction of range.

Suppose finally that the probability of an income value being missing increases with decreasing income. Then low incomes are under-represented in the sample, and the regression ignoring the missingness process will be systematically biased. In some 'biased' sample designs the same result may be achieved by design: we may deliberately over-sample high income respondents to ensure a sufficient sample with high incomes. Model fitting then requires either weighting by the inverse of the sample selection probabilities, or complex modelling of the sample distribution of the observed responses incorporating the design. Lawless *et al.* (1999) gave a recent comprehensive discussion.

When the response variable is fully observed but the explanatory variables are missing, *imputation* methods are widely used to 'fill in' the missing explanatory variable values, with the aim of providing a 'clean data set' for a standard analysis. *Single imputation* methods, in which a single value is provided, do not give MLEs of the model parameters from the standard analysis of the 'completed' data set even under MCAR assumptions, whatever form of imputation is used, and standard errors are always underestimated.

Multiple imputation methods (Rubin, 1987; Schafer, 1997) provide M multiple imputed values simulated from the conditional distribution of the variables with missing data given the observed variables, for each case with any missing data. These M multiple imputed values are used to construct M completed data sets, to each of which the standard analysis is applied; the M sets of estimates and their covariance matrices are then combined to give a single estimate and covariance matrix. Fay (1996) gave an alternative method of combining the M data sets, by weighting each set with weight $1/M$ in a single analysis.

Multiple imputation is implemented in the S-plus and R functions of Schafer (1997), and in the package SOLAS. Multiple imputation methods give a close approximation to MLEs and information-based standard errors without the difficult computation of the latter; their performance has been discussed in several review papers (see, e.g. Fay, 1996).

We conclude with a discussion of approximate methods available in statistical packages.

3.18 Approximate methods for missing data

Three approximate methods are in common use. The first omits all observations with missing data on any variables, and uses only complete records ('complete case' analysis). There are two serious disadvantages of this method. The first is the obvious one that the number of observations may be substantially reduced. In extreme cases there may be no complete records. The second is that serious bias in the model parameter estimates may result. When observation are missing on both response and explanatory variables, the parameter estimates from the complete cases are no longer consistent.

The second approximate method estimates the mean and variance of each variable from all the complete observations on that variable, and the covariance between each pair of variables from the complete observations on those pairs of variables ('available case' analysis). This does not avoid bias in the parameter estimates, and may result in a covariance matrix for the explanatory variables which is not positive definite.

The third approximate method fills in the missing values with single imputed values, with the object of creating a completed data set which can be used for standard analyses. As noted above, single imputation by any method does not provide MLEs, and always underestimates the standard errors of the estimates.

3.19 Modelling of variance heterogeneity

We have already seen in the `poison` example that variances may not be constant. There are several ways of allowing for variance heterogeneity.

The first is by adapting the variance expression of the usual regression estimator to allow for a general variance pattern. For the general regression model, suppose that

$$Y_i|\mathbf{x}_i \sim N(\mu_i, \sigma_i^2)$$

with

$$\mu_i = \boldsymbol{\beta}'\mathbf{x}_i,$$

where the variances σ_i^2 are different but unspecified.

Then the MLE $\hat{\boldsymbol{\beta}}$ of $\boldsymbol{\beta}$ under homogeneity has the repeated sampling distribution $\hat{\boldsymbol{\beta}} \sim N(\boldsymbol{\beta}, V)$ where

$$V = (X'X)^{-1}X'DX(X'X)^{-1}$$

and D is the diagonal matrix of variances σ_i^2. The variances are unknown, but can be estimated from the squared residuals from the regression as $\widetilde{s^2}_i = e_i^2$, leading to the *sandwich* or *robust* variance estimate

$$\tilde{V} = (X'X)^{-1}X'\tilde{D}X(X'X)^{-1},$$

according to White (1982). This estimate is provided in many statistical packages (including R via the package `sandwich`) in addition to the usual estimate assuming variance homogeneity.

Without knowledge of V we cannot improve on the least squares estimate. It might appear that we can use the estimated variances as inverse weights, and define the *weighted* least squares estimate

$$\tilde{\boldsymbol{\beta}} = (X'\tilde{D}^{-1}X)^{-1}X'\tilde{D}^{-1}\mathbf{y}$$

with variance

$$(X'\tilde{D}^{-1}X)^{-1},$$

or perhaps an iterated version of this, but it is easily seen that the estimate of $\boldsymbol{\beta}$ is inconsistent, because as $n \to \infty$ the variance estimates $\widetilde{s_i^2}$ do not converge to the true variances σ_i^2 – the number of these variances, each with (at most) 1 df, goes to infinity as well. To improve on the least squares estimate, we need to *model* the variances by a parsimonious structure of small dimension relative to the sample size.

We now extend the standard model by explicit modelling of the variance heterogeneity in terms of explanatory variables. Such models are called *dispersion* models, and are discussed in great generality in Jorgensen (1997). In Chapter 8 we allow for *random* variances, leading to the *t-distribution* for the observed response Y.

We generalize the regression model to the form:

$$Y_i|\mathbf{z}_i \sim N(\mu_i, \sigma_i^2)$$

with

$$\log \sigma_i^2 = \boldsymbol{\lambda}'\mathbf{z}_i,$$

a *log-linear model* for the variances. This ensures positivity of variances in the fitted model. The variables $\mathbf{z}_i$ may be a subset or superset of $\mathbf{x}_i$, or may be unrelated. We first consider the special case of *known* means μ_i, and write $e_i = y_i - \mu_i$. The likelihood is

$$L(\boldsymbol{\lambda}) = \prod_i \frac{1}{\sqrt{2\pi}\sigma_i} \exp\left\{-\frac{e_i^2}{2\sigma_i^2}\right\}.$$

Recall that $E_i^2/\sigma_i^2 \sim \chi_1^2$, and since the χ_ν^2 random variable is twice a gamma variable with degrees of freedom $\nu/2$, the likelihood $L(\boldsymbol{\lambda})$ is proportional to that of a set of n-scaled gamma variables, with scale factors $2\sigma_i^2$ and degrees of freedom 1/2.

The log-linear model for σ_i^2 can then be fitted in R directly by specifying the gamma error model for the squared residuals e_i^2, and a `glm` scale parameter of $1/\nu = 2$; the model fitted to the mean σ_i^2 is the log-linear model specified above.

Now consider the general case where μ_i is unknown, but specified by another linear model $\mu_i = \boldsymbol{\beta}'\mathbf{x}_i$. This *double generalized linear model* can be fitted by successive relaxation, by alternately fitting the mean and variance models (Smyth, 1986, 1989; Aitkin, 1987).

Given observations $(y_i, \mathbf{x}_i, \mathbf{z}_i),\ i = 1, \ldots, n$ the likelihood is

$$L(\boldsymbol{\beta}, \boldsymbol{\lambda}) = \prod_i \frac{1}{\sqrt{2\pi}\sigma_i} \exp\left\{-\frac{e_i^2}{2\sigma_i^2}\right\}$$

where now

$$e_i = y_i - \boldsymbol{\beta}'\mathbf{x}_i.$$

The log-likelihood is

$$\ell(\boldsymbol{\beta}, \boldsymbol{\lambda}) = -\frac{1}{2}\left[\sum \log \sigma_i^2 + \sum (y_i - \boldsymbol{\beta}'\mathbf{x}_i)^2/\sigma_i^2\right],$$

and the first and second derivatives with respect to the parameters are

$$\frac{\partial \ell}{\partial \boldsymbol{\beta}} = \sum \mathbf{x}_i (y_i - \boldsymbol{\beta}'\mathbf{x}_i)/\sigma_i^2$$

$$\frac{\partial^2 \ell}{\partial \boldsymbol{\beta} \partial \boldsymbol{\beta}'} = -\sum \mathbf{x}_i \mathbf{x}_i'/\sigma_i^2$$

$$\frac{\partial \ell}{\partial \boldsymbol{\lambda}} = \frac{1}{2}\left[-\sum \mathbf{z}_i + \sum (y_i - \boldsymbol{\beta}'\mathbf{x}_i)^2 \mathbf{z}_i/\sigma_i^2\right]$$

$$= \frac{1}{2}\sum (e_i^2/\sigma_i^2 - 1)\mathbf{z}_i$$

$$\frac{\partial^2 \ell}{\partial \boldsymbol{\lambda} \partial \boldsymbol{\lambda}'} = -\frac{1}{2}\sum (e_i^2/\sigma_i^2)\mathbf{z}_i \mathbf{z}_i'$$

$$\frac{\partial^2 \ell}{\partial \boldsymbol{\beta} \partial \boldsymbol{\lambda}'} = -\sum (e_i/\sigma_i^2)\mathbf{x}_i \mathbf{z}_i'.$$

Taking expectations, the expected information matrix is block-diagonal, with blocks

$$\mathcal{I}_{\boldsymbol{\beta}} = X'W_{11}X$$

$$\mathcal{I}_{\boldsymbol{\lambda}} = \frac{1}{2}Z'Z,$$

where

$$X' = [\mathbf{x}_1, \ldots, \mathbf{x}_n]$$

$$Z' = [\mathbf{z}_1, \ldots, \mathbf{z}_n]$$

$$W_{11} = \text{diag}(1/\sigma_i^2),$$

since $\mathrm{E}[E_i] = 0$ and $\mathrm{E}[E_i^2] = \sigma_i^2$. Thus a Fisher scoring algorithm for the simultaneous MLE of $\boldsymbol{\beta}$ and $\boldsymbol{\lambda}$ reduces to two separate algorithms for $\boldsymbol{\beta}$ and $\boldsymbol{\lambda}$. Since, however, W_{11} depends on $\boldsymbol{\lambda}$ and e_i depends on $\boldsymbol{\beta}$, it is simplest to formulate the scoring algorithm as a successive relaxation algorithm.

For given σ_i^2, $\hat{\boldsymbol{\beta}}$ is a weighted least squares estimate with weights $1/\sigma_i^2$, and for given $\boldsymbol{\beta}$, $\hat{\boldsymbol{\lambda}}$ is the MLE from a gamma model with response variable e_i^2 and a scale parameter 2.

The algorithm conveniently begins with an initial unweighted normal regression of y on $\mathbf{x}$, taking $\sigma_i^2 \equiv \sigma^2$. The squared residuals from the least squares fit are defined as a new response variable with a gamma distribution with scale parameter 2. The linear predictor $\boldsymbol{\lambda}'\mathbf{x}$ is then fitted using a log link function, and the disparity calculated for the initial estimate of $(\boldsymbol{\beta}, \boldsymbol{\lambda})$. A weighted normal regression of y on $\mathbf{x}$ is now fitted, with scale parameter 1 and weights given by the reciprocals of the fitted values from the gamma model. This process continues until the disparity converges. At this point the standard errors (based on the expected information) from both models are correct. However, the log-likelihood in the two sets of parameters may be very skewed, and so the standard errors are not a reliable indicator of variable importance. In addition, the loss of degrees of freedom in the variance model due to the estimation of the mean parameters may be serious, requiring a *marginal* or *restricted* likelihood maximization for the variance model. See Smyth and Verbyla (1999) for a discussion.

The above analysis assumes that the parameters $\boldsymbol{\beta}$ and $\boldsymbol{\lambda}$ are functionally independent, which will usually be the case.

This approach to variance heterogeneity modelling provides a routine analysis of the famous *Behrens–Fisher problem*, of how to test the hypothesis of equal means in a two-group problem when the variances are different. This is the simplest case of variance heterogeneity: we fit two models, the 'alternative hypothesis' model has the group factor in both mean and variance models, the 'null hypothesis' model has a null mean model and the same group variance model. A profile likelihood for the mean difference can be computed by defining this difference as an offset and fitting a null mean model. This test procedure however, does not provide a fixed size α test except in large samples (as $n \to \infty$): the test size depends on the true variance ratio.

The double modelling procedure is implemented in the `dglm` function available in the `dglm` package; its use is illustrated below with two examples, the Box–Cox poison data (Section 3.15) and the Minitab tree data (Section 3.6).

3.19.1 *Poison example*

We adopt the reciprocal scale for time, as in the previous analysis, but we now allow for general variance heterogeneity, though we found little evidence for this previously. With cross-classifications, we follow the general procedure of fitting a saturated model for the mean, and using this to find a suitable model for the dispersion, beginning with the saturated model and using backward elimination. When a final model for dispersion has been found, we simplify the mean model in a similar way.

```
> data(poison, package = "SMIR")

> library(dglm)
> poison$rate <- 1/poison$time
> poison.dglm <- dglm(rate ~ type * treat, dformula = ~type *
+     treat, data = poison)

> print(summary(poison.dglm, dispersion = 1), digits = 4)
```

```
Mean Coefficients:
                 Estimate Std. Error z value  Pr(>|z|)
(Intercept)       2.487       0.215  11.564  6.302e-31
typeII            0.782       0.416   1.879  6.026e-02
typeIII           2.316       0.314   7.366  1.761e-13
treatB           -1.323       0.232  -5.710  1.128e-08
treatC           -0.624       0.302  -2.067  3.871e-02
treatD           -0.797       0.269  -2.988  2.810e-03
typeII:treatB    -0.552       0.488  -1.131  2.580e-01
typeIII:treatB   -0.450       0.374  -1.205  2.281e-01
typeII:treatC     0.070       0.501   0.139  8.894e-01
typeIII:treatC    0.086       0.393   0.220  8.257e-01
typeII:treatD    -0.770       0.539  -1.429  1.532e-01
typeIII:treatD   -0.914       0.367  -2.487  1.288e-02
(Dispersion Parameters for gaussian family estimated as below )

    Scaled Null Deviance: 795.6 on 47 degrees of freedom
Scaled Residual Deviance: 48 on 36 degrees of freedom

Dispersion Coefficients:
                 Estimate Std. Error  z value Pr(>|z|)
(Intercept)      -1.687       0.707  -2.386    0.01702
typeII            1.008       1.000   1.008    0.31324
typeIII           0.129       1.000   0.129    0.89771
treatB           -1.824       1.000  -1.824    0.06812
treatC           -0.030       1.000  -0.030    0.97642
treatD           -0.618       1.000  -0.618    0.53684
typeII:treatB     1.031       1.414   0.729    0.46584
typeIII:treatB    1.367       1.414   0.967    0.33367
typeII:treatC    -1.326       1.414  -0.938    0.34841
typeIII:treatC   -1.597       1.414  -1.129    0.25872
typeII:treatD     0.301       1.414   0.213    0.83149
typeIII:treatD   -0.932       1.414  -0.659    0.50988
(Dispersion parameter for Gamma family taken to be 2 )

    Scaled Null Deviance: 44.96 on 47 degrees of freedom
Scaled Residual Deviance: 30.17 on 36 degrees of freedom

Minus Twice the Log-Likelihood: 39.14
Number of Alternating Iterations: 5
```

The disparity of the fitted model is held in the component labelled `m2loglik` of the dglm object and is reported in the summary of the `dglm` object (as `Minus Twice the Log-Likelihood`), however, the disparity does not correspond to that in the *boxcox* function as it is missing the term $\prod 1/y_i^2$ in the likelihood. The comparable value from the constant (null) variance model, `poison.rate.glm4` is 53.92, so the saturated variance model gives a disparity reduction of 14.79 on 11 df. This is not large, but we now examine the contribution of individual model terms to it by backward elimination.

None of the interaction parameters is large, and so the interaction can be omitted:

```
> poison.dglm2 <- dglm(rate ~ type * treat, dformula = ~type +
+      treat, data = poison)

> print(summary(poison.dglm2, dispersion = 1), digits = 4)

Mean Coefficients:
                 Estimate Std. Error z value  Pr(>|z|)
(Intercept)       2.487        0.234  10.608  2.748e-26
typeII            0.782        0.433   1.807  7.079e-02
typeIII           2.316        0.314   7.367  1.749e-13
treatB           -1.323        0.282  -4.699  2.612e-06
treatC           -0.624        0.281  -2.224  2.617e-02
treatD           -0.797        0.284  -2.805  5.025e-03
typeII:treatB    -0.552        0.520  -1.062  2.884e-01
typeIII:treatB   -0.450        0.378  -1.192  2.331e-01
typeII:treatC     0.070        0.518   0.134  8.931e-01
typeIII:treatC    0.086        0.376   0.230  8.183e-01
typeII:treatD    -0.770        0.524  -1.468  1.421e-01
typeIII:treatD   -0.914        0.381  -2.398  1.649e-02
(Dispersion Parameters for gaussian family estimated as below )

    Scaled Null Deviance: 505 on 47 degrees of freedom
Scaled Residual Deviance: 48 on 36 degrees of freedom

Dispersion Coefficients:
            Estimate Std. Error z value Pr(>|z|)
(Intercept)  -1.515       0.500  -3.030  0.002449
typeII        0.877       0.500   1.755  0.079312
typeIII      -0.226       0.500  -0.451  0.651715
treatB       -0.814       0.577  -1.410  0.158510
treatC       -0.836       0.577  -1.448  0.147547
treatD       -0.757       0.577  -1.311  0.189871
(Dispersion parameter for Gamma family taken to be 2 )

    Scaled Null Deviance: 44.96 on 47 degrees of freedom
Scaled Residual Deviance: 36.08 on 42 degrees of freedom

Minus Twice the Log-Likelihood: 45.05
Number of Alternating Iterations: 8
```

The disparity increases by 5.91 on 6 df. The parameter estimates in the saturated mean model are unchanged by changes in the variance model, but their standard errors are changed. For further reductions in the variance model we suppress the mean model output.

```
> poison.dglm3 <- dglm(rate ~ type * treat, dformula = ~type,
+       data = poison)

> summary(poison.dglm3$dispersion)

Coefficients:
             Estimate Std. Error t value  Pr(>|t|)
(Intercept) -2.091       0.336    -6.220  1.472e-07
typeII       0.917       0.475     1.930    0.05999
typeIII     -0.140       0.475    -0.295    0.76957

(Dispersion parameter for Gamma family taken to be 1.808)

    Null deviance: 44.9559  on 45  degrees of freedom
Residual deviance: 39.3152  on 45  degrees of freedom
AIC: NA

Number of Fisher Scoring iterations: 4
```

The omission of the treatment main effect gives a disparity increase of 3.24 on 3 df.

```
> poison.dglm4 <- dglm(rate ~ type * treat, dformula = ~1,
+       data = poison)
> summary(poison.dglm4$dispersion)
```

Omitting the poison type main effect gives a disparity increase of 5.64 on 2 df, and this is almost all concentrated in the contrast of type II with the other two types.

We define a dummy variable variable for type II:

```
> poison$type2 <- ifelse(poison$type == "II", 1, 0)
> poison.dglm5 <- dglm(rate ~ type * treat, dformula = ~type2,
+       data = poison)

> summary(poison.dglm5$dispersion)

Coefficients:
             Estimate Std. Error t value  Pr(>|t|)
(Intercept) -2.15865    0.23479 -9.1940 5.494e-12
type2        0.98498    0.40667  2.4221   0.01943

(Dispersion parameter for Gamma family taken to be 1.7640186)
```

```
    Null deviance: 44.9559  on 46  degrees of freedom
Residual deviance: 39.3937  on 46  degrees of freedom
AIC: NA

Number of Fisher Scoring iterations: 4
```

Replacing `type` by `type2` increases the disparity by only 0.08 on 1 df, the contrast of type II with the other two increasing it by 5.56 on 1 df. The final model on the rate scale thus specifies a common variance for types I and III but a larger variance for type II.

As noted above, the parameter estimates for the saturated mean model do not depend on the dispersion model fitted, but their standard errors do. The largest interaction of -0.9137 is for `typeIII:treatD`, and its standard error is 0.367 in the saturated dispersion model, 0.340 in the final dispersion model with a `type2` term, and 0.490 in the common dispersion model. The standard errors are smaller in the first two models because the variance estimate is based on the smaller type III variance. (When the null variance model is fitted, the variance estimate is the MLE of σ^2, not the unbiased estimate, so the standard errors are smaller than those reported by R in a standard analysis.)

We now reduce the mean model.

```
> poison.dglm6 <- dglm(rate ~ type + treat, data = poison,
+     dformula = ~type2)

> print(summary(poison.dglm6), digits = 4)

Mean Coefficients:
            Estimate Std. Error t value  Pr(>|t|)
(Intercept)    2.676      0.157  17.032 1.810e-20
typeII         0.469      0.186   2.521 1.557e-02
typeIII        1.996      0.149  13.359 1.022e-16
treatB        -1.612      0.190  -8.487 1.184e-10
treatC        -0.576      0.190  -3.033 4.144e-03
treatD        -1.314      0.190  -6.922 1.880e-08
(Dispersion Parameters for gaussian family estimated as below )

    Scaled Null Deviance: 363.9 on 47 degrees of freedom
Scaled Residual Deviance: 48 on 42 degrees of freedom

Dispersion Coefficients:
            Estimate Std. Error z value  Pr(>|z|)
(Intercept)   -1.856      0.250  -7.423 1.144e-13
type2          0.739      0.433   1.707 8.785e-02
(Dispersion parameter for Gamma family taken to be 2 )

    Scaled Null Deviance: 50.64 on 47 degrees of freedom
```

```
Scaled Residual Deviance: 47.54 on 46 degrees of freedom

Minus Twice the Log-Likelihood: 58.97
Number of Alternating Iterations: 4
```

Fitting the main effect model for the mean with the `type2` dispersion model gives a disparity increase of 10.60 on 6 df. We add a parameter for the type III, treatment D cell using a dummy variable:

```
> poison$d34 <- (poison$type == "III") * (poison$treat ==
+      "D")
> poison.dglm7 <- dglm(rate ~ type + treat + d34, data = poison,
+      dformula = ~type2)

> print(summary(poison.dglm7, disp = 1), digits = 4)

Mean Coefficients:
            Estimate Std. Error z value  Pr(>|z|)
(Intercept)   2.603       0.138   18.821 5.093e-79
typeII        0.469       0.175    2.679 7.387e-03
typeIII       2.155       0.144   14.958 1.377e-50
treatB       -1.598       0.165   -9.670 4.045e-22
treatC       -0.577       0.165   -3.491 4.807e-04
treatD       -1.033       0.202   -5.119 3.072e-07
d34          -0.632       0.273   -2.316 2.055e-02
(Dispersion Parameters for gaussian family estimated as below )

    Scaled Null Deviance: 425.8 on 47 degrees of freedom
Scaled Residual Deviance: 48 on 41 degrees of freedom

Dispersion Coefficients:
            Estimate Std. Error z value  Pr(>|z|)
(Intercept)   -2.050      0.250  -8.201 2.385e-16
type2          1.031      0.433   2.382 1.724e-02
(Dispersion parameter for Gamma family taken to be 2 )

    Scaled Null Deviance: 60.03 on 47 degrees of freedom
Scaled Residual Deviance: 53.93 on 46 degrees of freedom

Minus Twice the Log-Likelihood: 54.31
Number of Alternating Iterations: 4
```

The single interaction parameter reduces the disparity by 4.66 on 1 df, the remaining five interaction parameters accounting for a disparity of 5.94. The importance of this interaction term is not apparent in the common dispersion model, where the disparity change for `d34` is 2.29, and that for the remaining five interaction parameters is 5.73.

Table 3.12. Cell means for rate and [fitted values]

Type	Treat			
	A	C	D	B
III	4·80	4·26	3·09	3·03
	[4·76]	[4·18]	[3·09]*	[3·16]
II	3·27	2·71	1·70	1·39
	[3·07]	[2·49]	[2·04]	[1·47]
I	2·49	1·86	1·69	1·16
	[2·60]	[2·03]	[1·57]	[1·00]

Table 3.13. Cell variances for rate, and [fitted values]

Type	Treat			
	A	C	D	B
III	0·281	0·055	0·060	0·178
	[0·129]	[0·129]	[0·129]	[0·129]
II	0·676	0·174	0·493	0·306
	[0·361]	[0·361]	[0·361]	[0·361]
I	0·247	0·239	0·133	0·040
	[0·129]	[0·129]	[0·129]	[0·129]

Tables 3.12 and 3.13 show the cell means and variances (sums of squares divided by 4) for rate, and fitted values from the final model. The tables are arranged so that mean values decrease monotonically down and across the table. The d34 dummy variable reproduces the observed mean in the cell marked with an asterisk.

```
> poison.fitted <- predict(poison.dglm7)
> tab.means <- with(poison, tapply(rate, list(Type = type,
+     Treat = treat), mean, na.rm = TRUE))
> tab.fitted <- tapply(poison.fitted, list(Type = poison$type,
+     Treat = poison$treat), mean, na.rm = TRUE)

> poison.var.fitted <- predict(poison.dglm7$dispersion.fit)
> tab.var <- with(poison, tapply(rate, list(Type = type,
+     Treat = treat), var, na.rm = TRUE))
> tab.var.fitted <- tapply(poison.var.fitted,
+     list(Type = poison$type,
+     Treat = poison$treat), mean, na.rm = TRUE)
```

3.19.2 *Tree example*

We consider the 'solid of revolution' model with log transformations of `v`, `d` and `h`. With continuous explanatory variables there is no obvious 'saturated model' for the variance; we adopt here a second-degree response surface in `log h` and `log d` while maintaining the linear model in the mean; Cook and Weisberg (1982) noted some evidence of variance heterogeneity.

```
> data(trees, package = "SMIR")

> library(dglm)
> trees <- transform(trees, lh2 = lh^2, ld2 = ld^2,
+     lhld = lh * ld)
> trees.dglm <- dglm(lv ~ lh + ld, data = trees, dformula ~
+     lh + ld + lh2 + lhld + ld2)

> summary(trees.dglm, disp = 1)

Mean Coefficients:
             Estimate  Std. Error     z value        Pr(>|z|)
(Intercept) -6.415      0.223        -28.772      4.83297 6e-182
lh           1.094      0.052         20.775        7.316   2e-96
ld           1.936      0.014        134.366      0.00000 0e+00
(Dispersion Parameters for gaussian family estimated as below )

    Scaled Null Deviance: 18992.207 on 30 degrees of freedom
Scaled Residual Deviance: 30.999992 on 28 degrees of freedom

Dispersion Coefficients:
             Estimate  Std. Error     z value        Pr(>|z|)
(Intercept) -797.405    630.948      -1.264       2.06294 4e-01
lh           460.783    307.369       1.499       1.33841 0e-01
ld          -180.469     83.035      -2.173       2.97501 3e-02
lh2          -79.911     38.543      -2.073       3.814   7e-02
lhld          94.855     26.888       3.528       4.19157 2e-04
ld2          -44.681348   8.5465773 -5.2279815 1.71370 6e-07
(Dispersion parameter for Gamma family taken to be 2 )

    Scaled Null Deviance: 52.237701 on 30 degrees of freedom
Scaled Residual Deviance: 29.641284 on 25 degrees of freedom

Minus Twice the Log-Likelihood: -92.600489
Number of Alternating Iterations: 33
```

As noted above, the standard errors for the terms in the response surface model for the variance are very unreliable indicators of importance in this small sample and heavily parametrized model – the likelihood in these parameters is very skewed. The function gives a warning that the fitting algorithm did not converge for the variance modelling. Backward elimination of terms from the variance model

Table 3.14. Mean and variance models for log v, tree data

Mean			Variance						
1	*lh*	*ld*	1	*lh*	*ld*	lh^2	*lhld*	ld^2	*Disparity*
−6.42	1.09	1.94	−797	461	−180	−80	95	−45	−92.60
(0.27)	(0.07)	(0.02)	(847)	(412)	(111)	(52)	(36)	(11.5)	
−6.39	1.08	1.95	−180	0	134	0	0	−26	−88.31
(0.27)	(0.07)	(0.02)	(57)		(44)			(9)	
−6.15	1	2	−156	0	115	0	0	−22	−85.10
(0.00)			(36.89)		(28.83)			(5.60)	
−6.15	1	2	−156	0	0	0	0	0	−70.30
(0.01)			(0)						

leaves a quadratic in `ld` shown in Table 3.14, with the estimates and (standard errors) from the response surface variance model and the constant variance model; the mean model is consistent with the model `log h + 2log d`, and estimates for this model are also shown in the table, obtained by appropriately defining the offset `ofsm` as in Section 3.6.

```
> trees.dglm2 <- dglm(lv ~ lh + ld, data = trees, dformula = ~ld +
+     ld2)
> trees$z <- 2 * trees$ld + trees$lh
> trees.dglm3 <- dglm(lv - z ~ 1, data = trees, dformula = ~ld +
+     ld2)
> trees.dglm4 <- dglm(lv - z ~ 1, data = trees, dformula = ~1)
```

It is difficult to understand the meaning of the quadratic variance function from the estimates; Fig. 3.14 shows the squared residuals and fitted variance function on the log scale, and Fig. 3.15 shows them on the actual variance scale.

There is a striking increase in variability for intermediate values of tree diameter, with small variances for large and small trees. This unexpected result leads us to examine closely the influence of individual observations on the variance model, by extracting them following the variance model fit and graphing them against `ld` in Fig. 3.16.

```
> trees.dglm2 <- dglm(lv ~ lh + ld, data = trees, dformula = ~ld +
+     ld2)

> plot(trees$ld, hatvalues(trees.dglm2$dispersion.fit),
+     xlab = "log d", ylab = "variance influence",
+     las = 1)
```

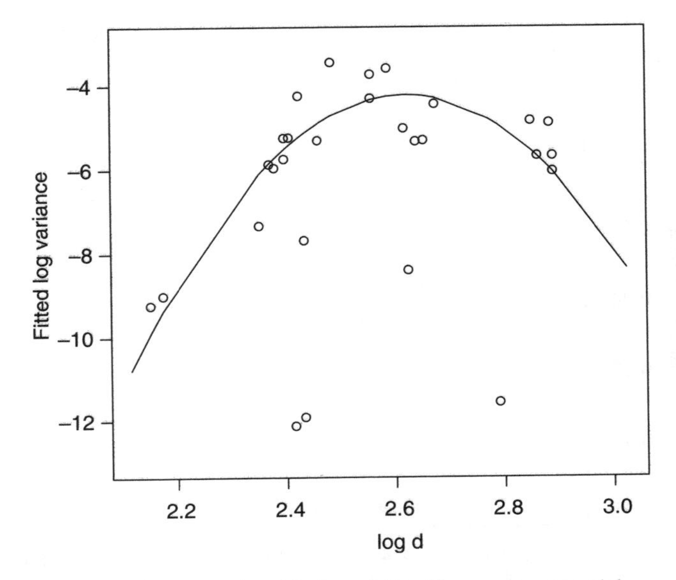

Fig. 3.14. Squared residuals and fitted log variance model

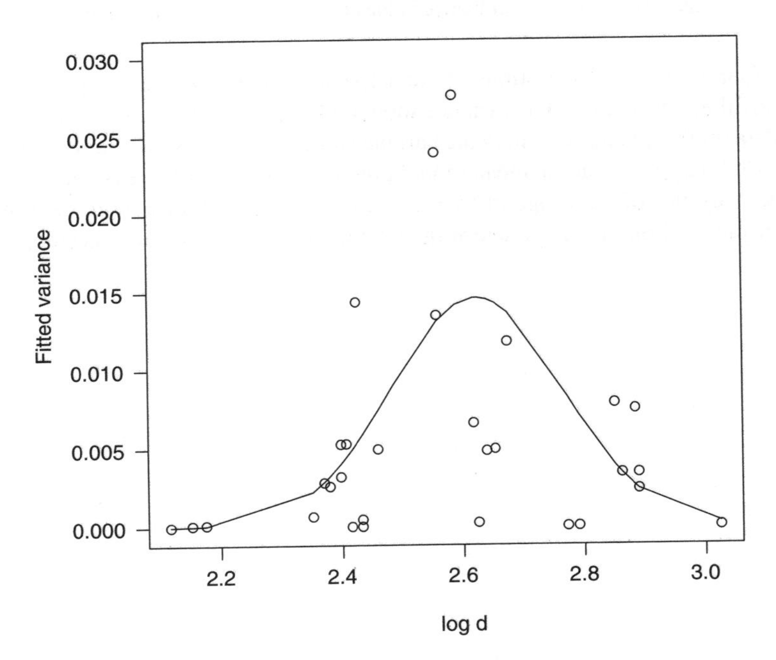

Fig. 3.15. Squared residuals and fitted variance model

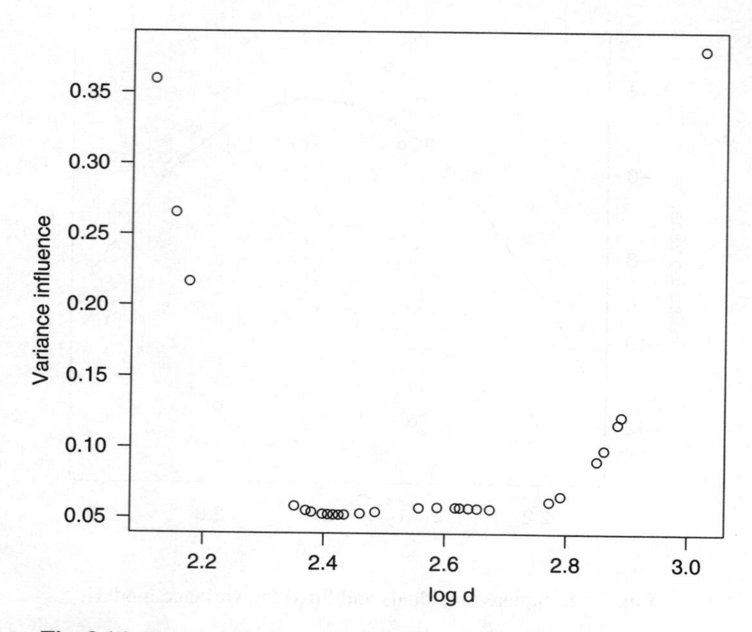

Fig. 3.16. Tree data: influence values from quadratic variance model

The four extreme observations, both large and small, appear to have a very powerful effect on the fitted variance model. However, if these are removed from the data set the quadratic term in the variance model still gives a disparity reduction of 4.16. It happens that the mean model prediction of tree volume is very accurate at the extremes of the range of tree diameter; it would require more data over a larger range of diameters to determine whether this effect is real or accidental.

4
Binary response data

4.1 Binary responses

Much social science data consist of *categorical* variables. Familiar examples are religion, nationality, residence (urban/rural), type of dwelling, level of education and social class. The categories may be unordered (religion, nationality) or ordered (degree of disablement, attitude to a social question).

The simplest categorical variable is one with just two categories. Simple Yes/No or Agree/Disagree responses to questionnaire items are examples, while in medical science Present/Absent, Susceptible/Not susceptible, Exposed/Not exposed are common examples. In Chapter 3 we considered the use of binary variables, coded 0 or 1, as explanatory variables in a regression model. Here, however, we are interested in the relationship between a binary response variable and other explanatory variables.

We begin with a simple example from Racine *et al.* (1986), of a small study of the effects of $m = 4$ increasing doses of a drug on the mortality of mice through acute toxicity on inhalation of the drug (Table 4.1).

At the i-th dose level d_i, $i = 1, \ldots, m$ (expressed in mg/mL), $n_i = 5$ mice are administered the drug dose, and r_i of them die. The small number of animals reflects the ethical need to minimize animal sacrifice; the dose levels are set to cover the full range of response probability, from 0 to 1. We want to model the relationship between the probability p_i of death and the dose d_i. We graph the *sample proportions* dying against dose in Fig. 4.1; because of the substantial skew in the dose values we use the log scale for dose.

```
> mice <- data.frame(dose = c(422, 744, 948, 2069),
+      r = c(0, 1, 3, 5), n = c(5, 5, 5, 5))
> mice <- transform(mice, ldose = log(dose), p = r/n)
> library(lattice)
> print(xyplot(p ~ ldose, data = mice, xlab = "log dose"))
```

For binary responses in general, we define the response variable

$$
\begin{aligned}
Y &= 1 \quad \text{'success' (occurrence of the response)} \\
&= 0 \quad \text{'failure' (non-occurrence)}
\end{aligned}
$$

Table 4.1. Toxicity of a drug in mice

i	1	2	3	4
d_i	422	744	948	2069
r_i	0	1	3	5
n_i	5	5	5	5

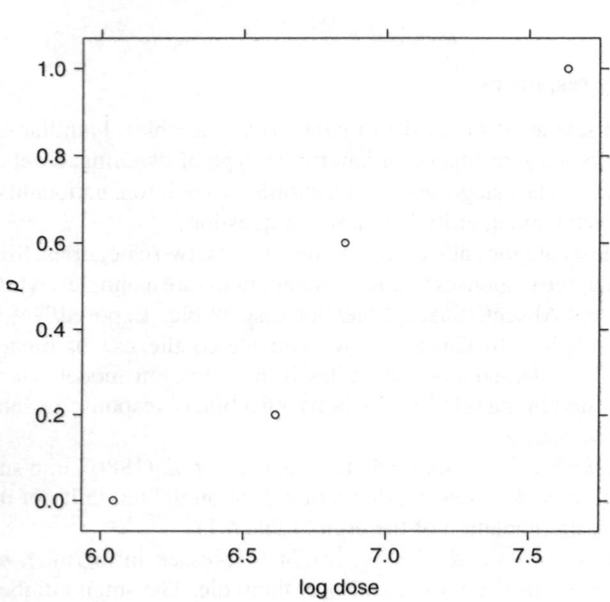

Fig. 4.1. Toxicity in mice – observed death proportions

and let p be the success probability for a randomly chosen individual at a given explanatory variable value x. Then Y has a *Bernoulli* distribution

$$\Pr[Y = y] = p^y(1 - p)^{1-y}, \quad y = 0 \text{ or } 1,$$

where p is related to x through a suitable regression function and link function. The random variable Y has mean p and variance $p(1 - p)$.

If n individuals are observed at the same value of x, the number of individuals Y with a success at this value of x has a *binomial* distribution

$$\Pr[Y = r] = \binom{n}{r} p^r(1 - p)^{n-r}, \quad r = 0, 1, \ldots, n$$

with mean np and variance $np(1 - p)$.

We now consider in detail the use of different link functions for probability parameters.

4.2 Transformations and link functions

The relation between the response variable Y and d will be through a regression model for p in which $x = \log d$ is used as the explanatory variable in the linear predictor. How is p to be related to the linear predictor?

In Chapter 3 the regression model was for the mean μ of the normal distribution, although as illustrated in Section 3.7 other functions of the mean could be used. Since the mean of the Bernoulli distribution is p, this suggests a linear regression model

$$p = \beta_0 + \beta_1 x$$

for the probability of death. Such linear models for proportions are sometimes used, but they suffer from obvious difficulties. If $\beta_1 > 0$, then for x sufficiently large, p will exceed 1, while for x sufficiently small, p may be negative. In practical data analysis these natural bounds for p are sometimes not exceeded, so that linear models may give sensible answers. For the mice data, it is clear that a linear model cannot be appropriate since at the extreme dose levels the response observed proportions are already 0 and 1.

Transformations of the success probability are the standard solution to the problem of the finite range for p. Let $H(\theta)$ be a strictly increasing function of θ, where $-\infty < \theta < \infty$, such that $H(-\infty) = 0$, $H(\infty) = 1$. This condition is satisfied if H is a cumulative distribution function (*cdf*) for any continuous random variable defined on $(-\infty, \infty)$. Noting that for each value of p there is a value of θ with $p = H(\theta)$, we can define a transformation of p from (0, 1) to $(-\infty, \infty)$ by

$$\theta = H^{-1}(p) = g(p).$$

A linear model for this transformed parameter

$$\theta = \beta_0 + \beta_1 x$$

now has the property that any finite value of the linear predictor will give a value of p in the range (0, 1).

Five such transformations are provided as standard options in R where they are defined by the link function $g(p)$, introduced in Section 2.9.2. They are the *logit* transformation or link

$$\theta = \operatorname{logit} p = \log\{p/(1-p)\}, \quad H(\theta) = \mathrm{e}^{\theta}/(1+\mathrm{e}^{\theta}),$$

for which $H(\theta)$ is the *cdf* for the logistic distribution and is the default link in R for binomial responses, the *probit* transformation or link

$$\theta = \operatorname{probit} p = \Phi^{-1}(p), \quad H(\theta) = \Phi(\theta),$$

for which $H(\theta)$ is the *cdf* for the standard normal distribution, the *cauchit* transformation or link

$$\theta = g(p) = \tan\{\pi(p - 1/2)\}, \quad p = 1/2 + 1/\pi \tan^{-1}(\theta),$$

the log transformation or link

$$\theta = \log(p), \quad H(\theta) = \exp\theta$$

and the *complementary log–log (*CLL*)* transformation or link

$$\theta = \log\{-\log(1 - p)\}, \quad H(\theta) = 1 - \exp(-e^{\theta}),$$

for which $H(\theta)$ is the *cdf* for the extreme value distribution (see Chapter 6).

The probit and logit links are similar, and can be made almost identical by a scale change to equate the variances of the underlying normal (1) and logistic ($\pi^2/3$) distributions. They generally give very similar fitted models, and are symmetric in p and $1 - p$. The cauchit link may be appropriate when the distribution of the data show heavier than expected tails. The CLL link is not symmetric, but for small p it is very close to the logit link, since in this case $\log(1 - p) \approx -p$, so that

$$\log(-\log(1 - p)) \approx \log p, \quad \log\{p/(1 - p)\} \approx \log p + p.$$

A sixth link function, the log–log link, can also be used in R by defining the response as $(1 - p)$ and using the CLL transformation. It is defined by

$$\theta = -\log(-\log p), \quad H(\theta) = \exp(-e^{-\theta}),$$

for which $H(\theta)$ is the *cdf* for the reversed extreme value distribution (Section 6.13): if z has the extreme value distribution with *cdf*

$$H(z) = 1 - \exp(-e^{z}),$$

then $w = -z$ has the reversed extreme value distribution with

$$H(w) = \exp(-e^{-w}).$$

For p near 1, the log–log link is very close to the logit link.

As $p \to 0$ or 1, all these link functions approach $\pm\infty$ apart from the *log* which approaches zero and ∞. The logistic transformation is also called the *log odds*, $p/(1 - p)$ being the *odds in favour* of $y = 1$.

To fit models to binary data in R, we have to specify that the probability distribution is binomial (`glm(..., family=binomial)`) and the response is either a single variable containing a binary or Bernoulli response, or a two level factor, or a two column matrix of which the first column holds the number of successes and the second holds the number of failures. The Bernoulli distribution

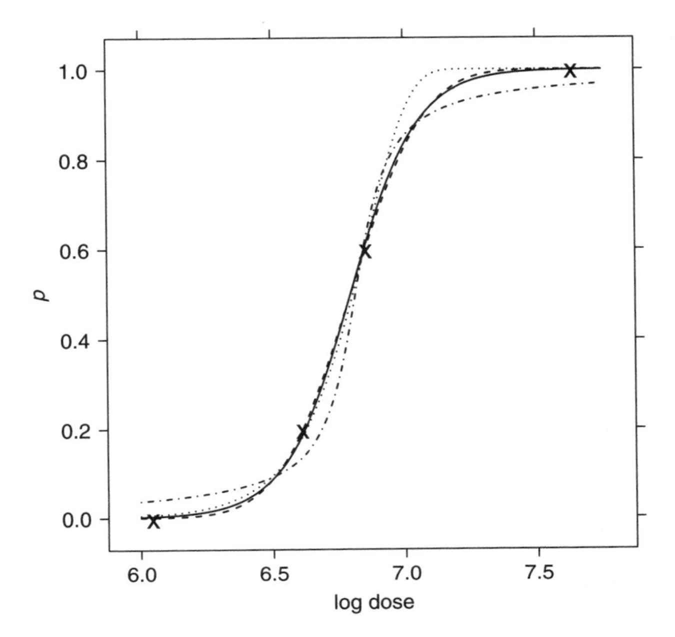

Fig. 4.2. Observed and fitted death probabilities

is not provided as a separate distribution, since it is the special case of the binomial when the number of trials n is equal to 1.

Alternatively for binomial outcomes the proportion of successes can be given as a response and the number of trials defined by the `weights` argument to the `glm` function.

We analyse the dose–response data using various link functions. For each link function we compute the fitted probabilities from the model over a fine grid of log dose values covering the observed range using `predict` to obtain a smooth fitted model (Fig. 4.2).

R's `glm` function does provide an identity link for the binomial family. However, only the null model which does not depend on the choice of link function provides a deviance. Adding `ldose` to the formula generates an error message about valid starting values. We were unsuccessful in supplying valid starting values.

```
> try(glm(cbind(r, (n - r)) ~ ldose, data = mice,
+     family = binomial(link = "identity")))
```

We use the links which give in-range probabilities.

```
> (mice.glm1 <- glm(cbind(r, (n - r)) ~ ldose, data = mice,
+     family = binomial))
```

```
Coefficients:
(Intercept)      ldose
     -53.90       7.93

Degrees of Freedom: 3 Total (i.e. Null);  2 Residual
Null Deviance:          15.79
Residual Deviance: 0.04669        AIC: 7.957

> gridx <- seq(6, 7.75, by = 0.01)
> fpg <- predict(mice.glm1, new = data.frame(ldose = gridx),
+      type = "response")
> (mice.glm2 <- update(mice.glm1,
+    family = binomial(link = "probit")))


Coefficients:
(Intercept)       ldose
    -31.127       4.579

Degrees of Freedom: 3 Total (i.e. Null);  2 Residual
Null Deviance:          15.79
Residual Deviance: 0.003763       AIC: 7.914

> fpp <- predict(mice.glm2, new = data.frame(ldose = gridx),
+   type = "response")
> (mice.glm3 <- update(mice.glm1,
+   family = binomial(link = "cloglog")))

Coefficients:
(Intercept)      ldose
     -43.45       6.33

Degrees of Freedom: 3 Total (i.e. Null);  2 Residual
Null Deviance:          15.79
Residual Deviance: 0.06667       AIC:    7.977

> fpc <- predict(mice.glm3, new = data.frame(ldose = gridx),
+      type = "response")
> (mice.glm4 <- update(mice.glm1,
+      family = binomial(link = "cauchit")))

Coefficients:
(Intercept)       ldose
     -73.05       10.72
```

```
Degrees of Freedom: 3 Total (i.e. Null);  2 Residual
Null Deviance:                   15.79
Residual Deviance: 0.9236             AIC: 8.834

> fpcauch <- predict(mice.glm4, new = data.frame(ldose = gridx),
+      type = "response")
> mice.new.df <- data.frame(gridx, fpg, fpp, fpc, fpcauch)

> library(lattice)
> print(xyplot(fpg + fpp + fpc + fpcauch ~ gridx,
+      data = mice.new.df, type = "l", xlab = "log dose",
+      ylab = "p", lwd = 1.5, lty = c(1, 2, 3, 4),
+      panel = function(x, y, ...) {
+          panel.xyplot(x, y, ...)
+          panel.text(mice$ldose, mice$p, "x", cex = 1.5,
+              type = "p", ...)
+      }))
```

The *logit, probit* and *cloglog* link functions all fit very closely, with 'scaled' deviances of almost zero – the differences among them are negligible. This is essentially because the parameters are determined by the two intermediate observations, as the extreme 0 and 1 proportions do not locate clearly the model parameters. The *cauchit* link model does not fit the probabilities close to the extremes of the p range well.

The R deviance is the LRTS for the model relative to the 'saturated' model with a parameter for each dose level. That is, the scaled deviance from the linear model is the LRTS for *linearity* of the log dose–response relationship, relative to a *cubic* log dose model, the 'saturated' polynomial model which exactly interpolates the data. (It is also the LRTS for the log dose model relative to a four-group model with arbitrary death probabilities at each dose level, but this is not of interest in the experimental context.)

The fitted probabilities closely interpolate the observed proportions, as is evident from the very small deviances. But the graph shows a discrepancy between the CLL link and the others, at the upper end of the dose range. Since there are no data in this region, the fit of the CLL link is as close as the others. We work with the logit link.

The standard error of the slope coefficient is quite large, with a Wald test statistic of $z^2 = (7.93/5.08)^2 = 1.56^2 = 2.44$, so it appears that the regression of death probability on log dose is not significant. However, the LRTS for the same hypothesis is the deviance difference 15.74, which is very large compared with χ^2_1. In this case the Wald test is not a reliable tool for assessing the importance of explanatory variables – the likelihood ratio test (LRT) shows clearly that dose is important. We discuss this important point further below.

4.2.1 *Profile likelihoods for functions of parameters*

In non-linear models with small samples, the likelihood function is often far from normality, and we cannot therefore rely on large-sample normality of parameter estimates, particularly for non-linear functions of the parameters. *Profile likelihoods* (Section 2.7.1) for the parameters allow for the non-linearity and provide correctly located regions and intervals for the parameters, though we are still dependent on asymptotic theory for the confidence coverage of these regions or intervals. We consider here several such functions for the very small drug example above.

(i) *The slope parameter*

This can be computed by computing the likelihood for β_0 with $\beta_1 x$ held as an offset and varying β_1 over a grid.

```
> beta1 <- seq(0, 25, by = 0.25)
> pl <- NULL
> for (b1 in beta1) {
+     z <- b1 * mice$ldose
+     fit <- glm(cbind(r, (n - r)) ~ offset(z), data = mice,
+          family = binomial)
+     logl <- with(mice, sum(dbinom(r, n, fitted(fit),
+          log = TRUE)))
+     pl <- c(pl, logl)
+ }
> pl <- pl - max(pl)
> print(xyplot(exp(pl) ~ beta1, type = "l",
+     xlab = expression(beta[1]), ylab = "relative likelihood",
+     lwd = 1.5, panel = function(...) {
+         panel.xyplot(...)
+         panel.abline(h = exp(3.84/-2), lty = 2, lwd = 1.5)
+     }))

> fc2(beta1[which.max(pl)])
[1] 8.00
```

The profile likelihood (Fig. 4.3) is substantially skewed. The asymptotic 95% confidence interval for β_1 can be found by linear interpolation in the deviance for the value 3.84.

We find the interval (2.27, 21.49), far away from zero. The 't' interval based on $\hat{\beta}_1 \pm 1.96 \cdot \mathrm{SE}(\hat{\beta}_1)$ is (−2.03, 17.89), misleadingly including zero.

The MASS package written for R provides the function `confint` for generalized linear models. For this example we obtain the profile intervals by

```
> confint(mice.glm1)
```

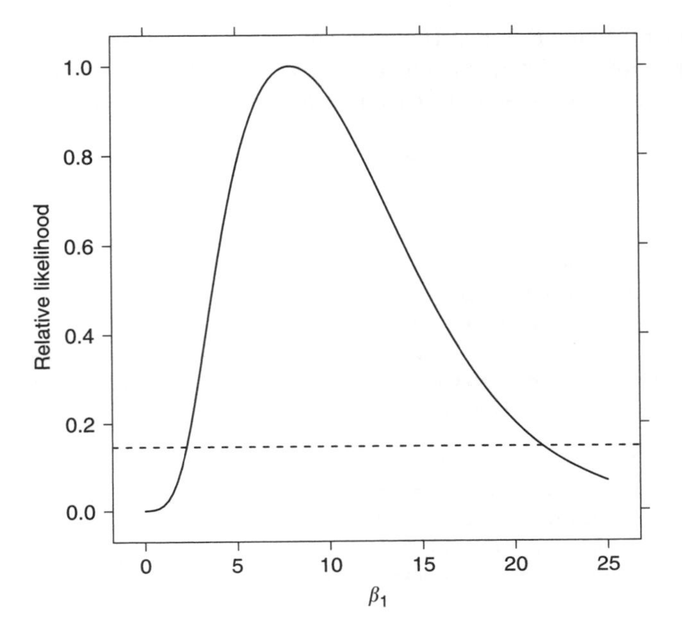

Fig. 4.3. Profile likelihood for slope

```
                   2.5 %     97.5 %
(Intercept) -146.055417 -15.94054
ldose          2.322979  21.49210
.
```

(ii) *The death probability at a given dose* x_0

We define the parameter of interest to be

$$\theta = \beta_0 + \beta_1 x_0$$

and express β_0 in terms of θ and β_1 in the model:

$$\operatorname{logit} p = \theta + \beta_1(x - x_0).$$

The model is fitted over a grid of θ with the explanatory variable x centred by x_0, an offset of θ and no intercept, and is finally converted to p by the anti-logit transformation

$$p = \frac{e^\theta}{1 + e^\theta}.$$

We illustrate with the dose $d = 1100$, corresponding to log dose $x = 7.00$. The grid interval parameters are set as for `logodds`, though they now refer to the probability scale rather than the β_1 scale. The values of p are held in the vector `pseq`. The `I` function inhibits the interpretation of operator '–' as a formula operator so that the value of `x0`, that is, log(1100), is subtracted from all *ldose* values.

```
> pseq <- seq(0.1, 0.99, by = 0.01)
> logl <- NULL
> xdose <- NULL
> x0 <- log(1100)
> for (p in pseq) {
+     theta <- rep(log(p/(1 - p)), 4)
+     fit <- glm(cbind(r, (n - r)) ~ offset(theta) +
+         I(ldose - x0) - 1, data = mice, family = binomial)
+     logl <- c(logl, with(mice, sum(dbinom(r, n, fitted(fit),
+         log = TRUE))))
+ }
> logl <- logl - max(logl)
> print(xyplot(exp(logl) ~ pseq, type = "l", xlab = "probability",
+     ylab = "relative likelihood", panel = function(...) {
+         panel.xyplot(...)
+         panel.abline(h = exp(3.84/-2), lty = 2, lwd = 1.5)
+     }))
```

The profile likelihood is very severely skewed, not surprisingly since the MLE of the death probability at log dose 7 is 0.84 (Fig. 4.4). The horizontal line is at a log relative likelihood of -1.92.

The 95% asymptotic confidence interval for p at dose 1100 is, to 2 dp, (0.41, 1.00).

(iii) *The dose needed for a given response probability* p_0

An important use of this model in drug trials is to establish the dose at which a desired response occurs. Two common doses of interest are the *medial lethal dose* or *median effective dose*, the LD50 or ED50, and the *90% effective dose*, the ED90. These dose levels are non-linear functions of the model parameters. For the given probability p_0, write x_0 for the corresponding log dose. Then

$$\theta_0 = \text{logit}\, p_0 = \beta_0 + \beta_1 x_0$$

and

$$\text{logit}\, p = \theta_0 + \beta_1(x - x_0).$$

This is formally similar to the previous case, but now x_0 is the parameter of interest and θ is the nuisance parameter. The macro fits the model over a grid of x_0 with the explanatory variable centred by x_0, a fixed offset of θ_0 and no intercept.

We illustrate with both the ED50 and the ED90. The grid interval parameters now refer to the log dose scale rather than the β_1 scale. The probability value is set

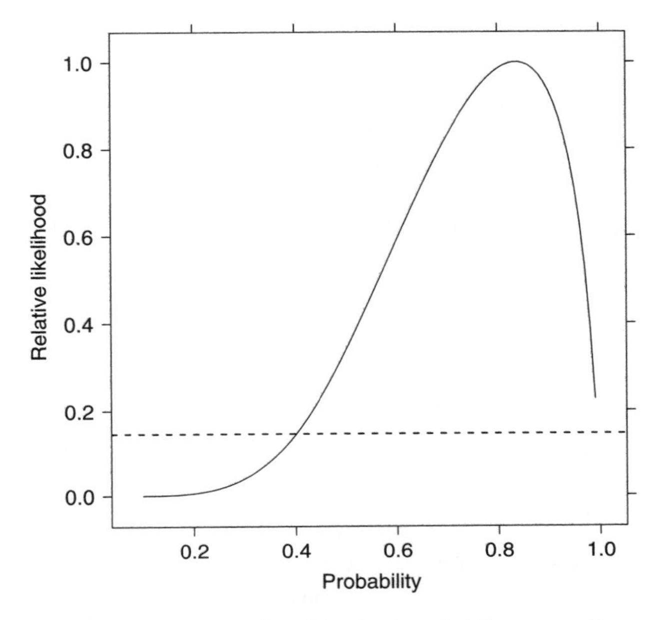

Fig. 4.4. Profile likelihood for death probability at $x = 7$.

in the scalar `p`. The explanatory variable (log dose) values are held in the vectors `ldseq50` and `ldseq90`.

```
> ldseq50 <- log(seq(600, 1600, len = 1001))
> logl <- NULL
> fit <- NULL
> p <- 0.5
> logitp <- rep(log(p/(1 - p)), 4)
> for (ld in ldseq50) {
+     fit <- glm(cbind(r, (n - r)) ~ I(ldose - ld) +
+         offset(logitp) - 1, data = mice, family = binomial)
+     logl <- c(logl, sum(with(mice, dbinom(r, n, fitted(fit),
+         log = TRUE))))
+ }
> logl <- logl - max(logl)
> plot(ldseq50, exp(logl), type = "l", xlab = "log dose",
+     ylab = "relative likelihood")
> abline(h = exp(3.84/-2), lty = 2, lwd = 1.5)
```

The profile likelihood for the ED50 (Fig. 4.5) is nearly symmetrical, that for the ED90 (Fig. 4.6) is severely skewed. Asymptotic 95% confidence intervals for the log ED50 and log ED90 are (6.59, 7.107) and (6.842, 7.92), respectively; the corresponding dose intervals are (728, 1221) and (937, 2751).

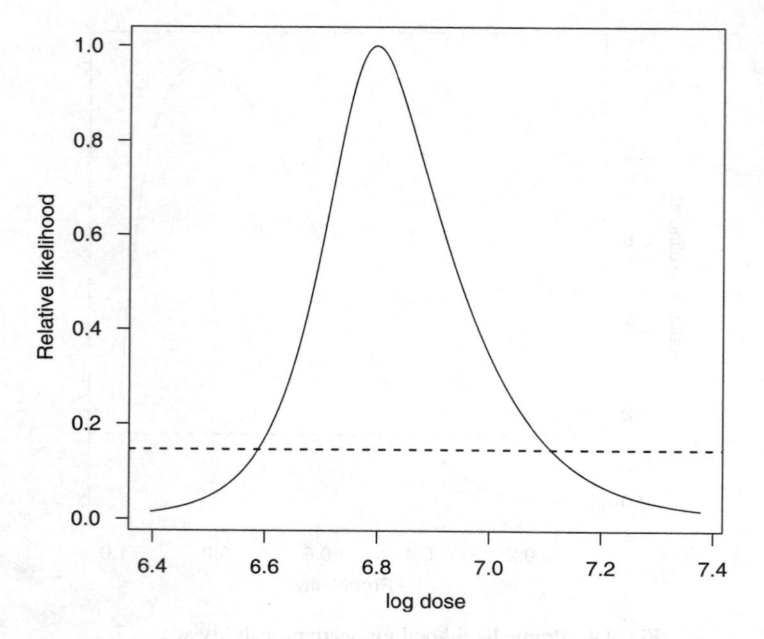

Fig. 4.5. Profile likelihood for ED50

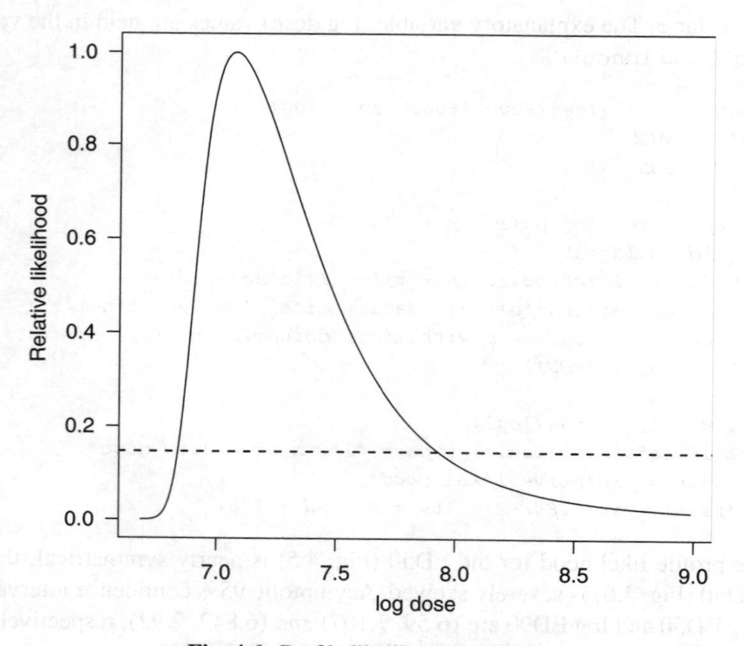

Fig. 4.6. Profile likelihood for ED90

4.3 Model criticism

In Section 3.4 we discussed model criticism for the normal distribution model. For discrete data the possibilities are more limited, but we follow the same sequence.

4.3.1 *Mis-specification of the probability distribution*

The `resid` function in R when used with binomial `glm` model prints the Pearson residuals

$$e_i = (y_i - n_i \hat{p}_i)/[n_i \hat{p}_i(1 - \hat{p}_i)]^{1/2}.$$

These are approximately standardized variables with mean 0 and variance approximately 1; the variance is approximate because of the estimation of $\boldsymbol{\beta}$ in the linear predictor. They are not, however, even approximately normally distributed unless the binomial sample sizes n_i are large, so a normal quantile plot is not appropriate and may be misleading. Large values of the Pearson (or other) residuals indicate failure of the model to fit at the corresponding points. The sum of squares of the Pearson residuals is the Pearson goodness-of-fit X^2, which can be calculated by `sum(resid(model)`^{2}`)`.

'Deviance residuals' have been defined by McCullagh and Nelder (1989); these are the signed square roots of the individual components of the deviance: their sum of squares is the deviance. Residual analysis in this chapter will be based on the Pearson residuals.

Mis-specification of the probability distribution resulting in *overdispersion* may arise from the omission of important variables in the model. This important issue is discussed at length in Chapter 8.

4.3.2 *Mis-specification of the link function*

R implements five standard links for the binomial distribution, but other more general families of links have been proposed and used for this distribution, by generalizing the log transformation of the odds to more general transformation families.

Stukel (1988) gave an extensive discussion of several families. We do not discuss these as we do not make use of them in this book.

4.3.3 *The occurrence of aberrant and influential observations*

With binomial data in which aggregation of individual responses has occurred across common explanatory variable values, with all $n_i > 1$, aberrant observations tend to be damped by the aggregation, whether they occur in the response or in the explanatory variables. With individual Bernoulli observations, however, the effect of single aberrant observations can be marked. We discuss this and other issues peculiar to binary data in the next section.

The simple results of Chapter 3 on regression diagnostics can be generalized in various ways (Pregibon, 1981) at the expense of additional computing. We will use the diagonal elements of the hat matrix from the iteratively reweighted least squares algorithm. At the final (r-th) iteration R computes the parameter estimates $\hat{\beta}$ using

$$X'W^{(r)}X\hat{\boldsymbol{\beta}} = X'W^{(r)}\mathbf{z}^{(r)},$$

where $W^{(r)}$ is the diagonal matrix of iterative weights and $\mathbf{z}^{(r)}$ the adjusted dependent variate at the final iteration. The hat matrix is then

$$H = (W^{(r)})^{\frac{1}{2}}X(X'W^{(r)}X)^{-1}X'(W^{(r)})^{\frac{1}{2}}.$$

The diagonal elements of H are obtained in R as the product of the variance of the linear predictor and the iterative weight variate, since

$$\begin{aligned}\mathrm{Var}(X\hat{\boldsymbol{\beta}}) &= X\mathrm{Var}(\hat{\boldsymbol{\beta}})X' \\ &= X(X'W^{(r)}X)^{-1}X'.\end{aligned}$$

These can be extracted from a model object using the `hatvalues` function. We illustrate their use in the next section.

4.4 Binary data with continuous covariates

We give an example from Finney (1947), see also Pregibon (1981). Table 4.2 gives the data obtained in a carefully controlled study of the effect of the rate `rate` and volume `vol` of air inspired by human subjects on the occurrence (coded 1) or non-occurrence (coded 0) of a transient vasoconstriction response `resp` in the skin of the fingers. These data are in the file `vaso`, in subject order. Three subjects were involved in the study: the first contributed 9 observations at different values of `rate` and `vol`, the second 8, and the third 22 observations. They are identified by S1, S2, and S3 in Table 4.2. The response is abbreviated to `r` in the table. The data set is available from the *SMIR* package.

Figure 1 in Finney graphs the correct data but Finney's data Table 1 has an error; the 32nd rate observation of 0.3 was given as 0.03. This observation had very high leverage, discussed below.

The experiment was designed to ensure as far as possible that successive observations obtained on each subject were independent, and the initial observations for each subject were discarded to allow for a training period in which they became familiar with the measurement procedure.

Correlation between successive observations on the same subject in such studies is always a possibility. We consider this issue in Chapter 9, but for the moment assume that the responses among and within individuals are independent, and that there are no differences in response level among individuals.

Table 4.2. Vasoconstriction response

S1			S2			S3					
vol	rate	r	vol	rate	r	vol	rate	r	vol	rate	r
3.7	0.825	1	0.9	0.45	0	0.85	1.415	1	1.9	0.95	1
3.5	1.09	1	0.8	0.57	0	1.7	1.06	0	1.6	0.4	0
1.25	2.5	1	0.55	2.75	0	1.8	1.8	1	2.7	0.75	1
0.75	1.5	1	0.6	3.0	0	0.4	2.0	0	2.35	0.3	0
0.8	3.2	1	1.4	2.33	1	0.95	1.36	0	1.1	1.83	0
0.7	3.5	1	0.75	3.75	1	1.35	1.35	0	1.1	2.2	1
0.6	0.75	0	2.3	1.64	1	1.5	1.36	0	1.2	2.0	1
1.1	1.7	0	3.2	1.6	1	1.6	1.78	1	0.8	3.33	1
0.9	0.75	0				0.6	1.5	0	0.95	1.9	0
						1.8	1.5	1	0.75	1.9	0
						0.95	1.9	0	1.3	1.625	1

We fit the two-variable logit model.

```
> data(vaso, package = "SMIR")

> (vaso.glm <- glm(Y ~ Rate + Volume, data = vaso,
+        family = binomial))


Coefficients:
(Intercept)          Rate        Volume
     -9.530         2.649         3.882

Degrees of Freedom: 38 Total (i.e. Null);  36 Residual
Null Deviance:              54.04
Residual Deviance: 29.77          AIC: 35.77
```

How is the scaled deviance to be interpreted in this model? The saturated model, with a parameter for every observation, will fit the data exactly and have a deviance of zero, so if we were to use the large sample theory of the LRT, we would treat the deviance as approximately χ^2_{36} if the fitted model is an adequate representation of the data. For binary data, however, large-sample theory does not apply to the distribution of the residual deviance from the fitted model, because the saturated model is equivalent to m separate models, one for each single observation: we are fitting m models to m samples of size 1.

For binary data the residuals are

$$e_i = (y_i - \hat{p}_i)/[\hat{p}_i(1 - \hat{p}_i)]^{\frac{1}{2}}$$

and since y_i is equal to 0 or 1, the e_i are equal to a_i or $-1/a_i$ for each i, where

$$a_i = -\{\hat{p}_i/(1-\hat{p}_i)\}^{\frac{1}{2}}.$$

Large residuals will still be evidence of model discrepancy from the data. We print them.

```
> round(resid(vaso.glm, type = "pearson"), 4)

      1       2       3       4       5       6       7       8
 0.0299  0.0310  0.3780  3.7515  0.3582  0.2923 -0.0738 -0.6853
      9      10      11      12      13      14      15      16
-0.1321 -0.0888 -0.0857 -0.9468 -1.4529  0.3538  0.1905  0.1538
     17      18      19      20      21      22      23      24
 0.0283  3.4578 -0.9409  0.3285 -0.2620 -0.3265 -0.7003 -0.9495
     25      26      27      28      29      30      31      32
 0.4973 -0.1992  0.4887 -0.6676  0.8340 -0.3233  0.2300 -0.8491
     33      34      35      36      37      38      39
-0.8141  0.7525  0.8077  0.3015 -0.6676 -0.4528  1.0931

> round(sum(resid(vaso.glm, type = "pearson")^2), 2)

[1] 39.01
```

There are two large residuals: 3.7515 for the fourth observation and 3.4578 for the 18th. These two observations are ‘successes’ at points where the fitted probability of success is very low: 0.072 for observation 4 and 0.083 for observation 18. They contribute a total of 26.03 to the Pearson X^2 of 39.01 (in large samples with binomial data the deviance and the Pearson X^2 agree closely, but they may differ substantially in small samples or with binary data).

A graph shows these points clearly. We graph in Fig. 4.7 successes and failures using plotting characters + and o, respectively.

```
> print(xyplot(Rate ~ Volume, data = vaso, group = Y))
```

The two ‘success’ observations with large residuals appear on the bottom left hand side of the cluster of ‘success’ observations.

What now? As with residual examination in normal regression models, we do not automatically remove these observations. They would be checked for correctness with the experimenter; if they are correct, then it is the model which needs modification, or else we accept the poor fit at these points as random variation.

One possible modification is the incorporation of a subject effect, to allow for differences in response level. The data come from three subjects who are not identified in the model. If subjects vary in their base level vaso-constriction response, then the fitted model is misspecified and a better fit may be obtained by the inclusion of the subject effect.

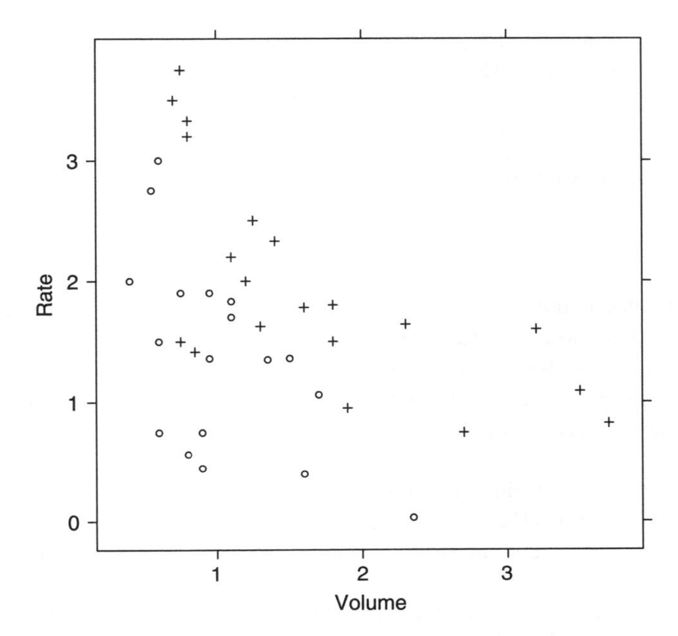

Fig. 4.7. Vasoconstriction response, + for an occurrence and **o** for a non-occurrence.

```
> vaso.glm1 <- update(vaso.glm, . ~ . + Subject)
> coef(vaso.glm1)

(Intercept)        Rate      Volume    Subject2    Subject3
 -10.548475    3.766308    4.810614   -4.297557   -1.920382

> anova(vaso.glm, vaso.glm1)

Analysis of Deviance Table

Model 1: Y ~ Rate + Volume
Model 2: Y ~ Rate + Volume + Subject
  Resid. Df Resid. Dev Df Deviance
1        36    29.7723
2        34    26.1612  2   3.6111

> round(resid(vaso.glm1, type = "pearson"), 4)

      1       2       3       4       5       6       7       8
 0.0056  0.0055  0.0871  1.9070  0.0688  0.0498 -0.0890 -1.7735
      9      10      11      12      13      14      15      16
-0.1832 -0.0121 -0.0120 -0.3980 -0.7187  0.7174  0.2363  0.3019
     17      18      19      20      21      22      23      24
 0.0374  4.5963 -0.8613  0.2265 -0.2218 -0.2495 -0.6408 -0.9367
```

```
     25      26      27      28      29      30      31      32
 0.3806 -0.1399  0.3986 -0.6898  0.8828 -0.1954  0.1878 -0.5913
     33      34      35      36      37      38      39
-0.8673  0.5744  0.6582  0.1408 -0.6898 -0.4264  1.0485
```

```
> round(sum(resid(vaso.glm1, type = "pearson")^2),
+     2)
```

```
[1] 36.79
```

The deviance is now 26.16, a reduction of 3.61 for 2 df, and the Pearson X^2 is 36.79, a reduction of 2.22. There is no strong evidence of a subject effect. The residual at the fourth observation has decreased to 1.907, but that at the eighteenth has increased to 4.5963. We cannot attribute the poor fit at these points to the absence of a subject effect, and we omit this term from subsequent analyses.

Could the form of the model be inappropriate? The analyses in Finney (1947) and Pregibon (1981) used log `rate` and log `vol` rather than `rate` and `vol`, though there were no strong scientific reasons given for choosing between them.

One data-based reason is the considerable skewness of both `rate` and `vol`, evident from Fig. 4.8. Such skewness is substantially reduced by a log transformation. Since these variables are explanatory, and not responses, this reason is not by itself compelling. Another reason is that the curves of constant logit, and therefore constant p, are straight lines cutting the `rate` and `vol` axes in our analysis above, but are hyperbolae which do not cut the axes if the log scales are used. We construct the straight lines – the level curves – for response probability values 0.05, 0.1, 0.3, 0.5, 0.7, 0.9, and 0.95, in Fig. 4.8 using the R code below. The corresponding logit values are −2.94, −2.02, −0.85, 0, 0.85, 2.02, and 2.94.

```
> print(xyplot(Rate ~ Volume, data = vaso, group = Y,
+     pch = c(1, 3), col = "black", panel = function(x,
+         y, ...) {
+         panel.xyplot(x, y, ...)
+         probs <- c(0.05, 0.1, 0.3, 0.5, 0.7, 0.9,
+             0.95)
+         logits <- log(probs/(1 - probs))
+         line.types <- c(4, 3, 2, 1, 2, 3, 4)
+         for (i in 1:7) panel.curve((logits[i] -
+             coef(vaso.glm)[1] -
+             coef(vaso.glm)[2] * x)/coef(vaso.glm)[3],
+             lty = line.types[i])
+         panel.text(c(2, 2.6, 3.2), c(0.6, 0.7, 0.8),
+             c("10", "50", "90"), font = 2)
+     }))
```

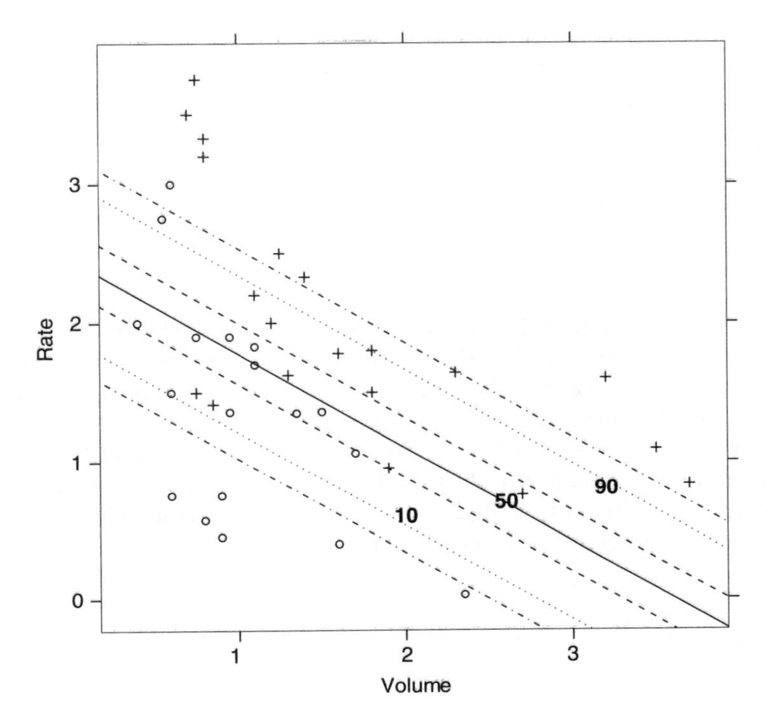

Fig. 4.8. Vasoconstriction response and linear model level curves

Now we transform the variable scales.

```
> vaso <- transform(vaso, lv = log(Volume), lr = log(Rate))
> vaso.glm2 <- glm(Y ~ lv + lr, data = vaso, family = binomial)
> anova(vaso.glm, vaso.glm2)

Analysis of Deviance Table

Model 1: Y ~ Rate + Volume
Model 2: Y ~ lv + lr
  Resid. Df Resid. Dev Df Deviance
1        36    29.7723
2        36    29.2274  0   0.5449

> sort(round(resid(vaso.glm2, type = "pearson"), 4))

     24      33      19      23       8      28      37      13
-1.3684 -1.2062 -1.0718 -1.0242 -1.0196 -0.8989 -0.8989 -0.7751
     12      38      22      26      21      30       9      11
-0.5073 -0.4874 -0.4193 -0.1595 -0.1076 -0.0992 -0.0938 -0.0370
      7      10      32      17       2      16       1      20
```

```
-0.0328 -0.0293 -0.0007  0.0709  0.1349  0.1576  0.2205  0.2405
     14       3      25      27      15      36       5      35
 0.2559  0.2923  0.3347  0.3645  0.4352  0.4828  0.5287  0.5404
     34       6      31      39      29      18       4
 0.5447  0.6090  0.6198  0.7053  0.8981  2.9062  3.5181
```

```
> round(sum(resid(vaso.glm2, type = "pearson")^2),
+     4)

[1] 34.2338
```

The deviance is now 29.23, 0.54 less than for the first model. The Pearson X^2 is reduced from 39.0128 to 34.2338, and the residuals at the fourth and eighteenth observations are reduced to 3.751 and 3.458. The formal LRT cannot discriminate between models with logged and unlogged explanatory variables, as Finney (1947) noted, but as discussed in Chapter 2 the MLR between the models is a good indicator of relative support as the models have the same number of parameters. The likelihood ratio of $\exp(1.20/2) = 1.82$ gives a very slight preference for the logged model. We graph the level curves in Fig. 4.9.

```
> logvolest <- function(x, model, p) {
+     logitp <- log(p/(1 - p))
+     exp((logitp - coef(model)[1] - coef(model)[2] *
+         log(x))/coef(model)[3])
+ }
```

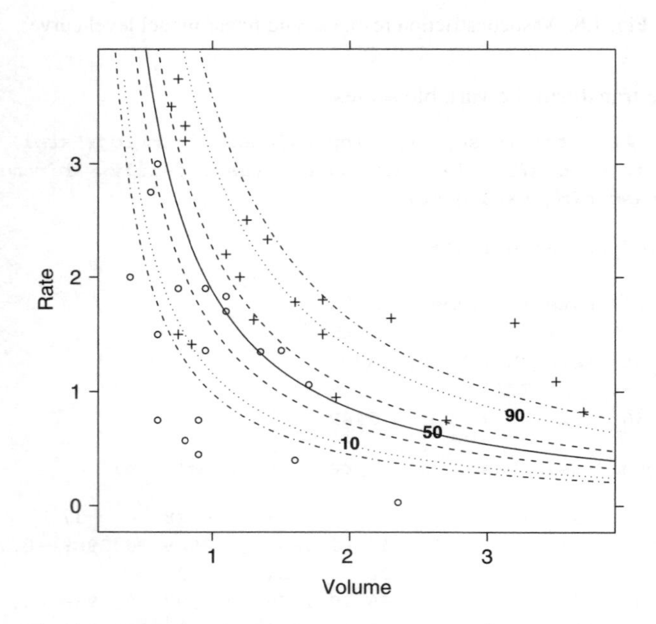

Fig. 4.9. Vasoconstriction response and log model level curves

```
> print(xyplot(Rate ~ Volume, data = vaso, group = Y,
+     pch = c(1, 3), col = "black", panel = function(x,
+         y, ...) {
+         panel.xyplot(x, y, ...)
+         probs <- c(0.05, 0.1, 0.3, 0.5, 0.7, 0.9,
+             0.95)
+         line.types <- c(4, 3, 2, 1, 2, 3, 4)
+         for (i in 1:7) panel.curve(logvolest(x, vaso.glm2,
+             p = probs[i]), lty = line.types[i])
+         panel.text(c(2, 2.6, 3.2), c(0.55, 0.65,
+             0.8), c("10", "50", "90"), font = 2)
+     }))
```

The level curves are quite different in the two models, but the models fit almost equally well, and agree on the two unusual responses.

We now examine the influence of observations in the two models. We first fit the linear model (Fig. 4.10).

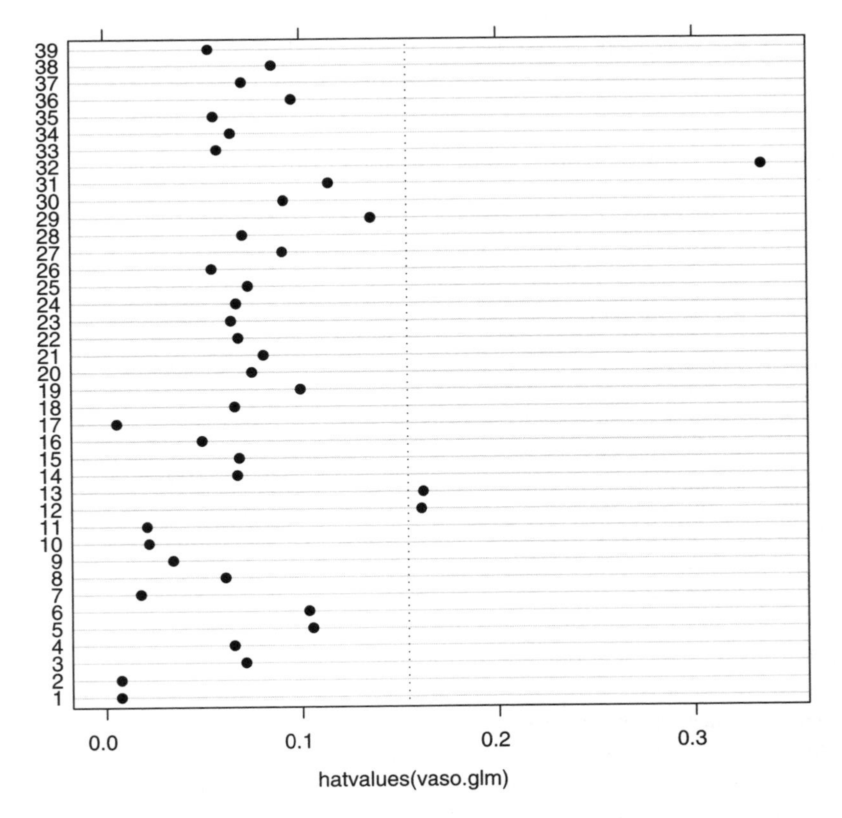

Fig. 4.10. Influence values, linear model

```
> round(hatvalues(vaso.glm), 4)
> print(dotplot(~hatvalues(vaso.glm), panel = function(...) {
+     panel.dotplot(...)
+     panel.abline(v = 0.154, lty = "dotted")
+ }))
```

The value of $2(p + 1)/n$ is 0.154. The 12th and 13th values exceed this marginally, and the 32nd considerably: it is 0.335. This point has the smallest value of `rate` and a large `vol`. Omitting it has a noticeable effect on the fitted model:

```
> vaso.glm3 <- update(vaso.glm, subset = row.names(vaso) !=
+     "32")
> summary(vaso.glm3)


Coefficients:
             Estimate Std. Error z value Pr(>|z|)
(Intercept)   -9.5668     3.2775  -2.919  0.00351
Rate           2.4681     0.8933   2.763  0.00573
Volume         4.2788     1.5810   2.706  0.00680

(Dispersion parameter for binomial family taken to be 1)

    Null deviance: 52.574  on 37  degrees of freedom
Residual deviance: 28.318  on 35  degrees of freedom
AIC: 34.318

     1      2      3      4      5      6      7      8      9
0.0076 0.0075 0.0712 0.0653 0.1054 0.1035 0.0176 0.0610 0.0345
    10     11     12     13     14     15     16     17     18
0.0221 0.0211 0.1609 0.1617 0.0673 0.0682 0.0494 0.0058 0.0661
    19     20     21     22     23     24     25     26     27
0.0997 0.0749 0.0809 0.0680 0.0643 0.0670 0.0732 0.0547 0.0908
    28     29     30     31     32     33     34     35     36
0.0705 0.1358 0.0914 0.1145 0.3347 0.0575 0.0646 0.0559 0.0958
    37     38     39
0.0705 0.0858 0.0534

Number of Fisher Scoring iterations: 6
```

The deviance changes by 1.45 and the `vol` parameter estimate changes by half a standard error, that for `rate` changing much less.

Omission of the two observations with large residuals, on the other hand, has a dramatic effect.

```
> vaso.glm4 <- update(vaso.glm, subset = !row.names(vaso) %in%
+     c("4", "18"))
> summary(vaso.glm4)

Coefficients:
             Estimate Std. Error z value Pr(>|z|)
(Intercept)  -41.989     22.200  -1.891   0.0586
Rate          10.744      5.670   1.895   0.0581
Volume        17.495      9.395   1.862   0.0626

(Dispersion parameter for binomial family taken to be 1)

    Null deviance: 51.266  on 36  degrees of freedom
Residual deviance: 10.700  on 34  degrees of freedom
AIC: 16.700

Number of Fisher Scoring iterations: 9
```

The deviance changes by more than 17 relative to the full model, and the parameter estimates have become much larger. The large estimates for the parameters show that a rapid change from 0 to 1 in the response probability occurs over a small interval of each variable: the successes and failures can be almost perfectly discriminated from the model. We illustrate in Fig. 4.11 with the level curves from this model.

```
> volest <- function(x, model, p) {
+     logitp <- log(p/(1 - p))
+     (logitp - coef(model)[1] - coef(model)[3] * x)/
+     coef(model)[2]}
> print(xyplot(Rate ~ Volume, data = vaso, group = Y,
+     pch = c(1, 3), col = "black", panel = function(x,
+         y, ...) {
+         panel.xyplot(x, y, ...)
+         probs <- c(0.05, 0.1, 0.3, 0.5, 0.7, 0.9,
+             0.95)
+         line.types <- c(4, 3, 2, 1, 2, 3, 4)
+         for (i in 1:7) panel.curve(volest(x, vaso.glm4,
+             p = probs[i]), lty = line.types[i])
+         panel.text(c(2, 2, 2), c(0.4, 0.7, 1), c("10",
+             "50", "90"), font = 2)
+     }))
```

We repeat the analysis for the logged model.

```
> vaso.glm5 <- glm(Y ~ lv + lr, data = vaso, family = binomial)
> summary(vaso.glm5)
```

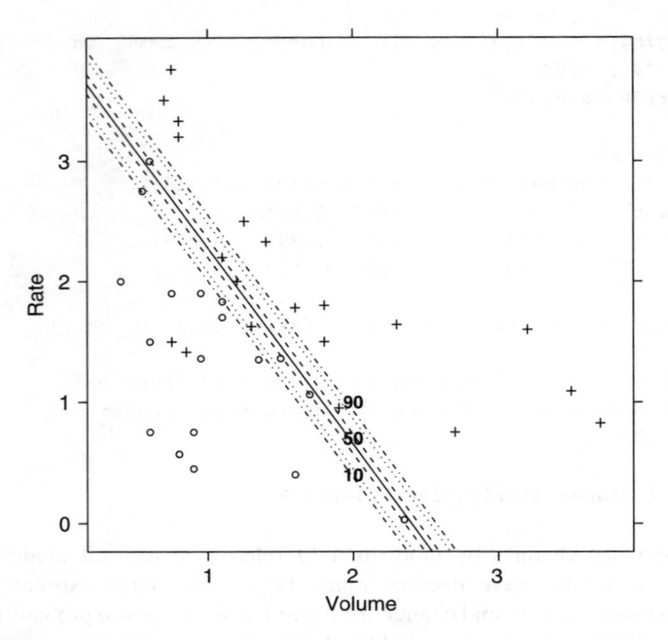

Fig. 4.11. Vaso-constriction response and linear model level curves

```
Coefficients:
             Estimate Std. Error z value Pr(>|z|)
(Intercept)    -2.875      1.321  -2.177  0.02946
lv              5.179      1.865   2.778  0.00547
lr              4.562      1.838   2.482  0.01306

(Dispersion parameter for binomial family taken to be 1)

    Null deviance: 54.040  on 38  degrees of freedom
Residual deviance: 29.227  on 36  degrees of freedom
AIC: 35.227

Number of Fisher Scoring iterations: 6
```

```
> sort(round(hatvalues(vaso.glm5), 4))

    32     10      7     11     17      9     21     16      2
0.0000 0.0072 0.0076 0.0097 0.0172 0.0342 0.0373 0.0402 0.0429
    30     33     20     39     26     14     35      8     25
0.0507 0.0510 0.0525 0.0531 0.0548 0.0551 0.0551 0.0559 0.0587
    34      3     28     37     27     24     23      4      1
0.0601 0.0612 0.0647 0.0647 0.0661 0.0717 0.0761 0.0868 0.0927
```

```
    18     38     22      5     36     19     15     12      6
0.0954 0.1000 0.1015 0.1158 0.1176 0.1315 0.1336 0.1481 0.1524
    13     29     31
0.1627 0.1682 0.2459
```

The 13th values and 29th values now exceed $2(p+1)/n$ marginally, and the 31st more substantially: it is 0.25. The 32nd observation does not have high leverage because of the log transformations.

```
> vaso.glm6 <- update(vaso.glm5, subset = row.names(vaso) !=
+     "31")
> summary(vaso.glm6)


Coefficients:
            Estimate Std. Error z value Pr(>|z|)
(Intercept)   -3.041      1.355  -2.244  0.02484
lv             4.966      1.824   2.723  0.00647
lr             4.765      1.885   2.528  0.01147

(Dispersion parameter for binomial family taken to be 1)

    Null deviance: 52.679  on 37  degrees of freedom
Residual deviance: 28.455  on 35  degrees of freedom
AIC: 34.455

Number of Fisher Scoring iterations: 6
```

Excluding the 31st observation gives a small deviance change and small changes in the parameter estimates, by less than 4% of a standard error.

```
> vaso.glm7 <- update(vaso.glm5, subset = !row.names(vaso) %in%
+     c("4", "18"))
> summary(vaso.glm7)


Coefficients:
            Estimate Std. Error z value Pr(>|z|)
(Intercept)   -24.58      14.02  -1.753   0.0796
lv             39.55      23.25   1.701   0.0889
lr             31.94      17.76   1.798   0.0721

(Dispersion parameter for binomial family taken to be 1)

    Null deviance: 51.266  on 36  degrees of freedom
Residual deviance:  7.361  on 34  degrees of freedom
AIC: 13.361
```

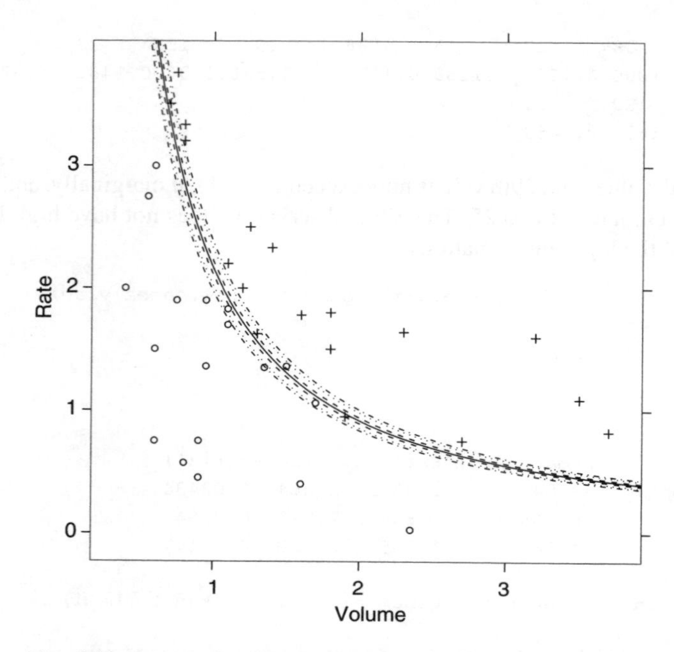

Fig. 4.12. Vasoconstriction response and log model level curves

```
Number of Fisher Scoring iterations: 11
```

Excluding observations 4 and 18 again has a dramatic effect, with a deviance change of 21.9 and very large parameter estimates for log `rate` and log `vol`, respectively. We graph in Fig. 4.12 the level curves for this model.

```
> print(xyplot(Rate ~ Volume, data = vaso, group = Y,
+     pch = c(1, 3), col = "black", subset = !row.names(vaso) %in%
+         c("4", "18"), panel = function(x, y, ...) {
+         panel.xyplot(x, y, ...)
+         probs <- c(0.05, 0.1, 0.3, 0.5, 0.7, 0.9,
+             0.95)
+         logits <- log(probs/(1 - probs))
+         line.types <- c(4, 3, 2, 1, 2, 3, 4)
+         for (i in 1:7) panel.curve(exp((logits[i] +
+             24.58 - 39.55 * log(x))/31.93), lty = line.types[i])
+     }))
```

The exclusion of the two observations results in a dramatically close fit of the model to the data – the level curves are so close together in Fig. 4.12 that they form a ‘cliff’ in the plane, with a ‘plateau’ of 0 on the lower left and a plateau of 1 on the upper right of the figure. A good fit to the data is, of course, the whole object of modelling, but when it is obtained by the removal of observations which do not fit the model, we should be very cautious indeed. If there is no experimental reason to doubt the correctness of these two observations, then the original model should be retained. As usual, more data might clarify the position.

We consider finally the simplification and interpretation of the log model. First we verify that both variables are needed in the model; as with the previous example, we need to use the LRT rather than the Wald test, by dropping each variable in turn from the model:

```
> drop1(vaso.glm5, test = "Chisq")

Single term deletions

Model:
Y ~ lv + lr
        Df Deviance    AIC     LRT   Pr(Chi)
<none>       29.227 35.227
lv       1   48.857 52.857 19.630 9.398e-06
lr       1   47.060 51.060 17.832 2.413e-05
```

The Wald test values for `lv` and `lr` taken from model *vaso.glm5* are $2.778^2 = 7.72$ and $2.482^2 = 6.16$. These are in very poor agreement with the deviance changes of 19.63 and 17.83, respectively.

We note that the coefficients of `lv` and `lr` are very similar, differing by only one third of a standard error of either. This suggests that they could be equated, as Finney (1947) noted, to give a single variable model using log (`vol`× `rate`).

```
> vaso.glm6 <- update(vaso.glm5, . ~ I(lv + lr))
> summary(vaso.glm6)


Coefficients:
             Estimate Std. Error z value Pr(>|z|)
(Intercept)    -3.031      1.280  -2.368  0.01791
I(lv + lr)      4.901      1.740   2.818  0.00484

(Dispersion parameter for binomial family taken to be 1)

    Null deviance: 54.040  on 38  degrees of freedom
Residual deviance: 29.518  on 37  degrees of freedom
AIC: 33.518

Number of Fisher Scoring iterations: 7
```

The deviance increases by 0.27 on 1 df. The corresponding fitted model for the odds ratio is

$$\frac{\hat{p}}{1-\hat{p}} = \exp(\hat{\theta}) = \mathrm{e}^{-3.051}\ (\mathtt{vol} * \mathtt{rate})^{4.928}$$

The power of vol× rate is very close to 5 – only 1/24 of a standard error away. We set it to 5 using an offset:

```
> vaso.glm7 <- update(vaso.glm5, . ~ offset(5 * I(lv + lr)))
> summary(vaso.glm7)

Coefficients:
            Estimate Std. Error z value Pr(>|z|)
(Intercept)  -3.0986     0.4588  -6.753 1.45e-11

(Dispersion parameter for binomial family taken to be 1)

    Null deviance: 29.521  on 38  degrees of freedom
Residual deviance: 29.521  on 38  degrees of freedom
AIC: 31.521

Number of Fisher Scoring iterations: 5
```

The device increases by 0.002, and we note that $e^{-3.1} = 0.04505$, very close to 0.05.

The simpler model

$$\frac{\hat{p}}{1-\hat{p}} = \exp(\hat{\theta}) = 0.05(\texttt{vol} * \texttt{rate})^5$$

can therefore be fitted by extending the offset, and fitting the model with no estimated parameters:

```
> vaso.glm7 <- update(vaso.glm7, . ~ offset(log(0.05) +
+     5 * I(lv + lr)) - 1)
> summary(vaso.glm7)

No Coefficients

(Dispersion parameter for binomial family taken to be 1)

    Null deviance: 29.571  on 39  degrees of freedom
Residual deviance: 29.571  on 39  degrees of freedom
AIC: 29.571

Number of Fisher Scoring iterations: 0
```

The df is now 39, equal to the number of observations.

The fitted model and observed data are now easily graphed in one dimension. First we sort the fitted probabilities into increasing order with lvr to give a smooth graph (Fig. 4.13).

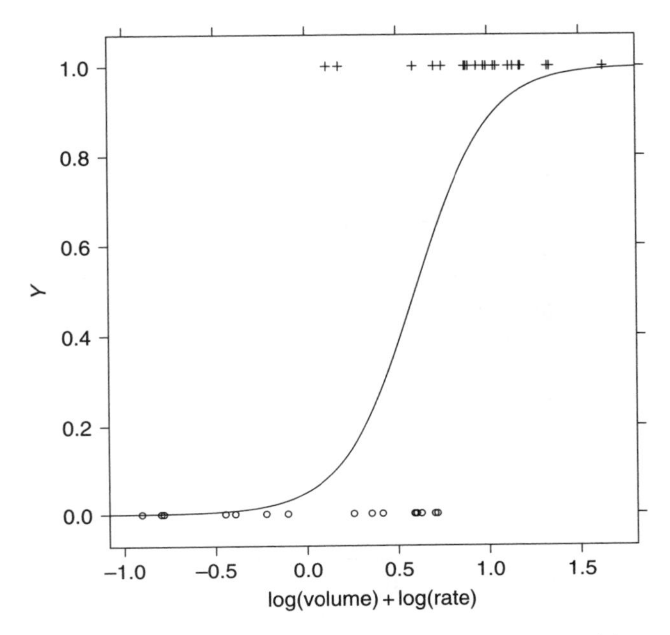

Fig. 4.13. Vasoconstriction response and fitted `lvr` model

```
> print(xyplot(Y ~ I(lv + lr), data = vaso, group = Y,
+      pch = c(1, 3), xlab = "log(Volume)+log(Rate)",
+      subset = !row.names(vaso) %in% c("32"),
+      panel = function(x, y, ...) {
+          panel.xyplot(x, y, ...)
+          panel.curve(exp(eta <- (log(0.05) + 5 * x))/(1 +
+              exp(eta)))
+      }))
```

The two observations 4 and 18 with large residuals are those at the top of the graph at the left-hand end of the row of '+'s. It can be seen that when they are removed from the fit, the slope of the regression will be much steeper as there is very little overlap in `lvr` between the 0 and 1 responses.

Note that the value of 5 for the power of `vol*rate` is chosen only for simplicity. No strong substantive interpretation should be drawn from it, in the absence of a physical model for an integer power.

4.5 Contingency table construction from binary data

In larger samples with binary data and continuous explanatory variables it is frequently possible to construct a contingency table by grouping the explanatory

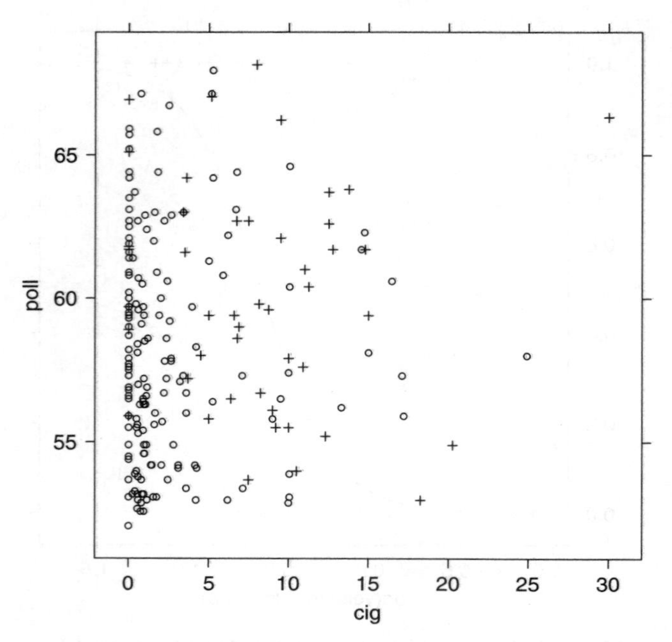

Fig. 4.14. Bronchitis response

variables into ordered categories. This allows more detailed model criticism than with the original binary data, at the expense of some loss of precision in parameter estimation. We illustrate with an example given by Wrigley (1976) of data from a 1974 study of chronic bronchitis in Cardiff conducted by Jones (1975). The data consist of observations on three variables for each of 212 men in a sample of Cardiff enumeration districts. The data are too extensive to be listed here; they are in the file `bronchit` (Fig. 4.14).

The variables are `cig`, the number of cigarettes smoked per day, `poll`, the smoke level in the locality of the respondent's home (obtained by interpolation from 13 air pollution monitoring stations in the city), and `r`, an indicator variable taking the value 1 if the respondent suffered from chronic bronchitis, and 0 if he did not. The presence or absence of chronic bronchitis was determined using a special questionnaire devised by the Medical Research Council; detection of chronic bronchitis using this questionnaire had been found to be almost completely consistent with the clinical diagnosis. (The text in Wrigley refers to `cig` as the total consumption of cigarettes ever smoked in hundreds, but the published data are either total consumption expressed in units of ten thousand, or are current rates of smoking in cigarettes per day. We have assumed the latter.)

Our aim is to develop and fit a statistical model relating `r` to `cig` and `poll`. We begin by graphing the data using + and o as plotting symbols as in the previous example.

```
> data(bronchit, package = "SMIR")

> print(xyplot(poll ~ cig, data = bronchit, group = r,
+     pch = c(1, 3), col = "black"))
```

There are 46 men with bronchitis, a sample proportion of 0.217. Seven of these are non-smokers; the remainder are more common at higher levels of cigarettes and pollution. Two very heavy smokers have different bronchitis outcomes.

We proceed to fit the logit model.

```
> bronchit.glm <- glm(r ~ cig + poll, data = bronchit,
+     family = binomial)
> summary(bronchit.glm)


Coefficients:
              Estimate Std. Error z value Pr(>|z|)
(Intercept) -10.08491    2.95100  -3.417 0.000632
cig           0.21169    0.03813   5.552 2.83e-08
poll          0.13176    0.04895   2.692 0.007113

(Dispersion parameter for binomial family taken to be 1)

    Null deviance: 221.78  on 211  degrees of freedom
Residual deviance: 174.21  on 209  degrees of freedom
AIC: 180.21

Number of Fisher Scoring iterations: 5

> round(sum(resid(bronchit.glm, type = "pearson")^2),
+     1)

[1] 192.6
```

Level curves for the fitted probability of bronchitis are shown in Fig. 4.15.

```
> print(xyplot(poll ~ cig, data = bronchit, group = r,
+     pch = c(1, 3), col = "black", panel = function(x, y, ...)
+         {panel.xyplot(x, y, ...)
+         probs <- c(0.05, 0.1, 0.3, 0.5, 0.7, 0.9,
+             0.95)
+         logits <- log(probs/(1 - probs))
+         line.types <- c(4, 3, 2, 1, 2, 3, 4)
+         for (i in 1:7) panel.curve((logits[i] -
+             coef(bronchit.glm)[1] -
+             coef(bronchit.glm)[2] * x)/coef(bronchit.glm)[3],
+             lty = line.types[i])
```

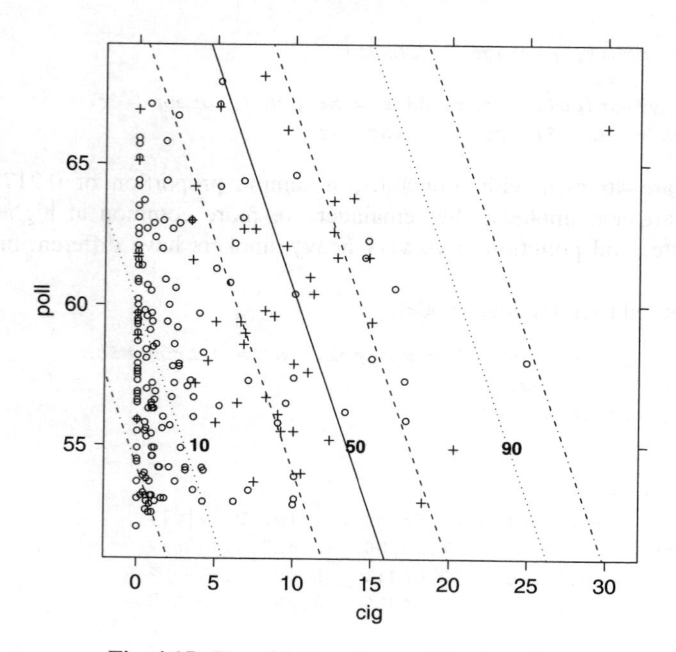

Fig. 4.15. Bronchitis response and level curves

```
+           panel.text(c(4, 14, 24), 55, c("10", "50",
+                "90"), font = 2)
+       }))
```

The deviance of 174.21 with 209 df does not necessarily indicate a good fit. The Pearson X^2 is 192.6. To display the residuals would take many screens of output; instead we display just the large residuals (greater than 2 in absolute value) by generating a logical vector *w* setting TRUE for the large residuals and FALSE for the rest:

```
> bronchit$w <- (abs((rbron <- resid(bronchit.glm,
+      type = "pearson"))) > 2)
> cbind(bronchit, fitted = round(fitted(bronchit.glm),
+      3), residual = round(rbron, 3))[bronchit$w, ]

     r  cig poll    w fitted residual
38   1  3.7 57.2 TRUE  0.146    2.417
84   1  0.0 55.9 TRUE  0.062    3.895
89   1  0.0 58.9 TRUE  0.089    3.197
108  1  0.0 65.1 TRUE  0.181    2.125
121  1  0.0 61.7 TRUE  0.124    2.658
128  1  0.0 59.7 TRUE  0.098    3.033
143  0 24.9 58.0 TRUE  0.944   -4.113
```

```
149 1  7.5 53.7 TRUE  0.194     2.036
155 1  5.0 55.8 TRUE  0.158     2.310
171 1  0.0 61.8 TRUE  0.125     2.641
192 1  4.5 58.0 TRUE  0.184     2.107
```

There are 11 residuals greater than 2 in absolute value. Where in the variable space are these observations?

```
> print(xyplot(poll ~ cig, data = bronchit, group = r,
+       pch = c(1, 3), col = 1, cex = 1.3, cex.axis = 1.3,
+       cex.lab = 1.3, subset = w == TRUE))
```

Of the 10 positive residuals, six occur for bronchitis sufferers at `cig` = 0 and the others for low values of `cig` and `poll`; the negative residual occurs for a non-occurrence for the heavy smoker, observation 143 with `cig` = 24.9 and `poll` = 58. The model is fitting badly for non- and light smokers (Fig. 4.16).

How should we amend the model? The ‘additive’ model assumes that the regression of logit p on `poll` is the same for smokers and non-smokers: is this reasonable? We might hypothesize that the irritating effects of air pollution might be greatly increased by heavy smoking, producing an interaction. Non-smokers might have a response to `poll` quite different from smokers.

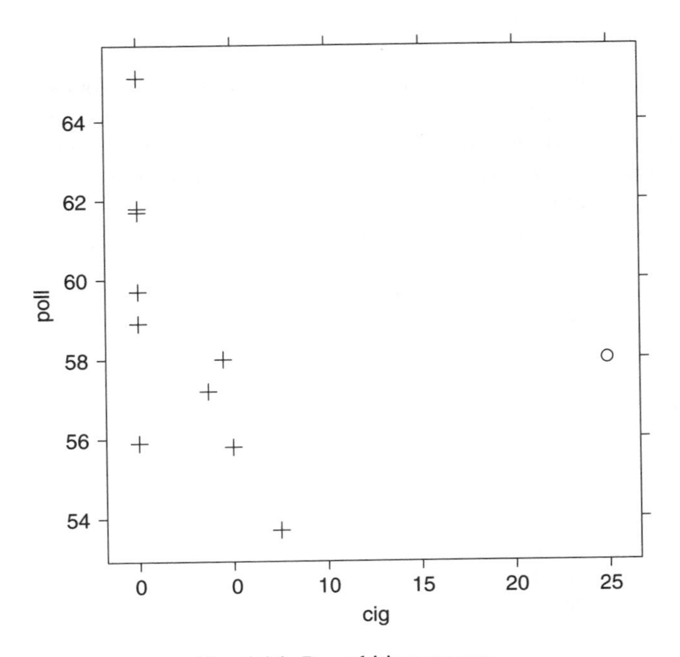

Fig. 4.16. Bronchitis response

This possibility is investigated by fitting models for smokers and nonsmokers separately:

```
> (glm(r ~ poll, data = bronchit, family = binomial,
+      subset = cig == 0))

Coefficients:
(Intercept)         poll
   -12.8284       0.1809

Degrees of Freedom: 53 Total (i.e. Null);  52 Residual
Null Deviance:              41.65
Residual Deviance: 39.38         AIC: 43.38

> (glm(r ~ cig + poll, data = bronchit, family = binomial,
+      subset = cig != 0))

Coefficients:
(Intercept)          cig         poll
    -9.3993       0.2288       0.1172

Degrees of Freedom: 157 Total (i.e. Null);  155 Residual
Null Deviance:              176.6
Residual Deviance: 134         AIC: 140
```

The `poll` slope is not very different from that for the two groups combined.

The total deviance for the two models is 173.38: compared with the deviance of 174.21 for the original model, almost nothing has been gained by a separate regression on `poll` for non-smokers. The poor fit for non-smokers is thus not due to a simple interaction.

We now extend the model to a second-degree response surface:

```
> bronchit.glm1 <- glm(r ~ cig + poll + I(poll^2) +
+     I(cig^2) + cig * poll, data = bronchit, family = binomial)
> summary(bronchit.glm1)

Coefficients:
             Estimate Std. Error z value Pr(>|z|)
(Intercept) -3.696e+01  4.350e+01  -0.850  0.39555
cig          4.620e-01  4.977e-01   0.928  0.35319
poll         1.006e+00  1.449e+00   0.694  0.48739
I(poll^2)   -7.215e-03  1.204e-02  -0.599  0.54910
I(cig^2)    -1.286e-02  4.005e-03  -3.210  0.00133
cig:poll    -3.895e-04  8.627e-03  -0.045  0.96399

(Dispersion parameter for binomial family taken to be 1)
```

```
    Null deviance: 221.78  on 211  degrees of freedom
Residual deviance: 163.72  on 206  degrees of freedom
AIC: 175.72

Number of Fisher Scoring iterations: 5
```

The deviance relative to the full linear model decreases by 10.49 on 3 df, almost all of which is due to the quadratic cigarette term, `I(cig^2)`.

```
> options(digits = 4)
> bronchit.glm2 <- update(bronchit.glm1, . ~ . - I(poll^2) -
+     cig:poll)
> summary(bronchit.glm2)

Coefficients:
             Estimate Std. Error z value Pr(>|z|)
(Intercept) -11.03608    3.06219   -3.60  0.00031
cig           0.44099    0.08299    5.31  1.1e-07
poll          0.13922    0.05038    2.76  0.00572
I(cig^2)     -0.01303    0.00376   -3.46  0.00054

(Dispersion parameter for binomial family taken to be 1)

    Null deviance: 221.78  on 211  degrees of freedom
Residual deviance: 164.08  on 208  degrees of freedom
AIC: 172.1

Number of Fisher Scoring iterations: 5
```

Reduction of the model by omitting `cig:poll` and `I(poll^2)` gives a deviance increase of only 0.36. The rate of increase of bronchitis logit falls off as `cig` increases, and becomes zero at `cig` = 0.4410/2(0.0130) = 17.0. This is a surprising result. We again examine the residuals.

```
> w <- abs(resid(bronchit.glm2, type = "pearson")) >
+     2
> cbind(bronchit, fitted = round(fitted(bronchit.glm2),
+     3), residual = round(resid(bronchit.glm2, type = "pearson"),
+     3))[w, ]

    r cig poll     w fitted residual
38  1 3.7 57.2  TRUE  0.165    2.247
54  1 0.0 66.9 FALSE  0.152    2.366
84  1 0.0 55.9  TRUE  0.037    5.088
89  1 0.0 58.9  TRUE  0.055    4.129
108 1 0.0 65.1  TRUE  0.122    2.682
121 1 0.0 61.7  TRUE  0.080    3.398
128 1 0.0 59.7  TRUE  0.062    3.906
```

```
155 1 5.0 55.8  TRUE  0.200    2.002
171 1 0.0 61.8  TRUE  0.081    3.375
```

Nine of the residuals remain large, and all are for bronchitis sufferers who are non-smokers or low on smoking and/or pollution. The response surface extension has not corrected the poor fit for non-smokers.

This complexity makes us suspect some systematic failure of the model to represent the data. To investigate the appropriateness of the model, we group both `cig` and `poll` into classes and construct a three way contingency table (actually a two way cross-classification of `r` and `n`). The number of bronchitis sufferers in each cell of the table is then modelled by a binomial distribution.

```
> bronchit <- transform(bronchit, fcig = factor(cut(bronchit$cig,
+      c(0, 0.1, 1, 3, 5, 8, Inf), include.lowest = TRUE),
+      ordered = FALSE), fpol = factor(cut(bronchit$pol,
+      c(50, 55, 57.5, 60, 62.5, 65, Inf)), ordered = FALSE))
```

The factors `fcig` and `fpol` take the values 1 to 6 : we have defined six categories for each of the two continuous variables `cig` and `poll`. Non-smokers are the first category of `fcig`. The category boundaries for `cig` are chosen by eye inspection to give roughly equal numbers in each marginal category; alternatively, `quantile` may be used to estimate approximate percentiles. The category boundaries for `poll` are equally spaced with steps of 2.5.

We form two new variates `tr` and `tn` to contain the grouped numbers of successes and sample sizes for the 36 cells.

```
> (tr <- with(bronchit, tapply(r, list(fcig, fpol),
+      sum, na.rm = TRUE)))

         (50,55] (55,57.5] (57.5,60] (60,62.5] (62.5,65] (65,Inf]
[0,0.1]        0         1         2         2         0        2
(0.1,1]        0         0         0         0         0        0
(1,3]          0         0         0         0         0        0
(3,5]          0         2         2         1         2       NA
(5,8]          1         1         3         0         2        2
(8,Inf]        3         5         5         5         3        2
```

```
> tn <- xtabs(~fcig + fpol, data = bronchit)

         (50,55] (55,57.5] (57.5,60] (60,62.5] (62.5,65] (65,Inf]
[0,0.1]        6        11        14        11         7        5
(0.1,1]       17        11         8         3         3        1
(1,3]          9         8         8         4         4        2
(3,5]          6         6         4         2         3        0
(5,8]          3         3         3         2         5        4
(8,Inf]        6        11         7         9         4        2
```

The table of frequencies can be converted to a data.frame very simply:

```
> tbronchit <- data.frame(tn)
> tbronchit$tr <- as.vector(tr)
```

The variates `fcig` and `fpol` each take values 1 to 6, but are of length 36. Cell sample sizes are spread fairly evenly from 2 to 11, with two larger values (14 and 17). One cell has no observations.

```
> bronchit.glm3 <- glm(cbind(tr, (Freq - tr)) ~ fcig +
+     fpol, data = tbronchit, family = binomial)
> summary(bronchit.glm3)


Coefficients:
                Estimate Std. Error z value Pr(>|z|)
(Intercept)       -3.236      0.753   -4.30  1.7e-05
fcig(0.1,1]      -19.614   5196.615 -0.0038   0.9970
fcig(1,3]        -19.585   5053.268 -0.0039   0.9969
fcig(3,5]          1.687      0.668    2.53   0.0115
fcig(5,8]          1.761      0.648    2.72   0.0066
fcig(8,Inf]        2.662      0.573    4.64  3.4e-06
fpol(55,57.5]      0.625      0.730    0.86   0.3917
fpol(57.5,60]      1.781      0.756    2.36   0.0185
fpol(60,62.5]      0.998      0.777    1.29   0.1986
fpol(62.5,65]      1.251      0.801    1.56   0.1182
fpol(65,Inf]       2.353      0.953    2.47   0.0135

(Dispersion parameter for binomial family taken to be 1)

    Null deviance: 98.027  on 34  degrees of freedom
Residual deviance: 14.877  on 24  degrees of freedom
  (1 observation deleted due to missingness)
AIC: 72.23

Number of Fisher Scoring iterations: 19

> round(resid(bronchit.glm3, type = "pearson"), 4)

      1       2       3       4       5       6       7       8
-0.4857  0.0000  0.0000 -1.1293  0.6545  0.7122  0.2948  0.0000
      9      10      11      12      13      14      15      16
 0.0000  0.2662  0.1276 -0.3875 -0.4437 -0.0001 -0.0001 -0.2334
     17      18      19      20      21      22      23      24
 1.4858 -0.3497  0.9595  0.0000  0.0000  0.3940 -1.1145 -0.3010
     25      26      27      28      29      30      31      32
-0.9809  0.0000  0.0000  0.8419 -0.1999  0.3671  0.5280  0.0000
     33      35      36
-0.0001 -0.9068  0.5809
```

The deviance of 14.88 on 24 df shows a good fit for binomial data, and the largest residual is 1.486 for the 17th cell.

```
> drop1(bronchit.glm3, test = "Chisq")

Single term deletions

Model:
cbind(tr, (Freq - tr)) ~ fcig + fpol
       Df Deviance   AIC   LRT Pr(Chi)
<none>        14.9  72.2
fcig    5     87.1 134.5  72.2 3.5e-14
fpol    5     25.0  72.3  10.1   0.073
```

Dropping each main effect in turn we find that cigarette consumption is much more important than smoke level. We assess further the need for `fpol`.

Inspection of the parameter estimates for the main effects model shows that the estimates for `fpol` generally increase with level, though the value for level 3 is considerably higher than for levels 2 or 4. This suggests a linear trend in logit with increasing `fpol`. We treat `fpol` as a variable by converting the factor into a variable using the *unclass* function which replaces the factor with a variable with values 1, . . . , 6.

```
> bronchit.glm4 <- update(bronchit.glm3, . ~ fcig +
+       unclass(fpol))
> anova(bronchit.glm4, bronchit.glm3, test = "Chisq")

Analysis of Deviance Table

Model 1: cbind(tr, (Freq - tr)) ~ fcig + unclass(fpol)
Model 2: cbind(tr, (Freq - tr)) ~ fcig + fpol
  Resid. Df Resid. Dev Df Deviance P(>|Chi|)
1        28      19.47
2        24      14.88  4     4.59      0.33

> round(coef(summary(bronchit.glm4)), 4)

              Estimate Std. Error z value Pr(>|z|)
(Intercept)     -3.048     0.6737 -4.5246   0.0000
fcig(0.1,1]    -19.711  5363.1176 -0.0037   0.9971
fcig(1,3]      -19.652  5241.5361 -0.0037   0.9970
fcig(3,5]        1.516     0.6456  2.3480   0.0189
fcig(5,8]        1.625     0.6213  2.6151   0.0089
fcig(8,Inf]      2.468     0.5447  4.5317   0.0000
unclass(fpol)    0.321     0.1403  2.2869   0.0222

> round(resid(bronchit.glm4), 4)
```

```
      1       2       3       4       5       6       7       8
-0.8719 -0.0001 -0.0001 -1.7687  0.3257  0.3165  0.0974 -0.0001
      9      10      11      12      13      14      15      16
-0.0001  0.2249  0.0716 -0.4048  0.3709 -0.0001 -0.0001  0.5655
     17      18      19      20      21      22      23      24
 2.3869  0.6589  0.3236 -0.0001 -0.0001  0.1753 -1.5823 -0.7074
     25      26      27      28      29      30      31      32
-1.7228 -0.0001 -0.0001  0.5206 -0.6513  0.0641  0.7597  0.0000
     33      35      36
-0.0001 -0.5000  0.9619
```

The deviance change is 4.59 on 4 df, and the largest residual is now 2.387 for the 17th cell. Inspection of the parameter estimates for `fcig` shows a very surprising pattern: for levels 2 and 3 the estimates are -19.7, corresponding to a fitted probability of almost zero, and we see from the tabled values of `tr` that there are no bronchitis sufferers at any air pollution level among smokers of not more than three cigarettes per day, though there are sufferers among non-smokers! It is for this reason that the response surface model failed to fit adequately: it was trying to reproduce both a zero proportion of bronchitis sufferers amongst light smokers and a non-zero proportion amongst non-smokers, as well as a high proportion amongst heavy smokers.

Further inspection of the parameter estimates suggests that the classification of `fcig` is unnecessarily detailed: the fourth and fifth levels, with estimates 1.516 and 1.625, can be collapsed, as can the second and third levels.

```
> tbronchit$mcig <- tbronchit$fcig
> levels(tbronchit$mcig) <- levels(tbronchit$mcig)[c(1,
+      2, 2, 3, 3, 4)]
> bronchit.glm5 <- update(bronchit.glm4, . ~ . - fcig + mcig)
> round(coef(summary(bronchit.glm5)), 4)
```

```
                Estimate Std. Error z value Pr(>|z|)
(Intercept)      -3.0691     0.6620 -4.6361   0.0000
unclass(fpol)     0.3265     0.1361  2.3981   0.0165
mcig(0.1,1]     -19.6768  3746.7096 -0.0053   0.9958
mcig(1,3]         1.5738     0.5325  2.9553   0.0031
mcig(3,5]         2.4735     0.5444  4.5438   0.0000
```

The interpretation of the model is that each step up the air pollution scale increases the logit by 0.327, that is, multiplies the odds on chronic bronchitis by 1.39. Categories 3 and 4 of `mcig`, smokers of 3–8 and more than eight cigarettes a day have their odds on bronchitis multiplied by 4.82 and 11.86, respectively, compared with non-smokers. Light smokers (1–3 a day) have no observed cases of bronchitis.

We finally return to the original data and refit the model.

```
> bronchit$c <- bronchit$fcig
> levels(bronchit$c) <- levels(bronchit$c)[c(1, 2,
+      2, 3, 3, 4)]
> bronchit.glm6 <- update(bronchit.glm1, . ~ c + unclass(poll))
> round(coef(summary(bronchit.glm6)), 4)

               Estimate  Std. Error z value Pr(>|z|)
(Intercept)     -9.3667      3.2821 -2.8539   0.0043
c(0.1,1]       -17.3944   1194.8680 -0.0146   0.9884
c(1,3]           1.5416      0.5296  2.9108   0.0036
c(3,5]           2.4657      0.5422  4.5477   0.0000
unclass(poll)    0.1241      0.0537  2.3121   0.0208
```

The equal intervals for the `unclass(poll)` scale correspond to 2.5 units of `poll`. Multiplying the `unclass(poll)` coefficient and standard error by 2.5 gives 0.3103 and 0.1343, which agree very closely with the `unclass(fpol)` values of 0.3210 and 0.1403. The estimate of -9.367 for `c(0.1,1]` tries to reproduce on the logit scale the observed proportion of zero bronchitis sufferers. The large standard error for this parameter estimate occurs because the likelihood in this parameter is nearly flat at the estimated value and hence the second derivative of the log-likelihood with respect to this parameter evaluated at the estimate is nearly zero. The constant bronchitis rate for smokers of more than eight cigarettes a day corresponds to the property of the earlier quadratic model, of a decreasing rate of linear increase with increasing `cig`.

Five of the previous eleven observations (84, 89, 121, 171), now have residuals larger than 2 in absolute value: these all correspond to non-smokers with bronchitis.

We do not show level curves for this model as they are not informative – for `cig` $= 0$ they are points, for `cig` between 0 and 3 they do not appear on the graph at all, and for `cig` greater than 3 they are lines parallel to the `cig` axis, but only the 30% and 50% lines appear.

It can be verified that the addition of a quadratic term in `poll` adds nothing to the model, and that using log `poll` gives almost the same deviance as `poll`. Further, the probit, CLL and log–log links all give almost the same deviances.

The paradoxical results for non- and light smokers require explanation. Do non-smokers also include those who have given up smoking? Are the non-smokers much older than the smokers? Without more details of the original survey, these questions remain open.

This example shows the value of categorizing continuous explanatory variables when the response variable is binary. In many cases, as in the last two examples in this chapter, such categorization is routinely carried out before the data are analysed. Even without categorization, however, binary responses can become binomial if the number of explanatory variable categories is limited. The next example illustrates both this feature and the use of binary models for prediction.

4.6 The prediction of binary outcomes

The data in the file ghq were published by Silvapulle (1981), and come from a psychiatric study of the relation between psychiatric diagnosis (as case or non-case) and the value of the score on a 12-item General Health Questionnaire (GHQ), for 120 patients attending a general practitioner's surgery. Each patient was administered the GHQ, resulting in a score between 0 and 12, and was subsequently given a full psychiatric examination by a psychiatrist who did not know the patient's GHQ score. The patient was classified by the psychiatrist as either a 'case', requiring psychiatric treatment, or a 'non-case'. The question of interest was whether the GHQ score, which could be obtained from the patient without the need for trained psychiatric staff, could indicate the need for psychiatric treatment. Specifically, given the value of GHQ score for a patient, what can be said about the probability that the patient is a psychiatric case? Sex of the patient is an additional variable. The data are shown in Table 4.3.

Both men and women patients are heavily concentrated at the low end of the GHQ scale, where the overwhelming majority are classified as non-cases. The small number of cases are spread over medium and high values of GHQ.

The ghq data file gives the number of cases c and non-cases nc at each ghq score, and the total number n = c + nc, classified by the factor sex. We first graph the proportion of cases against ghq for each sex separately:

```
> data(ghq, package = "SMIR")

> ghq <- transform(ghq, n = c + nc, p = c/(c + nc),
+       sex = factor(sex, labels = c("men", "women")))
> print(xyplot(p ~ ghq | sex, data = ghq, cex = 1.2))
```

Table 4.3. GHQ score for cases and non-cases

GHQ	Men		Women	
	Cases	Non-cases	Cases	Non-cases
0	0	18	2	42
1	0	8	2	14
2	1	1	4	5
3	0	0	3	1
4	1	0	2	1
5	3	0	3	0
6	0	0	1	0
7	2	0	1	0
8	0	0	3	0
9	0	0	1	0
10	1	0	0	0

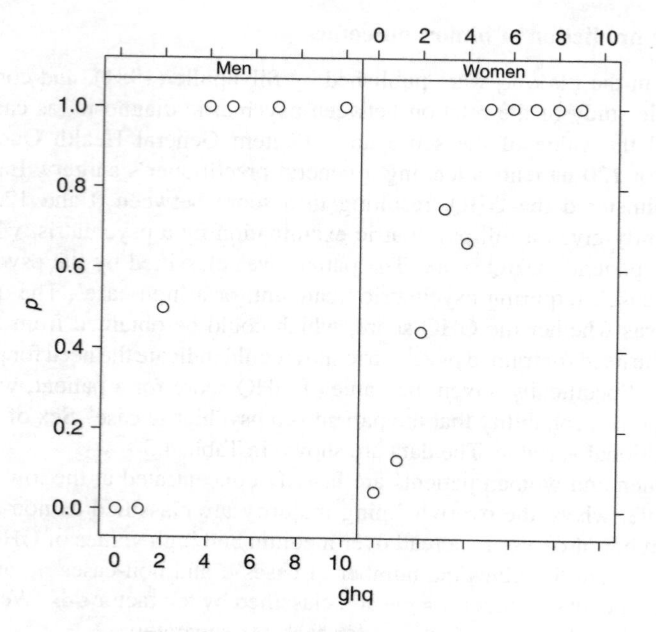

Fig. 4.17. Case response proportion

Males (Fig. 4.17) show a rapid change from non-cases to cases beween GHQ scores of 1 and 4.

Females also show a rapid change, though there is a small proportion of cases even for GHQ zero.

Thus GHQ score is a good indicator of being a psychiatric case. To make this statement more precise we fit a model:

```
> ghq.glm <- glm(cbind(c, nc) ~ sex + ghq, data = ghq,
+       family = binomial)
> summary(ghq.glm)


Coefficients:
              Estimate Std. Error z value Pr(>|z|)
(Intercept)    -4.072       0.976   -4.17  3.0e-05
sexwomen        0.794       0.929    0.85     0.39
ghq             1.433       0.291    4.93  8.3e-07

(Dispersion parameter for binomial family taken to be 1)

    Null deviance: 83.1763  on 16  degrees of freedom
Residual deviance:  4.9419  on 14  degrees of freedom
```

```
AIC: 23.42

Number of Fisher Scoring iterations: 6
```

The standard error of the sex parameter is large compared with the parameter estimate, suggesting that the sex effect is not needed in the model. This is confirmed by a formal test based on the deviance difference:

```
> (ghq.glm1 <- update(ghq.glm, . ~ . - sex))


Coefficients:
(Intercept)          ghq
      -3.45         1.44

Degrees of Freedom: 16 Total (i.e. Null);  15 Residual
Null Deviance:            83.2
Residual Deviance: 5.74         AIC: 22.2

> anova(ghq.glm, ghq.glm1, test = "Chisq")

Analysis of Deviance Table

Model 1: cbind(c, nc) ~ sex + ghq
Model 2: cbind(c, nc) ~ ghq
  Resid. Df Resid. Dev Df Deviance P(>|Chi|)
1        14       4.94
2        15       5.74 -1     -0.80      0.37
```

The change in deviance is quite small compared to χ_1^2. Thus as far as the relation between 'caseness' and GHQ is concerned, we can ignore the sex classification and use a single model for both sexes.

Our interest is in the probability of being a case, not in the logit of this probability, so we transform the fitted values back to the probability scale, and graph in Fig. 4.18 the observed and fitted proportions, pooled over sex:

```
> print(xyplot(p ~ ghq, data = ghq, col = 1, panel = function(x,
+       y, ...) {
+       panel.xyplot(x, y, ...)
+       panel.curve(exp(eta <- (coef(ghq.glm1)[1] +
+           coef(ghq.glm1)[2] * + x))/(1 + exp(eta)), lwd = 1.5)
+ }))
```

The analysis looks complete at this point, but if we examine the data more carefully a strange feature appears. Suppose we check the assumption of no interaction between `sex` and `ghq` by including their interaction in the model:

```
> ghq.glm2 <- update(ghq.glm, . ~ . + sex:ghq)
> summary(ghq.glm2)
```

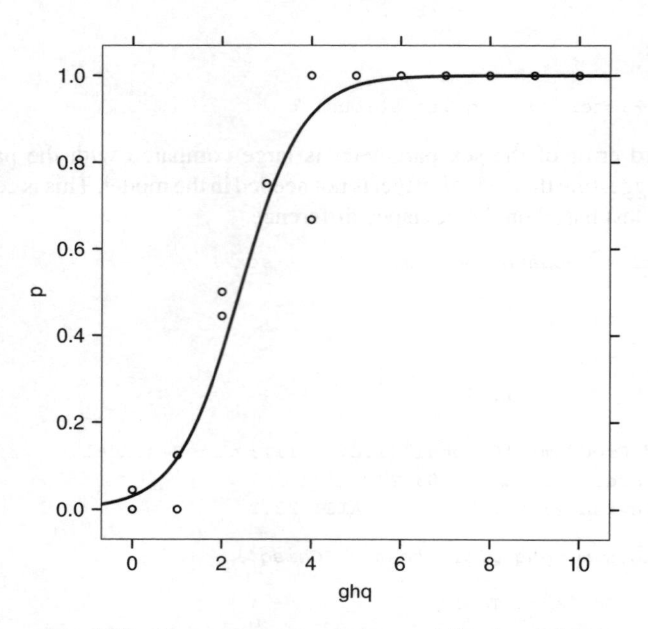

Fig. 4.18. Case response proportions and fitted probabilities

```
Coefficients:
                  Estimate Std. Error z value Pr(>|z|)
(Intercept)          -43.4    21986.7 -0.0020        1
sexwomen              40.4    21986.7  0.0018        1
ghq                   21.7    10993.4  0.0020        1
sexwomen:ghq         -20.4    10993.4 -0.0019        1

(Dispersion parameter for binomial family taken to be 1)

    Null deviance: 83.1763  on 16  degrees of freedom
Residual deviance:  1.5422  on 13  degrees of freedom
AIC: 22.02

Number of Fisher Scoring iterations: 23
```

Very large positive or negative estimates now appear for all the parameters, with even larger standard errors after 23 iterations of the scoring algorithm. Why is this happening?

The fitting of the full interaction model is equivalent to fitting separate `ghq` models for each sex, with different slopes and intercepts. This can also be achieved by fitting each sex separately using the *`subset`* argument.

```
> ghq.women.glm <- glm(cbind(c, nc) ~ ghq, data = ghq,
+      family = binomial, subset = sex == "women")
> summary(ghq.women.glm)


Coefficients:
            Estimate Std. Error z value Pr(>|z|)
(Intercept)   -2.999      0.567   -5.29  1.2e-07
ghq            1.242      0.297    4.18  2.9e-05

(Dispersion parameter for binomial family taken to be 1)

    Null deviance: 48.1983  on 9  degrees of freedom
Residual deviance:  1.5422  on 8  degrees of freedom
AIC: 16.63

Number of Fisher Scoring iterations: 6

> ghq.men.glm <- glm(cbind(c, nc) ~ ghq, data = ghq,
+      family = binomial, subset = sex == "men",
+      control = glm.control(maxit = 30))
> summary(ghq.men.glm)


Coefficients:
            Estimate Std. Error z value Pr(>|z|)
(Intercept)    -49.4    98529.9  -5e-04        1
ghq             24.7    49265.0   5e-04        1

(Dispersion parameter for binomial family taken to be 1)

    Null deviance: 3.4856e+01  on 6  degrees of freedom
Residual deviance: 3.0315e-10  on 5  degrees of freedom
AIC: 5.386

Number of Fisher Scoring iterations: 26
```

At convergence after 26 cycles, the deviance is less than 10^{-9}, and the model fits 'exactly'. The intercept and slope values are such that at GHQ = 2, the fitted logit from the model is zero, corresponding to the observed proportion of 0.5. Any large values of β_0 and β_1 with $\beta_1 > 0$ such that $\beta_0 + 2\beta_1 = 0$ will give an exact fit to the data, in the sense that the fitted values for `ghq` = 0 or 1 and for `ghq` > 2 can be made arbitrarily close to the observed values. Thus the R estimates in this example are determined solely by the convergence criterion in R: if this is made stricter then $\hat{\beta}_0$ and $\hat{\beta}_1$ will be larger, but still with $\hat{\beta}_0 + 2\hat{\beta}_1 = 0$. The likelihood function approaches its maximum of 1 as $\beta_0 \to -\infty$, $\beta_1 \to +\infty$ with $\beta_0 + 2\beta_1 = 0$. Thus the MLEs of β_0 and β_1 'do not exist', that is they are not finite. This causes R no difficulty, however, since the convergence criterion is satisfied for finite values of these parameters.

Since the evidence for interaction is not very strong – the deviance reduction on adding the interaction is 3.40 on 1 df – the simpler main effects model can be retained, and the sex effect can be omitted from this model, as we have already seen.

The psychiatric value of the model involving only GHQ is in prediction: for a new patient (assumed similar to those in the study) with a value of GHQ of 2 (say), what can we say about the probability that this patient is a psychiatric case?

The fitted logit from the model *(ghq.glm1)* at ghq $= 2$ is $-3.454 + 2 \times 1.44 = -0.573$ and the corresponding fitted probability is 0.361. For prediction purposes, however, this value is misleadingly precise. As in Section 4.2, we construct the profile relative likelihoods for the case probability, for the values 0 – 6 of GHQ. We do not show the separate graphs here, only the combined graph.

```
> pseq <- seq(from = 0.001, to = 0.998, by = 0.001)
> rprob <- function(x0, pseq) {
+      logl <- NULL
+      for (p in pseq) {
+           theta <- rep(log(p/(1 - p)), nrow(ghq))
+           xx0 <- ghq$ghq - rep(x0, nrow(ghq))
+           fit <- glm(cbind(c, nc) ~ offset(theta) +
+                xx0 - 1, data = ghq, family = binomial,
+                na.action = na.omit)
+           logl <- c(logl, with(ghq, sum(dbinom(c, (nc +
+                c), fitted(fit), log = TRUE))))
+      }
+      exp(logl - max(logl))
+ }
> print(xyplot(c(0, 1) ~ c(0, 1), type = "n", xlab = "",
+      ylab = "", panel = function(x, y, ...) {
+           for (i in 0:6) panel.curve(rprob(i, x), from = 0.001,
+                to = 0.999)
+           panel.text(c(0.05, 0.14, 0.4, 0.75, 0.94),
+                rep(1, 5), 0:6)
+      }))
```

The profile likelihoods for GHQ $= 0$ and 1 are quite concentrated, while those for GHQ $= 2$, 3, and 4 are quite diffuse. For GHQ 5 and above the likelihoods become very concentrated close to 1. Asymptotic confidence intervals for the case probability p are obtained as in Section 4.2 by interpolating in the deviance. We give the 95% intervals to 3 dp in Table 4.4, without calculation details. A GHQ score of zero means that the patient is very unlikely to be a case (only 2 out of 60 patients with this score were cases) and a score of 5 or more means that the patient

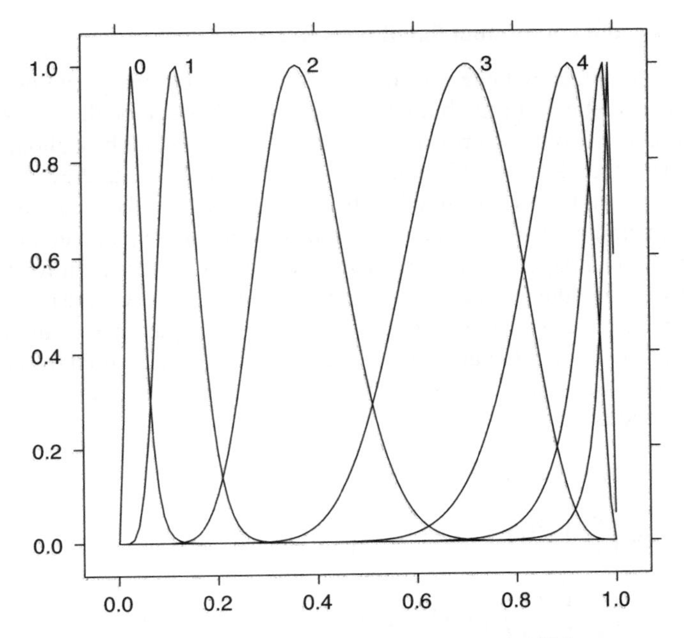

Fig. 4.19. Profile likelihoods for case probabilities

Table 4.4. Predicted case probability and 95% confidence interval

ghq	Lower limit	Predicted value	Upper limit
0	0.010	0.031	0.087
1	0.060	0.118	0.219
2	0.210	0.361	0.544
3	0.437	0.704	0.879
4	0.665	0.909	0.981
5	0.830	0.977	0.997
6	0.921	0.994	1.000

is very likely to be a case (all of the 15 patients with a score of 5 or more were cases). For values from 1 to 4, however, the GHQ score is not a definite indicator of 'caseness'. Note that for a score of 4, the fitted probability is 0.909, strongly suggesting that such a patient would be a case: however, the confidence interval includes values of p near 0.7, and we note that 1 out of 4 patients with this score was not a case (Fig. 4.20).

4.7 Profile and conditional likelihoods in 2 × 2 tables

The 2×2 contingency table arising from a randomized two-group design is an important special case of the binomial generalized linear model, and its analysis is still a matter of some controversy (see, e.g. Yates, 1984). Throughout the book we have used the LRT, or equivalently the profile likelihood, to draw conclusions about a model parameter. The likelihood is maximized over the values of the other (nuisance) parameters in the model, for a fixed value of the parameter of interest. As discussed in Chapter 2, in small samples both the justification for following this procedure and the exact coverage of the confidence interval are not altogether satisfactory, and other methods of eliminating nuisance parameters are of value.

Consider the 2×2 table in which a binary response is classified by a two-level group factor, with r_i successes in n_i trials with success probability p_i $(i = 1, 2)$. The likelihood is

$$\begin{aligned} L(p_1, p_2) &= \Pr[R_1 = r_1, R_2 = r_2] \\ &= \binom{n_1}{r_1} p_1^{r_1}(1-p_1)^{n_1-r_1} \cdot \binom{n_2}{r_2} p_2^{r_2}(1-p_2)^{n_2-r_2} \\ &= \binom{n_1}{r_1}\binom{n_2}{r_2}[p_1/(1-p_1)]^{r_1}[p_2/(1-p_2)]^{r_2}(1-p_1)^{n_1}(1-p_2)^{n_2}. \end{aligned}$$

We now define $r = r_1 + r_2$ and substitute for r_2 in the likelihood:

$$\begin{aligned} L(p_1, p_2) &= \binom{n_1}{r_1}\binom{n_2}{r-r_1}\{[p_1/(1-p_1)]/[p_2/(1-p_2)]\}^{r_1} \\ &\quad \cdot (1-p_1)^{n_1}(1-p_2)^{n_2}[p_2/(1-p_2)]^{r} \\ &= \binom{n_1}{r_1}\binom{n_2}{r-r_1}\theta^{r_1}(1-p_1)^{n_1}p_2^{r}(1-p_2)^{n_2-r}, \end{aligned}$$

where $\theta = \exp(\beta_1)$ is the odds ratio from the logistic model

$$\text{logit}\, p_i = \beta_0 + \beta_1 x_i$$

and x_i is a (0, 1) dummy variable taking the value 1 for the first level of the factor.

Now consider the distribution of the random variable $R = R_1 + R_2$, the total number of successes in both groups. We have

$$\Pr[R = r = r_1 + r_2] = \sum_{u=u_1}^{u=u_2} \Pr[R_1 = u]\Pr[R_2 = r - u]$$

where the limits of the sum are determined by the marginal totals. Since both $0 \le u \le n_1$ and $0 \le r - u \le n_2$, we have $u_1 = \max(0, r - n_2)$, $u_2 = \min(n_1, r)$.

Thus

$$\Pr[R = r] = \sum_{u=u_1}^{u=u_2} \binom{n_1}{u}\binom{n_2}{r-u}\theta^u (1-p_1)^{n_1} p_2^r (1-p_2)^{n_2-r},$$

and

$$\begin{aligned}\Pr[R = r_1, R_2 = r_2 \mid R = r] &= \Pr[R = r_1, R_2 = r_2]/\Pr[R = r] \\ &= \binom{n_1}{r_1}\binom{n_2}{r_2}\theta^{r_1} \Big/ \sum_{u=u_1}^{u=u_2} \binom{n_1}{u}\binom{n_2}{r-u}\theta^u.\end{aligned}$$

This conditional distribution depends only on θ – the nuisance parameter β_0 in the logistic model has been eliminated by the conditioning. Given the values of r_1 and r_2, this (non-central hypergeometric) conditional distribution defines the *conditional likelihood* (CL)

$$\mathrm{CL}(\theta \mid r) = \binom{n_1}{r_1}\binom{n_2}{r_2}\theta^{r_1} \Big/ \sum_{u=u_1}^{u=u_2} \binom{n_1}{u}\binom{n_2}{r-u}\theta^u.$$

This distribution was introduced by Fisher (1935), and the conditional likelihood was developed by Fisher as the appropriate analysis of the 2×2 table from this experimental design. Fisher argued that the marginal number of successes R was an *ancillary* statistic which provides no information about θ but does provide information about its precision, and is therefore an appropriate statistic on which to condition. Since the marginal distribution of R does depend on θ, R is not an exact ancillary and so there is some loss of information about θ in the conditioning, but Plackett (1977) showed that the information about θ in R is small compared to the information in R_1 unless the sample sizes n_1 and n_2 are very small.

The hypergeometric probabilities are easily calculated provided n_1 and r are not too large, since $\mathrm{CL}(\theta)$ is a polynomial in θ of degree u_2. The conditional likelihood in β_1 can then be obtained by log transforming the scale of θ.

The usual application of the conditional likelihood is to Fisher's 'exact' test of the hypothesis $\theta = 1$, which is equivalent to $p_1 = p_2$. The value CL(1) is the (Fisher) 'exact probability' of the observed table, that is, the hypergeometric probability of the observed table given fixed margins and $\theta = 1$. In the usual application of the 'exact' test this probability is assessed by comparison with the probabilities of 'more extreme' tables, and a p-value assigned to the observed table which is the cumulative probability of the observed table and all more extreme ones. Opinions differ over whether this probability should be doubled (corresponding to a two-tailed test) or the probabilities of extreme tables in the other direction should be cumulated as well (Yates, 1984 gives a discussion).

These procedures do not lend themselves easily to interval construction, but the conditional likelihood can be used directly for this purpose.

We illustrate with a small numerical extreme example. In samples of $n_1 = 11$ and $n_2 = 1$, $r_1 = 11$ and $r_2 = 0$ successes are observed. The likelihood is

$$L(p_1, p_2) = p_1^{11}(1 - p_2).$$

Assuming that it is appropriate to condition on the marginal total $r = 11$, the conditional likelihood is

$$\mathrm{CL}(\theta) = \frac{\theta^{11}}{11\theta^{10} + \theta^{11}} = \frac{\theta}{11 + \theta}.$$

This has a maximum of 1 as $\theta \to \infty$. We now compare the conditional likelihood with the profile likelihood. For the parameter β_1 in the logistic model, the profile likelihood is obtained as in Section 4.2.1. Since the full likelihood is maximized at 1 for $\hat{p}_1 = 1$, $\hat{p}_2 = 0$, the profile likelihood in β_1 will also have a maximum of 1 as $\beta_1 \to \infty$. We graph the two likelihoods over a grid $(-1(0.1)10)$ of $\beta_1 = \log \theta$. The values of $\log \theta$ are held in the vector *bseq* (Fig. 4.20).

```
> clog <- data.frame(r = c(11, 0), n = c(11, 1),
+     group = factor(c(1, 2)))
```

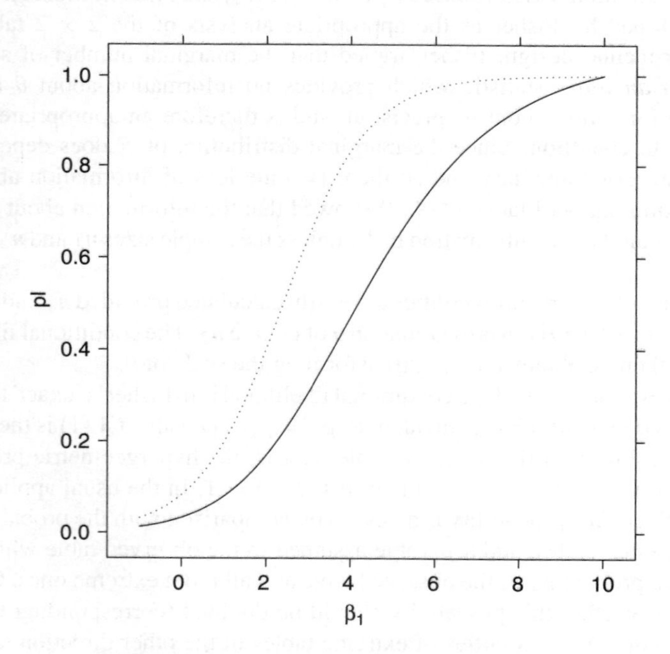

Fig. 4.20. Conditional and profile likelihoods for logodds

```
> beta1 <- seq(-1, 10, by = 0.1)
> pl <- NULL
> for (b in beta1) {
+     z <- b * c(1, 0)
+     logl <- logLik(glm(cbind(r, (n - r)) ~ offset(z),
+         family = binomial, data = clog))
+     pl <- c(pl, logl)
+ }
> pl <- exp(pl - max(pl))
> print(xyplot(pl ~ beta1, type = "l", panel = function(x,
+     y, ...) {
+     panel.xyplot(x, y, ...)
+     panel.curve(exp(x)/(11 + exp(x)), from = min(beta1),
+         to = max(beta1), lty = "dotted")
+ }))
```

Since both deviances go to zero as $\beta_1 \to \infty$, the deviance at $\beta_1 = 0$ is the LRTS for $\beta_1 = 0$ against a general alternative. The conditional LRTS is 4.97, the (unconditional) LRTS is 6.88. Both would lead, assuming the asymptotic distributions can be validly applied, to rejection of the hypothesis of equal probabilities, though at different levels: 0.026 for the CLRT and 0.009 for the LRT. The asymptotic 95% confidence intervals for $\beta_1 = \log\theta$ are quite different: $(0.64, \infty)$ for the CLRTS and $(1.84, \infty)$ for the LRTS.

The conclusion from Fisher's 'exact' test is expressed differently. The hypergeometric probabilities are evaluated under the null hypothesis $\theta = 1$ for the possible tables with the same margins as the observed table. For our example, there are only two possible tables with the same margins: the observed one and the table with $r_1 = 10, r_2 = 1$. The probabilities of these tables are $\theta^{11}/(11\theta^{10} + \theta^{11})$ and $11\theta^{10}/(11\theta^{10} + \theta^{11})$, respectively. At $\theta = 1$, these probabilities are 1/12 and 11/12, respectively. In this case there are no more 'extreme' tables than the observed one, since we already observe all the successes in one group and all the failures in the other. Thus the 'exact' probability of the observed table is 1/12 = 0.0833. This probability would not lead to rejection of the null hypothesis at conventional levels.

The 'exact' test assesses the probability of the observed event under the null hypothesis relative to those of other more extreme events under the same hypothesis. This is philosophically different from the conditional LRT which assesses the probability of the observed event under the null hypothesis relative to that of the same event under other hypotheses.

In this example with very small sample sizes, the conclusions about the log-odds ratio from the two analyses are not equivalent. In larger samples the conditional and profile likelihoods become equivalent, and so the computational difficulties of the conditional likelihood (the evaluation of high degree polynomials in θ) can be avoided in larger samples by computing the profile likelihood. The example above may seem contrived. In fact it is a real data set, taken from the ECMO (extracorporeal membrane oxygenation) study of Bartlett *et al.* (1985). In this study ECMO was compared with the then-standard treatment for neonatal respiratory failure, which had a very high mortality. To reduce mortality in the study for those assigned to the standard treatment, the treatment assignment was a form of 'play-the-winner' rule (Zelen, 1969) in which randomization to the new treatment group has a probability which increases with its success relative to the standard treatment on patients already treated. Under this form of assignment the conditioning argument used to derive the conditional likelihood is invalid, and a different conditional reference set is required for p-value calculations. In the actual study, the randomization was stopped after 10 patients had been treated, and the two additional patients included in the study were assigned non-randomly to the new treatment, but the stopping rule for termination of the trial is unknown. Formally this invalidates all p-value calculations since these require a complete statement of the sample design.

However, the likelihood is invariant to the sample design, so likelihood and Bayes analyses can be applied to the data without a knowledge of the stopping rule. Further discussion of this study, with many interesting comments on philosophical issues (see Royall's comments in particular) can be found in Begg (1990).

4.8 Three-dimensional contingency tables with a binary response

We now consider two examples of three-dimensional contingency tables with a binary response as one dimension.

4.8.1 *Prenatal care and infant mortality*

The data (Bishop *et al.*, 1975; p. 41) come from a study of two pre-natal clinics in a large US city. Mothers attended one of the pre-natal clinics for varying periods before the birth of the baby. The length of time the mother attended the clinic has been categorized for analysis into less than one month or more than one month.

The binary response variable of interest is the infant mortality outcome: infant died within one month, or survived at least one month. There are two explanatory variables: amount of prenatal care, and clinic (A or B); the data are given in Table 4.5. An additional column has been added giving the mortality rates for each group. Clinic B has a considerably higher mortality rate than Clinic A, but the duration of mother's attendance seems to be unrelated to mortality.

If the data for the two clinics are combined, a quite different picture appears (Table 4.6). Now mothers attending for less than one month seem to experience a higher mortality rate for their infants.

This example illustrates clearly two of the hazards of inference from observational data: the danger of aggregating data over important variables, and the inappropriate assumption of causality – that duration of pre-natal care causes or

Table 4.5. Survival experience of child

Mother's attendance	Died within one month	Survived at least one month	Total number	Mortality rate (%)
Clinic A				
<1 month	3	176	179	1.68
>1 month	4	293	297	1.35
Clinic B				
<1 month	17	197	214	7.94
>1 month	2	23	25	8.00

Table 4.6. Aggregate survival experience of child

Mother's attendance	Died within one month	Survived at least one month	Total number	Mortality rate (%)
Both clinics				
< 1 month	20	373	393	5.09
> 1 month	6	316	322	1.86

is responsible for the change in infant mortality. Such a conclusion is insupportable in a non-randomized study: it could be supported only if mothers were randomly assigned to the two classes of pre-natal care. In fact there is a systematic biasing factor visible from the complete data: of the mothers attending Clinic B, only 10% attended for more than one month, while this proportion was 62% for mothers attending Clinic A. Since Clinic B has a much higher mortality rate, when the clinic identification is suppressed the variation in mortality rate appears to be associated with duration of pre-natal care. This is a classic example of *Simpson's paradox* (Simpson, 1951).

It is equally invalid, however, to conclude from the full data that the difference in mortality rates is caused by different quality of prenatal advice and treatment in the two clinics. There is no randomization of mothers to clinics: each clinic serves the mothers in a local area of the city. The different mortality rates probably reflect different infant mortality rates in the subpopulations of the city served by each clinic. Without randomized assignment in such studies it is impossible to draw strong causal conclusions.

The analysis of the data through statistical models can, however, avoid the invalid conclusions produced by aggregating the data for both clinics. We proceed to model the data.

```
> prenatal <- data.frame(death = c(3, 4, 17, 2), total = c(179,
+      297, 214, 25), clinic = factor(c("A", "A", "B",
```

Table 4.7. Analysis of deviance tables

Source	Dev.	df	Source	Dev.	df
Attend	5.61	1	Clinic	17.75	1
Clinic	12.18	1	Attend	0.04	1
Interaction	0.04	1	Interaction	0.04	1

```
+      "B")), attend = factor(c("<1", ">1", "<1", ">1")))
> prenatal.glm <- glm(cbind(death, (total - death)) ~
+      attend + clinic, data = prenatal, family = binomial)
> prenatal.glm1 <- update(prenatal.glm, . ~ clinic +
+      attend)
```

The deviances are 17.83, 12.22, 0.08, 0.04, 0.00. Successive differencing of the deviances leads to two analysis of deviance tables (Table 4.7).
There is a striking interchange of deviance between Clinic and Attendance when their fitting order is reversed, reflecting the substantial correlation between the variables noted above. It is obvious from the table that the Clinic term is essential in the model, and that once it is included, the Attendance term is unnecessary. We refit the `clinic` model and calculate the fitted probabilities from the model: The fitted death rates are 0.0147 for clinic A and 0.0795 for clinic B, which are just the death rates for each clinic pooled over the attendance classification.

Thus statistical modelling of the three-way table has shown that the table can be collapsed over *attend* to give a simple representation, but it can not be collapsed over *clinic* without serious distortion.

4.8.2 *Coronary heart disease*

The dataset `chd` gives the number `r` of men diagnosed as having coronary heart disease (CHD) in an American study of 1329 men (the data are presented and analysed in Ku and Kullback (1974). The serum cholesterol level `chol` and blood pressure `bp` in mm mercury were recorded for each man, and are reported in one of four categories, giving a 4×4 cross-classified table in each cell of which the number `r` of men with CHD and the total number `n` of men examined are given as `r/n`. The data are reproduced in Table 4.8.

The explanatory variables were originally continuous, but have been categorized into ordered categories. We treat them first as unordered factors `bp` and `chol` are defined on the file.

How is the proportion of men suffering from CHD related to blood pressure and serum cholesterol levels? Before fitting any models, we find the proportion `p` suffering from CHD, and graph it against `bp` separately for each value of `chol` joined by lines (Fig. 4.21). For this purpose we use trellis graphics:

```
> data(chd, package = "SMIR")
```

Table 4.8. Blood pressure and cholesterol level in CHD

	bp			
	<127	127–146	147–166	>166
chol				
<200	2/119	3/124	3/50	4/26
200–219	3/88	2/100	0/43	3/23
220–259	8/127	11/220	6/74	6/49
>259	7/74	12/111	11/57	11/44

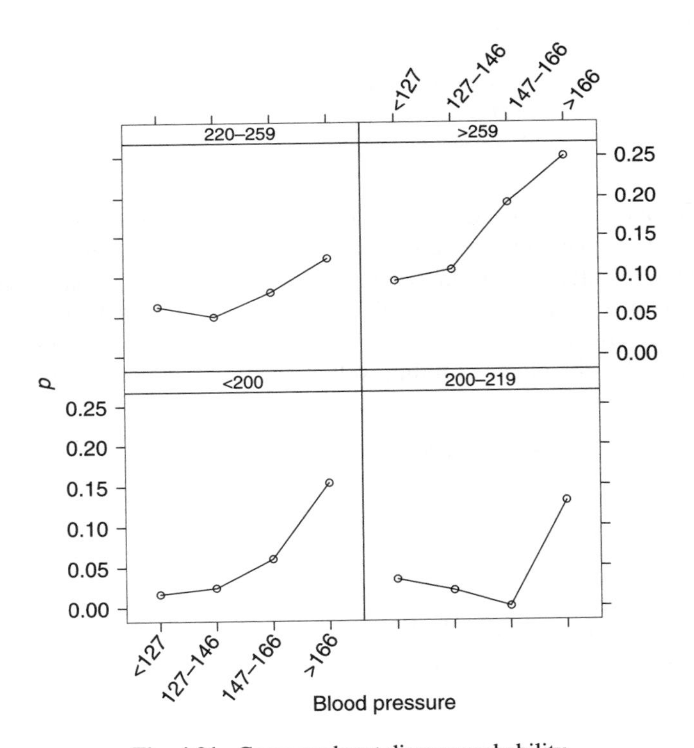

Fig. 4.21. Coronary heart disease probability

```
> chd <- transform(chd, p = chd$r/chd$t, bp = factor(chd$bp,
+     labels = c("<127", "127-146", "147-166", ">166")),
+     chol = factor(chd$chol, labels = c("<200", "200-219",
+         "220-259", ">259")))
> print(xyplot(p ~ bp | chol, data = chd, type = "b",
+     xlab = "blood pressure", scales = list(x = list(rot = 45))))
```

Apart from the zero proportion of cases in one cell, the pattern of increase is fairly consistent at each level of the other factor.

Now we try fitting models on the logit scale:

```
> chd.glm1 <- glm(cbind(r, (t - r)) ~ 1, data = chd,
+      family = binomial)
> chd.glm2 <- update(chd.glm1, . ~ . + bp)
> chd.glm3 <- update(chd.glm2, . ~ . + chol)
> anova(chd.glm1, chd.glm2, chd.glm3)

Analysis of Deviance Table

Model 1: cbind(r, (t - r)) ~ 1
Model 2: cbind(r, (t - r)) ~ bp
Model 3: cbind(r, (t - r)) ~ bp + chol
  Resid. Df Resid. Dev Df Deviance
1        15       58.7
2        12       35.2  3     23.6
3         9        8.1  3     27.1
```

The two analysis of deviance tables are given in Table 4.9 showing `chol` is the more important, but both main effects are necessary in the model. The interaction is not explicitly fitted, but we know that the interaction model is saturated and has a deviance of zero. The value of 8.076, near the mean (or median) of χ^2_9, shows that the main effects model with seven parameters provides a good fit to the data.

To check that the model is fitting well at each sample point, we examine the (Pearson) residuals.

```
> round(resid(chd.glm3, type = "pearson"), 4)

      1       2       3       4       5       6       7       8
-0.8351 -0.2977  0.3331  1.0759  0.5917 -0.2249 -1.3523  0.9708
      9      10      11      12      13      14      15      16
 0.6036  0.0541 -0.0949 -0.5755 -0.3047  0.2260  0.5212 -0.4636
```

For the main effects model the largest residual is −1.35, showing a good fit.

Now we examine the parameter estimates to interpret the model:

```
> round(summary(chd.glm3)$coeff, 4)

            Estimate Std. Error z value Pr(>|z|)
```

Table 4.9. Analysis of deviance tables

Source	Dev.	df	Source	Dev.	df
bp	23.56	3	chol	31.92	3
chol	27.09	3	bp	18.73	3
interaction	8.08	9	interaction	8.08	9

```
(Intercept)   -3.4819      0.3486 -9.9869    0.0000
bp127-146     -0.0415      0.3037 -0.1365    0.8914
bp147-166      0.5324      0.3324  1.6016    0.1093
bp>166         1.2004      0.3269  3.6723    0.0002
chol200-219   -0.2080      0.4664 -0.4459    0.6557
chol220-259    0.5622      0.3508  1.6027    0.1090
chol>259       1.3441      0.3430  3.9191    0.0001
```

The parameter estimates for both factors show a consistent increase as the factor level increases from 1 to 4, apart from level 2 which for each factor is very little different from level 1. This suggests that we equate levels 1 and 2 and then try smoothing the proportions further by using the factor level as though it were a continuous variable. We then fit a regression model using one parameter for each score by redefining the factors to be variates:

```
> chd$ch <- chd$chol
> levels(chd$ch) <- c(1, 1, 2, 3)
> chd$b <- chd$bp
> levels(chd$b) <- c(1, 1, 2, 3)
> chd.glm4 <- update(chd.glm3, . ~ ch + b)
> anova(chd.glm3, chd.glm4)

Analysis of Deviance Table

Model 1: cbind(r, (t - r)) ~ bp + chol
Model 2: cbind(r, (t - r)) ~ ch + b
  Resid. Df Resid. Dev Df Deviance
1         9       8.08
2        11       8.30 -2    -0.22

> round(summary(chd.glm4)$coeff, 4)

            Estimate Std. Error z value Pr(>|z|)
(Intercept)  -3.5923     0.2465 -14.574   0.0000
ch2           0.6478     0.2956   2.191   0.0284
ch3           1.4307     0.2865   4.994   0.0000
b2            0.5554     0.2812   1.975   0.0482
b3            1.2232     0.2743   4.460   0.0000

> round(resid(chd.glm4, type = "pearson"), 4)

      1       2       3       4       5       6       7       8
-0.6748 -0.1795  0.4809  1.2448  0.4237 -0.4209 -1.4364  0.7694
      9      10      11      12      13      14      15      16
 0.6719  0.0001 -0.0908 -0.5710 -0.2447  0.1683  0.5236 -0.4606
```

The deviance increases by 0.22 on 2 df, and the parameter estimates show nearly equal steps. Now we redefine ch and b as variables, and refit:

```
> chd.glm5 <- update(chd.glm4, . ~ unclass(ch) + unclass(b))
> anova(chd.glm5, chd.glm4)

Analysis of Deviance Table

Model 1: cbind(r, (t - r)) ~ unclass(ch) + unclass(b)
Model 2: cbind(r, (t - r)) ~ ch + b
  Resid. Df Resid. Dev Df Deviance
1        13       8.42
2        11       8.30  2     0.12

> round(summary(chd.glm5)$coeff, 4)

             Estimate Std. Error z value Pr(>|z|)
(Intercept)   -4.9625     0.3879 -12.792        0
unclass(ch)    0.7221     0.1427   5.062        0
unclass(b)     0.6069     0.1349   4.500        0

> round(resid(chd.glm5, type = "pearson"), 4)

      1       2       3       4       5       6       7       8
-0.6156 -0.1090  0.4634  1.3436  0.4944 -0.3627 -1.4438  0.8537
      9      10      11      12      13      14      15      16
 0.5806 -0.1070 -0.2897 -0.6257 -0.1752  0.2584  0.4520 -0.3540
```

The deviance increases by 0.12 on 2 df and there are no large residuals. The parameter estimates of 0.61 for b and 0.72 for ch (with standard errors about 0.14) suggest that these could be equated: if $\beta_1 = \beta_2 = \beta$, then the model

$$\theta = \beta_0 + \beta_1 x_1 + \beta_2 x_2$$

reduces to

$$\theta = \beta_0 + \beta(x_1 + x_2),$$

a simple linear regression on the total score $x_1 + x_2$.

```
> chd <- transform(chd, score = unclass(b) + unclass(ch))
> chd.glm6 <- update(chd.glm5, . ~ score)
> anova(chd.glm6, chd.glm5)

Analysis of Deviance Table

Model 1: cbind(r, (t - r)) ~ score
Model 2: cbind(r, (t - r)) ~ unclass(ch) + unclass(b)
  Resid. Df Resid. Dev Df Deviance
```

```
1          14        8.74
2          13        8.42  1      0.32
```

```
> round(summary(chd.glm6)$coeff, 4)

             Estimate Std. Error z value Pr(>|z|)
(Intercept)   -4.9235     0.3784 -13.011        0
score          0.6615     0.0936   7.066        0
```

```
> round(resid(chd.glm6, type = "pearson"), 4)

      1       2       3       4       5       6       7       8
-0.6628 -0.1653  0.3156  1.0686  0.4380 -0.4091 -1.5083  0.6185
      9      10      11      12      13      14      15      16
 0.6573 -0.0170 -0.3526 -0.8146  0.0477  0.5487  0.5536 -0.4133
```

The deviance has increased very little, and the model fits well at all points, with a largest residual of -1.51 for unit 7, where the observed proportion is zero. The regression coefficient of 0.66 is halfway between those for `b` and `ch`.

How do we interpret the model? We have effectively a five-point scale for `score` (2–6), and an increase of 1 point on this scale gives a predicted increase of 0.66 in the log–odds of having coronary heart disease, that is, it multiplies the odds in favour of heart disease by $e^{0.66} = 1.93$, so the odds approximately double for each point up the scale. If the regression coefficient were $\log 2 = 0.6931$, the odds would exactly double. We try smoothing further by fixing the slope β at 0.6931, and estimating only the intercept β_0. This is achieved by defining an offset of `log 2 × score`:

```
> chd$ofs <- log(2) * chd$score
> chd.glm7 <- update(chd.glm6, . ~ offset(ofs))
> anova(chd.glm7, chd.glm6)

Analysis of Deviance Table

Model 1: cbind(r, (t - r)) ~ offset(ofs)
Model 2: cbind(r, (t - r)) ~ score
  Resid. Df Resid. Dev Df Deviance
1        15       8.86
2        14       8.74  1     0.11
```

```
> round(summary(chd.glm7)$coeff, 4)

             Estimate Std. Error z value Pr(>|z|)
(Intercept)    -5.047      0.111  -45.47        0
```

Now we are fitting just the intercept as the `score` slope is specified. The deviance increases by 0.11 on 1 df and the intercept is -5.047. The fitted logits, odds and probabilities at each point on the scale are given in Table 4.10.

The odds values are very nearly 1/40, 1/20, 1/10, 1/5, 2/5. Can we smooth the model any further? If the intercept had been -5.075 instead of -5.047, we would have had exactly the rounded-off odds values above. Since the standard error of $\hat{\beta}_0$ is 0.11, we might as well do the extra smoothing to present easily interpreted values. We finally fit a model with no estimated parameters.

```
> chd$ofs2 <- -5.075 + chd$ofs
> chd.glm8 <- update(chd.glm7, . ~ offset(ofs2) - 1)
> anova(chd.glm8, chd.glm7)

Analysis of Deviance Table

Model 1: cbind(r, (t - r)) ~ offset(ofs2) - 1
Model 2: cbind(r, (t - r)) ~ offset(ofs)
  Resid. Df Resid. Dev Df Deviance
1        16       8.92
2        15       8.86  1     0.07
```

Table 4.10. Fitted logits, odds and probabilities

Score	2	3	4	5	6
Logit	−3.660	−2.967	−2.276	−1.581	−0.888
Odds	0.0257	0.0515	0.1029	0.2058	0.4116
Probability	0.0251	0.0489	0.0933	0.1707	0.2916

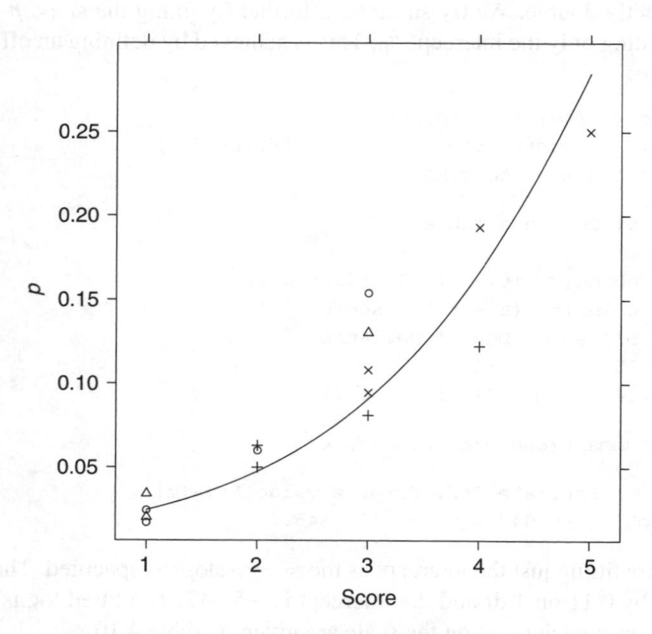

Fig. 4.22. CHD probability against blood pressure

In presenting the data finally we can simplify by defining the score on the range 1–5 instead of 2–6, by subtracting 1. The observed and (fitted) probabilities are graphed in Fig. 4.22 against the `score`, using the previous graph symbols for the levels of `chol`.

We do not quote standard errors for these fitted proportions (or fitted logits) because the purpose of presenting the smoothed fitted values is to give a simple interpretation to the data. Each step up the score scale doubles the odds on CHD, starting from odds of 1/40 at score 1. This gives a very simple and understandable relation between risk and the explanatory variables. A more precise statement might have been possible if the original values of blood pressure and serum cholesterol had been retained: the second example in this chapter showed how the model would then have been used. The categorization of the explanatory variables leads to a *simpler* interpretation, though we note that as far as the relation between CHD and the categorized variables is concerned, the lowest two categories for each variable do not need to be distinguished.

4.9 Multidimensional contingency tables with a binary response

The dataset `byssinosis` gives the number of workers in a survey of the US cotton industry suffering (`yes`) and not suffering (`no`) from the lung disease byssinosis, together with the values of five cross-classifying categorical explanatory variables: the `race`, `sex` and smoking habit `smok` of the worker, the length of employment `emp` in three categories, and the dustiness `dust` of the workplace in three categories. The data were presented and discussed in Higgins and Koch (1977).

The question of primary scientific interest was the relationship between the incidence of byssinosis and the dustiness of the workplace, but smoking habit and length of employment, and hence exposure to dust, are obviously relevant and need to be allowed for.

We first construct the completely cross-classified table of proportions suffering from byssinosis.

```
> data(byssinosis, package = "SMIR")
> byssinosis <- transform(byssinosis, n = yes + no,
+      p = yes/(yes + no))

> yestot <- with(byssinosis, tapply(yes, list(smok = smok,
+      sex = sex, emp = emp, race = race, dust = dust),
+      sum))
> ntot <- with(byssinosis, tapply(n, list(smok = smok,
+      sex = sex, emp = emp, race = race, dust = dust),
+      sum))
> round(with(byssinosis, ftable(yestot/ntot, row.vars = c(1,
+      2, 3), col.vars = c(4, 5))), 4)
```

```
                race  white                       non-white
                dust  most   less   least         most   less   least
smok   sex    emp
smoker male   <10     0.0750 0.0000 0.0077        0.1524 0.0000 0.0122
              10-20   0.2759 0.0196 0.0053        0.2105 0.0000 0.0000
              >20     0.2870 0.0070 0.0237        0.2439 0.0000 0.0000
       female <10     0.0000 0.0106 0.0164        0.0833 0.0136 0.0114
              10-20      NaN 0.0294 0.0208           NaN 0.0000 0.0000
              >20     0.0000 0.0319 0.0168        0.0000    NaN 0.0000
non    male   <10     0.0000 0.0000 0.0000        0.0741 0.0208 0.0081
              10-20   0.2000 0.0588 0.0000        0.1000    NaN 0.0000
              >20     0.0962 0.0000 0.0162        0.1667 0.0000 0.0000
       female <10     0.0000 0.0182 0.0117        0.0400 0.0207 0.0131
              10-20      NaN 0.0000 0.0110           NaN 0.0000 0.0000
              >20     0.0000 0.0158 0.0058           NaN 0.0000 0.0000
```

The table has a large number of zero values, but there is a noticeably high byssinosis rate at dust level 1. The *signif* function gives two significant digits in the smallest value of p. The value 'NaN' standing for 'Not a Number' is returned for results where division is by zero. Before modelling the table we examine marginal tabulations:

```
> for (i in 1:5) {
+     cat("\t", names(byssinosis)[i], "\n")
+     print(round(tapply(byssinosis$yes, byssinosis[,
+       i], sum, na.rm = TRUE)/tapply(byssinosis$n,
+        byssinosis[, i], sum, na.rm = TRUE), 4))
+ }

          dust
  most   less  least
0.1570 0.0138 0.0122
          race
  white non-white
 0.0262    0.0384
         sex
  male female
0.0439 0.0148
         smok
smoker    non
0.0392 0.0179
         emp
   <10  10-20    >20
0.0231 0.0365 0.0384

> with(byssinosis, round(sum(yes)/sum(n), 4))

[1] 0.0304
```

The highest dust level has an outstandingly high byssinosis rate. There appear to be clear racial, sex and smoking differences in byssinosis rate. Length of employment differences are less clear. The overall byssinosis rate for all workers is 3.0%.

As we saw in Section 4.8.1 there is a serious hazard in collapsing contingency tables over important explanatory variables: marginal tabulations may be misleading if explanatory variables are strongly associated. Such associations may be expected in survey data, and occur in this example, as is easily seen if we cross-tabulate by dust and race:

```
> with(byssinosis, round(tapply(yes, list(race = race,
+      dust = dust), sum)/tapply(n, list(race = race,
+      dust = dust), sum), 4))

           dust
race          most   less  least
  white     0.1835 0.0140 0.0129
  non-white 0.1393 0.0135 0.0104
```

Now we see that the higher marginal byssinosis rate for non-whites is spurious: at each dust level the sample rate is in fact lower for non-whites than for whites! The higher marginal rate results from the much higher proportion of non-whites than of whites working in the high dust condition, as can be verified by a further tabulation:

```
> with(byssinosis, tapply(n, list(race = race, dust = dust),
+      sum))

           dust
race        most less least
  white      267  855  2394
  non-white  402  445  1056
```

This is another example of Simpson's paradox.

We now proceed to model the full table. We fit a sequence of logit models, with the main effects and interactions added in an arbitrary (but hierarchical) order. We suppress the output and present only the changes in deviance in the analysis of deviance table.

```
> bys.glm1 <- glm(cbind(yes, no) ~ (dust + emp + smok +
+      sex + race)^3, data = byssinosis, family = binomial)
```

The models with `dust:emp:race` and more interactions take more than 20 iterations to converge, because they are attempting to reproduce the zero proportions with byssinois in the table. The standard errors for all these parameters, and for some of their marginal terms, become very large as a consequence and it is difficult to identify the important variables from the estimates from the full three-way interaction model. We use the analysis of deviance Table 4.11 to reduce the model.

Table 4.11. Analysis of deviance table, byssinosis

	df	Dev.	Resid. df	Resid. dev.
NULL			64	322.53
dust	2	252.11	62	70.42
emp	2	15.10	60	55.32
smok	1	11.43	59	43.88
sex	1	0.30	58	43.59
race	1	0.32	57	43.27
dust:emp	4	4.51	53	38.77
dust:smok	2	4.75	51	34.01
dust:sex	2	4.83	49	29.18
dust:race	2	0.25	47	28.94
emp:smok	2	1.94	45	27.00
emp:sex	2	2.04	43	24.96
emp:race	2	7.33	41	17.63
smok:sex	1	0.26	40	17.37
smok:race	1	1.26	39	16.11
sex:race	1	0.45	38	15.67
dust:emp:smok	4	0.91	34	14.76
dust:emp:sex	3	5.32	31	9.44
dust:emp:race	4	1.52	27	7.92
dust:smok:sex	2	1.47	25	6.44
dust:smok:race	2	0.07	23	6.37
dust:sex:race	2	1.31	21	5.06
emp:smok:sex	2	0.37	19	4.70
emp:smok:race	2	1.22	17	3.48
emp:sex:race	2	0.00	15	3.48
smok:sex:race	1	0.77	14	2.71

The immediately striking feature of this is the very large deviance for `dust`. However, this is fitted first in the model, and it may be substantially less important when fitted after the other main effects. Interactions are much smaller, though some may be important. Note that `dust:emp:sex` has only 3 df because one interaction parameter is not estimable.

The residual deviance for the estimable four- and five-way interactions is very small: the observed value of 2.71 is at the lower 0.06 percentage point of χ^2_{14}. This illustrates the failure of the asymptotic χ^2 distribution for the residual deviance when the table becomes very sparse, as the model more nearly reproduces the observed data. The failure of the residual deviance to follow the χ^2 distribution invalidates the usual test procedure for model selection based on the LRT, since this compares the size of the residual deviance from the current model to the χ^2 percentage point.

However, the same principle of model reduction by backward elimination can be applied to the three-way interaction model instead of the saturated model. We examine carefully the individual components of terms with multiple degrees of freedom, and continue elimination until a point is reached where the removal of any remaining term in the model produces a large change in deviance, so that no further simplification is possible without major distortion.

Inspection of the analysis of deviance table shows that all the three-way interactions are small and can be omitted, with the possible exception of `dust:emp:sex`. We omit all the three-way interactions except this one, and examine its contribution. Typing of large numbers of interactions in specifying the model can be avoided by using the product of the main effects model with itself; R interprets terms like `dust:dust` as main effects.

```
> bys.glm2a <- glm(cbind(yes, no) ~ (dust + emp + smok +
+      sex + race)^2, data = byssinosis, family = binomial)
> anova(bys.glm2a, bys.glm2 <- glm(cbind(yes, no) ~
+      (dust + emp + smok + sex + race)^2 + dust:emp:sex,
+      data = byssinosis, family = binomial))

Analysis of Deviance Table

Model 1: cbind(yes, no) ~ (dust + emp + smok + sex + race)^2
Model 2: cbind(yes, no) ~ (dust + emp + smok + sex + race)^2
                              + dust:emp:sex
  Resid. Df Resid. Dev Df Deviance
1        38      15.67
2        35      10.69  3     4.97
```

The deviance is 10.69 as compared with 15.67 for the model with all two-way interactions. The deviance for the `dust:emp:sex` interaction is thus 4.98 when it is fitted as the first of the three-way interactions, and this interaction can be omitted as well. We drop each of the two-way interactions in turn.

```
> bys.glm3 <- update(bys.glm2, . ~ . - dust:emp:sex)
> drop1(bys.glm3, test = "Chisq")

Single term deletions

Model:
cbind(yes, no) ~ dust + emp + smok + sex + race + dust:emp +
    dust:smok + dust:sex + dust:race + emp:smok + emp:sex +
    emp:race + smok:sex + smok:race + sex:race
          Df Deviance   AIC   LRT Pr(Chi)
<none>          15.7 176.3
dust:emp   4    20.4 173.1   4.8   0.312
dust:smok  2    19.3 176.0   3.7   0.160
dust:sex   2    20.8 177.4   5.1   0.078
```

```
dust:race 2     17.7 174.4   2.0    0.362
emp:smok  2     15.9 172.5   0.2    0.906
emp:sex   2     18.2 174.8   2.5    0.288
emp:race  2     23.4 180.1   7.8    0.021
smok:sex  1     15.9 174.6   0.2    0.639
smok:race 1     17.0 175.7   1.3    0.246
sex:race  1     16.1 174.8   0.4    0.503
```

Inspection of the table suggests that `emp:smok`, `smok:sex`, `smok:race` and `sex:race` can all be omitted, but `emp:race` is large, and some other terms may need to be retained. We refit the model omitting the first four interactions and examine the parameter estimates to see which other terms can be omitted.

```
> bys.glm4 <- update(bys.glm3, . ~ . - smok:(emp +
+       sex + race) - sex:race)
> round(summary(bys.glm4)$coeff, 4)[9:22, ]

                          Estimate Std. Error z value Pr(>|z|)
dustless:emp10-20          -0.7349     1.2978 -0.5663   0.5712
dustleast:emp10-20         -2.0187     1.0089 -2.0009   0.0454
dustless:emp>20            -0.6501     1.0900 -0.5964   0.5509
dustleast:emp>20           -1.0148     0.7770 -1.3061   0.1915
dustless:smoknon            1.0568     0.5684  1.8592   0.0630
dustleast:smoknon           0.4390     0.4466  0.9831   0.3255
dustless:sexfemale          2.1174     0.9450  2.2408   0.0250
dustleast:sexfemale         1.4298     0.7688  1.8597   0.0629
dustless:racenon-white     -0.8853     0.9755 -0.9075   0.3641
dustleast:racenon-white    -1.2218     0.7275 -1.6793   0.0931
emp10-20:sexfemale         -0.3540     0.9242 -0.3830   0.7017
emp>20:sexfemale           -0.9906     0.6140 -1.6135   0.1066
emp10-20:racenon-white     -1.7877     0.7860 -2.2743   0.0229
emp>20:racenon-white       -1.3962     0.6916 -2.0187   0.0435
```

The `emp:sex` parameter estimates are now the least important, and we omit them:

```
> bys.glm5 <- update(bys.glm4, . ~ . - emp:sex)
> round(summary(bys.glm5)$coeff, 4)[9:20, ]

                          Estimate Std. Error z value Pr(>|z|)
dustless:emp10-20          -1.0202     1.0917 -0.9345   0.3501
dustleast:emp10-20         -2.2175     0.8964 -2.4738   0.0134
dustless:emp>20            -1.3831     0.9811 -1.4097   0.1586
dustleast:emp>20           -1.4399     0.7195 -2.0012   0.0454
dustless:smoknon            1.0063     0.5668  1.7753   0.0758
dustleast:smoknon           0.3868     0.4462  0.8669   0.3860
dustless:sexfemale          1.7061     0.8552  1.9949   0.0461
dustleast:sexfemale         1.0322     0.7005  1.4735   0.1406
```

```
dustless:racenon-white    -0.8428     0.9743 -0.8650   0.3870
dustleast:racenon-white   -1.1510     0.7281 -1.5808   0.1139
emp10-20:racenon-white    -1.7849     0.7845 -2.2752   0.0229
emp>20:racenon-white      -1.3886     0.6932 -2.0032   0.0452
```

The `dust:race` terms are now the least important, and we omit them:

```
> bys.glm6 <- update(bys.glm5, . ~ . - dust:race)
> round(summary(bys.glm6)$coeff, 4)[9:18, ]

                        Estimate Std. Error z value Pr(>|z|)
dustless:emp10-20        -0.3495     0.7892 -0.4429   0.6579
dustleast:emp10-20       -1.3713     0.6807 -2.0147   0.0439
dustless:emp>20          -0.6710     0.5876 -1.1421   0.2534
dustleast:emp>20         -0.5744     0.4241 -1.3544   0.1756
dustless:smoknon          0.9957     0.5663  1.7585   0.0787
dustleast:smoknon         0.3782     0.4461  0.8477   0.3966
dustless:sexfemale        1.6852     0.8542  1.9727   0.0485
dustleast:sexfemale       1.0209     0.7001  1.4582   0.1448
emp10-20:racenon-white   -1.1458     0.6042 -1.8964   0.0579
emp>20:racenon-white     -0.7706     0.4896 -1.5738   0.1155
```

The `dust:smok` terms are now the least important and we omit them:

```
> bys.glm7 <- update(bys.glm6, . ~ . - dust:smok)
> round(summary(bys.glm7)$coeff, 4)[9:16, ]

                        Estimate Std. Error z value Pr(>|z|)
dustless:emp10-20        -0.3619     0.7892 -0.4585   0.6466
dustleast:emp10-20       -1.3847     0.6793 -2.0384   0.0415
dustless:emp>20          -0.5624     0.5855 -0.9605   0.3368
dustleast:emp>20         -0.5636     0.4234 -1.3311   0.1832
dustless:sexfemale        1.9095     0.8425  2.2665   0.0234
dustleast:sexfemale       1.0823     0.6929  1.5621   0.1183
emp10-20:racenon-white   -1.1405     0.6021 -1.8941   0.0582
emp>20:racenon-white     -0.7691     0.4884 -1.5749   0.1153
```

Inspection of the parameter estimates suggests that `dust:emp` can be omitted, except perhaps for `dustleast:emp10-20`:

```
> byssinosis$d3e2 <- ifelse((byssinosis$dust == "least") &
+       (byssinosis$emp == "10-20"), 1, 0)
> bys.glm8 <- update(bys.glm7, . ~ . - dust:emp + d3e2)
> anova(bys.glm8, bys.glm7)

Analysis of Deviance Table

Model 1: cbind(yes, no) ~ dust + emp + smok + sex + race + d3e2
    + dust:sex + emp:race
```

```
Model 2: cbind(yes, no) ~ dust + emp + smok + sex + race +
    dust:emp + dust:sex + emp:race
  Resid. Df Resid. Dev Df Deviance
1        52      29.88
2        49      27.72  3     2.16
```

```
> bys.glm9 <- update(bys.glm8, . ~ . - d3e2)
> anova(bys.glm9, bys.glm8)

Analysis of Deviance Table

Model 1: cbind(yes, no) ~ dust + emp + smok + sex + race +
    dust:sex + emp:race
Model 2: cbind(yes, no) ~ dust + emp + smok + sex + race + d3e2
    + dust:sex + emp:race
  Resid. Df Resid. Dev Df Deviance
1        53       33.2
2        52       29.9  1      3.4
```

```
> round(summary(bys.glm9)$coeff, 4)[9:12, ]

                          Estimate Std. Error z value Pr(>|z|)
dustless:sexfemale          2.0450     0.8341   2.452   0.0142
dustleast:sexfemale         1.2888     0.6826   1.888   0.0590
emp10-20:racenon-white     -0.7427     0.5577  -1.332   0.1829
emp>20:racenon-white       -0.5947     0.4740  -1.255   0.2096
```

The deviance change on omitting `dustleast:emp10-20` is only 3.36. The `emp.race` interactions are now quite small and can be omitted:

```
> bys.glm10 <- update(bys.glm9, . ~ . - emp:race)
> anova(bys.glm10, bys.glm9)

Analysis of Deviance Table

Model 1: cbind(yes, no) ~ dust + emp + smok + sex + race +
    dust:sex
Model 2: cbind(yes, no) ~ dust + emp + smok + sex + race +
    dust:sex + emp:race
  Resid. Df Resid. Dev Df Deviance
1        55       35.7
2        53       33.2  2      2.5
```

```
> summary(bys.glm10)


Coefficients:
                    Estimate Std. Error z value Pr(>|z|)
(Intercept)           -1.847      0.234   -7.89  2.9e-15
```

```
dustless                  -3.244       0.518   -6.26  3.8e-10
dustleast                 -2.849       0.248  -11.48  < 2e-16
emp10-20                   0.503       0.262    1.92  0.05447
emp>20                     0.698       0.216    3.23  0.00122
smoknon                   -0.658       0.195   -3.38  0.00072
sexfemale                 -1.001       0.611   -1.64  0.10129
racenon-white              0.113       0.207    0.55  0.58532
dustless:sexfemale         2.002       0.834    2.40  0.01634
dustleast:sexfemale        1.261       0.682    1.85  0.06462

(Dispersion parameter for binomial family taken to be 1)

    Null deviance: 322.527  on 64  degrees of freedom
Residual deviance:  35.721  on 55  degrees of freedom
AIC: 162.4

Number of Fisher Scoring iterations: 5
```

The `dust:sex` interaction is large and is retained, as are the marginal main effects of `dust` and `sex`. The `race` variable can clearly be omitted, but the `smok` and `emp` main effects are large and must be retained:

```
> bys.glm11 <- update(bys.glm10, . ~ . - race)
> anova(bys.glm11, bys.glm10)

Analysis of Deviance Table

Model 1: cbind(yes, no) ~ dust + emp + smok + sex + dust:sex
Model 2: cbind(yes, no) ~ dust + emp + smok + sex + race +
    dust:sex
  Resid. Df Resid. Dev Df Deviance
1        56       36.0
2        55       35.7  1      0.3

> summary(bys.glm11)


Coefficients:
                     Estimate Std. Error z value Pr(>|z|)
(Intercept)            -1.753      0.156  -11.21  < 2e-16
dustless               -3.277      0.514   -6.37  1.9e-10
dustleast              -2.878      0.242  -11.88  < 2e-16
emp10-20                0.464      0.251    1.85  0.06479
emp>20                  0.637      0.184    3.47  0.00053
smoknon                -0.658      0.195   -3.38  0.00072
sexfemale              -0.999      0.611   -1.64  0.10182
dustless:sexfemale      2.006      0.834    2.41  0.01612
dustleast:sexfemale     1.258      0.682    1.84  0.06530
```

```
(Dispersion parameter for binomial family taken to be 1)

    Null deviance: 322.527  on 64  degrees of freedom
Residual deviance:  36.019  on 56  degrees of freedom
AIC: 160.7

Number of Fisher Scoring iterations: 5
```

The two interaction terms differ by 0.75, or about one standard error, which suggests that they can be equated; the same is true for the main effect parameters. Equating `dustless` and `dustleast` is equivalent to collapsing these two categories into one.

```
> byssinosis$d2 <- byssinosis$dust
> levels(byssinosis$d2) <- c(1, 2, 2)
> bys.glm12 <- update(bys.glm11, . ~ . - dust:sex +
+      d2 + d2:sex)
> anova(bys.glm12, bys.glm11)
Analysis of Deviance Table

Model 1: cbind(yes, no) ~ dust + emp + smok + sex + d2 + sex:d2
Model 2: cbind(yes, no) ~ dust + emp + smok + sex + dust:sex
  Resid. Df Resid. Dev Df Deviance
1        57       37.4
2        56       36.0  1      1.4

> summary(bys.glm12)


Coefficients: (1 not defined because of singularities)
               Estimate Std. Error z value Pr(>|z|)
(Intercept)      -1.755      0.157  -11.21  < 2e-16
dustless         -2.873      0.317   -9.07  < 2e-16
dustleast        -2.973      0.236  -12.59  < 2e-16
emp10-20          0.464      0.251    1.85  0.06490
emp>20            0.641      0.184    3.49  0.00048
smoknon          -0.658      0.195   -3.38  0.00072
sexfemale        -0.998      0.611   -1.63  0.10228
d22                  NA         NA      NA       NA
sexfemale:d22     1.447      0.663    2.18  0.02903

(Dispersion parameter for binomial family taken to be 1)

    Null deviance: 322.527  on 64  degrees of freedom
Residual deviance:  37.443  on 57  degrees of freedom
AIC: 160.1
```

```
Number of Fisher Scoring iterations: 5
```

The aliased term d2 is included in the model to allow a direct comparison between the parameter estimates for d2:sex and dust:sex.

The deviance increase is very small, and the parameter estimates for dustless and dustleast are now almost equal. We remove the three-level dust factor and retain the two-level d2 factor.

```
> bys.glm13 <- update(bys.glm12, . ~ . - dust + d2)
> anova(bys.glm13, bys.glm12)

Analysis of Deviance Table

Model 1: cbind(yes, no) ~ emp + smok + sex + d2 + sex:d2
Model 2: cbind(yes, no) ~ dust + emp + smok + sex + d2 + sex:d2
  Resid. Df Resid. Dev Df Deviance
1        58       37.6
2        57       37.4  1      0.1

> summary(bys.glm13)


Coefficients:
              Estimate Std. Error z value Pr(>|z|)
(Intercept)     -1.754      0.156  -11.21  < 2e-16
emp10-20         0.462      0.251    1.84  0.06589
emp>20           0.641      0.184    3.49  0.00049
smoknon         -0.658      0.195   -3.38  0.00071
sexfemale       -0.998      0.611   -1.63  0.10217
d22             -2.951      0.227  -13.01  < 2e-16
sexfemale:d22    1.459      0.662    2.21  0.02742

(Dispersion parameter for binomial family taken to be 1)

    Null deviance: 322.527  on 64  degrees of freedom
Residual deviance:  37.564  on 58  degrees of freedom
AIC: 158.2

Number of Fisher Scoring iterations: 5
```

A similar recoding of emp can be used to equate levels 2 and 3:

```
> byssinosis$e2 <- byssinosis$emp
> levels(byssinosis$e2) <- c(1, 2, 2)
> bys.glm14 <- update(bys.glm13, . ~ . - emp + e2)
> anova(bys.glm14, bys.glm13)

Analysis of Deviance Table
```

```
Model 1: cbind(yes, no) ~ smok + sex + d2 + e2 + sex:d2
Model 2: cbind(yes, no) ~ emp + smok + sex + d2 + sex:d2
  Resid. Df Resid. Dev Df Deviance
1       59       38.1
2       58       37.6  1      0.5
> summary(bys.glm14)

Coefficients:
               Estimate Std. Error z value Pr(>|z|)
(Intercept)      -1.758      0.157  -11.23  < 2e-16
smoknon          -0.648      0.194   -3.34  0.00083
sexfemale        -0.993      0.611   -1.63  0.10382
d22              -2.949      0.227  -13.00  < 2e-16
e22               0.592      0.172    3.44  0.00059
sexfemale:d22     1.453      0.661    2.20  0.02807

(Dispersion parameter for binomial family taken to be 1)

    Null deviance: 322.527  on 64  degrees of freedom
Residual deviance:  38.105  on 59  degrees of freedom
AIC: 156.8

Number of Fisher Scoring iterations: 5
```

Before proceeding to interpret this model, a word of warning is in order. We equated levels 2 and 3 of emp because the difference between their parameter estimates was not large compared with the individual standard errors. Close examination of these parameter estimates suggests two other possible interpretations, however:

(1) since emp10-20 does not differ significantly from zero, levels 1 and 2 of emp could instead be equated;
(2) since the values 0, 0.462 and 0.641 are roughly linear, we could replace the factor emp by a variate taking the same values.

We illustrate the latter model:

```
> byssinosis$line <- as.numeric(unclass(byssinosis$emp))
> bys.glm15 <- update(bys.glm14, . ~ . - e2 + line)
> anova(bys.glm14, bys.glm15)

Analysis of Deviance Table

Model 1: cbind(yes, no) ~ smok + sex + d2 + e2 + sex:d2
Model 2: cbind(yes, no) ~ smok + sex + d2 + line + sex:d2
  Resid. Df Resid. Dev Df Deviance
```

```
1        59       38.1
2        59       37.9  0       0.2

> summary(bys.glm15)


Coefficients:
              Estimate Std. Error z value Pr(>|z|)
(Intercept)     -2.047      0.218   -9.38  < 2e-16
smoknon         -0.665      0.194   -3.42  0.00062
sexfemale       -1.021      0.609   -1.68  0.09389
d22             -2.950      0.227  -13.00  < 2e-16
line             0.319      0.091    3.50  0.00047
sexfemale:d22    1.478      0.661    2.24  0.02535

(Dispersion parameter for binomial family taken to be 1)

    Null deviance: 322.53  on 64  degrees of freedom
Residual deviance:  37.93  on 59  degrees of freedom
AIC: 156.6

Number of Fisher Scoring iterations: 5
```

The deviance is 37.93, slightly smaller than the value 38.11 when levels 2 and 3 of `emp` are equated. We cannot choose between these models: the data do not provide enough evidence to discriminate between the model in which risk is the same for the two longer-term exposure categories and that in which risk increases steadily with exposure. To make such a discrimination the actual number of years worked in the industry for each individual would be necessary, or at least a finer classification of `emp`. The parameter estimates for variables other than `emp` are very little affected by this ambiguity; we base our interpretation on the last `line` model.

The odds on having byssinosis for smokers are greater than for non-smokers by a factor of $e^{0.665} = 1.94$, or approximately 2, and increase steadily (in this model) with increasing duration of employment in the industry. Interpretation of other features of the model is simplified if we tabulate the observed and fitted proportions by the factors `smok`, `emp`, `sex` and `d2`. `race` can be omitted. We suppress the individual tables.

```
> fp <- fitted(bys.glm15)
> round(with(byssinosis, ftable(tapply(fp, list(smok = smok,
+     emp = emp, dust = d2, sex = sex), mean), row.vars = c(1,
+     2), col.vars = c(3, 4))), 3)
> round(with(byssinosis, ftable(tapply(yes, list(smok = smok,
+     emp = emp, dust = d2, sex = sex), sum)/tapply((yes +
+     no), list(smok = smok, emp = emp, dust = d2,
+     sex = sex), sum), row.vars = c(1, 2), col.vars = c(3,
+     4))), 3)
```

Table 4.12. Fitted and observed byssinosis proportions, and (sample sizes)

	Dust less-least		Dust most	
Emp	Female	Male	Female	Male
Non-smokers	0.007	0.005	0.032	0.084
<10	0.015	0.006	0.034	0.062
	(676)	(340)	(29)	(97)
10–20	0.010	0.007	0.043	0.112
	0.008	0.012	—	0.150
	(129)	(82)	(0)	(20)
>20	0.014	0.009	0.059	0.147
	0.009	0.012	0.000	0.114
	(537)	(248)	(2)	(70)
Smokers	0.014	0.009	0.060	0.151
<10	0.013	0.007	0.069	0.137
	(687)	(667)	(29)	(204)
10–20	0.020	0.013	0.081	0.196
	0.022	0.007	—	0.239
	(137)	(277)	(0)	(67)
>20	0.027	0.017	0.108	0.251
	0.022	0.019	0.000	0.275
	(275)	(695)	(2)	(149)

The sample sizes can also be tabulated in the same way.

```
> with(byssinosis, ftable(tapply(n, list(smok = smok,
+     emp = emp, dust = d2, sex = sex), sum), row.vars = c(1,
+     2)))
```

The fitted and observed proportions and sample sizes are shown in Table 4.12 to 3 dp (in vertical order fitted, observed, n) so that the fitted proportion with byssinosis increases across and down the table. Two cells have no observations.

It is now easily seen that there is a striking increase in the incidence of byssinosis for men in the dustiest working conditions. Another interesting feature is that smokers with less than 10 years' employment in the industry have the same incidence rates as non-smokers with more than 20 years' employment. The incidence of byssinosis is very high – one in four – amongst smoking men with more than 20 years' employment who are working in the dustiest conditions. Amongst smoking women under the same conditions the byssinois rate is 11%.

5

Multinomial and Poisson response data

5.1 The Poisson distribution

The dataset `faults` gives the number `n` of faults in 32 rolls of material of length ℓ metres; the data come from Bissell (1972), and are reproduced in Table 5.1.

The number of faults is a non-negative integer, and is naturally modelled by the *Poisson* distribution, the standard model for count data.

The number of faults N in a roll of material is modelled by

$$\Pr(N = n|\mu) = \mathrm{e}^{-\mu}\mu^n/n!, \quad \mu > 0, \quad n = 0, 1, 2, \ldots,$$

where μ is the mean number of faults, and is related to the linear predictor $\eta = \boldsymbol{\beta}'\mathbf{x}$ through

$$g(\mu) = \eta,$$

where the link function g is generally taken to be the (default) log, guaranteeing positive fitted values:

$$\log \mu = \eta = \boldsymbol{\beta}'\mathbf{x}$$

though the identity link is sometimes used. The fitting of the Poisson model in R is achieved by setting the *`family`* argument as `poisson` and the appropriate link and model specifications in the *`glm`* function.

For the fault data, the logical model would have the mean number of faults proportional to the length of the roll, so that $\mu = c\ell$ for some constant c. Using the log link gives $\eta = \log \mu = c' + \log \ell$; under the proportionality model the coefficient of $\log \ell$ should be 1.

We fit the regression on $\log \ell$ with log link:

Table 5.1. Fault data

ℓ	551	651	832	375	715	868	271	630	491	372	645
n	6	4	17	9	14	8	5	7	7	7	6
ℓ	441	895	458	642	492	543	842	905	542	522	122
n	8	28	4	10	4	8	9	23	9	6	1
ℓ	657	170	738	371	735	749	495	716	952	417	
n	9	4	9	14	17	10	7	3	9	2	

```
> faults <- transform(faults, ll = log(l))
> faults.glm <- glm(n ~ ll, data = faults, family = poisson)
> round(coef(summary(faults.glm)), 4)
            Estimate Std. Error z value Pr(>|z|)
(Intercept)   -4.173      1.135   -3.68    2e-04
ll             0.997      0.176    5.67    0e+00
```

The coefficient of $\log \ell$ is almost unity, and the proportionality model is well supported. We fit it explicitly by constraining the coefficient to be 1, using an offset:

```
> faults.glm1 <- glm(n ~ offset(ll), data = faults,
+     family = poisson)
> anova(faults.glm, faults.glm1, test = "Chisq")

Analysis of Deviance Table

Model 1: n ~ ll
Model 2: n ~ offset(ll)
  Resid. Df Resid. Dev Df Deviance P(>|Chi|)
1        30       64.5
2        31       64.5 -1 -0.00031       1.0
```

The deviance change is zero to three *dp*. The fitted log-linear model is shown as the solid straight line with the data on the original scales of mean and length in Fig. 5.1.

```
> xval <- with(faults, seq(min(l), max(l), len = 100))
> fval <- exp(coef(faults.glm1)[1] + log(xval))
> fval.sd <- sqrt(fval)
> tval <- qt(0.975, faults.glm1$df.residual)
> lb <- fval - tval * fval.sd
> ub <- fval + tval * fval.sd
> print(xyplot(fval + lb + ub ~ xval, col = "black",
+     type = "l", xlab = "length", ylab = "number",
+     lty = c(1, 3, 3), ylim = c(0, 30), panel = function(...) {
+         panel.xyplot(...)
+         lpoints(faults$l, faults$n, col = "black")
+     }))
```

There is no sign of non-linearity in the plot, but the deviance from the fitted model (i.e. the LRTS for the linear model relative to the saturated model) is very large (64.54) compared with its degrees of freedom (31), and examination of the residuals shows several which are very large. We have clear evidence of *overdispersion*: the most plausible explanation is that there are other factors varying over the data which are affecting the chance of a fault. Since these factors are not included in the model, they inflate the variability of the

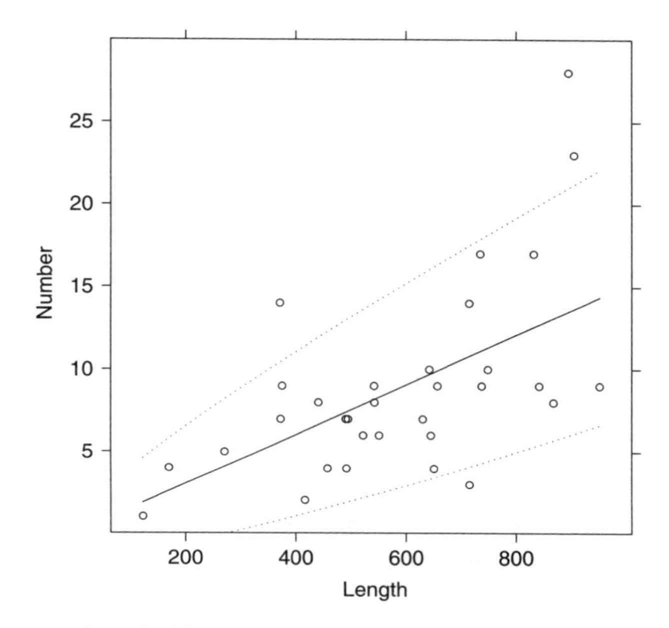

Fig. 5.1. Observed and fitted values with 2 SD bounds

data beyond the Poisson variance, and appear in the large residuals. We can illustrate this further by graphing the fitted Poisson variance, in the form of 2 SD bounds above and below the mean, also shown in Fig. 5.1 by dashed curves:

Four points fall outside the 2 SD limits, two of them far outside. Although the log-linear model appears appropriate for the relationship between length of roll and number of faults, the Poisson distribution understates the extent of variability in the data. The precision of the regression coefficient estimates is overstated from the Poisson model, and the estimates themselves may be biased. We deal with these problems in detail in Chapter 8, and postpone further discussion of this example until then.

5.2 Cross-classified counts

The dataset `claims` gives the number of policyholders `n` of an insurance company who were 'exposed to risk', and the number `c` of car insurance claims made in the third quarter of 1973 by these policyholders, arranged as a contingency table cross-classified by three four-level factors: `dist`, the district in which the policyholder lived (1: rural, 2: small towns, 3: large towns, 4: major cities), `car`, the engine capacity of the car (1: $<$ 1 litre, 2: 1–1.5 litres, 3: 1.5–2 litres, 4: $>$ 2 litres), and

Table 5.2. Claims data showing the number of claims `c` and policyholders `n` by `car`, `dist` and `age`

dist	car	age							
		1		2		3		4	
		n	c	n	c	n	c	n	c
1	1	197	38	264	35	246	20	1680	156
	2	284	63	536	84	696	89	3582	400
	3	133	19	286	52	355	74	1640	233
	4	24	4	71	18	99	19	452	77
2	1	85	22	139	19	151	22	931	87
	2	149	25	313	51	419	49	2443	290
	3	66	14	175	46	221	39	1110	143
	4	9	4	48	15	72	12	322	53
3	1	35	5	73	11	89	10	648	67
	2	53	10	155	24	240	37	1635	187
	3	24	8	78	19	121	24	692	101
	4	7	3	29	2	43	8	245	37
4	1	20	2	33	5	40	4	316	36
	2	31	7	81	10	122	22	724	102
	3	18	5	39	7	68	16	344	63
	4	3	0	16	6	25	8	114	33

`age`, the age of the policyholder (1: < 25, 2: 25–29, 3: 30–35, 4: > 35). The data are shown in Table 5.2, adapted from Baxter *et al.* (1980).

```
> data(claims, data = "SMIR")
```

We wish to model the relation between the frequency of claims and the explanatory variables. At first sight the appropriate probability model appears to be binomial: we might reasonably suppose that each policyholder has a constant probability p of making a claim, so that the number C would be binomial $b(n, p)$, with p modelled as a logistic function of the explanatory variables. However, it is possible (though unlikely) for any policyholder to submit several claims during the quarter, so the binomial distribution may not be appropriate: we examine the use of the binomial distribution at the end of this section.

In the i-th cell of the contingency table, there are n_i policyholders. We model the number of claims C_{ij} made by the j-th policyholder in the i-th cell as a Poisson distribution with mean μ_i, and for each cell, the C_{ij} are assumed independent. The total number of claims

$$C_i = \sum_{j=1}^{n_i} C_{ij}$$

submitted by all the policyholders in the i-th cell is the sum of n_i independent Poisson variables each with mean μ_i, and has a Poisson distribution with mean $\theta_i = n_i\mu_i$.

If the c_{ij} were observed, the model $\log \mu_i = \boldsymbol{\beta}'\mathbf{x}_i$ could be fitted directly. Since only c_i is observable, we can fit a model only to θ_i. We have

$$\begin{aligned}\log \theta_i &= \log n_i + \log \mu_i \\ &= \log n_i + \boldsymbol{\beta}'\mathbf{x}_i.\end{aligned}$$

Thus to specify the model correctly we must include the term $\log n_i$ as an explanatory variable with a coefficient of 1, that is, $\log n_i$ must be taken as an offset for the model. We suppress the output.

```
> claims <- transform(claims, ln = log(n))
> claims.glm <- glm(c ~ offset(ln) + dist + car + age +
+       car:age + dist:car + dist:age, data = claims,
+       family = poisson)
> anova(claims.glm)
```

Here `car:age` is fitted before the other interactions as the one most likely to be important. We construct an analysis of deviance table (Chapter 4) in Table 5.3 for the factors in the order fitted.

The observed values of c are generally large, so the standard asymptotic χ^2 distribution theory for deviance differences should hold adequately. The three-way interaction term is not explicitly fitted as its deviance is just the residual deviance from the model with all two-way interactions. All the interactions are close to their degrees of freedom, suggesting that the main effects model with a residual deviance of 51.42 and 54 df provides a good fit. This model was used by Baxter *et al.* (1980). We check the fit at each sample point and present the Pearson residuals ordered by their absolute value.

```
> claims.glm1 <- update(claims.glm, . ~ offset(ln) +
+       dist + car + age)
> claims.res <- round(resid(claims.glm1, type = "pearson"),
+       3)
```

Table 5.3. Analysis of deviance table for the claims data

	df	Dev.	Resid. df	Resid. dev.
NULL			63	236.26
dist	3	12.73	60	223.53
car	3	87.24	57	136.29
age	3	84.87	54	51.42
car:age	9	10.51	45	40.91
dist:car	9	7.38	36	33.53
dist:age	9	6.24	27	27.29

```
> claims.res[rev(order(abs(claims.res)))]
     3      5     28     64     35     23     33     30     55
-2.279  2.101 -1.908  1.853  1.774  1.754 -1.541 -1.515 -1.311
     2     38      1      4     16     13     24     37     31
 1.228 -1.167  1.087 -1.080 -1.038 -1.033  1.011  1.008 -0.882
     8     11     60      6     45     40     27     54     48
 0.849  0.828 -0.815 -0.753 -0.744 -0.744  0.738  0.683  0.651
    12     19     42     43     32      7     34     57     47
 0.649 -0.612  0.591  0.577  0.575 -0.553 -0.485  0.415  0.371
     9     53     44     20     61     46     25     63     26
-0.364 -0.353 -0.331  0.327 -0.288  0.273  0.271  0.269 -0.257
    56     29     49     52     51     36     15     41     14
-0.256 -0.243 -0.228  0.220  0.217 -0.207 -0.193 -0.183 -0.164
    10     58     50     39     22     17     62     59     18
-0.145 -0.133  0.097  0.084  0.080 -0.046  0.031  0.026 -0.017
    21
-0.014
```

The Pearson residuals for the Poisson distribution are

$$e_i = (y_i - \hat{\theta}_i)/\sqrt{\hat{\theta}_i}.$$

These are approximately standardized variables with mean 0 and variance approximately 1; the variance is approximate because of the estimation of $\boldsymbol{\beta}$ in the linear predictor. There are two residuals exceeding 2 in absolute value: −2.279 at observation 3 and 2.101 at observation 5. Since the observed numbers of claims c_i are fairly large, the residuals e_i can be treated as approximately standard normal variables, and a normal quantile plot provides a check on their assumed distribution. Try *qqnorm(claims.res)*.

The fit to a straight line (not shown) is very close, so the assumption of a Poisson model is well supported.

In fitting the model, we have specified that `ln` has a fixed coefficient of 1. This is implicit in the structure of the model, but it may not be in accordance with the data. To check this, we fit `ln` explicitly, removing the offset.

```
> claims.glm2 <- update(claims.glm1, . ~ ln + dist +
+      car + age)
> print(anova(claims.glm1, claims.glm2), digits = 4)

Analysis of Deviance Table

Model 1: c ~ dist + car + age + offset(ln)
Model 2: c ~ ln + dist + car + age
  Resid. Df Resid. Dev Df Deviance
1        54      51.42
2        53      49.45  1     1.97
```

```
> round(coef(summary(claims.glm2))[2, , drop = FALSE],
+     4)

   Estimate Std. Error z value Pr(>|z|)
ln     1.20      0.144    8.34        0
```

The deviance decreases by only 1.97 on 1 df, and the parameter estimate for `ln` is within 1.4 standard errors of the offset value 1. The value of 1 is in accordance with the data. Now we examine the other parameter estimates.

```
> round(coef(summary(claims.glm1)), 4)

                     Estimate Std. Error z value Pr(>|z|)
(Intercept)           -1.8217     0.0768 -23.724   0.0000
distsmall towns        0.0259     0.0430   0.601   0.5476
distlarge towns        0.0385     0.0505   0.763   0.4457
distmajor cities       0.2342     0.0617   3.797   0.0001
car1-1.5               0.1613     0.0505   3.193   0.0014
car1.5-2               0.3928     0.0550   7.142   0.0000
car>2                  0.5634     0.0723   7.791   0.0000
age25-29              -0.1910     0.0829  -2.305   0.0211
age30-35              -0.3450     0.0814  -4.239   0.0000
age  >35              -0.5367     0.0700  -7.672   0.0000
```

The parameters for `distsmall towns` and `distlarge towns` are smaller than their standard errors, but that for `distmajor cities` is nearly four times its standard error. It appears that the first three districts can be amalgamated, only district four – London and other major cities – being kept separate.

```
> claims <- transform(claims, d4 = factor(dist == "major cities"))
> claims.glm4 <- update(claims.glm1, . ~ . - dist +
+     d4)
> anova(claims.glm4, claims.glm1)

Analysis of Deviance Table

Model 1: c ~ car + age + d4 + offset(ln)
Model 2: c ~ dist + car + age + offset(ln)
  Resid. Df Resid. Dev Df Deviance
1        56       52.1
2        54       51.4  2      0.7

> round(coef(summary(claims.glm4)), 4)

            Estimate Std. Error z value Pr(>|z|)
(Intercept)   -1.810     0.0753  -24.03   0.0000
car1-1.5       0.162     0.0505    3.21   0.0013
```

```
car1.5-2        0.394    0.0550    7.16   0.0000
car>2           0.565    0.0723    7.82   0.0000
age25-29       -0.189    0.0828   -2.28   0.0225
age30-35       -0.342    0.0813   -4.21   0.0000
age  >35       -0.533    0.0698   -7.63   0.0000
d4TRUE          0.218    0.0585    3.73   0.0002
```

The car estimates and the age estimates show similar patterns of nearly linear increase or decrease with factor level, suggesting that each factor can be replaced by the corresponding variable.

```
> claims <- transform(claims, lcar = unclass(car))
> claims.glm5 <- update(claims.glm4, . ~ . - car +
+     lcar)
> anova(claims.glm5, claims.glm4)

Analysis of Deviance Table

Model 1: c ~ age + d4 + lcar + offset(ln)
Model 2: c ~ car + age + d4 + offset(ln)
  Resid. Df Resid. Dev Df Deviance
1        58       53.0
2        56       52.1  2      0.9

> round(coef(summary(claims.glm5)), 4)
            Estimate Std. Error z value Pr(>|z|)
(Intercept) -2.026    0.0797   -25.43  0.0000
age25-29    -0.190    0.0828    -2.30  0.0217
age30-35    -0.344    0.0813    -4.24  0.0000
age  >35    -0.535    0.0698    -7.67  0.0000
d4TRUE       0.218    0.0585     3.73  0.0002
lcar         0.198    0.0208     9.48  0.0000

> claims <- transform(claims, lage = unclass(age))
> claims.glm6 <- update(claims.glm5, . ~ . - age +
+     lage)
> anova(claims.glm6, claims.glm5)

Analysis of Deviance Table

Model 1: c ~ d4 + lcar + lage + offset(ln)
Model 2: c ~ age + d4 + lcar + offset(ln)
  Resid. Df Resid. Dev Df Deviance
1        60       53.1
2        58       53.0  2      0.1

> round(coef(summary(claims.glm6)), 4)
```

```
            Estimate Std. Error z value Pr(>|z|)
(Intercept)   -1.853     0.0799  -23.18    0e+00
d4TRUE         0.219     0.0585    3.74    2e-04
lcar           0.198     0.0208    9.51    0e+00
lage          -0.177     0.0185   -9.56    0e+00
```

The parameter estimates for `lcar` and `lage` differ in magnitude by one standard error: it appears that we can simplify the model further by equating their magnitudes.

```
> claims <- transform(claims, score = lcar - lage)
> claims.glm7 <- update(claims.glm6, . ~ offset(ln) +
+      d4 + score)
> round(coef(summary(claims.glm7)), 4)
```

```
            Estimate Std. Error z value Pr(>|z|)
(Intercept)   -1.794     0.0241  -74.39    0e+00
d4TRUE         0.220     0.0585    3.76    2e-04
score          0.186     0.0139   13.35    0e+00
```

The coefficient of `d4TRUE` is close to that for score: being in district 4 is nearly equivalent to one point on the `score` scale. We try making them equivalent.

```
> claims <- transform(claims, s2 = as.numeric(d4) -
+      1 + score)
> claims.glm8 <- update(claims.glm7, . ~ offset(ln) +
+      s2)
> round(coef(summary(claims.glm8)), 4)
```

```
            Estimate Std. Error z value Pr(>|z|)
(Intercept)   -1.789     0.0223   -80.3        0
s2             0.188     0.0136    13.8        0
```

Each point up the `score` scale adds 0.188 to the fitted mean log number of claims, so the fitted mean is multiplied by $\exp(0.188) = 1.207$ – the mean number of claims increases by 20%. The `score` scale runs from -3 to $+3$, with an extra point for district 4; the fitted mean number of claims for an individual with scale score -3 is $\exp(-1.789 - 3 \times 0.188) = 0.095$.

We present finally the observed mean numbers of claims per individual, c_i/n_i, by `car` and `age`, for districts 1–3, and for district 4 separately, together with the fitted means. These are arranged in a table with the age classification reversed, so that `score` increases from the top left-hand to the bottom right-hand corner. Within each cell of the table, the order of districts is

```
1 2
3
```

with district 3 below district 1. We label the cells with the value of `score` + 4, which runs from 1 to 8.

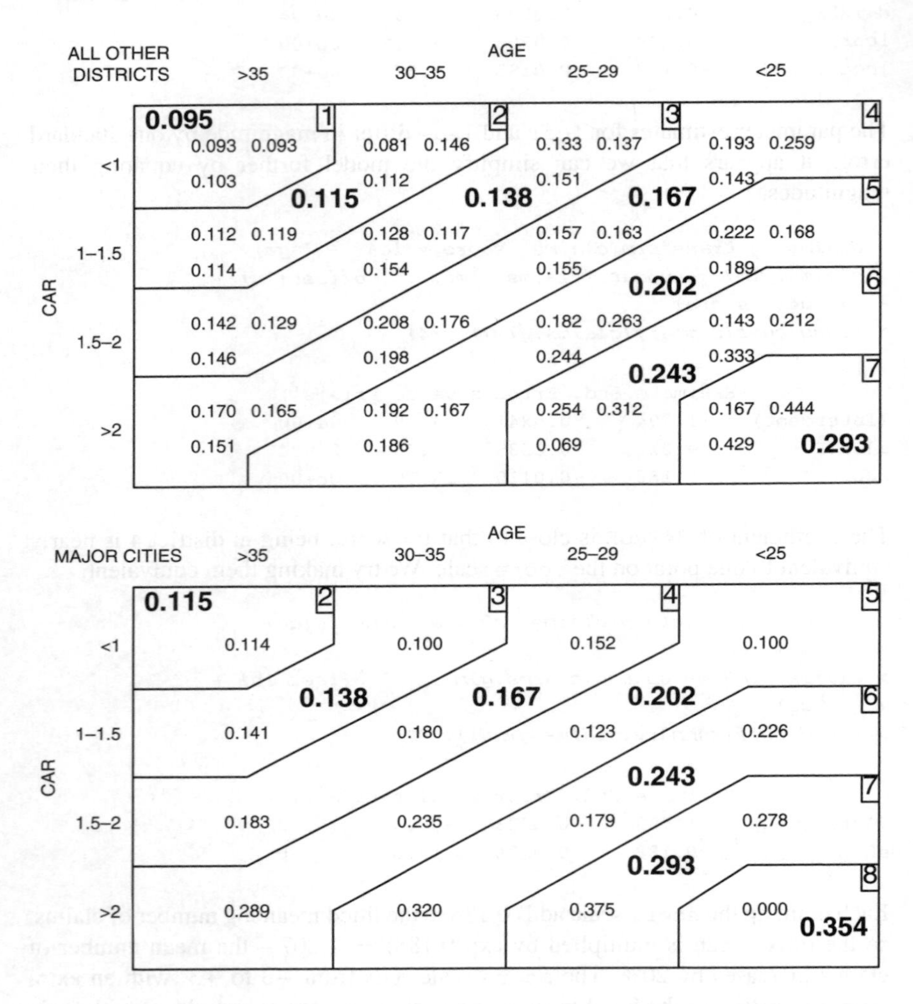

The fitted values accurately reproduce the observed means where the numbers exposed are large. The simplification of the table, as in the example in Chapter 4 on coronary heart disease, is very great: only two numbers are necessary to describe the whole 64-cell table.

It can be verified that a very similar main effects model results, with a deviance of 63.09 on 54 df, if c is assumed to be binomial $b(n, p)$ with a logit link for p (Baxter *et al.*, 1980). This is because the mean claim frequency is low, even in the highest score category, and so the probability of more than one claim in the quarter, which is $\{1 - (\mu + 1)e^{-\mu}\}$, is very small, 0.049 for the largest mean value of 0.353.

5.3 Multicategory responses

The dataset `miners` gives the numbers of coalminers classified by radiological examination into one of three categories of pneumoconiosis (`n` – normal, `m` – mild pneumoconiosis, `s` – severe pneumoconiosis), and by `period` spent working at the coalface (interval midpoint). Period is declared as a factor with eight levels, the levels corresponding to midpoints of class intervals: 5.8, 15.0, 21.5, 27.5, 33.5, 39.5, 46.0, and 51.5 `years` worked at the coalface. The data are discussed in Ashford (1959) and Mccullagh and Nelder (1989; p. 170), and are given in Table 5.4.

```
> data(miners)
> miners <- transform(miners, t = n + m + s)
> miners <- transform(miners, np = n/t, mp = m/t, sp = s/t)
```

We first graph the proportions of miners in each category against years worked (Fig. 5.2).

```
> miners.plot <- xyplot(np + mp + sp ~ years, data = miners,
+       col = 1, ylab = "proportion", type = "p", pch = c("n",
+           "m", "s"))
> print(update(miners.plot, type = "b", cex = 1.5,
+       lty = 1:3))
```

The proportions free of pneumoconiosis decline steadily with years, and those for the other two categories increase; there is a noticeable increase in the proportion with severe pneumoconiosis in the 7th and 8th periods.

How should multiple response proportions be modelled? One simple way would be to convert the three categories to two by collapsing two adjacent categories into one. For example, the second and third categories could be collapsed, and we could model the logit of the probability of no pneumoconiosis.

Table 5.4. Pneumoconiosis level

Years	n	m	s
5.8	98	0	0
15.0	51	2	1
21.5	34	6	3
27.5	35	5	8
33.5	32	10	9
39.5	23	7	8
46.0	12	6	10
51.5	4	2	5

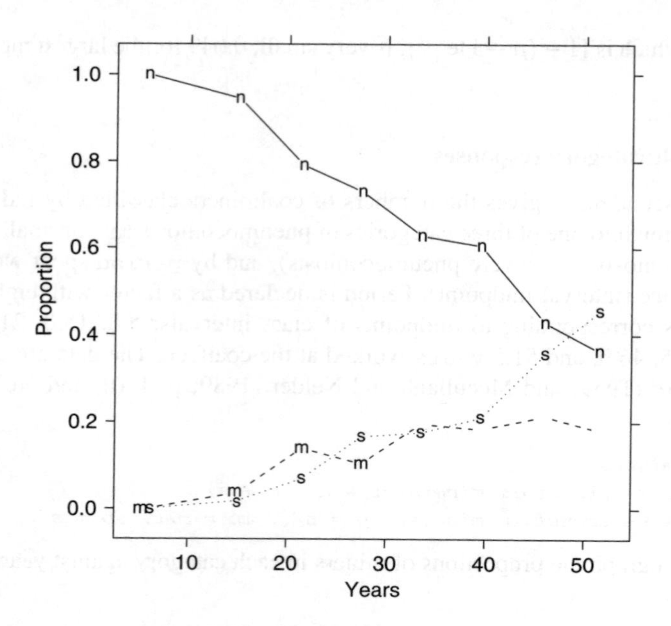

Fig. 5.2. Proportions

```
> miners <- transform(miners, period = factor(years))
> miners.glm <- glm(cbind(n, (t - n)) ~ period, family = binomial,
+       data = miners)
> summary(miners.glm)

Coefficients:
              Estimate Std. Error z value Pr(>|z|)
(Intercept)      27.3    51695.1  0.00053        1
period15        -24.5    51695.1 -0.00047        1
period21.5      -26.0    51695.1 -0.00050        1
period27.5      -26.3    51695.1 -0.00051        1
period33.5      -26.8    51695.1 -0.00052        1
period39.5      -26.9    51695.1 -0.00052        1
period46        -27.6    51695.1 -0.00053        1
period51.5      -27.9    51695.1 -0.00054        1

(Dispersion parameter for binomial family taken to be 1)

    Null deviance: 9.7564e+01  on 7  degrees of freedom
Residual deviance: 2.7531e-10  on 0  degrees of freedom
AIC: 41.84

Number of Fisher Scoring iterations: 22
```

The very large values on the logit scale are a consequence of the 100% normal response in the first group. The exact deviance for this model is zero as we are fitting a parameter for each observation.

The fitted logits decline steadily with period, as we have already seen. Their large negative values reflect the zero proportion with pneumoconiosis at the first period. Could this be expressed as a linear decline with year?

```
> miners.glm1 <- update(miners.glm, . ~ years)
> print(anova(miners.glm1), digits = 4)

Analysis of Deviance Table

Model: binomial, link: logit

Response: cbind(n, (t - n))

Terms added sequentially (first to last)

      Df Deviance Resid. Df Resid. Dev
NULL                      7      97.56
years  1    86.00         6      11.57

> round(resid(miners.glm1, type = "pearson"), 3)

     1      2      3      4      5      6      7      8
 1.801  0.525 -1.534 -1.016 -0.761  0.797  0.457  0.694
```

The linear model does not fit well, with a deviance of 11.57 on 6 df, and the residuals are not satisfactory: although the largest is only 1.801, there is a systematic pattern of positive, negative and then positive residuals as period increases. This suggests a systematic failure of the model. We could attempt to improve the fit by using a higher-order polynomial, but we first examine the complementary log–log link.

```
> miners.glm1a <- update(miners.glm1,
+        family = binomial(link = "cloglog"))
> print(anova(miners.glm1a), digits = 4)

Analysis of Deviance Table

Model: binomial, link: cloglog

Response: cbind(n, (t - n))

Terms added sequentially (first to last)
```

```
      Df Deviance Resid. Df Resid. Dev
NULL                      7      97.56
years  1    92.60         6       4.96

> round(resid(miners.glm1a, type = "pearson"), 3)
     1      2      3      4      5      6      7      8
 1.000  0.165 -1.325 -0.496 -0.242  0.978  0.211  0.284
```

The fit is considerably improved, with a deviance of 4.96, and the largest residual is only -1.325, but the same pattern of positive, negative and positive residuals appears. We try to remove the systematic sign changes in the residuals by a log transformation of years.

```
> miners <- transform(miners, ly = log(years))
> miners.glm2 <- update(miners.glm1a, . ~ ly)
> summary(miners.glm2)

Coefficients:
            Estimate Std. Error z value Pr(>|z|)
(Intercept)    4.582      0.709    6.46  1.0e-10
ly            -1.312      0.213   -6.15  7.6e-10

(Dispersion parameter for binomial family taken to be 1)

    Null deviance: 97.5639  on 7  degrees of freedom
Residual deviance:  1.4099  on 6  degrees of freedom
AIC: 31.25

Number of Fisher Scoring iterations: 5

> round(resid(miners.glm2, type = "pearson"), 3)

     1      2      3      4      5      6      7      8
 0.076  0.169 -0.593  0.188  0.073  0.761 -0.482 -0.414
```

The fit is greatly improved again over the linear model, with a deviance of 1.41, and no systematic pattern now appears in the residuals. We calculate the fitted probabilities from the model for later plotting.

```
> miners$cnp <- fitted(miners.glm2)
```

We now try the log transformation of years with the logit link.

```
> miners.glm3 <- update(miners.glm2, . ~ ly, family = binomial)
> summary(miners.glm3)

Coefficients:
            Estimate Std. Error z value Pr(>|z|)
(Intercept)    9.609      1.339    7.18  7.2e-13
ly            -2.576      0.386   -6.67  2.6e-11
```

```
(Dispersion parameter for binomial family taken to be 1)

    Null deviance: 97.5639  on 7  degrees of freedom
Residual deviance:  3.1313  on 6  degrees of freedom
AIC: 32.98

Number of Fisher Scoring iterations: 4

> round(resid(miners.glm3, type = "pearson"), 3)

     1      2      3      4      5      6      7      8
 0.780  0.338 -1.010 -0.250 -0.144  0.873 -0.089 -0.024
```

The fit is not as good, with a deviance of 3.13, but the difference is not large. A similar residual pattern of alternating signs is evident, but less pronounced. The CLL link appears to be preferable to the logit. We graph the observed proportions and the two sets of fitted proportions against years in Fig. 5.3.

```
> miners$lnp <- fitted(miners.glm3)
> print(xyplot(np + cnp + lnp ~ years, type = c("p",
+      "l", "l"), data = miners, ylab = "proportion",
+      panel = panel.superpose.2, lty = c(1, 2)))
```

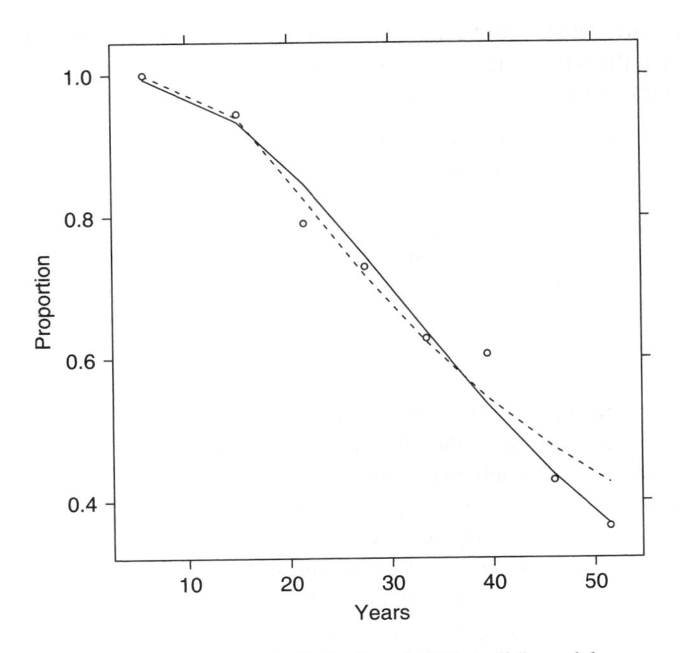

Fig. 5.3. Fitted logit (dashed) and CLL (solid) models

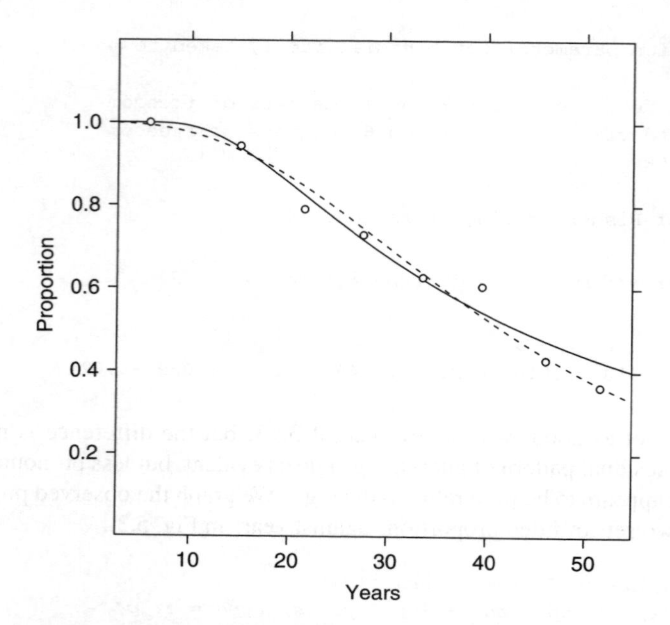

Fig. 5.4. Fitted logit (solid) and CLL (dashed) models

The graphs are not smooth because of the linear interpolation between the points. We improve the smoothness of the fitted proportions by applying the respective inverse link function in each `panel.curve` (Fig. 5.4) function.

```
> print(xyplot(np ~ years, data = miners, ylab = "proportion",
+     ylim = c(0, 1.2), col = "black", panel = function(x,
+         y) {
+         panel.xyplot(x, y)
+         panel.curve(1 - exp(-exp(4.582 - 1.312 *
+             log(x))))
+         panel.curve((eta <- exp(9.609 - 2.576 * log(x)))/(1 +
+             eta), lty = 2)
+     }))
```

The fitted proportions agree closely except for large years. The complementary log–log model flattens out at the 8th period, while the logistic model continues to decline. However, the practical conclusions are the same.

While this analysis could be repeated for `m` and `s`, the resulting models for the proportions in the three categories would not in general give fitted probabilities summing to 1, and the models themselves might involve different functions of `years` or different link functions. We need to model *simultaneously* the probabilities for all the categories. This can be achieved in R using the *multinomial logit model*.

5.4 Multinomial logit model

Suppose the response variable has r categories, with category probabilities p_j, $j = 1, 2, \ldots, r$. For the moment we ignore any *ordering* which may exist in the response categories; models for ordered categories are considered in Section 5.7. The probability that in a sample of n observations we obtain n_j in the j-th category ($j = 1, 2, \ldots, r$) is

$$p(n_1, n_2, \ldots, n_r) = \frac{n!}{n_1!n_2!\ldots n_r!} p_1^{n_1} p_2^{n_2} \ldots p_r^{n_r}$$

with

$$\sum_{j=1}^{r} p_j = 1,$$

that is there are only $r - 1$ distinct probabilities.

The *multinomial logit* transformation of $p_1, \ldots, p_r$ is the set of parameters $\theta_1, \theta_2, \ldots, \theta_r$, called multinomial logits, defined by

$$\theta_j = \log(p_j/p_1), \quad j = 1, \ldots, r.$$

Note that $\theta_1 \equiv 0$, so there are only $r - 1$ distinct logit parameters. This definition uses the convention that the first category is the reference category for the others: this is convenient when using R because the dummy variable coding for factors in R uses the same convention. However, any other category could be used as the reference category without changing the fit of the model: if we define

$$\theta_j^* = \log(p_j/p_r), \quad j = 1, \ldots, r,$$

for example, with $\theta_r^* = 0$, then the θ^*s are just a reparametrization of the θs, since

$$\theta_j^* = \theta_j + \log(p_1/p_r).$$

The multinomial logits play the same role in the analysis of multi-category response data as the binomial logit does in two-category response data. We could also define *multinomial probits* analogously to the binomial probits in Chapter 4, but we will not use these as they cannot be fitted in R using generalized linear models. For the `miners` data, taking the normal category as the reference category, we can construct the empirical logits $\log(m_i/n_i)$ and $\log(s_i/n_i)$:

```
> miners <- transform(miners, logit2 = log(m/n),
+        logit3 = log(s/n))
> print(miners[c("years", "logit2", "logit3")])

  years logit2 logit3
1   5.8    -Inf   -Inf
2  15.0 -3.239 -3.932
```

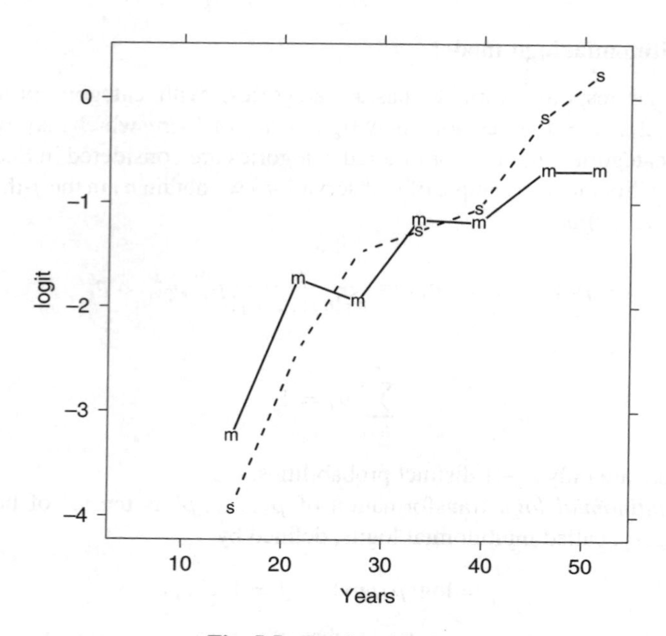

Fig. 5.5. Empirical logits

```
3  21.5 -1.735 -2.428
4  27.5 -1.946 -1.476
5  33.5 -1.163 -1.269
6  39.5 -1.190 -1.056
7  46.0 -0.693 -0.182
8  51.5 -0.693  0.223
```

We graph the logits in Fig. 5.5, removing the infinite points and connecting points with lines using the previously defined styles.

```
> print(xyplot(logit2 + logit3 ~ years, type = "b",
+      pch = c("m", "s"), data = miners, cex = 1.5,
+      ylab = "logit", lty = c(1, 2), lwd = 1.8))
```

The logits show a rapid increase with `years`. Curvature in the plots is very marked, clearly indicating the need for a log transformation of the `years` scale (Fig. 5.6).

```
> print(xyplot(logit2 + logit3 ~ years, type = "b",
+     pch = c("m", "s"), col = "black", lty = c(2,
+          3), lwd = 1.5, data = miners, cex = 1.5,
+     ylab = "logit", xlab = "years (log scale)",
+     scale = list(x = list(log = TRUE,
+          at = c(10, 20, 30, 40, 50), labels = c("10",
+               "20", "30", "40", "50")))))
```

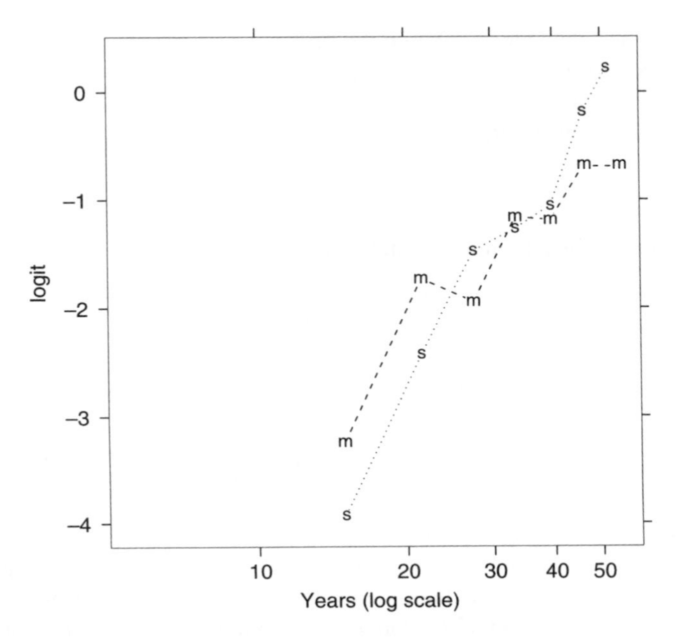

Fig. 5.6. Empirical logits

The trends are now nearly linear, with a steeper slope for the severe category. We want now to fit simultaneously the multinomial logit models

$$\begin{aligned} \theta_{2i} &= \beta_{20} + \beta_{21} \log(\text{years}_i) \\ \theta_{3i} &= \beta_{30} + \beta_{31} \log(\text{years}_i) \end{aligned} \qquad i = 1, \ldots, 8,$$

assess the goodness of fit, and convert the fitted models back to fitted multinomial probabilities. This cannot be done directly, since R does not include the multinomial distribution in its family of standard distributions. The model can be fitted indirectly by using the relation between the multinomial and Poisson distributions, which we now describe.

5.5 The Poisson-multinomial relation

We begin with the simple example of Cox (1958). Two radioactive sources are being compared. We count the number of particles n_1 and n_2 emitted from the sources in the same fixed time. The question of interest is – are the sources composed of the same radioactive material, or more generally, are the sources emitting particles at the same rate?

We assume the Poisson model for radioactive disintegrations, so the numbers N_1 and N_2 of particles emitted in a fixed time will be Poisson with means μ_1

and μ_2 for the two sources. Define the parameter of interest to be $\theta = \mu_2/\mu_1$, and define a *nuisance parameter* $\phi = \mu_1 + \mu_2$. If the sources are the same, then $\mu_1 = \mu_2$, or $\theta = 1$.

The likelihood in μ_1, μ_2 from observed counts n_1 and n_2 is

$$L(\mu_1, \mu_2) = \mathrm{e}^{-\mu_1 - \mu_2} \mu_1^{n_1} \mu_2^{n_2} / (n_1! n_2!).$$

Transforming to the parameters θ and ϕ, we have

$$L(\theta, \phi) = \mathrm{e}^{-\phi} \left(\frac{\phi}{1+\theta} \right)^{n_1} \left(\frac{\theta\phi}{1+\theta} \right)^{n_2} / (n_1! n_2!).$$

Write $n = n_1 + n_2$; then the likelihood can be re-written in the form

$$L(\theta, \phi) = \left[\mathrm{e}^{-\phi} \phi^n / n! \right] \cdot \left[\frac{n!}{n_1! n_2!} \left(\frac{1}{1+\theta} \right)^{n_1} \left(\frac{\theta}{1+\theta} \right)^{n_2} \right].$$

This striking result shows that the likelihood factors into two components, corresponding to independent random variables: the total count N is Poisson (ϕ), and conditional on $N = n$, N_2 has independently the binomial distribution $b(n, p)$ with $p = \theta/(1+\theta)$. Further, if the Poisson distributions themselves have log-linear models

$$\log \mu_j = \boldsymbol{\beta}_j' \mathbf{x}, \quad j = 1, 2$$

relating their means to explanatory variables $\mathbf{x}$, then

$$\theta = \exp(\boldsymbol{\beta}_2 - \boldsymbol{\beta}_1)' \mathbf{x}$$

and

$$\operatorname{logit} p = \log \theta = (\boldsymbol{\beta}_2 - \boldsymbol{\beta}_1)' \mathbf{x},$$

and so the binomial distribution has a logistic linear model, whose regression coefficient is the *difference* betweeen those for the Poisson models for sources 2 and 1.

This result holds quite generally between the multinomial and Poisson distributions for any number of response categories. Suppose $N_1, N_2, \ldots, N_r$ are independent Poisson random variables with means $\mu_1, \mu_2, \ldots, \mu_r$. Then the sum $N = N_1 + N_2 + \cdots + N_r$ also has a Poisson distribution with mean $\mu = \mu_1 + \mu_2 + \cdots + \mu_r$, and the *joint* distribution of the N_j, conditional on $N = n$, is the multinomial distribution

$$\Pr(N_1 = n_1, \ldots, N_r = n_r | N = n) = \frac{n!}{n_1! \ldots n_r!} p_1^{n_1} \ldots p_r^{n_r}$$

with

$$p_j = \mu_j/\mu.$$

This is easily shown analytically as above: we have

$$\begin{aligned}\Pr(n_1, \ldots, n_r) &= \prod_{j=1}^{r} \exp(-\mu_j)\mu_j^{n_j}/n_j! \\ &= \exp(-\mu)\prod_j \mu_j^{n_j}/n_j!\end{aligned}$$

and

$$\Pr(n) = \exp(-\mu)\mu^n/n!$$

so that

$$\begin{aligned}\Pr(n_1, \ldots, n_r|n) &= \Pr(n_1, \ldots, n_r)/\Pr(n) \\ &= \prod_j (\mu_j/\mu)^{n_j} \left(n! \Big/ \prod_j n_j!\right)\end{aligned}$$

as required. The multinomial logit parameters θ_j are then

$$\begin{aligned}\theta_j &= \log(p_j/p_1) \\ &= \log(\mu_j/\mu_1).\end{aligned}$$

We want to fit a regression model to each θ_j parameter (apart from θ_1):

$$\theta_{ji} = \boldsymbol{\beta}'_j \mathbf{x}_i,$$

where i denotes the observation index, for the given set of explanatory variables $\mathbf{x}$. The regression coefficients $\boldsymbol{\beta}_j$ will in general be different for each j. The multinomial logit models are equivalent to

$$\log(\mu_{ji}/\mu_{1i}) = \boldsymbol{\beta}'_j \mathbf{x}_i$$

so that

$$\phi_{ji} = \log(\mu_{ji}) = \psi_i + \boldsymbol{\beta}'_j \mathbf{x}_i, \quad j = 1, \ldots, r; \;\; i = 1, \ldots, n,$$

where $\boldsymbol{\beta}_1 = 0$, and $\phi_{1i} = \psi_i = \log \mu_{1i}$. The ψ_i are a set of nuisance parameters. To fit the model in R, we need to define a response factor with r levels corresponding to the multinomial response categories. The parameter $\boldsymbol{\beta}_j$ is then the coefficient of the *interaction* of the j-th level of the response factor with the explanatory variables $\mathbf{x}$. The fitted model must contain in addition a parameter for each observation; a 'nuisance' factor with n levels has to be fitted as well. This would be exceptionally demanding of computer time unless the explanatory variables form a contingency table of reasonably small dimension; in this case the nuisance factor can be written as the full cross-classification of the explanatory factors.

In R the function *gnm* available in the package *gnm* includes an `eliminate` feature which allows for the fitting of fixed multinomial totals efficiently and reducing the output from printed summaries (see Section 5.8). We show the numerical equivalence of multinomial and Poisson models with a simple two-category response for which the binomial logit model can be fitted explicitly, as shown above. In Section 4.8.1 we considered a 2×2 cross-classification of a binary survival response. We repeat the data in a slightly different form and the R analysis below.

```
> clinic <- data.frame(death = c(3, 4, 17, 2), survive = c(176,
+     293, 197, 23), clinic = gl(2, 2), attend = gl(2,
+     1))
> clinic <- transform(clinic, total = death + survive)
> clinic.glm <- glm(cbind(death, survive) ~ 1, data = clinic,
+     family = binomial)
> clinic.glma <- update(clinic.glm, . ~ clinic)
> clinic.glmb <- update(clinic.glma, . ~ . + attend)
> clinic.glmc <- update(clinic.glm, ~clinic:attend -
+     1)
```

The deviances and parameter estimates are reproduced below for the models.

Model	Deviance	Estimate (s.e.)
1	17.828	-3.277
		(0.200)
clinic	0.082	-4.205 + 1.756 clinic2
		(0.381) (0.450)
clinic + attend	0.043	
clinic * attend	0	

To use the Poisson analysis, we define a response factor of length 8 and a `count` vector which contains the `survive` and `death` counts, and block the explanatory variables by duplicating their values in factors of length 8. This is achieved easily using the `reshape` function.

```
> pclinic <- reshape(clinic, varying = list(c("survive",
+     "death")), v.names = "count", timevar = "resp",
+     times = c("survive", "death"), direction = "long")
> pclinic <- transform(pclinic, resp = factor(resp))
> clinic.glm1 <- glm(count ~ resp + clinic:attend,
+     family = poisson, data = pclinic)
> round(coef(summary(clinic.glm1)), 4)
```

```
                 Estimate Std. Error z value Pr(>|z|)
(Intercept)       -0.0953      0.278  -0.343    0.731
respsurvive        3.2771      0.200  16.404    0.000
clinic1:attend1    1.9685      0.213   9.220    0.000
clinic2:attend1    2.1471      0.211  10.159    0.000
clinic1:attend2    2.4749      0.208  11.884    0.000
```

```
> clinic.glm2 <- update(clinic.glm1, . ~ . + resp:clinic)
> round(coef(summary(clinic.glm2)), 4)
```

```
                      Estimate Std. Error z value Pr(>|z|)
(Intercept)              0.687      0.297    2.31   0.0209
respsurvive              4.205      0.381   11.04   0.0000
clinic1:attend1          0.281      0.485    0.58   0.5619
clinic2:attend1          2.147      0.211   10.16   0.0000
clinic1:attend2          0.787      0.482    1.63   0.1025
respsurvive:clinic2     -1.756      0.450   -3.90   0.0001
```

```
> clinic.glm3 <- update(clinic.glm2, . ~ . + resp:attend)
> clinic.glm4 <- update(clinic.glm3, . ~ . + resp:attend:clinic)
> anova(clinic.glm3)
```

```
Analysis of Deviance Table

Model: poisson, link: log

Response: count

Terms added sequentially (first to last)


              Df Deviance Resid. Df Resid. Dev
NULL                              7       1066
resp           1      768         6        299
clinic:attend  3      281         3         18
resp:clinic    1       18         2      0.082
resp:attend    1    0.039         1      0.043
```

The coefficient and standard error for `respsurvive` in the Poisson model are identical to the intercept estimate and standard error in the null logit model. The nuisance parameters for `clinic:attend` serve only to reproduce the marginal totals over the `resp` factor in each clinic/attendance cell. If these factors are omitted, the Poisson model does not correspond to the multinomial logit model because the marginal totals are being treated as random variables rather than fixed, as assumed in the logit model. It can be verified that the fitted counts added over the two response categories do reproduce the observed margins:

```
> print(with(pclinic, tapply(fitted(clinic.glm1),
+      list(clinic = clinic, attend = attend), sum)))

      attend
clinic   1   2
     1 179 297
     2 214  25

> print(with(pclinic, tapply(count, list(clinic = clinic,
+      attend = attend), sum)))

      attend
clinic   1   2
     1 179 297
     2 214  25
```

Similarly, for the second Poisson model above, the coefficients and standard errors of `respsurvive` and `respsurvive:clinic2` are identical to those of the intercept and `clinic2` in the logit model.

If the vector of counts is constructed with `death` above `survive`, the signs of the parameter estimates for terms involving `resp` are reversed, since we are then modelling the logit of the survival instead of the death probability. The need to fit nuisance parameters for the complete cross-classification of the explanatory variables makes the computational burden of the Poisson model very considerable in large contingency tables with a multinomial response. To demonstrate the elimination of nuisance parameters from the model the analysis will be repeated using the *gnm* function.

```
> library(gnm)
> clinic.gnm <- gnm(count ~ resp, eliminate = clinic:attend,
+      data = pclinic, family = poisson, verbose = FALSE)
> summary(clinic.gnm)

Coefficients of interest:
            Estimate Std. Error z value Pr(>|z|)
respsurvive     3.28       0.20    16.4   <2e-16

(Dispersion parameter for poisson family taken to be 1)

Residual deviance: 17.828 on 3 degrees of freedom
AIC: 68.01

Number of iterations: 4

> clinic.gnm1 <- update(clinic.gnm, . ~ resp + resp:clinic)
> summary(clinic.gnm1)
```

```
Coefficients of interest:
                      Estimate Std. Error z value Pr(>|z|)
respsurvive              4.205      0.381    11.0  < 2e-16
clinic2:respsurvive     -1.756      0.450    -3.9  9.4e-05

(Dispersion parameter for poisson family taken to be 1)

Residual deviance: 0.08229 on 2 degrees of freedom
AIC: 52.26

Number of iterations: 4
```

The deviances are identical as before. For the first model, only the `respsurvive` effect is displayed, which is the same as before. For the second model, both parameter estimates are identical to those found before.

5.6 Fitting the multinomial logit model

We return now to the `miners` example, and construct the necessary vectors for the Poisson model using the *reshape* function.

```
> miners.long <- reshape(miners, drop = names(miners)[6:11],
+      varying = list(names(miners)[2:4]), timevar = "resp",
+      times = c("n", "m", "s"), v.names = "count",
+      direction = "long")
> miners.long <- transform(miners.long, resp = factor(resp))
> miners.long <- transform(miners.long, resp = relevel(resp,
+      ref = "n"), lyr = log(years), period = factor(years))
> miners.glm5 <- glm(count ~ period + resp, data = miners.long,
+      family = poisson)
> coef(summary(miners.glm5))[9:10, , drop = FALSE]

      Estimate Std. Error z value Pr(>|z|)
respm    -2.03      0.173   -11.8 6.45e-32
resps    -1.88      0.162   -11.6 2.83e-31

> miners.glm6 <- update(miners.glm5, . ~ . + resp:years)
> anova(miners.glm5, miners.glm6)

Analysis of Deviance Table

Model 1: count ~ period + resp
Model 2: count ~ period + resp + resp:years
  Resid. Df Resid. Dev Df Deviance
1        14      101.6
2        12       13.9  2     87.7

> round(summary(miners.glm6)$coef[9:12, ], 4)
```

```
             Estimate Std. Error z value Pr(>|z|)
respm         -4.2917     0.5214   -8.23    0.000
resps         -5.0598     0.5964   -8.48    0.000
respn:years   -0.1093     0.0165   -6.64    0.000
respm:years   -0.0257     0.0198   -1.30    0.193
```

The first model is equivalent to the null multinomial model, and the second is equivalent to the multinomial model using `years`. The parameter estimates appear confusing because the coefficient of the *last* category of `resp` has been set to zero instead of the first, as we might have expected. This is because the main effect of `years` has not been included in the model. The standard parametrization can be recovered by adding `years` to the model:

```
> miners.glm7 <- update(miners.glm6, . ~ . + years +
+     resp:years)
> print(anova(miners.glm7, miners.glm6), digits = 5)

Analysis of Deviance Table

Model 1: count ~ period + resp + years + resp:years
Model 2: count ~ period + resp + resp:years
  Resid. Df Resid. Dev Df  Deviance
1        12     13.928
2        12     13.928  0 1.421e-14

> round(coef(summary(miners.glm7))[9:12, ], 4)

            Estimate Std. Error z value Pr(>|z|)
respm        -4.2917     0.5214   -8.23        0
resps        -5.0598     0.5964   -8.48        0
respm:years   0.0836     0.0153    5.47        0
resps:years   0.1093     0.0165    6.64        0

> miners.long$fval7 <- fitted(miners.glm7)
> miners.long$resid7 <- resid(miners.glm7)
> options(digits = 3)
> cbind(miners.long[1:8, c("count", "fval7", "resid7")],
+     miners.long[9:16, c("count", "fval7", "resid7")],
+     miners.long[17:24, c("count", "fval7", "resid7")],
+     row.names = NULL)

    count fval7 resid7 count fval7  resid7 count fval7  resid7
1.n    98 94.76  0.331     0  2.11 -2.0519     0  1.13 -1.5057
2.n    51 49.97  0.145     2  2.39 -0.2626     1  1.63 -0.5346
3.n    34 37.42 -0.568     6  3.09  1.4657     3  2.49  0.3133
4.n    35 37.96 -0.488     5  5.17 -0.0755     8  4.87  1.2991
5.n    32 34.65 -0.456    10  7.79  0.7574     9  8.56  0.1507
6.n    23 20.57  0.525     7  7.64 -0.2344     8  9.79 -0.5899
```

```
7.n      12 10.74  0.377      6  6.86 -0.3373     10 10.39 -0.1233
8.n       4  2.91  0.603      2  2.95 -0.5861      5  5.14 -0.0624
```

Note that `years` is aliased, because it can be expressed as a linear function of the dummy variables for the levels of `period`. The fit of the model is consequently unchanged.

The deviance changes by 87.71 for the 2 df, and the residual deviance of 13.93 with 12 df looks reasonable. The largest residual is 1.466 at the 11th observation, but the residuals for the first and third categories show the same pattern of alternating signs as in the binomial logit analysis. We try the log year model.

```
> miners.glm8 <- update(miners.glm6, . ~ period + resp +
+      lyr + resp:lyr)
> print(anova(miners.glm6, miners.glm8), digits = 4)

Analysis of Deviance Table

Model 1: count ~ period + resp + resp:years
Model 2: count ~ period + resp + lyr + resp:lyr
  Resid. Df Resid. Dev Df Deviance
1        12     13.928
2        12      5.347  0    8.581

> print(coef(summary(miners.glm8))[9:12, , drop = FALSE],
+      digits = 5)

           Estimate Std. Error z value   Pr(>|z|)
respm       -8.9360    1.58044 -5.6541 1.5662e-08
resps      -11.9751    2.00045 -5.9862 2.1478e-09
respm:lyr    2.1654    0.45749  4.7332 2.2102e-06
resps:lyr    3.0675    0.56521  5.4272 5.7258e-08

> miners.long$fval8 <- fitted(miners.glm8)
> miners.long$resid8 <- resid(miners.glm8)
> options(digits = 3)
> cbind(miners.long[1:8, c("count", "fval8", "resid8")],
+      miners.long[9:16, c("count", "fval8", "resid8")],
+      miners.long[17:24, c("count", "fval8", "resid8")],
+      row.names = NULL)

    count fval8  resid8 count fval8 resid8 count  fval8 resid8
1.n    98 97.29  0.0719     0 0.576 -1.073     0  0.135 -0.519
2.n    51 50.38  0.0871     2 2.334 -0.224     1  1.286 -0.262
3.n    34 36.50 -0.4190     6 3.687  1.103     3  2.811  0.112
4.n    35 35.93 -0.1558     5 6.184 -0.493     8  5.886  0.826
5.n    32 32.61 -0.1069    10 8.605  0.464     9  9.787 -0.255
6.n    23 20.27  0.5930     7 7.643 -0.236     8 10.085 -0.681
```

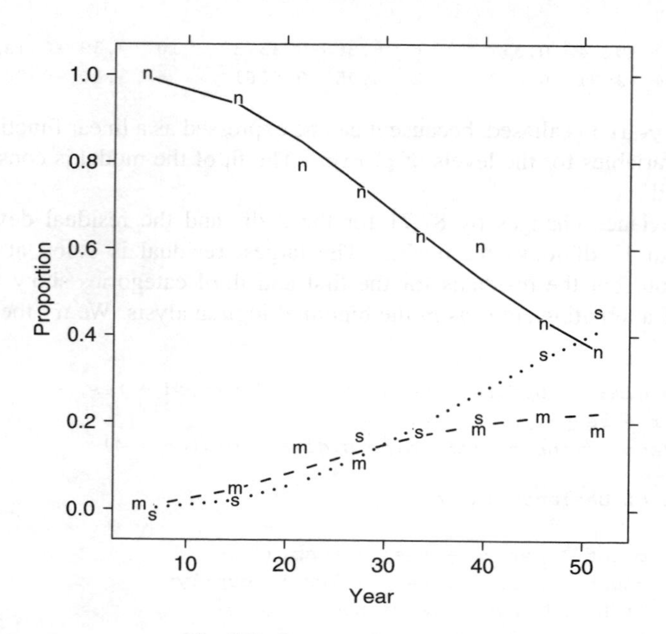

Fig. 5.7. Separate slope model

```
7.n      12 12.08 -0.0226       6 6.333 -0.134     10   9.588   0.132
8.n       4  3.94  0.0303       2 2.638 -0.411      5   4.422   0.269
```

The deviance drops to 5.35, and the slopes are 2.165 (s.e. 0.457) for the second category and 3.067 (s.e. 0.565) for the third. The alternating sign pattern in the residuals is much weaker, and the model provides a good fit. We construct and graph (Fig. 5.7) the observed and fitted proportions from the model.

```
> miners.long$fv8 <- fitted(miners.glm8)
> miners.long <- transform(miners.long, f2p8 = fv8/t,
+     op = count/t)
> print(miners.plot <- xyplot(op ~ years, data = miners.long,
+     groups = resp, ylab = "proportion", xlab = "years",
+     pch = c("n", "m", "s"), cex = 1.5))
> print(update(miners.plot, panel = function(x, y,
+     ...) {
+     panel.xyplot(x, y, ...)
+     panel.xyplot(x, miners.long$f2p8, type = "l",
+         ...)
+ }))
```

The severe proportion increases considerably for large values of `years`.

Since the estimated slopes differ by about twice either standard error, it seems unlikely that the true slopes could be equal. We investigate this question by repeating the analysis but setting the reference level of the `resp` factor to `m`.

```
> miners.long$respm <- relevel(miners.long$resp, ref = "m")
> coef(summary(update(miners.glm8, . ~ period + respm *
+     lyr)))[12, , drop = FALSE]

            Estimate Std. Error z value Pr(>|z|)
respms:lyr     0.902      0.669    1.35    0.178
```

The standard error of the difference between the moderate and severe slopes is 0.669; the observed difference of 0.902 is only 1.35 standard errors, so the slopes can be set equal. The estimated slopes have a large negative covariance, so the variance of the difference is substantially larger than the individual variances.

We now fit the model using a single dummy variable to estimate the common slope.

```
> miners.long <- transform(miners.long, r23 = ifelse(resp !=
+     "n", 1, 0))
> miners.long <- transform(miners.long, r23lyr = r23 *
+     lyr)
> miners.glm9 <- update(miners.glm8, . ~ factor(years) +
+     resp + r23lyr)
> anova(miners.glm9, miners.glm8)

Analysis of Deviance Table

Model 1: count ~ factor(years) + resp + r23lyr
Model 2: count ~ period + resp + lyr + resp:lyr
  Resid. Df Resid. Dev Df Deviance
1        13       7.21
2        12       5.35  1     1.86

> print(coef(summary(miners.glm9))[9:11, , drop = FALSE],
+     5)

       Estimate Std. Error z value   Pr(>|z|)
respm   -10.378    1.34433 -7.7199 1.1645e-14
resps   -10.231    1.34299 -7.6184 2.5687e-14
r23lyr    2.576    0.38632  6.6680 2.5926e-11
```

The deviance increases by 1.86 on 1 df. The compromise slope is 2.58 and the largest residual is 1.481 at the 11th observation.

We finally construct and graph (Fig. 5.8) the fitted proportions from this model.

```
> miners.long$fval9 <- fitted(miners.glm9)
> miners.long <- transform(miners.long, f2p9 = fval9/t,
+     op = count/t)
```

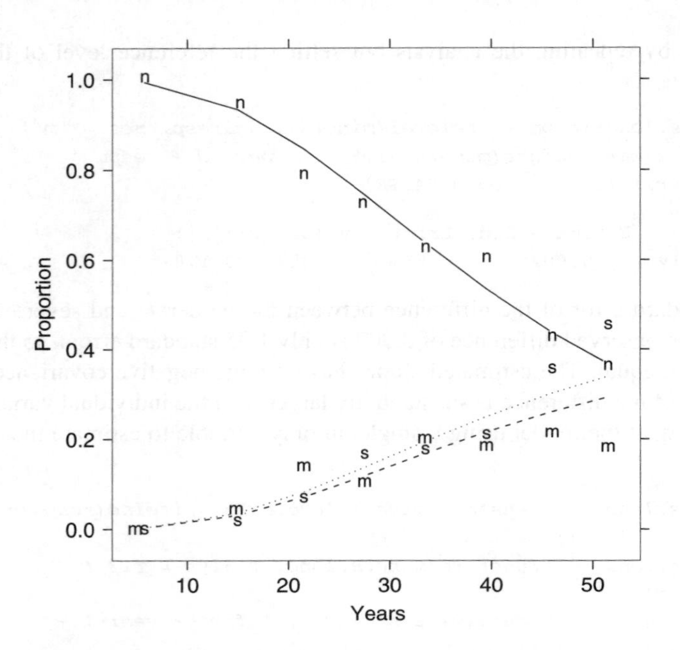

Fig. 5.8. Common slope model

```
> print(update(miners.plot, panel = function(x, y,
+       ...) {
+       panel.xyplot(x, y, ...)
+       panel.xyplot(x, miners.long$f2p9, type = "l",
+           ...)
+ }))
```

In this case the fitted probabilities of the normal category are the same as those from the binomial logit model, but this happens only because the slopes for the mild and severe categories have been equated. In the `resp.lyr` interaction model (*`miners.glm9`*), the fitted probabilities from the binomial and multinomial models are different.

In the common slope model the fitted probabilities for the mild and severe categories are very close to each other, but they are not close to the observed proportions at the upper end, where the sample sizes are small. We do not have enough sample information to discriminate the two models at the upper end of the `years` range.

5.7 Ordered response categories

The three-category response considered in Sections 5.3 and 5.6 had a natural order to the categories: normal, mild, and severe. This is a common feature of categorical responses, and such responses often have many categories: between five and nine is quite common. The number of parameters in the multinomial models then becomes very large, and it is natural to look for simpler models with fewer parameters. Such simpler models make various assumptions about the relations among the category probabilities. We now consider a number of possibilities, in all of which only one regression coefficient vector $\boldsymbol{\beta}$ is used, rather than $(r-1)$ such vectors.

5.7.1 *Common slopes for the regressions*

The general model, in the notation of Section 5.5, is

$$\theta_1 = 0$$
$$\theta_j = \log(p_j/p_1) = \boldsymbol{\beta}'_j\mathbf{x}, \quad j = 2, \ldots, r.$$

In the `miners` example of Section 5.6 we found that the slope estimates $\hat{\beta}_{21}$ and $\hat{\beta}_{31}$ could be equated. If the slopes are equal across categories for *all* the explanatory variables, excluding the intercepts, we can write

$$\theta_1 = 0$$
$$q_j = \gamma_j + \boldsymbol{\beta}'\mathbf{x}, \quad j = 2, \ldots, r.$$

Then

$$p_1 = 1/(1 + \delta e^{\boldsymbol{\beta}'\mathbf{x}})$$
$$p_j = \frac{e^{\gamma_j + \boldsymbol{\beta}'\mathbf{x}}}{1 + \delta e^{\boldsymbol{\beta}'\mathbf{x}}}, \quad j = 2, \ldots, r,$$

where

$$\delta = \sum_{j=2}^{r} \exp(\gamma_j).$$

Since δ can be expressed as e^α, it can be absorbed as an additional intercept term into $\boldsymbol{\beta}'\mathbf{x}$. The model for p_1 is then an ordinary binomial logit model for category 1 relative to the other pooled categories. This explains why the fitted logit model for the normal proportion in Section 5.3 gave the same fitted probabilities as those for the multinomial logit model in Section 5.6 in which the regression coefficients β_{21} and β_{31} were equated. The models for p_2 and p_3 are, however, not binomial logit models since the exponents in the numerator and denominator differ.

If the intercepts are also equated, then

$$\theta_1 = 0$$
$$\theta_j = \boldsymbol{\beta}'\mathbf{x}, \quad j = 2, \ldots, r$$

and

$$p_1 = 1/\{1 + (r-1)e^{\boldsymbol{\beta}'\mathbf{x}}\}$$

$$p_j = e^{\boldsymbol{\beta}'\mathbf{x}}/\{1 + (r-1)e^{\boldsymbol{\beta}'\mathbf{x}}\}, \quad j = 1, \ldots, r.$$

Identical models are now being fitted for all but the first category: the second and third categories are indistinguishable and are effectively being collapsed.

We continue the analysis of the `miners` example of Section 5.6, where we have already fitted a common slope for the response categories in the regression on `lyr`. Equal intercepts are fitted by replacing the `resp` factor in this model by `r23`:

```
> miners.glm10 <- update(miners.glm9, . ~ . - resp +
+       r23)
> anova(miners.glm10, miners.glm9)

Analysis of Deviance Table

Model 1: count ~ factor(years) + r23lyr + r23
Model 2: count ~ factor(years) + resp + r23lyr
  Resid. Df Resid. Dev Df Deviance
1        14       7.65
2        13       7.21  1     0.44

> print(coef(summary(miners.glm10))[9:10, , drop = FALSE],
+       digits = 4)

       Estimate Std. Error z value  Pr(>|z|)
r23lyr    2.576     0.3863   6.668 2.594e-11
r23     -10.302     1.3391  -7.693 1.433e-14

> miners.long <- transform(miners.long,
+       fval10 = fitted(miners.glm10)/t,
+       resid10 = resid(miners.glm10))
> print(update(miners.plot, panel = function(x, y,
+       ...) {
+       panel.xyplot(x, y, ...)
+       panel.xyplot(x, miners.long$fval10, type = "l",
+           ...)
+ }))
```

The deviance increases by only 0.44 on 1 df, not surprising since the estimates for `respm` and `resps` in the previous model, *miners.glm9*, were very close. Note that this deviance does not correspond to the deviance of 3.13 with 6 df for the logit model (*miners.glm3*) in Section 5.3 because the variation between the second and third categories is suppressed in that model. The difference of 4.52 with 8 df is the variation due to constraining these two categories to have the same response probabilities. The fitted proportions from this model are shown in Fig. 5.9.

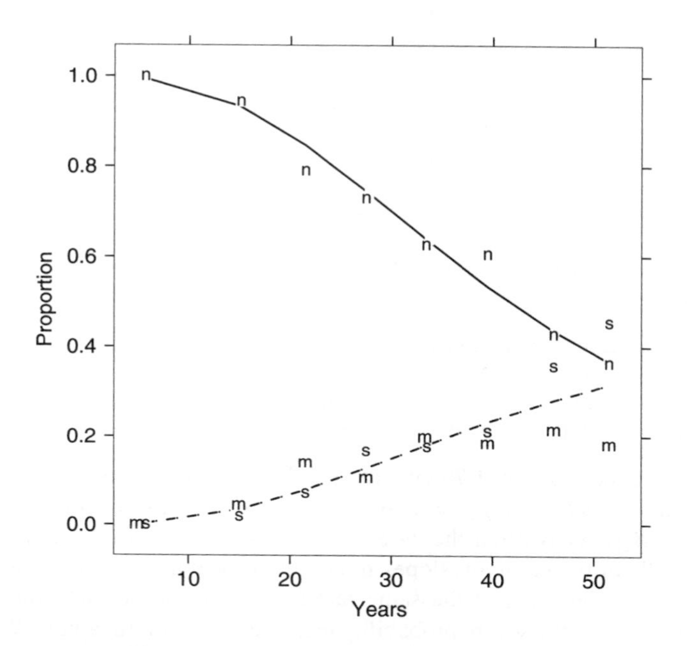

Fig. 5.9. Common slope and intercept model

5.7.2 *Linear trend over response categories*

Suppose that the regression coefficients $\boldsymbol{\beta}_j$, rather than being equal, increase in equal steps with the category index (apart from the intercepts):

$$\theta_1 = 0$$
$$\theta_j = \gamma_j + (j-1)\boldsymbol{\beta}'x, \quad j = 2, \ldots, r.$$

Thus the regression coefficients $\boldsymbol{\beta}_j$ have a *linear trend* over the given ordering of the response categories. In using the Poisson representation to fit this model, we replace the interaction between response factor and explanatory variables by its linear component.

We now define the linear component of `resp`:

```
> miners.long$lres <- unclass(miners.long$resp) - 1
```

If the slopes 2.165 and 3.067 were estimates of β and 2β, then β would be roughly $5.232/3 = 1.744$.

```
> miners.glm11 <- update(miners.glm10, . ~ factor(years) +
+       resp + lyr + lres + lyr:lres)
> anova(miners.glm11, miners.glm8)
```

```
Analysis of Deviance Table

Model 1: count ~ factor(years) + resp + lyr + lres + lyr:lres
Model 2: count ~ period + resp + lyr + resp:lyr
  Resid. Df Resid. Dev Df Deviance
1        13       7.05
2        12       5.35  1     1.70

> print(coef(summary(miners.glm11))[9:11, , drop = FALSE],
+      4)

          Estimate Std. Error z value  Pr(>|z|)
respm       -7.429     0.8771  -8.470 2.451e-17
resps      -13.362     1.8325  -7.292 3.059e-13
lyr:lres     1.726     0.2584   6.678 2.427e-11
```

The deviance increases by 1.70 on 1 df, and the estimate of β is 1.726.

Thus the model with $\beta_{31} = 2\beta_{21}$ fits just as well as the previous model with $\beta_{31} = \beta_{21}$. As is often the case, different smoothed structures fit the data equally well. In the common slopes model the probabilities of mild and severe pneumoconiosis increase at the same rate relative to the normal group; in the linear trend model, the severe probability increases much more rapidly. We cannot distinguish between these two interpretations.

If the *intercepts* γ_j also increase in equal steps (Haberman, 1974; Goodman, 1979), we can write

$$\theta_j = (j-1)\boldsymbol{\beta}'\mathbf{x} = b(j-1),$$

where

$$b = \boldsymbol{\beta}'\mathbf{x},$$

and

$$p_j = p_1 \mathrm{e}^{b(j-1)}, \quad j = 2, \ldots, r$$

which is equivalent to

$$p_j = (1-\theta)\theta^{j-1}/(1-\theta^r), \quad j = 1, \ldots, r$$

with

$$\theta = \mathrm{e}^b.$$

Thus the category probabilities $p_1, \ldots, p_r$, follow a *truncated geometric distribution* with parameter $\theta = \mathrm{e}^{\boldsymbol{\beta}'\mathbf{x}}$.

The *practical* interpretation of this distribution is less interesting in general than the model simplification resulting: the regression coefficients are simply proportional.

Is the linear trend in the intercept model reasonable for the `miners` data? The intercept for category 3 is very close to twice that for category 2: it looks as though the model will fit well.

```
> miners.glm12 <- update(miners.glm11, . ~ . - resp +
+      lres)
> anova(miners.glm12, miners.glm11)

Analysis of Deviance Table

Model 1: count ~ factor(years) + lyr + lres + lyr:lres
Model 2: count ~ factor(years) + resp + lyr + lres + lyr:lres
  Resid. Df Resid. Dev Df Deviance
1         14      24.14
2         13       7.05  1    17.09

> round(coef(summary(miners.glm12))[9:10, , drop = FALSE],
+      4)

          Estimate Std. Error z value Pr(>|z|)
lres         -8.09      0.955   -8.47        0
lyr:lres      2.09      0.271    7.69        0
```

Surprisingly, the deviance increases by 17.09. Why does this happen? The answer becomes clear if we examine the correlation between the `respm` and `resps` estimates from model *miners.glm6*.

```
> round(summary(update(miners.glm6, . ~ . - resp:years +
+      lyr:lres), corr = TRUE)$corr[9:11, 9:11, drop = FALSE],
+      4)

           respm  resps lyr:lres
respm      1.000  0.979   -0.979
resps      0.979  1.000   -0.995
lyr:lres -0.979 -0.995    1.000
```

The correlation is extremely high: 0.9785. Thus the (asymptotic) variance of $\hat{\gamma}_3 - 2\hat{\gamma}_2$ is $(1.832^2 - 4 \times 0.9785 \times 1.832 \times 0.877 + 4 \times 0.877^2)$, that is, 0.1442, so the standard error is 0.380, and $(\hat{\gamma}_3 - 2\hat{\gamma}_2)/\text{s.e.}(\hat{\gamma}_3 - 2\hat{\gamma}_2) = 3.95$.

Thus the truncated geometric distribution is untenable for the proportions of normal, mild and severe pneumoconiosis, though the model with a linear trend in the slopes fits well.

5.7.3 *Proportional slopes*

A weaker constraint than the linear trend in Section 5.7.2 is proportional slopes. Suppose

$$\theta_1 = 0$$
$$\theta_j = \gamma_j + \delta_j \boldsymbol{\beta}'\mathbf{x}, \quad j = 2, \ldots, r;$$

then the regression coefficients $\boldsymbol{\beta}$ are proportional across categories (except for the intercepts). This model is called the *stereotype model* by Anderson (1984); If there is only one explanatory variable, the model imposes no constraint on the full multinomial logit model.

5.7.4 *The continuation ratio model*

A somewhat different model expresses the multinomial model as a succession of binomial logit models, each of which can be fitted independently. The reduction in the number of parameters comes about through connections between the logit models.

Consider the three-category multinomial model, with response counts n_1, n_2 and n_3:

$$\Pr(N_1 = n_1, N_2 = n_2, N_3 = n_3) = \frac{n!}{n_1!n_2!n_3!} p_1^{n_1} p_2^{n_2} p_3^{n_3}.$$

We define $n_{23} = n_2 + n_3$, and write the above probability function as

$$\frac{n!}{n_1!n_{23}!} p_1^{n_1}(1 - p_1)^{n_{23}} \cdot \frac{n_{23}!}{n_2!n_3!} \left(\frac{p_2}{1 - p_1}\right)^{n_2} \left(\frac{p_3}{1 - p_1}\right)^{n_3}.$$

The first term is the marginal binomial distribution of N_1, and the second is the conditional binomial distribution of N_2, given that the first response did not occur. In our example p_1 is the probability that the response is 'normal', while $p_2/(1-p_1)$ is the conditional probability that the response is 'mild' rather than 'severe', given that it is not 'normal'. Now consider binomial logit models for p_1 and $p_2/(1-p_1)$. We see immediately that the model for p_1 is exactly that considered at the beginning of Section 5.3. This model also arises, as we saw in Section 5.7.1, from the multinomial logit model if the regression coefficients β_2 and β_3 are equal, including the intercepts. However, the binomial logit model parameters do not correspond in general to those of the multinomial logit model, since

$$\text{logit}\, p_1 = \log[(p_1/(p_2 + p_3)] = -\log(e^{\boldsymbol{\beta}_2'\mathbf{x}} + e^{\boldsymbol{\beta}_3'\mathbf{x}}),$$

and only if $\boldsymbol{\beta}_2'\mathbf{x} = \boldsymbol{\beta}_3'\mathbf{x} + c$, where c is a constant, is this equivalent to a linear regression on $\mathbf{x}$.

However,

$$\text{logit } p_2/(1-p_1) = \log(p_2/p_3) = (\boldsymbol{\beta}_2 - \boldsymbol{\beta}_3)'\mathbf{x}.$$

So the conditional category 2/category 3 comparison is both an ordinary logit model and equivalent to (part of) the multinomial logit model.

To fit the two logit models, we require a derived data structure, in order to relate the two models. We again use the *reshape* function to do this.

```
> miners$notn <- miners$t - miners$n
> miners.cr <- reshape(miners, drop = names(miners)[c(6:8,
+      10, 11)], varying = list(names(miners)[2:3]),
+      times = c("n", "m"), timevar = "resp", v.names = "count",
+      direction = "long")
> miners.cr$other <- miners.cr$t - miners.cr$count
> miners.cr$other[miners.cr$resp == "m"] <-
+      miners.cr$other[miners.cr$resp == "n"]
+      - miners.cr$count[miners.cr$resp == "m"]
> miners.cr <- transform(miners.cr, lyears = log(years),
+      resp = relevel(factor(resp), ref = "n"))
> miners.cr[c("years", "resp", "count", "other")]

    years resp count other
1.n   5.8    n    98     0
2.n  15.0    n    51     3
3.n  21.5    n    34     9
4.n  27.5    n    35    13
5.n  33.5    n    32    19
6.n  39.5    n    23    15
7.n  46.0    n    12    16
8.n  51.5    n     4     7
1.m   5.8    m     0     0
2.m  15.0    m     2     1
3.m  21.5    m     6     3
4.m  27.5    m     5     8
5.m  33.5    m    10     9
6.m  39.5    m     7     8
7.m  46.0    m     6    10
8.m  51.5    m     2     5

> miners.cr.glm <- glm(cbind(count, other) ~ resp,
+      data = miners.cr, family = binomial)
> summary(miners.cr.glm)

Coefficients:
            Estimate Std. Error z value Pr(>|z|)
(Intercept)    1.260      0.125   10.07  < 2e-16
respm         -1.406      0.254   -5.53  3.2e-08
```

```
(Dispersion parameter for binomial family taken to be 1)

    Null deviance: 132.10  on 14  degrees of freedom
Residual deviance: 101.64  on 13  degrees of freedom
AIC: 150.8

Number of Fisher Scoring iterations: 5

> miners.cr.glm2 <- update(miners.cr.glm, . ~ . + lyears)
> summary(miners.cr.glm2)

Coefficients:
            Estimate Std. Error z value Pr(>|z|)
(Intercept)    8.734      1.128    7.74  9.9e-15
respm         -0.682      0.275   -2.48    0.013
lyears        -2.321      0.327   -7.10  1.2e-12

(Dispersion parameter for binomial family taken to be 1)

    Null deviance: 132.0998  on 14  degrees of freedom
Residual deviance:   7.6268  on 12  degrees of freedom
AIC: 58.78

Number of Fisher Scoring iterations: 4

> miners.cr.glm3 <- update(miners.cr.glm2, . ~ . +
+     resp:lyears)
> summary(miners.cr.glm3)

Coefficients:
             Estimate Std. Error z value Pr(>|z|)
(Intercept)     9.609      1.339    7.18  7.2e-13
respm          -5.745      3.003   -1.91    0.056
lyears         -2.576      0.386   -6.67  2.6e-11
respm:lyears    1.440      0.852    1.69    0.091

(Dispersion parameter for binomial family taken to be 1)

    Null deviance: 132.0998  on 14  degrees of freedom
Residual deviance:   4.8784  on 11  degrees of freedom
AIC: 58.03

Number of Fisher Scoring iterations: 4
```

The full interaction model gives unrelated logistic regressions on log years for the normal category relative to abnormal (9.609 − 2.576 `lyears`), the same as for our original logit analysis (*miners.glm3*) with the mild/severe categories combined, and for the mild category relative to severe (3.864 − 1.136 `lyears`). The deviance for this model is 4.88 with 11 df, one df less than for the multinomial logit model; this is because for the mild/severe data the first observation is 0/0 and is automatically omitted from the analysis.

The fit is similar to the multinomial logit, but slightly closer. The interaction term is not needed; omitting it gives a deviance of 7.63, similar to the common slope multinomial logit model deviance of 7.21 (*miners.glm9*). The common slope of −2.321 (*miners.cr.glm2*) is similar, but not identical, in magnitude (but of opposite sign) to the common slope of 2.576 in the multinomial logit model. For this model the odds on abnormality versus normality increase at the same rate with years as do the odds on severe versus mild pneumoconiosis: the odds ratio 'continues' with increasing severity of the response.

We now convert the three models back to the equivalent multinomial model. We first tabulate the fitted probabilities (to 3 dp) from the three models.

```
> miners.cr$fval <- fitted(miners.cr.glm3)
> round(cr.tab <- with(miners.cr, tapply(fval,
+      list(Response = resp, Period = period), mean)), 3)

         Period
Response   5.8    15  21.5  27.5  33.5  39.5    46  51.5
       n 0.994 0.933 0.846 0.745 0.637 0.535 0.437 0.367
       m 0.866 0.687 0.593 0.524 0.468 0.422 0.381 0.351
```

The first row of the table gives the fitted probabilities for response `n` in the multinomial model. The second row gives the fitted conditional probabilities for response `m`, conditional on `m` + `s`. To obtain the fitted multinomial probabilities for responses `m` and `s`, we multiply the `n` entries by the corresponding probabilities $(1 - p_{\text{n}})$ to give the multinomial probabilities for category `m`; those for category `s` are obtained by difference from 1. The observed and fitted proportions are shown in Fig. 5.10.

```
> fn <- cr.tab[1, ]
> fm <- cr.tab[2, ] * (1 - fn)
> fs <- 1 - fn - fm
> round(rbind(fn, fm, fs), 3)

     5.8    15  21.5  27.5  33.5  39.5    46  51.5
fn 0.994 0.933 0.846 0.745 0.637 0.535 0.437 0.367
fm 0.005 0.046 0.091 0.134 0.170 0.197 0.214 0.222
fs 0.001 0.021 0.063 0.121 0.193 0.269 0.349 0.411

> print(xyplot(np + mp + sp + fn + fm + fs ~ years,
+     data = miners, type = c("p", "p", "p", "l", "l",
+         "l"), cex = 1.5, panel = panel.superpose.2,
+     ylab = "proportion", xlab = "year", pch = c("n",
+         "m", "s"), lty = 1:3, lwd = 1.5))
```

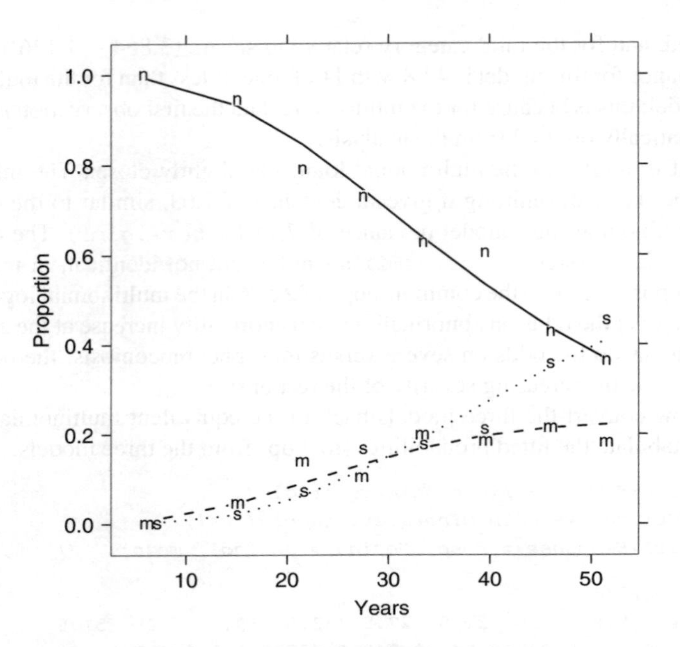

Fig. 5.10. Continuation ratio model

The fitted model for the 'mild' and 'severe' categories is quite different from the common slope multinomial logit model though it has nearly the same deviance, and is very similar to the separate slopes model. The sparseness of data for these categories at high values of `years` allows different models to fit the data equally well.

Note that the unrelated models can be fitted without 'blocking' the data into vectors of length 16, but the no-interaction continuation-ratio model cannot be fitted without this device.

5.7.5 *Other models*

Cumulative probabilities may be modelled directly using the same link functions as for the binomial. Define the cumulative probability P_j for the j-th category by

$$P_j = \sum_{s=1}^{j} p_s, \quad j = 1, \ldots, r-1,$$

$$P_r = 1.$$

Then

$$g(P_j) = \boldsymbol{\beta}'_j\mathbf{x}, \quad j \neq r$$

is a model for the cumulative probability, with g the logit, probit or CLL link function. This model has as many parameters as the multinomial, so is of interest only if restricted. A useful restriction is

$$g(P_j) = \theta_j + \boldsymbol{\beta}'\mathbf{x},$$

where θ_j is an intercept parameter, since then $\boldsymbol{\beta}$ is invariant under the collapsing together of adjacent categories. If g is the logit link, this model is called the *proportional odds model*. If g is the CLL link, the model is called the *proportional hazards model*.

This class of models can be expressed in terms of a *latent variable representation*. If

$$g(P_j) = \theta_j + \boldsymbol{\beta}'\mathbf{x}$$

for some link function g, then

$$P_j = H(\theta_j + \boldsymbol{\beta}'\mathbf{x})$$

where H is the cumulative distribution function inverse to the link function g (see Section 4.2). Then the category probabilities p_j can be represented as

$$p_j = H(\theta_j + \boldsymbol{\beta}'\mathbf{x}) - H(\theta_{j-1} + \beta'\mathbf{x}).$$

Suppose Z is an unobserved continuous random variable with a location/scale parameter distribution with cumulative distribution function $H\{(z - \mu)/\sigma\}$, and let $\phi_1, \ldots, \phi_{r-1}$ be fixed cut-point values of Z. We observe only the interval (ϕ_{j-1}, ϕ_j) in which Z lies, with probability

$$p_j = H((\phi_j - \mu)/\sigma) - H((\phi_{j-1} - \mu)/\sigma), \quad j = 1, \ldots, r$$

with $\phi_0 = -\infty$, $\phi_r = +\infty$. This is identical to the above model if

$$\phi_j/\sigma = \theta_j$$

and

$$-\mu/\sigma = \boldsymbol{\beta}'\mathbf{x}.$$

Thus an ordered cumulative probability model always has an interpretation in terms of modelling an underlying latent variable, though no such interpretation is necessary to use the model.

These models are discussed by (Bock, 1975, Chapter 8), McCullagh and Nelder (1989) and Anderson (1984). The cumulative distribution function H is usually taken to be normal or logistic. Programs for fitting some of these models include MULTIQUAL (Bock and Yates, 1973) and PLUM (McCullagh, 1980) and the function `polr` in the *MASS* package of Venables and Ripley (2002).

5.8 An Example

A survey of student opinion on the Vietnam War was taken at the University of North Carolina in Chapel Hill in May 1967 and published in the student newspaper. Students were asked to fill in 'ballot papers', available in the Student Council building, stating which policy out of A, B, C, or D they supported. Responses were cross-classified by sex and by undergraduate year or graduate status. The policies were:

```
A:    The US should defeat the power of North Vietnam by
      widespread bombing of its industries, ports and
      harbours and by land invasion.

B:    The US should follow the present policy in Vietnam.

C:    The US should de-escalate its military activity,
      stop bombing North Vietnam, and intensify its efforts
      to begin negotiation.

D:    The US should withdraw its military forces from Vietnam
      immediately.
```

Much more detail of this survey and a complementary correspondence analysis of the data are given in Aitkin (1996*c*).

The dataset `vietnam` is reproduced in Table 5.5.

```
> data(vietnam, package = "SMIR")
```

What can we conclude about students' attitude to the war? A first question concerns the response rates: what proportion of students actually responded in the survey? The numbers of students enrolled at the University, and the numbers responding (summed over the four categories A, B, C, and D) are shown in Table 5.6, with the response rates.

Overall, only 26% of male and 17% of female students expressed an opinion by filling in a ballot paper. The response rate is somewhat higher than average for fourth-year males and third-year females, and slightly lower than average for the other groups. We could model these response rates, treating the populations as samples, but the pattern of response is clear.

Given such low response rates, it seems difficult to draw any conclusion about the view of the student population. We have no way of knowing whether response

Table 5.5. Vietnam survey sample sizes

Year	A	B	C	D	Total
Males					
1	175	116	131	17	439
2	160	126	135	21	442
3	132	120	154	29	435
4	145	95	185	44	469
G	118	176	345	141	780
Females					
1	13	19	40	5	77
2	5	9	33	3	50
3	22	29	110	6	167
4	12	21	58	10	101
G	19	27	128	13	187

Table 5.6. Survey response rates

Year	Response	Enrolled	Response rate
Males			
1	439	1768	0.248
2	442	1792	0.247
3	435	1693	0.257
4	469	1522	0.308
G	780	3005	0.260
Total	2565	9780	0.262
Females			
1	77	487	0.158
2	50	326	0.153
3	167	772	0.216
4	101	608	0.166
G	187	1221	0.153
Total	582	3414	0.170

is itself related to intensity of feeling, and therefore possibly to the policy chosen. We can, however, investigate how the policy chosen varies by sex and year for those who responded. This is a different question; whether the conclusions are applicable to the population as a whole is unresolvable.

We show in Table 5.7 the observed proportions choosing each policy. For males, the proportions choosing policies A and B decrease, and those choosing C and D

Table 5.7. Observed proportions

Year	Policy			
	A	B	C	D
Males				
1	0.399	0.264	0.298	0.039
2	0.362	0.285	0.305	0.048
3	0.303	0.276	0.354	0.067
4	0.309	0.203	0.394	0.094
G	0.151	0.226	0.442	0.181
Females				
1	0.169	0.247	0.519	0.065
2	0.100	0.180	0.660	0.060
3	0.132	0.174	0.659	0.036
4	0.119	0.208	0.574	0.099
G	0.102	0.144	0.684	0.070

increase steadily with year. For females, policy C is consistently favoured but there is substantial variation with year.

```
> oval <- aperm(array(vietnam$count, dim = c(4, 5,
+     2), dimnames = list(Policy = levels(vietnam$policy),
+     Year = levels(vietnam$year), Sex = levels(vietnam$sex)[c(2,
+         1)])), c(3, 2, 1))
> vietnam.ob.tab <- prop.table(oval, c(1, 2))
> round(ftable(vietnam.ob.tab, row.vars = c(1, 2)),
+     3)
```

We proceed to model the proportions as though the respondents were randomly sampled from the student population. The variable response rates noted above are irrelevant in the analysis since we are including sex and year in the model (see Section 2.9 for a discussion of unequal sampling fractions).

We give two analyses, one based on the multinomial logit model and one on the continuation ratio model. In both analyses, we give a different analysis from that in Aitkin *et al.* (1989), using a suggestion of M. Green (in Aitkin, 1996*c*), to deal with the awkward Graduate group – we want to model trends over year of enrolment, but Graduates are a mixed group, whose actual years of University study may range from 5 to 9 or more. Green suggested using year 7 as a compromise, and we derive the very simple model resulting from this approach. We give first the multinomial logit model.

5.8.1 *Multinomial logit model*

The data are read in as 40 counts classified by factors `policy`, `year` and `sex`.

```
> vietnam.gnm1 <- gnm(count ~ policy, eliminate = year:sex,
+       data = vietnam, family = poisson, verbose = FALSE)
```

This is the null model on the multinomial logit scale: the `policy` and (eliminated) `year:sex` terms are included to reproduce the marginal totals. The fit is very poor: strong differences in response proportions are present. We now add the sex and year interactions with response:

```
> vietnam.gnm2 <- update(vietnam.gnm1, . ~ . + policy:sex)
> vietnam.gnm3 <- update(vietnam.gnm1, . ~ . + policy:year)
> vietnam.gnm4 <- update(vietnam.gnm3, . ~ . + policy:sex)
> print(anova(vietnam.gnm1, vietnam.gnm2, vietnam.gnm4),
+       digits = 5)

Analysis of Deviance Table

Model 1: count ~ policy
Model 2: count ~ policy + sex:policy
Model 3: count ~ policy + year:policy + sex:policy
  Resid. Df Resid. Dev Df Deviance
1        27     361.72
2        24     216.31  3   145.40
3        12      19.19 12   197.12

> print(anova(vietnam.gnm1, vietnam.gnm3, vietnam.gnm4),
+       digits = 5)

Analysis of Deviance Table

Model 1: count ~ policy
Model 2: count ~ policy + year:policy
Model 3: count ~ policy + year:policy + sex:policy
  Resid. Df Resid. Dev Df Deviance
1        27     361.72
2        15     153.94 12   207.78
3        12      19.19  3   134.74
```

Large changes in deviance show that both interactions are necessary. The only term not yet fitted is the three-way interaction; the deviance for this term is 19.19 on 12 df. Although this value does not indicate a notably bad fit, it is possible that a large contribution to this deviance comes from only a few degrees of freedom of the interaction term. We have already seen a different response pattern for males and females in the observed proportions and this should be investigated further. We model the responses for each sex separately.

```
> vietnam.males <- subset(vietnam, sex == "male")
```

Only males are now being analysed, so no sex terms need be fitted (if they are, they will be aliased and ignored).

```
> vietnam.males.gnm1 <- gnm(count ~ policy, data = vietnam.males,
+     eliminate = year, family = poisson, verbose = FALSE)
> vietnam.males.gnm2 <- update(vietnam.males.gnm1,
+     . ~ . + policy:year)
> summary(vietnam.males.gnm2)

Coefficients of interest:
              Estimate Std. Error z value Pr(>|z|)
policyB        -0.4112     0.1197   -3.43  0.00059
policyC        -0.2896     0.1155   -2.51  0.01219
policyD        -2.3316     0.2540   -9.18  < 2e-16
year2:policyB   0.1723     0.1689    1.02  0.30761
year3:policyB   0.3159     0.1739    1.82  0.06931
year4:policyB  -0.0117     0.1782   -0.07  0.94783
yearG:policyB   0.8110     0.1688    4.80  1.6e-06
year2:policyC   0.1197     0.1643    0.73  0.46641
year3:policyC   0.4437     0.1656    2.68  0.00736
year4:policyC   0.5332     0.1602    3.33  0.00087
yearG:policyC   1.3624     0.1572    8.67  < 2e-16
year2:policyD   0.3009     0.3441    0.87  0.38184
year3:policyD   0.8161     0.3265    2.50  0.01244
year4:policyD   1.1390     0.3069    3.71  0.00021
yearG:policyD   2.5096     0.2830    8.87  < 2e-16

(Dispersion parameter for poisson family taken to be 1)

Residual deviance: 3.2272e-15 on 0 degrees of freedom
AIC: 169.6

Number of iterations: 3
```

The large deviance for the interaction term (207.78) shows that proportions adopting each policy differ significantly over years. How can we describe these differences simply from the parameter estimates?

We have two orderings: by year and by policy. Within each policy, the logit increases with year, but not smoothly: there is a large jump in the logit in each case from year 4 to year 5, and for policy B the logit for year 4 is small and negative, whereas it is large and positive for the other years. Thus a linear trend over year will not fit all years, though it might fit years 1–4. We include the graduates in the model by assigning them to year 7 in the linear modelling of years. The year pattern of logits becomes stronger with increasing policy, suggesting that a linear

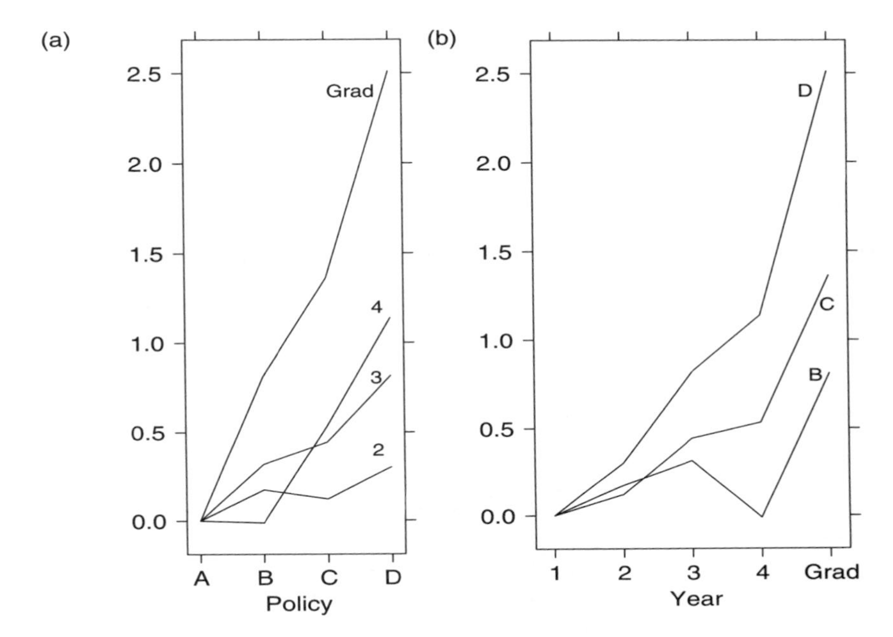

Fig. 5.11.

trend over policy might fit as well, that is, the regression coefficients on year are proportional over policy (Fig. 5.11).

We try these two model simplifications.

```
> vietnam.males <- transform(vietnam.males,
+       linp = unclass(policy))
> vietnam.males.gnm3 <- update(vietnam.males.gnm2,
+      . ~ . - policy:year + linp + linp:year)
> summary(vietnam.males.gnm3)

Coefficients of interest:
           Estimate Std. Error z value Pr(>|z|)
policyB     -0.4130     0.0692   -5.97  2.4e-09
policyC     -0.3542     0.1035   -3.42  0.00062
policyD     -2.1134     0.1625  -13.00  < 2e-16
linp             NA         NA      NA       NA
year2:linp   0.0688     0.0715    0.96  0.33547
year3:linp   0.2274     0.0713    3.19  0.00141
year4:linp   0.3227     0.0700    4.61  4.0e-06
yearG:linp   0.7378     0.0653   11.29  < 2e-16

(Dispersion parameter for poisson family taken to be 1)
```

```
Std. Error is NA where coefficient has been
         constrained or is unidentified

Residual deviance: 13.914 on 8 degrees of freedom
AIC: 167.5

Number of iterations: 4
```

The deviance shows a reasonable fit. We now assign year 7 to the graduates.

```
> vietnam.males <- transform(vietnam.males, liny = ifelse(year ==
+     "G", 7, unclass(year)))
> vietnam.males.gnm4 <- update(vietnam.males.gnm3,
+     . ~ . - year:linp + liny:linp)
> summary(vietnam.males.gnm4)

Coefficients of interest:
           Estimate Std. Error z value Pr(>|z|)
policyB    -0.57211    0.06141   -9.32   <2e-16
policyC    -0.67272    0.08424   -7.99   <2e-16
policyD    -2.58919    0.14072  -18.40   <2e-16
linp             NA         NA      NA       NA
linp:liny   0.12673    0.00963   13.16   <2e-16

(Dispersion parameter for poisson family taken to be 1)

Std. Error is NA where coefficient has been constrained
         or is unidentified

Residual deviance: 14.935 on 11 degrees of freedom
AIC: 162.6

Number of iterations: 4
```

The linear model across years fits very well. We examine the residuals.

```
> round(residuals(vietnam.males.gnm4, "pearson"), 3)

     1      2      3      4      5      6      7      8      9
-0.542 -0.073  1.016 -0.675 -0.127  0.783 -0.164 -0.973 -0.512
    10     11     12     13     14     15     16     17     18
 0.570  0.272 -0.600  1.578 -2.215  0.455  0.045 -0.260  0.807
    19     20
-0.977  0.974
```

There is a large residual of −2.22 at the 14th observation, where the observed number 95 of fourth-year students choosing policy B – the current policy – is not close to the model fitted value of 120.04. We refit the above model excluding the 14th observation using the *gnm* subset option.

```
> vietnam.males.gnm5 <- update(vietnam.males.gnm4,
+     subset = (row.names(vietnam) != "14"))
> summary(vietnam.males.gnm5)

Coefficients of interest:
          Estimate Std. Error z value Pr(>|z|)
policyB   -0.51643    0.06402   -8.07   <2e-16
policyC   -0.68685    0.08496   -8.08   <2e-16
policyD   -2.61173    0.14191  -18.40   <2e-16
linp            NA         NA      NA       NA
linp:liny  0.12845    0.00973   13.21   <2e-16

(Dispersion parameter for poisson family taken to be 1)

Std. Error is NA where coefficient has been constrained
          or is unidentified

Residual deviance: 6.4964 on 10 degrees of freedom
AIC: 147.7

Number of iterations: 4
```

The model fits very closely. Should we look for a further model which fits better, or is there a sensible interpretation of this model, and the outlying 14th observation?

A linear trend in policy means a linear increase in the log odds against policy A with each step along the policy scale: the policy options have been expressed in a smoothly ordered way. The outlying observation 14 does not fit this trend: fourth-year male students choose option B much less often than option A, instead of slightly less often as would be predicted from the linear increase in the odds against A.

We can give a plausible explanation for this. Fourth-year male students faced the draft and the possibility of serving in the war at the end of their academic year, unless they obtained deferment, for example by becoming graduate students. A policy of continuing the present conflict, rather than either bringing it to a presumably rapid end by invasion, or withdrawing to relatively safe positions, might have been viewed as the policy most likely to endanger their own lives. It is not surprising to find that this policy was much less favoured than A or C among the fourth-year students. Students in other undergraduate years did not face this immediate prospect.

We now turn to the females.

```
> vietnam.females <- subset(vietnam, sex == "female")
> vietnam.females.gnm1 <- gnm(count ~ policy,
+     data = vietnam.females, eliminate = year,
+     family = poisson)

> summary(vietnam.females.gnm1)
```

```
Coefficients of interest:
        Estimate Std. Error z value Pr(>|z|)
policyB    0.391      0.154    2.55   0.0109
policyC    1.648      0.130   12.72   <2e-16
policyD   -0.652      0.203   -3.21   0.0013

(Dispersion parameter for poisson family taken to be 1)

Residual deviance: 13.263 on 12 degrees of freedom
AIC: 124.0

Number of iterations: 2

> sort(round(residuals(vietnam.females.gnm1, "pearson"),
+     3))

    32     23     38     37     35     25     30     28     33
-1.417 -1.262 -1.160 -0.798 -0.754 -0.445 -0.206 -0.100 -0.092
    26     24     27     40     29     31     34     39     21
-0.007  0.047  0.231  0.322  0.360  0.400  0.651  0.867  1.177
    22     36
 1.371  1.412
```

The null logit model gives a good fit; the largest residual is -1.42. There are no differences among the proportions choosing the four policies, other than those attributable to sampling variation. The fitted proportions from the model are obtained from

$$\hat{p}_j = e^{\hat{\theta}_j} \Bigg/ \sum_{j=1}^{r} e^{\hat{\theta}_j}, \quad \hat{\theta}_1 = 0,$$

and are 0.122, 0.180, 0.634, and 0.064. These are simply the marginal proportions giving each response, collapsed over year.

The device of fitting separate models for each sex is useful as the model structures are very different. The same result can be achieved by fitting the following single model to the complete data.

```
> vietnam <- transform(vietnam, liny = ifelse(year ==
+     "G", 7, unclass(year)), linp = unclass(policy),
+     female = sex == "female")
> vietnam <- transform(vietnam, v = (sex == "male") *
+     liny * linp)
> vietnam.gnm5 <- gnm(count ~ policy + female:policy +
+     v, data = vietnam, eliminate = year:sex, family = poisson,
+     subset = (row.names(vietnam) != "14"), verbose = FALSE)
> summary(vietnam.gnm5)
```

```
Coefficients of interest:
                   Estimate Std. Error z value Pr(>|z|)
policyB            -0.51643    0.06402   -8.07   <2e-16
policyC            -0.68685    0.08496   -8.08   <2e-16
policyD            -2.61173    0.14191  -18.40   <2e-16
policyA:femaleTRUE -2.17895         NA      NA       NA
policyB:femaleTRUE -1.27124         NA      NA       NA
policyC:femaleTRUE  0.15602         NA      NA       NA
policyD:femaleTRUE -0.21898         NA      NA       NA
v                   0.12845    0.00973   13.21   <2e-16

(Dispersion parameter for poisson family taken to be 1)

Std. Error is NA where coefficient has been
      constrained or is unidentified

Residual deviance: 19.759 on 22 degrees of freedom
AIC: 271.8

Number of iterations: 4
```

The deviance of 19.76 on 22 df is the sum of the values 6.50 on 10 df for males (from *vietnam.males.gnm5*) and 13.26 on 12 df for females (*vietnam.females.gnm1*). The 14th observation is still excluded. The corresponding Pearson residual can be directly calculated: it can not be extracted using the *resid* function since this observation is left out.

We now construct the fitted proportions from the model. The fitted models for each sex are shown in Fig. 5.12.

```
> vietnam$v <- as.numeric(vietnam$v)
> vietnam.gnm5.14 <- update(vietnam.gnm5, data = vietnam,
+     subset = row.names(vietnam))
> vietnam$fv <- fitted(vietnam.gnm5.14)
> vietnam$t <- rep(t(tapply(vietnam$count, list(vietnam$sex,
+     vietnam$year), sum)[c(2, 1), ]), rep(4, 10))
> vietnam <- transform(vietnam, fp = fv/t, op = count/t)
> print(xyplot(fp ~ liny | sex, data = vietnam, groups = policy,
+     panel = function(x, y, type, subscripts, ...) {
+         lpoints(vietnam$liny[subscripts], vietnam$op[subscripts],
+             pch = LETTERS[vietnam$policy[subscripts]],
+             cex = 1.5)
+         panel.superpose(x, y, type = "l", subscripts,
+             ...)
+     }))
```

The model fits the observed proportions closely except at observation 14.

In conclusion, 63% of female students responding choose policy C, while 18% choose B, 12% A and 6% D. These patterns are consistent over years. For males,

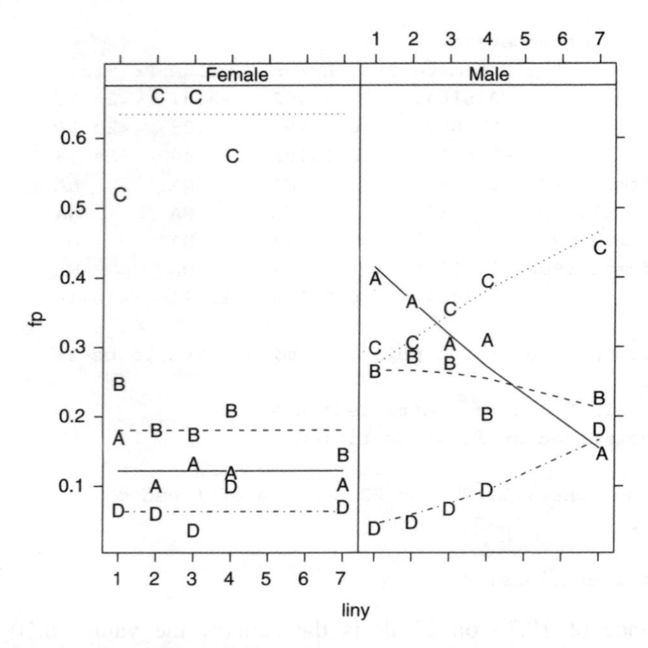

Fig. 5.12. Fitted and observed proportions of males and females supporting each of the policies

support for policy A declines steadily with year, while that for policy B increases slightly and then decreases. Support for C increases rapidly, but that for D only slowly.

We now consider the continuation ratio model.

5.8.2 *Continuation ratio model*

We write the model in the form

$$\frac{n!}{n_{\mathrm{A}}!(n_{\mathrm{B}}+n_{\mathrm{C}}+n_{\mathrm{D}})!}p_{\mathrm{A}}^{n_{\mathrm{A}}}(1-p_{\mathrm{A}})^{n_{\mathrm{B}}+n_{\mathrm{C}}+n_{\mathrm{D}}},$$

$$\frac{(n_{\mathrm{B}}+n_{\mathrm{C}}+n_{\mathrm{D}})!}{n_{\mathrm{B}}!(n_{\mathrm{C}}+n_{\mathrm{D}})!}\left(\frac{p_{\mathrm{B}}}{1-p_{\mathrm{A}}}\right)^{n_{\mathrm{B}}}\left(\frac{p_{\mathrm{C}}+p_{\mathrm{D}}}{1-p_{\mathrm{A}}}\right)^{n_{\mathrm{C}}+n_{\mathrm{D}}},$$

$$\frac{(n_{\mathrm{C}}+n_{\mathrm{D}})!}{n_{\mathrm{C}}!n_{\mathrm{D}}!}\left(\frac{p_{\mathrm{C}}}{1-p_{\mathrm{A}}-p_{\mathrm{B}}}\right)^{n_{\mathrm{C}}}\left(\frac{p_{\mathrm{D}}}{1-p_{\mathrm{A}}-p_{\mathrm{B}}}\right)^{n_{\mathrm{D}}}.$$

We fit three binomial logit models by constructing two new variables of length 30 containing in blocks of 10 the counts for A versus (B+C+D), B versus (C+D), and

C versus D. The five-category year factor is again replaced by its linear term, with Graduates coded to year 7.

```
> policy.tab <- with(vietnam, tapply(count, list(year,
+     sex, policy), sum))
> Atab <- apply(policy.tab[, , 1], c(1, 2), sum)
> Btab <- apply(policy.tab[, , 2], c(1, 2), sum)
> Ctab <- apply(policy.tab[, , 3], c(1, 2), sum)
> Dtab <- apply(policy.tab[, , 4], c(1, 2), sum)
> BCDtab <- apply(policy.tab[, , 2:4], c(1, 2), sum)
> CDtab <- apply(policy.tab[, , 3:4], c(1, 2), sum)
> vietnam.cr <- expand.grid(year = factor(c(1:4, "G")),
+     sex = c("female", "male"), policy = c("A", "B",
+         "C"))
> vietnam.cr$d1 <- c(Atab, Btab, Ctab)
> vietnam.cr$d2 <- c(BCDtab, CDtab, Dtab)
> vietnam.cr.glm <- glm(cbind(d1, d2) ~ policy + sex +
+     year + sex:year, data = vietnam.cr, family = binomial)
> anova(vietnam.cr.glm)

Analysis of Deviance Table

Model: binomial, link: logit

Response: cbind(d1, d2)

Terms added sequentially (first to last)


         Df Deviance Resid. Df Resid. Dev
NULL                        29       1927
policy    2     1565        27        362
sex       1       44        26        318
year      4      188        22        130
sex:year  4       22        18        108

> vietnam.cr.glm1 <- update(vietnam.cr.glm, . ~ . +
+     policy:sex)
> summary(vietnam.cr.glm1)

Coefficients:
                Estimate Std. Error z value Pr(>|z|)
(Intercept)      -1.6275     0.2180   -7.47  8.3e-14
policyB           0.6291     0.1679    3.75  0.00018
policyC           4.3124     0.2159   19.98  < 2e-16
sexmale           1.1555     0.2317    4.99  6.1e-07
year2            -0.4266     0.3242   -1.32  0.18817
year3            -0.2528     0.2403   -1.05  0.29286
```

```
year4            -0.4032      0.2685   -1.50   0.13320
yearG            -0.5606      0.2405   -2.33   0.01974
sexmale:year2     0.3344      0.3402    0.98   0.32570
sexmale:year3    -0.0731      0.2616   -0.28   0.77979
sexmale:year4    -0.1092      0.2873   -0.38   0.70396
sexmale:yearG    -0.5005      0.2580   -1.94   0.05242
policyB:sexmale  -0.2855      0.1808   -1.58   0.11441
policyC:sexmale  -1.8742      0.2328   -8.05   8.1e-16

(Dispersion parameter for binomial family taken to be 1)

    Null deviance: 1927.170  on 29  degrees of freedom
Residual deviance:   23.482  on 16  degrees of freedom
AIC: 203.2

Number of Fisher Scoring iterations: 4

> vietnam.cr.glm2 <- update(vietnam.cr.glm1, . ~ . +
+      policy:year)
> summary(vietnam.cr.glm2)

Coefficients:
                Estimate Std. Error z value Pr(>|z|)
(Intercept)      -1.5448     0.2281   -6.77  1.3e-11
policyB           0.4849     0.2174    2.23   0.0257
policyC           4.1810     0.3195   13.08  < 2e-16
sexmale           1.1286     0.2305    4.90  9.8e-07
year2            -0.5116     0.3464   -1.48   0.1397
year3            -0.3616     0.2661   -1.36   0.1742
year4            -0.2816     0.2884   -0.98   0.3289
yearG            -0.8121     0.2653   -3.06   0.0022
sexmale:year2     0.3546     0.3414    1.04   0.2990
sexmale:year3    -0.0488     0.2628   -0.19   0.8526
sexmale:year4    -0.1262     0.2852   -0.44   0.6582
sexmale:yearG    -0.4683     0.2584   -1.81   0.0699
policyB:sexmale  -0.2734     0.1830   -1.49   0.1351
policyC:sexmale  -1.8489     0.2354   -7.85  4.1e-15
policyB:year2     0.1517     0.2105    0.72   0.4712
policyC:year2     0.1258     0.3438    0.37   0.7143
policyB:year3     0.1537     0.2036    0.76   0.4502
policyC:year3     0.2685     0.3207    0.84   0.4025
policyB:year4    -0.1911     0.2071   -0.92   0.3562
policyC:year4    -0.1592     0.3038   -0.52   0.6002
policyB:yearG     0.4295     0.1919    2.24   0.0252
policyC:yearG     0.2824     0.2789    1.01   0.3114

(Dispersion parameter for binomial family taken to be 1)
```

```
    Null deviance: 1927.170  on 29  degrees of freedom
Residual deviance:   10.576  on  8  degrees of freedom
AIC: 206.3

Number of Fisher Scoring iterations: 4
```

The three-way interaction deviance is 10.58 on 8 df. Inspection of the parameter estimates for the saturated model (not shown) shows that the three-way interactions all have z-statistics around 1 or less. We may eliminate them, and then examine the no-three-factor interaction model above. We first recode the Graduates to year 7 and define a variate with these values.

```
> vietnam.cr$linyear <- with(vietnam.cr, ifelse(year ==
+      "G", 7, unclass(year)))
```

Examining the two-way interactions, we find that only appreciable terms are `policyC:sexmale`, `policyB:yearG` and `sexmale:yearG`. We replace the five-year `year` factor by the `linyear` variate.

```
> vietnam.cr.glm3 <- update(vietnam.cr.glm2, . ~ . -
+      policy:year + policy:linyear)
> anova(vietnam.cr.glm3, vietnam.cr.glm2)

Analysis of Deviance Table

Model 1: cbind(d1, d2) ~ policy + sex + year + sex:year
     + policy:sex + policy:linyear
Model 2: cbind(d1, d2) ~ policy + sex + year + sex:year
     + policy:sex + policy:year
  Resid. Df Resid. Dev Df Deviance
1        14      18.86
2         8      10.58  6     8.29

> summary(vietnam.cr.glm3)

Coefficients: (1 not defined because of singularities)
                Estimate Std. Error z value Pr(>|z|)
(Intercept)      -1.5025     0.2276   -6.60  4.0e-11
policyB           0.3920     0.2010    1.95    0.051
policyC           4.1926     0.2670   15.70  < 2e-16
sexmale           1.1275     0.2300    4.90  9.5e-07
year2            -0.4236     0.3259   -1.30    0.194
year3            -0.2523     0.2479   -1.02    0.309
year4            -0.4049     0.2834   -1.43    0.153
yearG            -0.5736     0.2933   -1.96    0.051
sexmale:year2     0.3411     0.3398    1.00    0.315
sexmale:year3    -0.0560     0.2612   -0.21    0.830
sexmale:year4    -0.0854     0.2872   -0.30    0.766
```

```
sexmale:yearG     -0.4671     0.2585   -1.81     0.071
policyB:sexmale   -0.2737     0.1812   -1.51     0.131
policyC:sexmale   -1.8690     0.2337   -8.00   1.3e-15
policyA:linyear   -0.0322     0.0365   -0.88     0.378
policyB:linyear    0.0291     0.0364    0.80     0.423
policyC:linyear        NA         NA      NA        NA

(Dispersion parameter for binomial family taken to be 1)

    Null deviance: 1927.170  on 29  degrees of freedom
Residual deviance:   18.862  on 14  degrees of freedom
AIC: 202.6

Number of Fisher Scoring iterations: 4
```

The deviance increase is consistent with the df. We examine the residuals.

```
> round(resid(vietnam.cr.glm3), 3)

     1      2      3      4      5      6      7      8      9
-0.196 -0.451 -0.150  0.105  0.496 -0.040 -0.453 -0.491  1.879
    10     11     12     13     14     15     16     17     18
-0.842  0.789  0.238 -0.540  0.879 -0.838 -0.129  0.563  0.473
    19     20     21     22     23     24     25     26     27
-2.076  0.881 -1.190  0.222  1.207 -1.447  0.600  0.367 -0.022
    28     29     30
 0.145 -0.132 -0.131

> vres <- resid(vietnam.cr.glm3)
```

The largest residual is -2.1 – the variate model fits well. The interaction terms for `policy:linyear` look small. We remove them.

```
> vietnam.cr.glm4 <- update(vietnam.cr.glm3, . ~ . -
+      policy:linyear)
> anova(vietnam.cr.glm4, vietnam.cr.glm3, test = "Chisq")

Analysis of Deviance Table

Model 1: cbind(d1, d2) ~ policy + sex + year + sex:year +
    policy:sex
Model 2: cbind(d1, d2) ~ policy + sex + year + sex:year +
    policy:sex + policy:linyear
  Resid. Df Resid. Dev Df Deviance P(>|Chi|)
1        16      23.48
2        14      18.86  2     4.62      0.10
```

The deviance change is at the 10% point of χ^2_2. We continue the factor-variate replacement.

```
> vietnam.cr.glm5 <- update(vietnam.cr.glm4, . ~ . -
+     sex:year + sex:linyear)
> anova(vietnam.cr.glm5, vietnam.cr.glm4)

Analysis of Deviance Table

Model 1: cbind(d1, d2) ~ policy + sex + year + policy:sex
     + sex:linyear
Model 2: cbind(d1, d2) ~ policy + sex + year + sex:year
     + policy:sex
  Resid. Df Resid. Dev Df Deviance
1        19      25.30
2        16      23.48  3     1.82

> vietnam.cr.glm6 <- update(vietnam.cr.glm5, . ~ . -
+     year + linyear)
> anova(vietnam.cr.glm6, vietnam.cr.glm5)

Analysis of Deviance Table

Model 1: cbind(d1, d2) ~ policy + sex + linyear + policy:sex
      + sex:linyear
Model 2: cbind(d1, d2) ~ policy + sex + year + policy:sex
      + sex:linyear
  Resid. Df Resid. Dev Df Deviance
1        22      25.59
2        19      25.30  3     0.29

> coef(summary(vietnam.cr.glm6))

                Estimate Std. Error z value Pr(>|z|)
(Intercept)      -1.6709     0.1848   -9.04 1.54e-19
policyB           0.6267     0.1677    3.74 1.87e-04
policyC           4.3055     0.2156   19.97 1.04e-88
sexmale           1.4223     0.1960    7.26 3.96e-13
linyear          -0.0762     0.0350   -2.18 2.94e-02
policyB:sexmale  -0.2830     0.1807   -1.57 1.17e-01
policyC:sexmale  -1.8686     0.2325   -8.04 9.22e-16
sexmale:linyear  -0.1067     0.0376   -2.84 4.53e-03
```

The effect of linyear for females is small −0.076. We try to remove it.

```
> vietnam.cr <- transform(vietnam.cr, male.linyear = ifelse(sex ==
+     "male", linyear, 0))
> vietnam.cr.glm7 <- update(vietnam.cr.glm6, . ~ . -
+     linyear - sex:linyear + male.linyear)
> anova(vietnam.cr.glm7, vietnam.cr.glm6)

Analysis of Deviance Table
```

```
Model 1: cbind(d1, d2) ~ policy + sex + male.linyear + policy:sex
Model 2: cbind(d1, d2) ~ policy + sex + linyear + policy:sex
     + sex:linyear
  Resid. Df Resid. Dev Df Deviance
1        23       30.4
2        22       25.6  1      4.8
```

The linear trend for females is small, but very significant. We use the previous model (*vietnam.cr.glm6*).

The final continuation ratio model (*vietnam.cr.glm6*) has a deviance of 25.59 on 22 df compared with the final multinomial logit model (*vietnam.gnm5*) deviance of 19.76, also with 22 df. The deviance difference of 5.83 looks very substantial, with a corresponding likelihood ratio of 0.0542 against the continuation ratio model. However, the deviance reported for the multinomial logit model excludes observation 14 – if this is included the model deviance increases by 8.44 to 28.20, and the continuation ratio model fits better, without any need to declare observation 14 an outlier!

It may be puzzling, however, that the continuation ratio model finds a linear trend over years for females where the multinomial logit model does not. This difference occurs because we did not in fact fit the linear trend model for females, because the null model appeared to fit well. We re-examine this point below, but first present the final continuation ratio model on the original scale.

We first tabulate the fitted probabilities (shown to 4 dp) from the three logit models.

```
> fval <- array(fitted(vietnam.cr.glm6), dim = c(5,
+      2, 3), dimnames = list(Year = levels(vietnam.cr$year),
+      Sex = levels(vietnam.cr$sex)[c(2, 1)],
+      Policy = levels(vietnam.cr$policy)))
> round(ftable(fval, row.vars = c(2, 1), col.vars = 3),
+      4)

                Policy      A      B      C
Sex    Year
male   1               0.1484 0.2459 0.9281
       2               0.1390 0.2321 0.9229
       3               0.1302 0.2188 0.9173
       4               0.1218 0.2060 0.9113
       G               0.0994 0.1711 0.8910
female 1               0.3938 0.4781 0.8814
       2               0.3510 0.4327 0.8609
       3               0.3106 0.3885 0.8375
       4               0.2728 0.3460 0.8110
       G               0.1781 0.2341 0.7125
```

The first row of each sex of the table also give the fitted probabilities for response A in the multinomial model. The remaining rows give the fitted conditional probabilities for responses B and C, conditional on B+C+D and C+D, respectively. To obtain the fitted multinomial probabilities for responses B, C, and D, we multiply the `policyB` entries by the corresponding probabilities $(1 - p_{\mathrm{A}})$ to give the multinomial probabilities for category B, and multiply the `policyC` entries by $(1 - p_{\mathrm{A}} - p_{\mathrm{B}})$ to give the multinomial probabilities for category C; those for category D are obtained by difference from 1. These are easily calculated in R, and are given below rounded to 3 dp:

```
> fvalA <- fval[, , 1]
> fvalB <- fval[, , 2] * (1 - fvalA)
> fvalC <- fval[, , 3] * (1 - fvalA - fvalB)
> fvalD <- (1 - fvalA - fvalB - fvalC)
> fvalABCD <- array(c(fvalA, fvalB, fvalC, fvalD),
+      dim = c(5, 2, 4))
> dimnames(fvalABCD) <- list(Year = levels(vietnam$year),
+      Sex = levels(vietnam$sex), Policy = levels(vietnam$policy))
> round(ftable(fvalABCD, row.vars = c(2, 1), col.vars = 3),
+      3)

              Policy      A     B     C     D
Sex    Year
female 1                0.148 0.209 0.596 0.046
       2                0.139 0.200 0.610 0.051
       3                0.130 0.190 0.623 0.056
       4                0.122 0.181 0.635 0.062
       G                0.099 0.154 0.665 0.081
male   1                0.394 0.290 0.279 0.038
       2                0.351 0.281 0.317 0.051
       3                0.311 0.268 0.353 0.069
       4                0.273 0.252 0.386 0.090
       G                0.178 0.192 0.449 0.181
```

We note again the quite different models for females from the constant model given previously. In particular, the female proportions supporting the 'hawk' policies A and B decline steadily with year, as they do for males. For the continuation ratio model, the females behave quite similarly to the males, in terms of trend over year. Note also that the 'discrepancy' we found in the under-use of Policy B by males in year 4 is not present in this model. Thus our interpretation of this discrepancy is speculative.

We finally return to the multinomial logit analysis of the females, and try the effect of linear trends across policies and years.

```
> vietnam.females.glm <- glm(count ~ year + policy,
+      data = vietnam, subset = sex == "female",
+      family = "poisson")
> vietnam.females.glm1 <- update(vietnam.females.glm,
+      . ~ . + policy:liny)
```

```
> anova(vietnam.females.glm, vietnam.females.glm1)

Analysis of Deviance Table

Model 1: count ~ year + policy
Model 2: count ~ year + policy + policy:liny
  Resid. Df Resid. Dev Df Deviance
1        12      13.26
2         9       7.70  3     5.56

> round(coef(summary(vietnam.females.glm1))[-c(1:5),
+     ], 4)

              Estimate Std. Error z value Pr(>|z|)
policyB          0.410     0.3163   1.296   0.1950
policyC          1.269     0.2701   4.700   0.0000
policyD         -1.108     0.4392  -2.522   0.0117
policyA:liny    -0.112     0.0940  -1.195   0.2322
policyB:liny    -0.117     0.0885  -1.325   0.1851
policyC:liny    -0.018     0.0789  -0.228   0.8194

> vietnam.females.glm2 <- update(vietnam.females.glm1,
+     . ~ . - policy:liny + linp:liny)
> anova(vietnam.females.glm2, vietnam.females.glm1)

Analysis of Deviance Table

Model 1: count ~ year + policy + liny:linp
Model 2: count ~ year + policy + policy:liny
  Resid. Df Resid. Dev Df Deviance
1        11       8.75
2         9       7.70  2     1.05

> round(coef(summary(vietnam.females.glm2))[-c(1:5),
+     ], 4)
          Estimate Std. Error z value Pr(>|z|)
policyB      0.188     0.1779    1.06   0.2898
policyC      1.229     0.2311    5.32   0.0000
policyD     -1.300     0.3668   -3.54   0.0004
liny:linp    0.053     0.0252    2.10   0.0354
> vietnam.females.glm3 <- update(vietnam.females.glm2,
+     . ~ . - linp:liny)
> anova(vietnam.females.glm3, vietnam.females.glm2)

Analysis of Deviance Table

Model 1: count ~ year + policy
```

```
Model 2: count ~ year + policy + liny:linp
  Resid. Df Resid. Dev Df Deviance
1        12      13.26
2        11       8.75  1     4.51
```

The replacement of year by its linear component in the interaction fits quite well ($\chi^2_3 = 5.56$) as does the replacement of policy by its linear component. The single df for interaction is then important, giving the same pattern of response as for males.

We show in Fig. 5.13 the fitted proportions for males and females from the continuation ratio model. Note the use of the `as.data.frame.table` function to convert the array `fvalABCD` to a dataframe ready for the lattice function `xyplot`.

```
> vietnam.cr.df <- as.data.frame.table(aperm(fvalABCD,
+     c(2, 1, 3)))
> vietnam.cr.df$pobs <- as.vector(vietnam.ob.tab[c(2,
+     1), , ])
> vietnam.cr.df$year <- with(vietnam.cr.df, ifelse(Year ==
+     "G", 7, unclass(vietnam.cr.df$Year)))
> print(xyplot(Freq ~ year | Sex, data = vietnam.cr.df,
+     groups = Policy, panel = function(x, y, type,
+         subscripts, ...) {
+         lpoints(vietnam.cr.df$year[subscripts],
+                 vietnam.cr.df$pobs[subscripts],
```

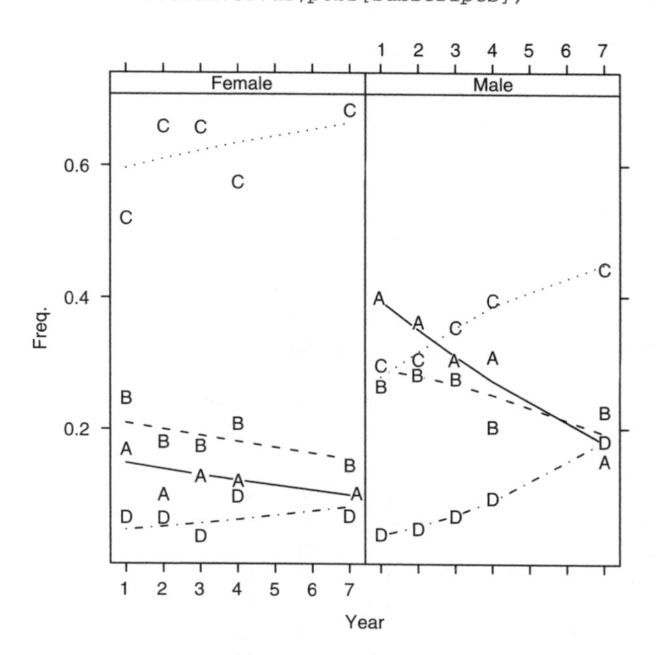

Fig. 5.13. Fitted continuation ratio model

```
+            pch = LETTERS[vietnam.cr.df$Policy[subscripts]],
+            cex = 1.5)
+        panel.superpose(x, y, type = "l", subscripts,
+            ...)
+    }))
```

5.9 Structured multinomial responses

In many studies with observational data, several categorical responses may be observed. The associations between these categorical response variables may be of interest, and particularly the effect of explanatory variables on the strength of these associations. The dataset `toxaemia` contains the number of women giving birth to their first child who showed toxaemic signs (hypertension and/or proteinurea, classified as Yes or No) during pregnancy. The data may be found in Brown *et al.* (1983) and were collected in Bradford between 1968 and 1977 on 13,384 UK women.

The four possible response patterns are cross-classified by social class (I–V), giving five categories of a factor `class`, and a three-level factor `smoke`, relating to the number of cigarettes smoked per day during pregnancy (0, 1–19, 20+). The counts in the four response categories are given by the variables `HU`, `HN`, `NU` and `NN`, representing those with hypertension and proteinurea, hypertension only, proteinurea only and neither symptom. We show these counts in Table 5.8, and the proportions in all four categories in Table 5.9.

```
> data(toxaemia, package = "SMIR")

> tox.prop.table1 <- with(toxaemia, prop.table(tapply(count,
+     list(class = class, response = response, smoke = smoke),
+     sum), c(1, 3))[, c(2, 1, 4, 3), 1:2])
> tox.prop.table2 <- with(toxaemia, prop.table(tapply(count,
+     list(class = class, response = response, smoke = smoke),
+     sum), c(1, 3))[, c(2, 1, 4, 3), 3, drop = FALSE])
```

How do the prevalence and association of these symptoms vary with `class` and `smoke`? We first consider each binary outcome separately, and then examine their association and the effect this has on the analysis.

Table 5.8. Hypertension and proteinurea counts

Smoke	0				1–19				20+			
class	HU	HN	NU	NN	HU	HN	NU	NN	HU	HN	NU	NN
I	28	82	21	286	5	24	5	71	1	3	0	13
II	50	266	34	785	13	92	17	284	0	15	3	34
III	278	1101	164	3160	120	492	142	2300	16	92	32	383
IV	63	213	52	656	35	129	46	649	7	40	12	163
V	20	78	23	245	22	74	34	321	7	14	4	65

Table 5.9. Hypertension and proteinurea proportions

Class	Smoke response											
	0				1-19				20+			
	HU	HN	NU	NN	HU	HN	NU	NN	HU	HN	NU	NN
I	0.067	0.197	0.050	0.686	0.048	0.229	0.048	0.676	0.059	0.176	0.000	0.765
II	0.044	0.234	0.030	0.692	0.032	0.227	0.042	0.700	0.000	0.288	0.058	0.654
III	0.059	0.234	0.035	0.672	0.039	0.161	0.046	0.753	0.031	0.176	0.061	0.732
IV	0.064	0.216	0.053	0.667	0.041	0.150	0.054	0.756	0.032	0.180	0.054	0.734
V	0.055	0.213	0.063	0.669	0.049	0.164	0.075	0.712	0.078	0.156	0.044	0.722

5.9.1 *Independent outcomes*

We begin with hypertension, and tabulate the observed proportions with hypertension. The four outcome combinations are held in variates of length 15.

```
> toxaemia$H <- factor(toxaemia$response %in% c("HN",
+     "HU"))
> round(ftable(with(toxaemia, prop.table(tapply(count,
+     list(class, H, smoke), sum), margin = c(1, 3)))[,
+     2, ]), 4)

         0  1-19   20+

I    0.264 0.276 0.235
II   0.278 0.259 0.288
III  0.293 0.200 0.206
IV   0.281 0.191 0.212
V    0.268 0.213 0.233
```

The observed proportions are quite variable over the range 0.2–0.3 and show a generally lower hypertension rate for the smokers. Class differences are hard to identify. We fit models.

```
> toxaemia1 <- data.frame(xtabs(count ~ class + H +
+     smoke, data = toxaemia))
> hypertension <- cbind(toxaemia1[toxaemia1$H == "TRUE",
+     -2], t = toxaemia1[toxaemia1$H == "FALSE", 4])
> hypertension.glm <- glm(cbind(Freq, t) ~ 1, data = hypertension,
+     family = binomial)
> hypertension.glm1 <- update(hypertension.glm, . ~
+     . + class)
> hypertension.glm2 <- update(hypertension.glm1, . ~
+     . + smoke)
> hypertension.glm3 <- update(hypertension.glm2, . ~
+     . + class * smoke)
```

```
> anova(hypertension.glm, hypertension.glm1, hypertension.glm2,
+     hypertension.glm3)

Analysis of Deviance Table

Model 1: cbind(Freq, t) ~ 1
Model 2: cbind(Freq, t) ~ class
Model 3: cbind(Freq, t) ~ class + smoke
Model 4: cbind(Freq, t) ~ class + smoke + class:smoke
  Resid. Df Resid. Dev Df Deviance
1        14      126.2
2        10      117.7  4      8.5
3         8       14.6  2    103.1
4         0    2.9e-14  8     14.6

> round(coef(summary(hypertension.glm3)), 4)
                      Estimate Std. Error z value Pr(>|z|)
(Intercept)            -1.0264      0.111 -9.2363    0.000
classII                 0.0740      0.129  0.5722    0.567
classIII                0.1466      0.116  1.2673    0.205
classIV                 0.0843      0.132  0.6396    0.522
classV                  0.0203      0.162  0.1255    0.900
smoke1-19               0.0629      0.245  0.2569    0.797
smoke20+               -0.1523      0.583 -0.2615    0.794
classII:smoke1-19      -0.1637      0.278 -0.5892    0.556
classIII:smoke1-19     -0.5670      0.251 -2.2578    0.024
classIV:smoke1-19      -0.5649      0.269 -2.0972    0.036
classV:smoke1-19       -0.3647      0.295 -1.2353    0.217
classII:smoke20+        0.2018      0.661  0.3051    0.760
classIII:smoke20+      -0.3140      0.593 -0.5294    0.597
classIV:smoke20+       -0.2203      0.609 -0.3615    0.718
classV:smoke20+        -0.0313      0.644 -0.0485    0.961
```

The two-way interaction deviance is large and cannot be omitted. The interaction terms allow several different interpretations. We note that the standard errors for the `smoke20+` category are much larger than those for the `smoke1-19` category because of the much smaller sample sizes. A careful inspection shows that the differences between the levels 2 and 3 of `smoke` in the interactions is nearly constant, between 0.253 and 0.366. This suggests that the interaction structure can be simplified by setting these differences equal in the model. To clarify the simplification, write `s` for `smoke` and `c` for `class`. The model

$$\beta_1 c2 \cdot s2 + \beta_2 c2 \cdot s3 + \beta_4 c3 \cdot s2 + \beta_5 c3 \cdot s3 + \cdots$$

under the restriction $\beta_2 - \beta_1 = \beta_4 - \beta_3 = \cdots = \delta$ reduces to

$$\beta_1 c2 \cdot (s2 + s3) + \beta_3 c3 \cdot (s2 + s3) + \cdots + \delta(c2 + c3 + \cdots) \cdot s3.$$

The variables $s2 + s3$ and $c2 + c3 + \cdots . + c5$ are simply expressed through their logical complements:

```
> hypertension <- transform(hypertension, sn1 = ifelse(smoke ==
+      "0", 0, 1), cn1 = ifelse(class == "I", 0, 1),
+      s3 = ifelse(smoke == "20+", 1, 0))
> hypertension.glm4 <- update(hypertension.glm, . ~
+      class + smoke + sn1 + cn1 + class:sn1 + cn1:s3)
> summary(hypertension.glm4)

Coefficients: (2 not defined because of singularities)
              Estimate Std. Error z value Pr(>|z|)
(Intercept)    -1.0264     0.1111   -9.24   <2e-16
classII         0.0740     0.1294    0.57    0.567
classIII        0.1466     0.1156    1.27    0.205
classIV         0.0843     0.1318    0.64    0.522
classV          0.0203     0.1621    0.13    0.900
smoke1-19       0.0629     0.2449    0.26    0.797
smoke20+       -0.1523     0.5825   -0.26    0.794
sn1                 NA         NA      NA       NA
cn1                 NA         NA      NA       NA
classII:sn1    -0.1549     0.2753   -0.56    0.574
classIII:sn1   -0.5727     0.2509   -2.28    0.022
classIV:sn1    -0.5534     0.2670   -2.07    0.038
classV:sn1     -0.3573     0.2917   -1.22    0.221
cn1:s3          0.2909     0.6185    0.47    0.638

(Dispersion parameter for binomial family taken to be 1)

    Null deviance: 126.18185  on 14  degrees of freedom
Residual deviance:   0.26458  on 3   degrees of freedom
AIC: 115.3

Number of Fisher Scoring iterations: 3
```

The aliased terms are included as they are marginal to the interactions; without them R reparametrizes the interactions. Further inspection shows that the interaction parameter estimates for classes III, IV, and V are very similar. If this were true for class II as well we could replace the `class` variable in the interaction by `cn1` defined above. The estimate for class II is considerably smaller, so we include a separate term in the model for the interaction of class II with `sn1`. The marginal main effect for class II is not needed.

```
> hypertension <- transform(hypertension, c2 = ifelse(class ==
+      "II", 1, 0))
> hypertension.glm5 <- update(hypertension.glm4, . ~
+      . - class:sn1 + cn1:sn1 + c2:sn1)
```

```
> coef(summary(hypertension.glm5))

             Estimate Std. Error z value Pr(>|z|)
(Intercept)  -1.0264      0.111  -9.236 2.55e-20
classII       0.0740      0.129   0.572 5.67e-01
classIII      0.1387      0.115   1.204 2.29e-01
classIV       0.0835      0.124   0.670 5.03e-01
classV        0.1269      0.138   0.919 3.58e-01
smoke1-19     0.0629      0.245   0.257 7.97e-01
smoke20+     -0.1523      0.582  -0.261 7.94e-01
cn1:s3        0.2921      0.619   0.472 6.37e-01
sn1:cn1      -0.5517      0.249  -2.212 2.70e-02
sn1:c2        0.3967      0.133   2.980 2.88e-03
```

The smoke terms are small and can be omitted. We continue with model simplification.

```
> hypertension.glm6 <- update(hypertension.glm5, . ~
+       . - smoke)
> summary(hypertension.glm6)

Coefficients: (1 not defined because of singularities)
             Estimate Std. Error z value Pr(>|z|)
(Intercept)  -1.0264     0.1111   -9.24   <2e-16
classII       0.0740     0.1294    0.57   0.5672
classIII      0.1387     0.1152    1.20   0.2287
classIV       0.0835     0.1245    0.67   0.5026
classV        0.1269     0.1382    0.92   0.3583
sn1           0.0342     0.2321    0.15   0.8827
cn1               NA         NA      NA       NA
cn1:s3        0.0769     0.0894    0.86   0.3898
sn1:cn1      -0.5230     0.2369   -2.21   0.0273
sn1:c2        0.3967     0.1331    2.98   0.0029

(Dispersion parameter for binomial family taken to be 1)

    Null deviance: 126.1818  on 14  degrees of freedom
Residual deviance:   2.0751  on  6  degrees of freedom
AIC: 111.1

Number of Fisher Scoring iterations: 3

> hypertension.glm7 <- update(hypertension.glm6, . ~
+       . - cn1:s3)
> summary(hypertension.glm7)

Coefficients: (1 not defined because of singularities)
             Estimate Std. Error z value Pr(>|z|)
```

```
(Intercept)  -1.0264      0.1111   -9.24   <2e-16
classII       0.0740      0.1294    0.57   0.5672
classIII      0.1383      0.1152    1.20   0.2300
classIV       0.0851      0.1245    0.68   0.4940
classV        0.1272      0.1382    0.92   0.3570
sn1           0.0342      0.2321    0.15   0.8827
cn1               NA          NA      NA       NA
sn1:cn1      -0.5106      0.2364   -2.16   0.0308
sn1:c2        0.3931      0.1330    2.96   0.0031

(Dispersion parameter for binomial family taken to be 1)

    Null deviance: 126.1818  on 14  degrees of freedom
Residual deviance:   2.8074  on  7  degrees of freedom
AIC: 109.9

Number of Fisher Scoring iterations: 3

> hypertension.glm8 <- update(hypertension.glm7, . ~
+     . - class)
> summary(hypertension.glm8)

Coefficients:
            Estimate Std. Error z value Pr(>|z|)
(Intercept)  -1.0264      0.1111   -9.24   <2e-16
sn1           0.0342      0.2321    0.15   0.8827
cn1           0.1205      0.1141    1.06   0.2912
sn1:cn1      -0.5048      0.2361   -2.14   0.0325
sn1:c2        0.3409      0.1117    3.05   0.0023

(Dispersion parameter for binomial family taken to be 1)

    Null deviance: 126.1818  on 14  degrees of freedom
Residual deviance:   4.2377  on 10  degrees of freedom
AIC: 105.3

Number of Fisher Scoring iterations: 3

> hypertension.glm9 <- update(hypertension.glm8, . ~
+     . - sn1)
> summary(hypertension.glm9)

Call:
glm(formula = cbind(Freq, t) ~ cn1 + sn1:cn1 + sn1:c2,
    family = binomial, data = hypertension)

Deviance Residuals:
    Min       1Q   Median       3Q      Max
-0.8547  -0.3973  -0.0702   0.4017   0.8128
```

```
Coefficients:
            Estimate Std. Error z value Pr(>|z|)
(Intercept)  -1.0186     0.0976  -10.44   <2e-16
cn1           0.1127     0.1010    1.12   0.2645
cn1:sn1      -0.4706     0.0433  -10.87   <2e-16
sn1:c2        0.3409     0.1117    3.05   0.0023

(Dispersion parameter for binomial family taken to be 1)

    Null deviance: 126.1818  on 14  degrees of freedom
Residual deviance:   4.2594  on 11  degrees of freedom
AIC: 103.3

Number of Fisher Scoring iterations: 3

> hypertension.glm10 <- update(hypertension.glm9, . ~
+     . - cn1)
> summary(hypertension.glm10)

Coefficients:
            Estimate Std. Error z value Pr(>|z|)
(Intercept)  -0.9136     0.0252  -36.30   <2e-16
cn1:sn1      -0.4629     0.0428  -10.83   <2e-16
sn1:c2        0.3409     0.1117    3.05   0.0023

(Dispersion parameter for binomial family taken to be 1)

    Null deviance: 126.1818  on 14  degrees of freedom
Residual deviance:   5.5233  on 12  degrees of freedom
AIC: 102.6

Number of Fisher Scoring iterations: 3

> hypertension$fval <- fitted(hypertension.glm10)
> print(with(hypertension, tapply(fval, list(Class = class,
+     Smoke = smoke), mean)), digits = 3)

     Smoke
Class     0  1-19   20+
  I   0.286 0.286 0.286
  II  0.286 0.262 0.262
  III 0.286 0.202 0.202
  IV  0.286 0.202 0.202
  V   0.286 0.202 0.202
```

For non-smokers there are no social class differences, and for class I there are no smoking differences. For the other classes smokers have a slightly lower hypertension rate in class II, and a much lower rate in classes III, IV, and V. Note that for this hypertension response, the smoking categories can be collapsed to smokers/non-smokers, and classes III, IV, and V can be collapsed together.

We now examine the proteinurea response.

```
> toxaemia$U <- factor(toxaemia$response %in% c("NU",
+     "HU"))
> round(ftable(with(toxaemia, prop.table(tapply(count,
+     list(class = class, hypertension = U, smoke = smoke),
+     sum), c(1, 3)))), row.vars = 1, col.vars = c(3,
+     2)), 4)[, c(2, 4, 6)]

      [,1]   [,2]   [,3]
[1,] 0.117 0.0952 0.0588
[2,] 0.074 0.0739 0.0577
[3,] 0.094 0.0858 0.0918
[4,] 0.117 0.0943 0.0856
[5,] 0.117 0.1242 0.1222
```

The proteinurea rate appears to decline with increasing smoking and shows some class differences. We fit models.

```
> proteinurea <- data.frame(xtabs(count ~ class + U +
+     smoke, data = toxaemia))
> proteinurea2 <- cbind(proteinurea[proteinurea$U ==
+     "TRUE", -2], t = proteinurea[proteinurea$U ==
+     "FALSE", 4])
> proteinurea.glm <- glm(cbind(Freq, t) ~ class + smoke,
+     data = proteinurea2, family = binomial)
> anova(proteinurea.glm)

Analysis of Deviance Table

Model: binomial, link: logit

Response: cbind(Freq, t)

Terms added sequentially (first to last)


      Df Deviance Resid. Df Resid. Dev
NULL                     14      27.25
class  4    20.99        10       6.26
smoke  2     3.18         8       3.08

> round(coef(summary(proteinurea.glm)), 4)
```

```
            Estimate Std. Error z value Pr(>|z|)
(Intercept) -2.0536     0.1376 -14.927  0.0000
classII     -0.4512     0.1673  -2.697  0.0070
classIII    -0.2043     0.1427  -1.432  0.1522
classIV     -0.0427     0.1558  -0.274  0.7840
classV       0.1370     0.1721   0.796  0.4258
smoke1-19   -0.1070     0.0642  -1.667  0.0956
smoke20+    -0.1180     0.1231  -0.958  0.3380
```

The `class` main effect model is a good fit.

```
> proteinurea.glm1 <- update(proteinurea.glm, . ~ . -
+     smoke)
> summary(proteinurea.glm1)

Coefficients:
            Estimate Std. Error z value Pr(>|z|)
(Intercept)  -2.0774     0.1369  -15.17   <2e-16
classII      -0.4576     0.1673   -2.74   0.0062
classIII     -0.2263     0.1422   -1.59   0.1115
classIV      -0.0749     0.1547   -0.48   0.6282
classV        0.0970     0.1706    0.57   0.5697

(Dispersion parameter for binomial family taken to be 1)

    Null deviance: 27.2544  on 14  degrees of freedom
Residual deviance:  6.2603  on 10  degrees of freedom
AIC: 95.55

Number of Fisher Scoring iterations: 3
```

Classes I, IV, and V have very similar responses, while II and III are different. We combine the former classes, and display the fitted proportions.

```
> proteinurea2 <- transform(proteinurea2, c2 = ifelse(class ==
+     "II", 1, 0), c3 = ifelse(class == "III", 1, 0))
> proteinurea.glm2 <- update(proteinurea.glm1, . ~
+     . - class + c2 + c3)
> summary(proteinurea.glm2)

Coefficients:
            Estimate Std. Error z value Pr(>|z|)
(Intercept)  -2.0943     0.0540  -38.77  < 2e-16
c2           -0.4407     0.1102   -4.00  6.4e-05
c3           -0.2094     0.0662   -3.16   0.0016

(Dispersion parameter for binomial family taken to be 1)
```

```
    Null deviance: 27.2544  on 14  degrees of freedom
Residual deviance:  8.1525  on 12  degrees of freedom
AIC: 93.45

Number of Fisher Scoring iterations: 4

> proteinurea2$fval <- fitted(proteinurea.glm2)
> round(with(proteinurea2, tapply(fval, class, mean)),
+     3)

    I    II   III    IV     V
0.110 0.073 0.091 0.110 0.110
```

Class II has the lowest rate, and class III is lower than the others.

We now consider the association between the two binary responses.

5.9.2 *Correlated outcomes*

In each of the 5×3 cell of the table classified by `class` and `smoke`, we have a 2×2 classification of H and U. We compute the log-odds ratio in each table, and present the table of 'raw' log-odds ratios:

```
> table1 <- with(toxaemia, tapply(count, list(class,
+     smoke, response), sum))
> lodds <- log(table1[, , 1] * table1[, , 4]/(table1[,
+     , 2] * table1[, , 3]))
> round(-lodds, 2)

       0 1-19  20+
I   1.54 1.08  Inf
II  1.47 0.86 -Inf
III 1.58 1.37 0.73
IV  1.32 1.34 0.87
V   1.00 1.03 2.09
```

The `Inf` values are due to the two zero observed counts, which are both associated with very small sample sizes. The log-odds ratios appear fairly constant at about 1.2.

To perform a full analysis of these data with the four response categories we need to use the Poisson model representation.

```
> toxaemia.glm11 <- glm(count ~ class:smoke + H * U,
+     data = toxaemia, family = poisson)
> print(anova(toxaemia.glm11), digits = 7)

Analysis of Deviance Table

Model: poisson, link: log
```

```
Response: count

Terms added sequentially (first to last)


             Df Deviance Resid. Df Resid. Dev
NULL                            59   33142.60
H             1  3427.33        58   29715.27
U             1 10229.26        57   19486.01
class:smoke  14 18813.16        43     672.85
H:U           1   493.82        42     179.03

> coef(summary(toxaemia.glm11))[c(2, 3, 18), ]

             Estimate Std. Error z value  Pr(>|z|)
HTRUE           -1.24     0.0218   -57.1  0.00e+00
UTRUE           -2.77     0.0425   -65.3  0.00e+00
HTRUE:UTRUE      1.36     0.0606    22.5 3.22e-112
```

The `class:smoke` term reproduces the marginal totals of the women in each class/smoking combination. The `H*U` term reproduces the 'saturated model' pattern of symptom dependence. Consequently, this model is equivalent to the null model on the multinomial logit scale, the marginal totals being fixed for each explanatory variable combination and a common response pattern being fitted. The deviance is 179.03 on 42 df: the null model is a poor fit as we know and there are strong patterns of association to be modelled. The interaction between `H` and `U` is positive and substantial.

We proceed by removing the `H:U` interaction terms from the saturated model as far as possible, to identify the association structure between the symptoms, and then follow the analysis for the separate hypertension and proteinurea responses. To save space we suppress the parameter estimates in the early models.

```
> toxaemia.glm12 <- update(toxaemia.glm11, . ~ H *
+     U + class:smoke + H:class + H:smoke + H:class:smoke +
+     U:class + U:smoke + U:class:smoke + H:U:class +
+     H:U:smoke)
> toxaemia.glm13 <- update(toxaemia.glm12, . ~ . -
+     H:U:class)
> toxaemia.glm14 <- update(toxaemia.glm13, . ~ . -
+     H:U:smoke)
> anova(toxaemia.glm12, toxaemia.glm13, toxaemia.glm14)

Analysis of Deviance Table

Model 1: count ~ H + U + H:U + class:smoke + H:class + H:smoke
    + U:class + U:smoke + H:class:smoke + U:class:smoke
    + H:U:class + H:U:smoke
```

```
Model 2: count ~ H + U + H:U + class:smoke + H:class + H:smoke
    + U:class + U:smoke + H:class:smoke + U:class:smoke
    + H:U:smoke
Model 3: count ~ H + U + H:U + class:smoke + H:class + H:smoke
    + U:class + U:smoke + H:class:smoke + U:class:smoke
  Resid. Df Resid. Dev Df Deviance
1         8      12.68
2        12      15.65 -4    -2.97
3        14      22.29 -2    -6.64
```

The deviance change is rather large. We refit this interaction and examine the parameter estimates for it.

```
> round(coef(summary(toxaemia.glm13))[47:48, ], 4)

                          Estimate Std. Error z value Pr(>|z|)
HTRUE:UTRUE:smoke1-19      -0.207      0.132   -1.58   0.1150
HTRUE:UTRUE:smoke20+       -0.587      0.258   -2.28   0.0227
```

The strength of the association appears to decrease with increasing smoking. We try replacing smoke by its linear component.

```
> toxaemia$lsmoke <- unclass(toxaemia$smoke)
> toxaemia.glm15 <- update(toxaemia.glm13, . ~ . -
+     H:U:smoke + H:U:lsmoke)
> anova(toxaemia.glm13, toxaemia.glm15)

Analysis of Deviance Table

Model 1: count ~ H + U + H:U + class:smoke + H:class + H:smoke +
    U:class + U:smoke + H:class:smoke + U:class:smoke + H:U:smoke
Model 2: count ~ H + U + H:U + class:smoke + H:class + H:smoke +
    U:class + U:smoke + H:class:smoke + U:class:smoke + H:U:lsmoke
  Resid. Df Resid. Dev Df Deviance
1        12      15.65
2        13      15.92 -1    -0.27

> coef(summary(toxaemia.glm15))[c(4, 47), ]

                          Estimate Std. Error z value Pr(>|z|)
HTRUE:UTRUE                  1.752      0.161   10.92  9.6e-28
HFALSE:UFALSE:lsmoke        -0.252      0.100   -2.51  1.2e-02
```

The linear trend term is clearly important, with a deviance change of 6.36 if it is omitted. The association between the two symptoms decreases with increasing level of smoking, from a log-odds ratio of 1.5 ($1.75 - 0.25$ – an odds ratio of 4.5) for non-smokers to 1.00 ($1.75 - 3 * 0.25$ – an odds ratio of 2.7) for heavy smokers. We proceed with simplification of the single interaction terms with H or U.

```
> toxaemia.glm16 <- update(toxaemia.glm15, . ~ . -
+     U:class:smoke)
> toxaemia.glm17 <- update(toxaemia.glm16, . ~ . -
+     U:smoke)
> anova(toxaemia.glm15, toxaemia.glm16, toxaemia.glm17)

Analysis of Deviance Table

Model 1: count ~ H + U + H:U + class:smoke + H:class + H:smoke
    + U:class + U:smoke + H:class:smoke + U:class:smoke
    + H:U:lsmoke
Model 2: count ~ H + U + H:U + class:smoke + H:class + H:smoke
    + U:class + U:smoke + H:class:smoke + H:U:lsmoke
Model 3: count ~ H + U + H:U + class:smoke + H:class + H:smoke
    + U:class + H:class:smoke + H:U:lsmoke
  Resid. Df Resid. Dev Df Deviance
1       13      15.92
2       21      19.39 -8    -3.47
3       22      19.61 -1    -0.22
```

We clarify the model by including the marginal terms without changing the model fit.

```
> toxaemia.glm18 <- update(toxaemia.glm17, . ~ . +
+     lsmoke + H:lsmoke + U:lsmoke)
> round(coef(summary(toxaemia.glm18))[23:24, ], 4)

                 Estimate Std. Error z value Pr(>|z|)
HTRUE:smoke1-19    0.0999      0.246   0.407    0.684
HTRUE:smoke20+    -0.0799      0.583  -0.137    0.891
```

The H:smoke interaction terms can be omitted, while the marginal H.lsmoke interaction is retained.

```
> toxaemia.glm19 <- update(toxaemia.glm18, . ~ . -
+     H:smoke)
> anova(toxaemia.glm18, toxaemia.glm19)

Analysis of Deviance Table

Model 1: count ~ H + U + lsmoke + H:U + class:smoke + H:class
    + H:smoke + U:class + H:lsmoke + U:lsmoke + H:class:smoke
    + H:U:lsmoke
Model 2: count ~ H + U + lsmoke + H:U + class:smoke + H:class
    + U:class + H:lsmoke + U:lsmoke + H:class:smoke + H:U:lsmoke
  Resid. Df Resid. Dev Df Deviance
1       22       19.6
2       22       19.6  0 -2.2e-13
```

```
> round(coef(summary(toxaemia.glm19))[29:37, ], 4)

                         Estimate Std. Error z value Pr(>|z|)
HTRUE:classI:smoke1-19     0.1511      0.281  0.5374   0.5910
HTRUE:classII:smoke1-19   -0.0253      0.190 -0.1326   0.8945
HTRUE:classIII:smoke1-19  -0.4222      0.149 -2.8394   0.0045
HTRUE:classIV:smoke1-19   -0.4150      0.178 -2.3345   0.0196
HTRUE:classV:smoke1-19    -0.2091      0.180 -1.1646   0.2442
HTRUE:classI:smoke20+      0.0224      0.645  0.0348   0.9722
HTRUE:classII:smoke20+     0.1993      0.418  0.4773   0.6332
HTRUE:classIII:smoke20+   -0.3041      0.298 -1.0199   0.3078
HTRUE:classIV:smoke20+    -0.2002      0.329 -0.6086   0.5428
```

The `class:smoke` interactions for `H` show the same pattern as before so we define the same terms, and replace the `class:smoke` interaction as before.

```
> toxaemia <- transform(toxaemia, cn1 = ifelse(class !=
+     "I", 1, 0), sn1 = ifelse(smoke != "0", 1, 0),
+     s3 = ifelse(smoke == "20+", 1, 0))
> toxaemia.glm20 <- update(toxaemia.glm19, . ~ . -
+     H:class:smoke + cn1 + sn1 + s3 + class:sn1 +
+     cn1:s3 + H:(cn1 + sn1 + class:sn1 + cn1:s3))
> anova(toxaemia.glm19, toxaemia.glm20)

Analysis of Deviance Table

Model 1: count ~ H + U + lsmoke + H:U + class:smoke + H:class +
    U:class + H:lsmoke + U:lsmoke + H:class:smoke + H:U:lsmoke
Model 2: count ~ H + U + lsmoke + cn1 + sn1 + s3 + H:U
    + class:smoke + H:class + U:class + H:lsmoke + U:lsmoke
    + class:sn1 + cn1:s3 + H:cn1 + H:sn1 + H:U:lsmoke
    + H:class:sn1 + H:cn1:s3
  Resid. Df Resid. Dev Df Deviance
1        22      19.61
2        25      19.90 -3    -0.29

> round(coef(summary(toxaemia.glm20))[35, , drop = F],
+     4)

            Estimate Std. Error z value Pr(>|z|)
HTRUE:cn1:s3    0.287      0.619   0.463    0.643

> toxaemia.glm21 <- update(toxaemia.glm20, . ~ . -
+     H:cn1:s3)
> anova(toxaemia.glm20, toxaemia.glm21)

Analysis of Deviance Table

Model 1: count ~ H + U + lsmoke + cn1 + sn1 + s3 + H:U
```

```
     + class:smoke + H:class + U:class + H:lsmoke + U:lsmoke
     + class:sn1 + cn1:s3 + H:cn1 + H:sn1 + H:U:lsmoke
     + H:class:sn1 + H:cn1:s3
Model 2: count ~ H + U + lsmoke + cn1 + sn1 + s3 + H:U
     + class:smoke + H:class + U:class + H:lsmoke + U:lsmoke
     + sn1:class + cn1:s3 + H:cn1 + H:sn1 + H:U:lsmoke
     + H:sn1:class
  Resid. Df Resid. Dev Df Deviance
1       25      19.90
2       26      20.12 -1    -0.22

> round(coef(summary(toxaemia.glm21))[31:34, ], 4)

                    Estimate Std. Error z value Pr(>|z|)
HTRUE:sn1:classII     -0.130      0.264  -0.492   0.6228
HTRUE:sn1:classIII    -0.540      0.238  -2.270   0.0232
HTRUE:sn1:classIV     -0.515      0.255  -2.020   0.0433
HTRUE:sn1:classV      -0.312      0.281  -1.113   0.2657

> toxaemia <- transform(toxaemia, c2 = ifelse(class ==
+     "II", 1, 0))
> toxaemia.glm22 <- update(toxaemia.glm21, . ~ . -
+     H:class:sn1 + H:cn1 + H:c2 + cn1:sn1 + c2:sn1 +
+     H:cn1:sn1 + H:c2:sn1)
> anova(toxaemia.glm21, toxaemia.glm22)

Analysis of Deviance Table

Model 1: count ~ H + U + lsmoke + cn1 + sn1 + s3 + H:U
     + class:smoke + H:class + U:class + H:lsmoke + U:lsmoke
     + sn1:class + cn1:s3 + H:cn1 + H:sn1 + H:U:lsmoke
     + H:sn1:class
Model 2: count ~ H + U + lsmoke + cn1 + sn1 + s3 + H:U
     + class:smoke + H:class + U:class + H:lsmoke + U:lsmoke
     + sn1:class + cn1:s3 + H:cn1 + H:sn1 + H:c2 + cn1:sn1
     + sn1:c2 + H:U:lsmoke + H:cn1:sn1 + H:sn1:c2
  Resid. Df Resid. Dev Df Deviance
1       26      20.12
2       28      22.00 -2    -1.88

> round(coef(summary(toxaemia.glm22))[23:26, ], 4)

               Estimate Std. Error z value Pr(>|z|)
UTRUE:classII   -0.4953      0.172  -2.888   0.0039
UTRUE:classIII  -0.2263      0.146  -1.545   0.1224
UTRUE:classIV   -0.0398      0.160  -0.249   0.8034
UTRUE:classV     0.1381      0.176   0.783   0.4337

> toxaemia <- transform(toxaemia, c3 = ifelse(class ==
```

```
+      "III", 1, 0))
> toxaemia.glm23 <- update(toxaemia.glm22, . ~ . -
+      U:class + +c3 + U:(c2 + c3))
> anova(toxaemia.glm22, toxaemia.glm23)

Analysis of Deviance Table

Model 1: count ~ H + U + lsmoke + cn1 + sn1 + s3 + H:U
    + class:smoke + H:class + U:class + H:lsmoke + U:lsmoke
    + sn1:class + cn1:s3 + H:cn1 + H:sn1 + H:c2 + cn1:sn1
    + sn1:c2 + H:U:lsmoke + H:cn1:sn1 + H:sn1:c2
Model 2: count ~ H + U + lsmoke + cn1 + sn1 + s3 + c3
    + H:U + class:smoke + H:class + H:lsmoke + U:lsmoke
    + sn1:class + cn1:s3 + H:cn1 + H:sn1 + H:c2 + cn1:sn1
    + sn1:c2 + U:c2 + U:c3 + H:U:lsmoke + H:cn1:sn1
    + H:sn1:c2
  Resid. Df Resid. Dev Df Deviance
1        28      22.00
2        30      23.93 -2    -1.92

> round(coef(summary(toxaemia.glm23))[19:22, ], 4)

                 Estimate Std. Error z value Pr(>|z|)
HTRUE:classII      0.1480      0.130   1.136    0.256
HTRUE:classIII     0.1764      0.116   1.523    0.128
HTRUE:classIV      0.0862      0.124   0.693    0.489
HTRUE:classV       0.1299      0.138   0.940    0.347

> toxaemia.glm24 <- update(toxaemia.glm23, . ~ . -
+      H:class + H:cn1)
> anova(toxaemia.glm23, toxaemia.glm24)

Analysis of Deviance Table

Model 1: count ~ H + U + lsmoke + cn1 + sn1 + s3 + c3
    + H:U + class:smoke + H:class + H:lsmoke + U:lsmoke
    + sn1:class + cn1:s3 + H:cn1 + H:sn1 + H:c2 + cn1:sn1
    + sn1:c2 + U:c2 + U:c3 + H:U:lsmoke + H:cn1:sn1 + H:sn1:c2
Model 2: count ~ H + U + lsmoke + cn1 + sn1 + s3 + c3
    + H:U + class:smoke + H:lsmoke + U:lsmoke + sn1:class
    + cn1:s3 + H:cn1 + H:sn1 + H:c2 + cn1:sn1 + sn1:c2
    + U:c2 + U:c3 + H:U:lsmoke + H:cn1:sn1 + H:sn1:c2
  Resid. Df Resid. Dev Df Deviance
1        30      23.93
2        32      26.37 -2    -2.45

> round(coef(summary(toxaemia.glm24))[c(2, 3, 10, 19:28),
+      ], 4)
```

```
                        Estimate Std. Error z value Pr(>|z|)
HTRUE                    -1.3251     0.1442  -9.192   0.0000
UTRUE                    -2.7849     0.1242 -22.428   0.0000
classII:smoke0            1.1703     0.0814  14.376   0.0000
HTRUE:lsmoke              0.0953     0.0894   1.066   0.2862
UTRUE:lsmoke              0.1270     0.0663   1.917   0.0553
HTRUE:cn1                 0.1563     0.1150   1.359   0.1742
HTRUE:sn1                -0.0340     0.2532  -0.134   0.8932
HFALSE:c2                 0.0109     0.0732   0.149   0.8815
UTRUE:c2                 -0.4945     0.1128  -4.386   0.0000
UTRUE:c3                 -0.2203     0.0666  -3.309   0.0009
HTRUE:UTRUE:lsmoke       -0.2454     0.0991  -2.476   0.0133
HTRUE:cn1:sn1            -0.5226     0.2364  -2.210   0.0271
HTRUE:sn1:c2              0.3932     0.1331   2.953   0.0031
```

We conclude the reduction at this point, though some unnecessary terms remain, and could be removed by further definitions of explicit variables for these terms and other terms to which they are marginal.

As for the separate symptom analyses, for hypertension there are no significant social class differences for non-smokers, and for class I there are no significant smoking differences. For the other classes smokers have a slightly lower hypertension rate in class III, and a much lower rate in classes III, IV and V.

For proteinurea, class II has a lower incidence than the others.

We give finally separate fitted models for proteinurea and hypertension from this model.

```
> H.tab <- with(toxaemia, tapply(fitted(toxaemia.glm24),
+     list(Class = class, Smoke = smoke, H), mean))
> H.ptab <- H.tab[, , 2]/(H.tab[, , 1] + H.tab[, ,
+     2])
> round(H.ptab, 3)

     Smoke
Class     0  1-19   20+
  I   0.264 0.269 0.281
  II  0.278 0.260 0.275
  III 0.288 0.198 0.209
  IV  0.295 0.203 0.213
  V   0.295 0.203 0.213

> U.tab <- with(toxaemia, tapply(fitted(toxaemia.glm24),
+     list(Class = class, U), mean))
> U.ptab <- U.tab[, 2]/(U.tab[, 1] + U.tab[, 2])
> round(U.ptab, 3)

    I    II   III    IV     V
0.111 0.073 0.091 0.110 0.109
```

6 Survival data

6.1 Introduction

Over the last 30 years there has been a rapid development of probability models and statistical analysis for technological and medical survival data. Many studies have been made of the length of life or of periods of remission of animal or human subjects being treated for serious diseases. These studies have used probability models for duration of life which originated in engineering reliability studies of the operating lifetimes of electrical and mechanical systems. We will use the terms 'death' and 'failure' interchangeably to represent the events and 'survival time' to represent the times to these events. In this chapter we give an account of the main probability models and their use in data analysis. The recognition of the special form of the likelihood in an important class of these distributions allows their simple fitting as generalized linear models. Many books are now available on survival analysis; a concise account can be found in Cox and Oakes (1984), detailed discussions in Kalbfleisch and Prentice (1980) and Lawless (1982), and more applied treatments in Collett (2003) and Hosmer and Lemeshow (1999).

This chapter is restricted to a discussion of single events: we do not deal with models for *recurrent events*, like repeated spells of remission or of unemployment.

6.2 The exponential distribution

The exponential distribution plays a central role in survival analysis. Although few systems have exponentially distributed lifetimes, most of the useful survival distributions are directly related to the exponential distribution. We first consider its properties as a survival distribution. In this chapter, we use the notation T for a non-negative random variable.

The density function of the exponential random variable T for survival time t is

$$\begin{aligned} f(t) &= \Pr(t < T < t + \mathrm{d}t)/\mathrm{d}t \\ &= (1/\mu)\mathrm{e}^{-t/\mu} \quad \mu,\ t > 0, \end{aligned}$$

where μ is the mean of the distribution; the variance is μ^2. The cumulative distribution function is

$$\Pr(T \leq t) = F(t) = 1 - e^{-t/\mu}, \quad \mu,\ t > 0.$$

A fundamental concept in survival analysis is that of the *hazard function* $h(t)$ which is the conditional density function at time t given survival up to time t:

$$\begin{aligned} h(t)\mathrm{d}t &= \Pr(t < T \leq t + \mathrm{d}t \mid T > t) \\ &= f(t)\mathrm{d}t/[1 - F(t)]. \end{aligned}$$

It is convenient to introduce a notation for the upper tail probability: we define the *survivor function*

$$S(t) = \Pr(T > t) = 1 - F(t).$$

Then

$$h(t) = f(t)/S(t).$$

The hazard function can be interpreted as the instantaneous failure rate at time t. Any continuous probability distribution can be specified equivalently by its density, survivor or hazard function. In particular the density and survivor functions can be obtained from the hazard function by

$$\begin{aligned} S(t) &= \exp(-H(t)) \\ f(t) &= h(t)\exp(-H(t)), \end{aligned}$$

where the function

$$H(t) = \int_0^t h(u)\,\mathrm{d}u$$

is the *integrated hazard function*. The usefulness of the hazard function is clear in the servicing and repair of mechanical or electrical systems. When the system has been in operation for some time, the appropriate servicing or repair schedule depends on the probability of failure, conditional on the previous operating history. Historically, the effect of explanatory variables on survival time has been expressed by modelling the hazard function; as we shall see, such models are models for functions of the parameters of the probability distribution.

For the exponential distribution

$$\begin{aligned} h(t) &= 1/\mu \\ H(t) &= t/\mu. \end{aligned}$$

Thus the exponential distribution has *constant hazard*: the probability of death at time t is not dependent on the length of previous lifetime. The exponential distribution represents the lifetime distribution of an item which does not age or wear: the instantaneous probability of failure is the same no matter how long the item has already survived. In most applications this strong property does not hold, and so the exponential distribution has limited application as a survival distribution.

In a more general context, if 'death' is replaced by an event which can occur repeatedly in time, then these events occur in a time-homogeneous Poisson process with rate $\lambda = 1/\mu$ if the times between events are independent exponential variables with the same mean μ. Thus the probability of r events in time t has the Poisson distribution

$$\Pr(r \mid \lambda, t) = e^{-\lambda t}(\lambda t)^r/r!, \quad r = 0, 1, 2, \ldots.$$

This relation between the exponential and Poisson distributions is used to advantage in Section 6.7 for ML estimation with censored data.

6.3 Fitting the exponential distribution

The exponential distribution is available as a standard distribution in R as it is a special case of the *gamma* distribution,.The error distribution and link function are specified as `family=Gamma(link="log")` in the `glm` function and the dispersion parameter set to one when extracting the standard errors from the `glm` object.

We illustrate with a simple example from Feigl and Zelen (1965). The dataset `feigl` contains the survival times (the variable `time`) in weeks of 33 patients suffering from acute myelogeneous leukaemia, and the values of two explanatory variables, white blood cell count `wbc` in thousands and a positive or negative factor `ag`, positive values (`ag`=+) being defined by the presence of Auer rods and/or significant granulation of the leukaemic cells in the bone marrow at diagnosis, and negative values (`ag`=−) if both Auer rods and granulation are absent.

```
> data(feigl, package = "SMIR")
```

We first graph `time` against `wbc` in Fig. 6.1.

```
> print(xyplot(time ~ wbc, data = feigl))
```

There is a heavy concentration of points along both the horizontal and vertical axes. We repeat the graph (Fig. 6.2) this time using log scales and separate plotting symbols for the two levels of `ag` (pluses for `ag` +, circles for `ag`−).

```
> print(xyplot(time ~ wbc, data = feigl, xlab = "wbc (log scale)",
+      groups = ag, ylab = "time (log scale)", pch = c(3,
+          1), col = "black", scales = list(x = list(log = TRUE,
+          at = (xv <- c(2, 5, 10, 20, 40, 60, 80)),
```

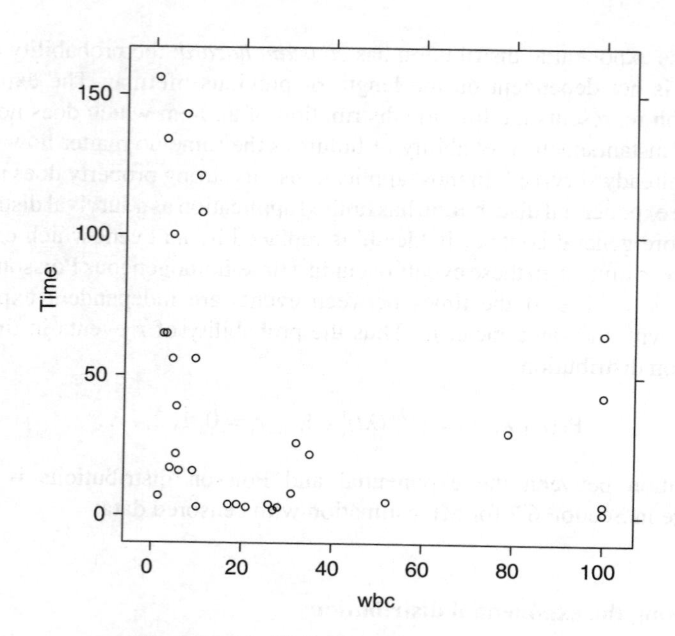

Fig. 6.1. Feigl data: survival time vs wbc

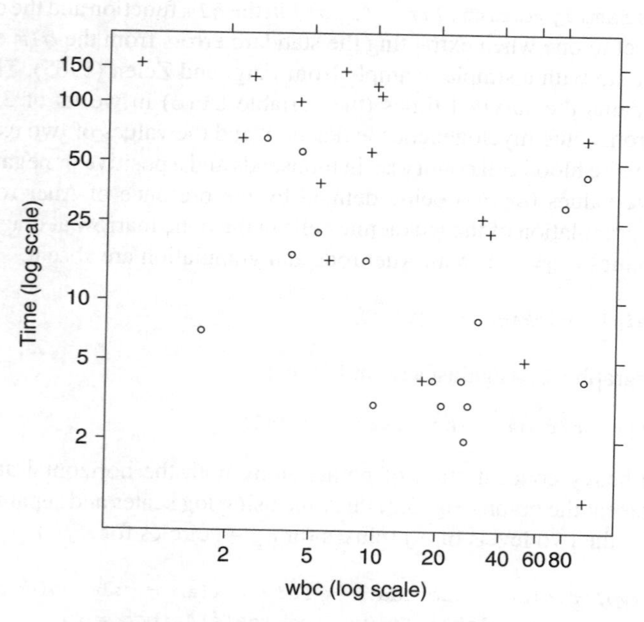

Fig. 6.2. Feigl data: log survival time vs log wbc

```
+         labels = xv), y = list(log = TRUE, at = (yv <- c(2,
+         5, 10, 25, 50, 100, 150)), labels = yv))))
```

There is a general decline in log survival time with increasing `lwbc` and the negative `ag` group appears to have shorter survival times. We try fitting the exponential distribution using a log link function for the mean with `lwbc` as the explanatory variable interacting with the factor `ag`.

Thus the model is $\log \mu = -\log h(t) = \boldsymbol{\beta}'\mathbf{x}$, so a log-linear model for the mean is also a log-linear model for the hazard function.

```
> feigl <- transform(feigl, lwbc = log(wbc))
> feigl.glm1 <- glm(time ~ ag * lwbc, data = feigl,
+      family = Gamma(link = "log"))
> summary(feigl.glm1, dispersion = 1)

Coefficients:
             Estimate Std. Error z value Pr(>|z|)
(Intercept)   5.1499     0.5008  10.283  < 2e-16
ag+          -1.8704     0.7848  -2.383  0.01716
lwbc         -0.4818     0.1736  -2.775  0.00552
ag+:lwbc      0.3278     0.2669   1.228  0.21938

(Dispersion parameter for Gamma family taken to be 1)

    Null deviance: 58.138  on 32  degrees of freedom
Residual deviance: 38.555  on 29  degrees of freedom
AIC: 301.74

Number of Fisher Scoring iterations: 11
```

It might be thought that the residual deviance from the model could be used for testing the goodness of fit, since the standard application of large sample theory would suggest that the deviance is distributed as χ^2_{n-p}. However, as in Section 4.4, as $n \to \infty$ the number of parameters in the saturated model also goes to infinity, and this violates one of the regularity conditions for the validity of this asymptotic distribution.

The interaction appears non-significant and might be omitted from the model. We note also that the slope of the regression in the negative `ag` group is $-0.4818 + 0.3278$, that is, -0.154, so that an alternative simplification of the model might be achieved by setting this slope to zero, as in the `solv` example in Section 3.1.

```
> feigl.glm2 <- update(feigl.glm1, . ~ . - ag:lwbc)
> print(coef(summary(feigl.glm2, dispersion = 1)),
+      digits = 5)
```

```
             Estimate Std. Error z value  Pr(>|z|)
(Intercept)   4.7303    0.41176 11.4880 1.5161e-30
ag+          -1.0176    0.34922 -2.9140 3.5680e-03
lwbc         -0.3044    0.13186 -2.3085 2.0974e-02
```

```
> feigl <- transform(feigl, z = ifelse(ag == "+", 0,
+        lwbc))
> feigl.glm3 <- update(feigl.glm2, . ~ ag + z)
> anova(feigl.glm2, feigl.glm3)
```

```
Analysis of Deviance Table

Model 1: time ~ ag + lwbc
Model 2: time ~ ag + z
  Resid. Df Resid. Dev Df Deviance
1        30     40.319
2        30     39.373  0    0.946
```

```
> print(coef(summary(feigl.glm3, dispersion = 1)),
+      3)
```

```
            Estimate Std. Error z value Pr(>|z|)
(Intercept)    5.150      0.501   10.28 8.41e-25
ag+           -2.263      0.560   -4.04 5.28e-05
z             -0.482      0.174   -2.77 5.52e-03
```

The deviances of the two reduced models differ by only 0.95. We cannot choose between these two models on the basis of the data, because the slope of the regression in the negative `ag` group is poorly defined.

The log link, while a natural choice for positive survival times, is not the only possible choice. The mean survival time can also be related to the linear predictor using the (default) *reciprocal link*, giving a *linear hazard model*

$$\mu^{-1} = \lambda = \boldsymbol{\beta}'\mathbf{x}.$$

Here λ is the mean rate of dying, or failure rate (per unit time). This choice of link function is sometimes thought natural because sufficient statistics exist for the model parameters on this scale. This property does not, however, assist the model fitting in any way, and it does not prevent the occurrence of negative fitted values from the linear model.

```
> feigl.glm4 <- update(feigl.glm1, family = Gamma)
> summary(feigl.glm4, dispersion = 1)
```

```
Coefficients:
            Estimate Std. Error z value Pr(>|z|)
(Intercept) 0.005976   0.003592   1.664   0.0961
```

```
ag+           0.021953   0.026435   0.830    0.4063
lwbc          0.005848   0.002383   2.453    0.0141
ag+:lwbc      0.005807   0.011224   0.517    0.6049

(Dispersion parameter for Gamma family taken to be 1)

    Null deviance: 58.138  on 32  degrees of freedom
Residual deviance: 39.779  on 29  degrees of freedom
AIC: 302.96

Number of Fisher Scoring iterations: 6
```

The interaction is small and can be omitted.

```
> feigl.glm5 <- update(feigl.glm4, . ~ . - ag:lwbc)
> summary(feigl.glm5, dispersion = 1)

Coefficients:
            Estimate Std. Error z value Pr(>|z|)
(Intercept) 0.005795   0.003400   1.704  0.08830
ag+         0.034415   0.014689   2.343  0.01914
lwbc        0.006105   0.002326   2.625  0.00867

(Dispersion parameter for Gamma family taken to be 1)

    Null deviance: 58.138  on 32  degrees of freedom
Residual deviance: 40.044  on 30  degrees of freedom
AIC: 301.22

Number of Fisher Scoring iterations: 6
```

With the reciprocal link an equally good fit is obtained if `wbc` instead of `lwbc` is used as the explanatory variable.

Finally, we consider the identity link. The original analysis by Feigl and Zelen used this link, with

$$\mu = 1/\lambda = \boldsymbol{\beta}'\mathbf{x}.$$

```
> feigl.glm6 <- update(feigl.glm1,
+        family = Gamma(link = "identity"))
> summary(feigl.glm6, dispersion = 1)

Coefficients:
            Estimate Std. Error z value Pr(>|z|)
(Intercept)  110.414     36.022   3.065  0.00218
ag+          -80.055     37.210  -2.151  0.03144
lwbc         -20.015      8.654  -2.313  0.02073
ag+:lwbc      15.425      9.280   1.662  0.09647
```

```
(Dispersion parameter for Gamma family taken to be 1)

    Null deviance: 58.138  on 32  degrees of freedom
Residual deviance: 40.848  on 29  degrees of freedom
AIC: 304.00

Number of Fisher Scoring iterations: 25
```

The deviance after 25 iterations is 40.85, and the interaction, though large, does not appear significant. Omitting the interaction gives a warning:

```
> feigl.glm7 <- update(feigl.glm6, . ~ . - ag:lwbc)

Warning messages:
1: NaNs produced in: log(x)
2: step size truncated due to divergence
3: algorithm did not converge in: glm.fit(x = X, y = Y,
    weights = weights, start = start, etastart = etastart,
```

The three interaction models have very similar deviances, so it is clear that we cannot choose among them from statistical considerations. All three models can be expressed in a Box–Cox form for their link functions:

$$\boldsymbol{\beta}'\mathbf{x} = \eta = (\mu^{\theta} - 1)/\theta, \qquad \mu = (1 + \theta\eta)^{1/\theta}$$

with $\theta = -1,\ 0$, and 1 for the inverse, log and identity links. The corresponding deviances are 39.78, 38.55, and 40.85: the log link has the smallest deviance, but all three fit the data equally well.

The choice of a final model is made difficult, as in the `solv` example of Section 3.1, by the small sample size. We choose the log link and the parallel slopes model `ag + lwbc` as a reasonable description.

We now consider model criticism for the exponential distribution.

6.4 Model criticism

How do we know that the exponential distribution is a reasonable choice? Probability plotting of approximately standardized exponential variables provides a guide, as for the normal distribution. For many of the probability distributions in this chapter, the survivor function has a simple form which facilitates plotting. We therefore use, in preference to the empirical *cdf*, the *empirical survivor function* $\hat{S}(t)$, defined by

$$\hat{S}(t) = \frac{\text{no. of } t_i > t}{n}.$$

If T_i is exponential with mean $\mu_i = 1/\lambda_i$, then $T_i/\mu_i = \lambda_i T_i$ is standard exponential with mean 1. If $\mu_i = g(\eta_i)$ with η_i the linear predictor, define

$$u_i = t_i/\hat{\mu}_i$$

with

$$\hat{\mu}_i = g(\hat{\boldsymbol{\beta}}'\mathbf{x}_i).$$

The u_i have approximately standard exponential distributions, though they are not independent, as with normal residuals.

The survivor function for the standard exponential is

$$S(u) = \mathrm{e}^{-u}$$

so that

$$\log S(u) = -u$$

and

$$-\log(-\log S(u)) = -\log u.$$

We first graph the empirical survivor function $\hat{S}(u)$ of the u_i against u, using the interaction model.

```
> feigl <- transform(feigl, u = time/fitted(feigl.glm1))
```

The u_i are closely related to the Pearson residuals. Theres are

$$e_i = \frac{t_i - \hat{\mu}_i}{\hat{\mu}_i} = u_i - 1.$$

We calculate the empirical survivor function, by ranking the scaled survival times in increasing order.

```
> feigl <- transform(feigl, s = 1 - rank(u)/(length(u) +
+     1))
```

The survivor function estimate uses $(n + 1)$ in the denominator to avoid a zero value for the last observation . We graph $\hat{S}(u)$ against `u`, together with the fitted survivor function `exp(-u)` in Fig. 6.3.

```
> print(xyplot(s ~ u, xlab = "u", ylab = "S(u)", data = feigl,
+     panel = function(x, y) {
```

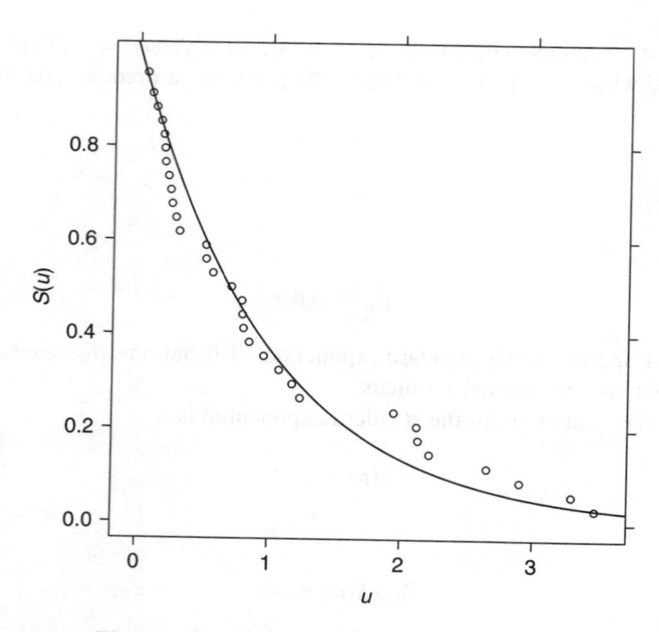

Fig. 6.3. Feigl data: residual survivor functions

```
+          panel.xyplot(x, y)
+          panel.curve(exp(-x), lwd = 1.5)
+      }))
```

As with other *cdf* and probability plots, and especially with small samples, it is difficult to decide whether a real failure of the probability model is occurring. Since $\hat{S}(u)$ is essentially an estimated binomial probability for each u, 95% simultaneous confidence bounds can be placed on the graph in Fig. 6.4 using the function `NPL.bands` from the `SMIR` package as in Section 2.3. We ignore here the dependence between the residuals caused by the model fitting; the bounds are used only qualitatively to indicate model and data agreement. A formal consideration of the dependence between observations would suggest that the bounds should be widened somewhat but we do not consider this further. We suppress in this graph, and the two following, the empirical survivor function to focus attention on the adequacy of the model fit.

The bounds on the true survivor function in Fig. 6.4 are very wide because of the small sample size, and the exponential model clearly fits well. We now log transform the scale of $S(u)$, since the graphs on these scales are linear for the exponential distribution, and may indicate the form of the departure from the exponential distribution if this is incorrect (see Section 6.14).

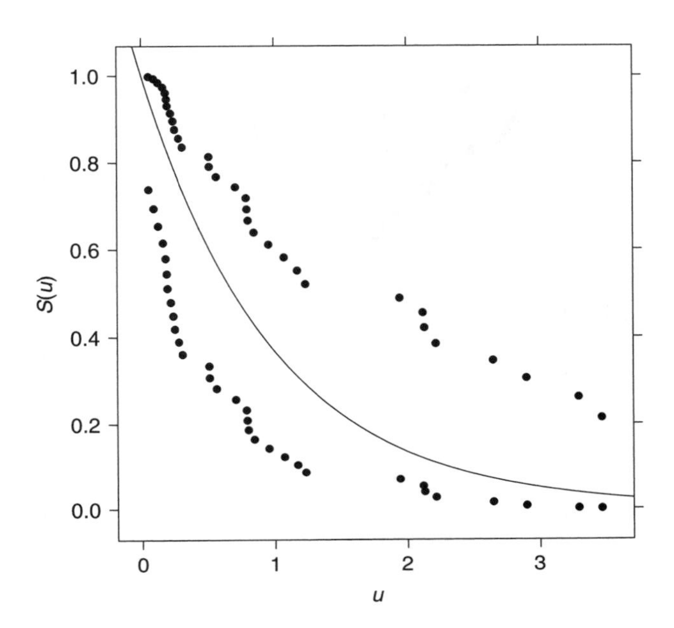

Fig. 6.4. Feigl data: fitted survivor function and bounds

```
> library("SMIR")

> print(xyplot((1 - lower) + (1 - upper) ~ x,
+     data = NPL.bands(feigl$u),
+     xlab = "u", ylab = "S(u)",
+     panel = function(x, y) {
+         panel.xyplot(x, y, pch = 19)
+         panel.curve(exp(-x))
+     }))

> print(xyplot(log(1 - lower) + log(1 - upper) ~ x,
+     data = NPL.bands(feigl$u), xlab = "u", ylab = "log S(u)",
+     panel = function(x, y) {
+         panel.xyplot(x, y, pch = 19)
+         panel.curve(-x)
+     }))
```

The bounds in Fig. 6.5 are very condensed for small values of u. We now repeat the log transformations on both scales in Fig. 6.6.

```
> print(xyplot((-log(-log(1 - lower))) + (-log(-log(1 -
+     upper))) ~ log(x), data = NPL.bands(feigl$u),
+     xlab = "log u", ylab = "-log[-log S(u)]",
```

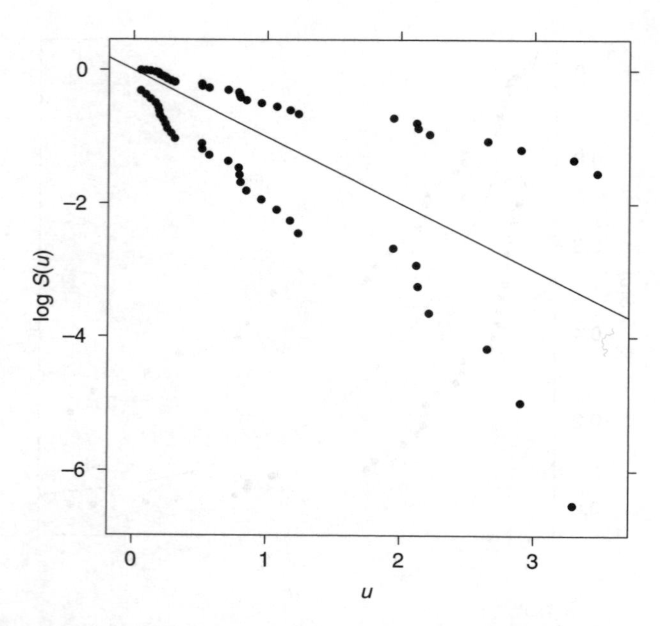

Fig. 6.5. Feigl data: fitted log survivor function and bounds

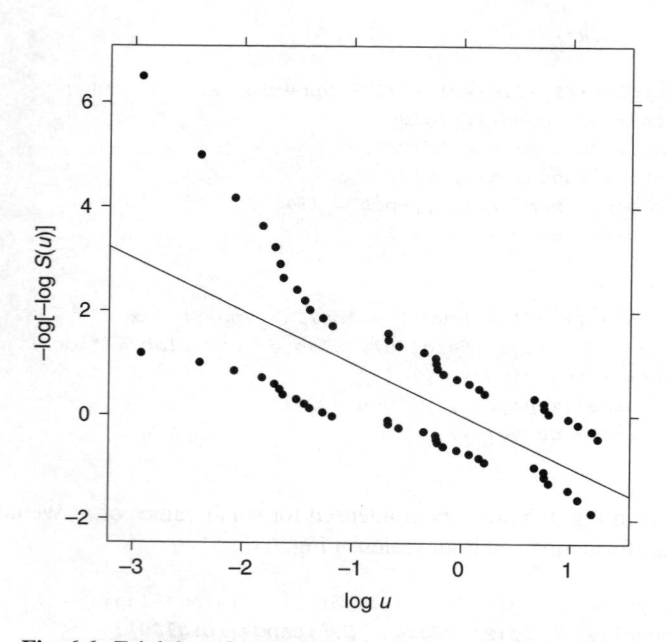

Fig. 6.6. Feigl data: fitted log–log survivor function and bounds

```
+          panel = function(x, y) {
+          panel.xyplot(x, y, pch = 19)
+          panel.curve(-x)
+      }))
```

Figure 6.6 is more informative, with a more nearly constant boundwidth, and is more useful in detecting departures from the exponential distribution, particularly for the Weibull distribution (Section 6.10). The figure shows a good agreement of the fitted straight line with the bounds. The exponential distribution assumption seems satisfactory.

Now we assess the possibility that particularly influential observations have affected the fitted model. The assessment of influence is based on the hat matrix from the iteratively reweighted least squares algorithm, as in Section 4.3. The diagonal elements of this matrix can be extracted from the model using the `influence` function.

```
> print(dotplot(influence(feigl.glm1)$hat, type = "h",
+      scales = list(y = list(cex = 0.8)),
+      ylab = "observation number",
+      xlab = "leverage", panel = function(...) {
+          panel.dotplot(...)
+          panel.abline(v = 2 * 4/33, lty = 2)
+      }))
```

The leverage values are shown in Fig. 6.7. Using the value $2(p+1)/n = 0.242$ as a rough guide, we find that observations $2 (h_2 = 0.297)$ and $21 (h_{21} = 0.282)$ are potentially influential. Both are on the edge of the variable space, with the two smallest values of `wbc`. Deleting these observations in turn makes no substantial difference to the fitted model. Returning to the main effect model, we find only one potentially influential observation, (2), which has no substantial effect on the fitted model. We conclude that the data are represented adequately by the exponential distribution with parallel regressions on the log scale (although other regression models also fit the data adequately).

Figure 6.8 shows the complete data on the log `t`–log `wbc` scale and the parallel slopes models for survival time using the log (straight lines) and reciprocal (curves) links.

```
> feigl.fitted.base <- xyplot(time ~ wbc, data = feigl,
+      groups = ag, pch = c(3, 1),
+      scales = list(x = list(log = "e",
+          at = (xv <- c(1, 2, 5, 10, 20, 50, 100)),
+          labels = xv), y = list(log = "e", at = (yv <- c(1,
+          2, 5, 10, 20, 50, 100, 200, 400)), labels = yv)))
> print(update(feigl.fitted.base, panel = function(...) {
+      panel.xyplot(...)
```

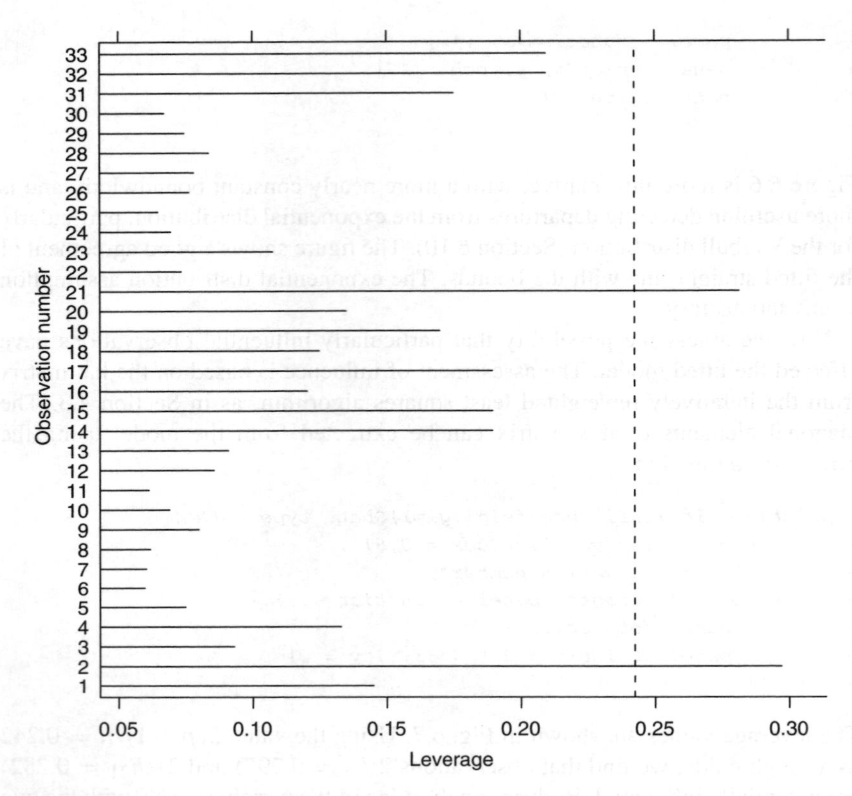

Fig. 6.7. Feigl data : leverage values, interaction model

```
+     panel.curve(4.7303 - 1.0176 - 0.3044 * x, lwd = 1.5,
+       )
+     panel.curve(4.7303 - 0.3044 * x, lwd = 1.5)
+     panel.curve(log(1/(0.005795 + 0.006105 * x)),
+       lty = 3, lwd = 1.5)
+     panel.curve(log(1/(0.005795 + 0.034415 + 0.006105 *
+          x)), lty = 3, lwd = 1.5)
+ }))
```

The fitted models agree closely relative to the random variation in the data. One aspect of this figure may seem strange: the fitted models for each group seem to lie above most of the points for that group. This is a consequence of the exponential assumption: on the log scale of time the distribution is extreme value (Section 6.12) with scale parameter 1. This distribution has negative skew: large negative and small positive values are characteristic of it (see Fig. 6.19).

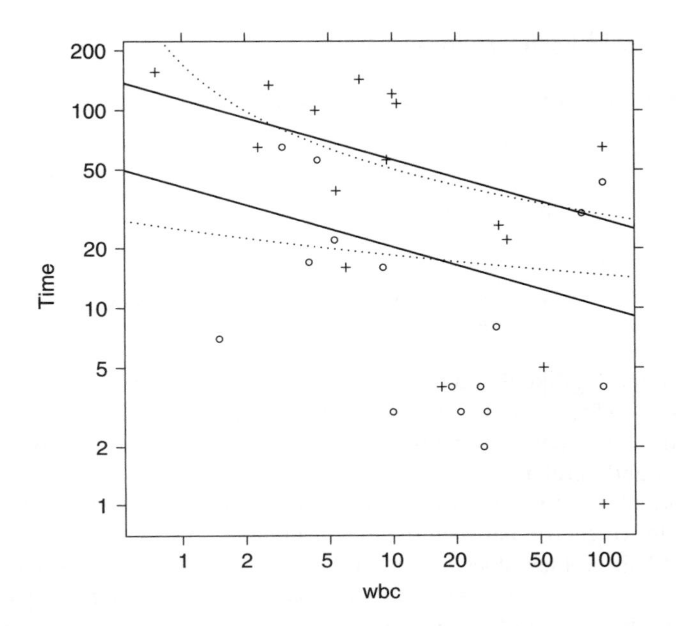

Fig. 6.8. Feigl data: fitted log (solid) and reciprocal (dotted) link function models

6.5 Comparison with the normal family

The Box–Cox normal transformation family discussed in Chapter 3 provides an alternative family of possible survival distributions. Does this family provide a better representation than the exponential distribution? We first explore the Box–Cox transformation without the log transformation of wbc:

```
> library(MASS)
> boxcox(time ~ ag * wbc, data = feigl, lambda = seq(-2,
+      2, 0.5), plotit = FALSE)

$x
[1] -2.0 -1.5 -1.0 -0.5  0.0  0.5  1.0  1.5  2.0

$y
[1] -254.9396 -218.9341 -188.0606 -165.9068 -155.9290 -159.6253
[7] -174.7848 -196.9978 -223.0891
```

Specifying a grid for λ of $-2(0.5)2$, we find the maximum of the log-likelihood $p\ell(\lambda)$ near $\lambda = 0$. Recalculating over a finer grid we find the maximum log-likelihood of -155.6 occurs at $\lambda = 0.11$, with an approximate 95% confidence interval of $(-0.15, 0.38)$. The value at $\lambda = 0$ is -155.93. The log scale for time is strongly indicated.

Should wbc be log transformed as well?

```
> boxcox(time ~ ag * lwbc, data = feigl, lambda = seq(-2,
+     2, 0.5), plotit = FALSE)

$x
[1] -2.0 -1.5 -1.0 -0.5  0.0  0.5  1.0  1.5  2.0

$y
[1] -261.1728 -224.4366 -192.0687 -167.5736 -155.0523 -156.6631
[7] -170.5804 -192.2566 -218.1347
```

The maximum log-likelihood of -154.0 now occurs at $\lambda = 0.19$, and the value at $\lambda = 0$ is -155.1. The log transformation of wbc at $\lambda = 0$ increases the log-likelihood by 0.8: there is not much to choose between the models, but the lwbc model is slightly preferred.

The normal quantile plot of the raw residuals shows a reasonable straight-line fit. How do we choose between the exponential and the lognormal distributions for time? This problem does not fit into the standard LRT theory because the two distributions are not in the same family and have different numbers of parameters. We discuss the problem further in Section 6.19, where the two distributions are embedded in a larger family: here we simply compare the disparities for the two models.

Further problems arise in this comparison, because for the exponential model there are constants omitted from the calculation of the deviance which have to be included if comparisons of disparity are to be made between different probability models.

For the exponential distribution, the deviance is

$$D = -2\log\Big[L(\hat{\boldsymbol{\beta}})/L\{(\hat{\mu}_i)\}\Big],$$

where

$$L\{(\hat{\mu}_i)\} = \prod_i \frac{1}{\hat{\mu}_i} \mathrm{e}^{-t_i/\hat{\mu}_i}$$

and $\hat{\mu}_i = t_i$ giving

$$L\{(\hat{\mu}_i)\} = \mathrm{e}^{-n} / \prod_i t_i.$$

Thus

$$\begin{aligned} -2\ell(\hat{\boldsymbol{\beta}}) &= -2\log L(\hat{\boldsymbol{\beta}}) \\ &= D + 2(n + \sum \log t_i). \end{aligned}$$

For the exponential model with log link, $D = 38.55$, and we calculate the disparity using

```
> -2 * logLik(feigl.glm1)
```

which gives a disparity of 291.74, compared with 288.37 for the lognormal.

Thus the exponential distribution provides a slightly worse fit, possibly because it has no free scale parameter. For the lognormal distribution the fitted model is

$$\underset{(0.603)}{5.425} - \underset{(0.945)}{2.525}\,\text{ag-} - \underset{(0.209)}{0.818}\,\text{lwbc} + \underset{(0.321)}{0.583}\,\text{ag-:lwbc}$$

Again the interaction term can be omitted (a disparity increase of 3.56), giving a final model

$$\underset{(0.514)}{4.802} - \underset{(0.436)}{0.988}\,\text{ag-} - \underset{(0.165)}{0.571}\,\text{lwbc}$$

The slope estimate is greater in magnitude for the lognormal than for the exponential model (-0.304), but the `ag` difference is very similar. Survival appears to decline more rapidly with `wbc` under the lognormal distribution than under the exponential, but since these distributions fit nearly equally well, we cannot be precise about this rate of decline.

The fitted models for the lognormal (`feigl.lognormal`) shown as dotted lines and exponential (`feigl.glm1`) shown as solid lines are given in Fig. 6.9.

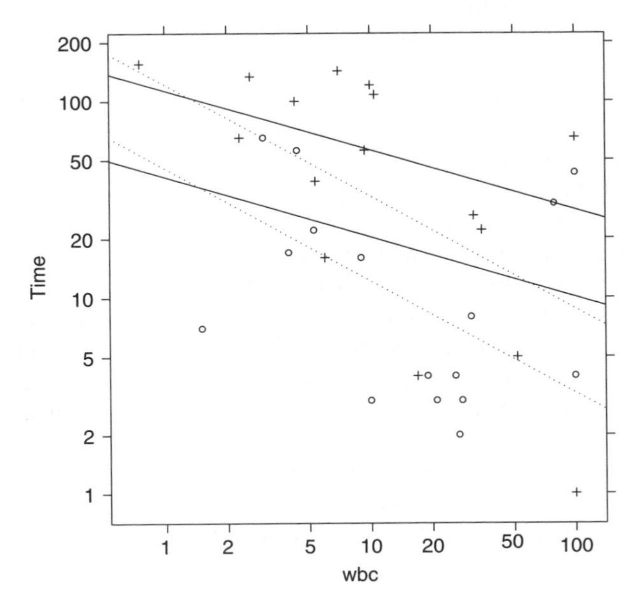

Fig. 6.9. Feigl data: fitted lognormal (dotted) and exponential log link (solid) models

```
> print(update(feigl.fitted.base, panel = function(...) {
+     panel.xyplot(...)
+     panel.curve(4.8022 - 0.9883 - 0.5709 * x, lty = 3,
+         lwd = 1.5)
+     panel.curve(4.8022 - 0.5709 * x, lty = 3, lwd = 1.5)
+     panel.curve(4.7303326 - 0.3044004 * x, lty = 1,
+         lwd = 1.5)
+     panel.curve(4.7303326 - 1.0176452 - 0.3044004 *
+         x, lty = 1, lwd = 1.5)
+ }))
```

We consider this dataset further in Section 6.9, 6.11, and 6.17, where we compare these models with those produced assuming gamma, Weibull and piecewise exponential distributions.

6.6 Censoring

A characteristic feature of survival data is the occurrence of *censored* observations, that is, observations on which the complete lifetime or survival time is not observed. Censoring occurs for a number of different reasons. In industrial experiments on destructive testing of components, testing is usually terminated after a fixed time, or after the failure of some fixed proportion of components on test. The remaining components have not failed and their complete lifetimes are not observed. In medical studies censoring arises from the withdrawal of patients from the study because they have moved away or have stopped returning for follow-up, and from the need to analyse the data at various stages of the study before complete lifetimes are observed on all patients. Censoring is generally on the right; that is, the observed time is less than the actual survival time. Much less commonly it may be on the left; that is, the observed time is *greater* than the actual survival time. We will assume that censoring is on the right for most of this chapter; left-censoring is considered in Section 6.18.

Throughout this chapter we shall assume that censoring is *uninformative*, that is, that censoring of a lifetime does not occur because of factors associated with the survival time which would have been observed. This assumption is similar to that of data missing at random in Section 3.17. Censoring would be informative if, for example, patients were withdrawn from a drug treatment study because they were suffering unexpected or severe effects of the treatment which might have reduced their survival times.

As with missing data, it is prudent to compare by logistic regression the group of censored observations with the group of uncensored observations to see if censoring is systematically related to the explanatory variables. Systematic effects would be a warning of the possible failure of the assumption of uninformative censoring, though they do not imply it logically.

6.7 Likelihood function for censored observations

When some observations have censored lifetimes, the likelihood function is changed, and, in general, MLE can no longer be carried out using the standard GLM fitting algorithm (Section 2.9.3). Nevertheless, for survival distributions related to the exponential (and also for the logistic and log-logistic distributions in Section 6.18) it is still possible to use GLM model fitting procedures. We illustrate with the exponential distribution.

For each individual in the sample we observe the vector of explanatory variables $\mathbf{x}_i$, and a pair of variables (t_i, w_i). The *censoring indicator* w_i takes the value 1 if the survival time t_i for the i-th observation is *uncensored* and zero if it is censored. Thus when $w_i = 1$, the true survival time $T_i = t_i$, while when $w_i = 0$, it is known only that $T_i > t_i$. The contribution of a censored survival time to the likelihood is thus the probability $S(t_i)$.

The joint distribution of T_i and the Bernoulli censoring indicator W_i can be expressed as

$$\begin{aligned} \Pr(T_i, W_i) &= \Pr(T_i \mid W_i) \Pr(W_i) \\ &= \Pr(W_i \mid T_i) \Pr(T_i). \end{aligned}$$

The critical assumption of uninformative censoring is that $\Pr(W_i \mid T_i) = \Pr(W_i)$ is independent of T_i, and that the Bernoulli distribution of W_i is free of the model parameters in the distribution of T_i. In this case the censoring process can be omitted from the likelihood, which can then be expressed as

$$\begin{aligned} L(\boldsymbol{\beta}) &= \prod_{i=1}^{n} [f(t_i)]^{w_i} [S(t_i)]^{1-w_i} \\ &= \prod_{i=1}^{n} [h(t_i)]^{w_i} S(t_i). \end{aligned}$$

For the exponential distribution with mean μ_i, we have

$$\begin{aligned} h(t_i) &= 1/\mu_i = \lambda_i \\ S(t_i) &= \mathrm{e}^{-\lambda_i t_i} \end{aligned}$$

so that

$$L(\boldsymbol{\beta}) = \prod_i \lambda_i^{w_i} e^{-\lambda_i t_i}$$

$$= \prod_i (\lambda_i t_i)^{w_i} e^{-\lambda_i t_i} \Big/ \prod_i t_i^{w_i}$$

The term in the denominator is not a function of the parameter vector $\boldsymbol{\beta}$ and can be omitted for ML calculations, though it must be retained for the computation of the disparity. The remaining term is identical to the likelihood function for a set of n observations w_i having independent Poisson distributions with mean $\lambda_i t_i$ (the term $1/w_i!$ is missing from the Poisson likelihood, but this is a constant and can also be omitted). Since w_i is only either 0 or 1, this result looks strange. Where does the Poisson distribution come from? The Poisson result is easily understood if we visualize a set of n independent time-homogeneous Poisson processes as in Section 6.2, the i-th having rate λ_i. The number of events for the i-th process in a time interval of length t_i then has a Poisson distribution with mean $\lambda_i t_i$. We observe the i-th process until either the first event occurs, or a fixed time has elapsed without the event occurring. In the first case, one event occurs at time t_i; in the second, no event has occurred by time t_i.

The Poisson likelihood for the censored and uncensored exponential survival times allows us to fit models to the hazard rate λ using the Poisson model (Aitkin and Clayton, 1980). Write θ_i for the Poisson mean:

$$\theta_i = \lambda_i t_i.$$

If the *linear hazard model* is used, then

$$\lambda_i = \boldsymbol{\beta}' \mathbf{x}_i$$

and

$$\theta_i = \boldsymbol{\beta}'(t_i \mathbf{x}_i).$$

To fit the model, each explanatory variable $\mathbf{x}_i$ (including the unit vector $\mathbf{1}$) has to be multiplied by t_i, w_i is declared as the response variable with Poisson error and identity link function, and the intercept term is omitted.

If the *log linear hazard model* is used, then

$$-\log \mu_i = \log \lambda_i = \boldsymbol{\beta}' \mathbf{x}_i$$

and

$$\log \theta = \log \lambda_i + \log t_i$$

$$= \boldsymbol{\beta}' \mathbf{x}_i + \log t_i.$$

This model is particularly simple to fit: the same error and response variable specifications are used, but now with log link function and an offset of log t_i.

This approach can also be used for uncensored observations, by defining the censoring indicator as a vector of ones. We illustrate with the example of Section 6.3:

```
> feigl <- transform(feigl, w = rep(1, length(time)),
+       lt = log(time))
> feigl.glm8 <- glm(w ~ offset(lt) + ag * lwbc, data = feigl,
+       family = poisson)
> summary(feigl.glm8)

Coefficients:
            Estimate Std. Error z value Pr(>|z|)
(Intercept)  -5.1499     0.5142 -10.014  < 2e-16
ag+           1.8705     0.7318   2.556  0.01059
lwbc          0.4818     0.1797   2.681  0.00733
ag+:lwbc     -0.3278     0.2462  -1.332  0.18298

(Dispersion parameter for poisson family taken to be 1)

    Null deviance: 58.138  on 32  degrees of freedom
Residual deviance: 38.555  on 29  degrees of freedom
AIC: 112.55

Number of Fisher Scoring iterations: 6
```

We obtain the same parameter estimates and deviance as in Section 6.3 (model `feigl.glm1`), though with opposite signs for the parameters, but the standard errors are different (e.g. 0.732 for the s.e. of `ag-`, rather than 0.785). This is because the expected information matrix for the Poisson model treats t_i as a constant explanatory variable, while in the exponential model t_i is treated as a random variable and is replaced in the expected information matrix by its expected value μ_i. Thus the standard errors reported from the Poisson model are based on the observed, rather than the expected, information matrix for the censored exponential model. See Section 2.9.3 for details.

It may seem strange that we can fit a model to a constant response variable, but the offset introduces the variation to be modelled by the explanatory variables.

If the *reciprocal hazard model* is used, then

$$\lambda_i = 1/\boldsymbol{\beta}'\mathbf{x}_i$$

giving

$$\mu_i = \boldsymbol{\beta}'\mathbf{x}_i$$

and the model is a linear model for the *mean*. Then

$$\theta_i = \lambda_i t_i = t_i / \boldsymbol{\beta}' \mathbf{x}_i,$$

so that

$$\theta_i^{-1} = \boldsymbol{\beta}'(\mathbf{x}_i / t_i).$$

The model can be fitted by dividing each explanatory variable (including the unit vector **1**) by t_i, and using the reciprocal link function without an intercept.

The Poisson representation of the likelihood applies more generally to *proportional hazard* models, as we shall see in Section 6.15.

6.8 Probability plotting with censored data: the Kaplan–Meier estimator

The presence of censoring complicates our assessment of the probability distribution assumption for the response variable. If T_i is exponential with mean μ_i, then T_i/μ_i is standard exponential, but if the true lifetime is censored at t_i, then t_i/μ_i is a censored value from the standard exponential distribution. Thus when a model has been fitted to mixed censored and uncensored observations, the $\mu_i = t_i/\hat{\mu}_i$ are themselves mixed censored and uncensored values from the (approximately) standard exponential distribution. The u_i are the Poisson-fitted values for any link: $u_i = \hat{\theta}_i = \hat{\lambda}_i t_i = t_i/\hat{\mu}_i$. However, it is not clear how to construct the empirical survivor function of the u_i to assess the correctness of the probability distribution. The *product-limit* or *Kaplan–Meier* estimator (Kaplan and Meier, 1958) of the survivor function is used for this purpose.

We begin by considering the set of all discrete times $0 < a_1 < a_2 < \cdots < a_N$ at which deaths or censorings may occur. There is no real loss of generality in assuming that these times are discrete, because the finite precision of measurement means that the values of survival time actually recorded can take only a finite (though possibly large) number of values (see Chapter 2 for a discussion of measurement precision). Let A_j denote the time-interval $(a_{j-1}, a_j]$ with $a_0 = 0$. For the continuous survival distribution with density $f(t)$, survivor function $S(t)$ and hazard $h(t)$, we define the corresponding functions for the resulting grouped discrete distribution:

$$\begin{aligned} f_j &= \Pr(T \in A_j) \\ &= S(a_{j-1}) - S(a_j) \\ &= s_j - s_{j+1}, \end{aligned}$$

where

$$s_j = \Pr(T > a_{j-1}) = f_j + f_{j+1} + \cdots + f_N,$$

and

$$\begin{aligned} h_j &= \Pr(T \in A_j \mid T > a_{j-1}) \\ &= f_j / s_j. \end{aligned}$$

Then

$$h_j = (s_j - s_{j+1})/s_j$$

so that

$$s_{j+1}/s_j = 1 - h_j$$

whence

$$s_{r+1} = \prod_{j=1}^{r} s_{j+1}/s_j = \prod_{j=1}^{r}(1 - h_j)$$

since $s_1 = 1$. The survivor function at time t can therefore be expressed as

$$S(t) = \prod_{j:a_j<t} (1 - h_j).$$

We estimate the survivor function by considering the passage of each individual through time, shown in the table below:

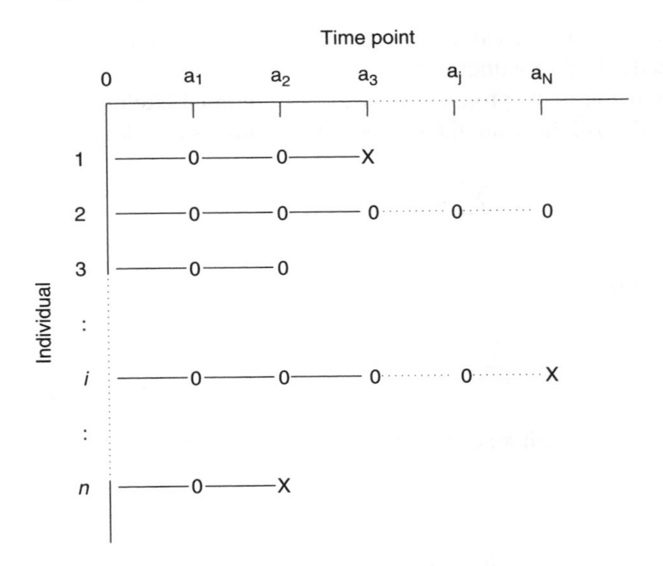

The circles represent censoring and the crosses death. The i-th individual survives to the beginning of the interval A_{j_i} and is either finally censored or dies at the end of the interval. Let w_{ij} be the censoring indicator for the i-th individual at a_j; the hazard in A_j is h_j, and the survivor function at a_{j-1} is s_j, and at a_j is s_{j+1}.

The contribution L_i to the likelihood for the i-th individual is f_{j_i} if the individual dies in A_{j_i}, and s_{j_i+1} if he survives beyond the end a_j of the interval. Thus

$$\begin{aligned}
L_i &= f_{j_i}{}^{w_{ij_i}} s_{j_i+1}{}^{1-w_{ij_i}} \\
&= (h_{j_i} s_{j_i})^{w_{ij_i}} [s_{j_i}(1-h_{j_i})]^{1-w_{ij_i}} \\
&= s_{j_i} h_{j_i}{}^{w_{ij_i}} (1-h_{j_i})^{1-w_{ij_i}} \\
&= h_{j_i}{}^{w_{ij_i}} (1-h_{j_i})^{1-w_{ij_i}} \prod_{j=1}^{j_i-1} (1-h_j) \\
&= \prod_{j=1}^{j_i} h_j{}^{w_{ij}} (1-h_j)^{1-w_{ij}},
\end{aligned}$$

since all the w_{ij} for $j < j_i$ are zero. The likelihood function over all individuals is then

$$L = \prod_{i=1}^{n} L_i = \prod_{j=1}^{N} \prod_{i \in R_j} h_j{}^{w_{ij}} (1-h_j)^{1-w_{ij}},$$

where R_j is the *risk set* of individuals in the j-th interval, that is the set of individuals not already dead or withdrawn (censored).

The likelihood function is the product of a set of N binomial likelihoods, one for each interval. In the j-th interval, the number of individuals at risk is

$$\sum_{i \in R_i} 1 = r_j,$$

and the number dying is

$$\sum_{i \in R_j} w_{ij} = d_j.$$

The constant probability of death for each individual is h_j, so we have immediately that the MLE of h_j is

$$\hat{h}_j = d_j / r_j$$

and therefore that of $S(t)$ is

$$\hat{S}(t) = \prod_{j:a_j<t} (1 - d_j/r_j).$$

This is the *product-limit* or *Kaplan–Meier* estimator of the survivor function. When $d_j = 0$, $\hat{h}_j$ is zero: a change in the estimated survivor function occurs only at an observed death time. Thus in computing the Kaplan–Meier estimator we need to consider only the times at which deaths occur: all other times can be omitted from the set of a_j. When there is no censoring, the Kaplan–Meier estimator reduces to the usual empirical survivor function. Thus the usual empirical cumulative distribution function is in fact the non-parametric maximum likelihood (NPML) estimator of the cumulative distribution function $F(t)$ when no assumption is made about the form of F.

The Kaplan–Meier estimator is a *step function*, with steps only at the observed death times. It is usually graphed as a step function, but we will follow the convention of graphing it as a point function only at the observed death times.

We illustrate with a simple example. The dataset `gehan` contains 42 observations of remission times in weeks of child patients with acute leukaemia. A randomized treatment group was treated with 6-mercaptopurine, the other group was a control. The data are presented and analysed in Freireich *et al.* (1963) and Gehan (1965). We first examine the treatment group, in which censoring is heavy. The counts of observations t for this group are given below, in increasing order, with the censoring indicator w.

```
> data(gehan, package = "SMIR")

> xtabs(~cens + time, data = gehan, subset = treat ==
+     "6-MP")

    time
cens 6 7 9 10 11 13 16 17 19 20 22 23 25 32 34 35
   0 1 0 1  1  1  0  0  1  1  1  0  0  1  2  1  1
   1 3 1 0  1  0  1  1  0  0  0  1  1  0  0  0  0
```

The first censored observation at 6 weeks we take to just *exceed* 6 weeks, and similarly for that at 10 weeks. The survivor function takes the value 1 at $t_0 = 0$ and changes value only at the observed remission times $t_j = 6, 7, 10, 13, 16, 22$, and 23 weeks.

In R the Kaplan–Meier estimator can be computed using the function `survfit` found in the package `survival`. The data must be first coerced to a survival object using the `Surv` function.

```
time n.risk n.event survival std.err lower 95% CI upper 95% CI
   6     21       3    0.857  0.0764        0.720        1.000
```

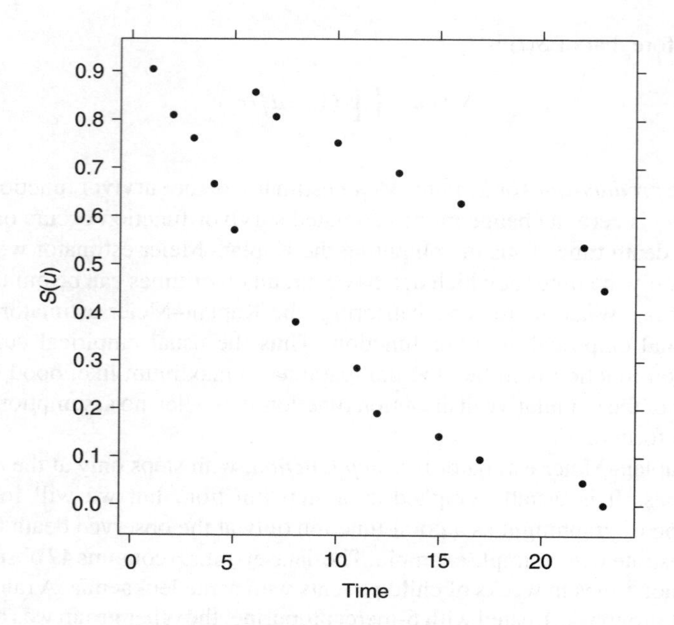

Fig. 6.10. Gehan data: survivor functions

```
    7     17          1     0.807   0.0869          0.653          0.996
   10     15          1     0.753   0.0963          0.586          0.968
   13     12          1     0.690   0.1068          0.510          0.935
   16     11          1     0.627   0.1141          0.439          0.896
   22      7          1     0.538   0.1282          0.337          0.858
   23      6          1     0.448   0.1346          0.249          0.807
```

The plot of the survival curves for each group is shown in Fig. 6.10. The plot can be generated using the plot method for a survival curve (by default, the Kaplan–Meier estimate) with the `survfit` function but here we plot the survivor function explicitly.

We illustrate with the full set of data from `gehan` in which the response variable is `time`, the censoring indicator is `cens` and the treatment group factor is `treat`.

```
> library(survival)
> gehan.surv <- with(gehan, Surv(time, cens))
> gehan.survfit <- survfit(gehan.surv ~ treat, data = gehan,
+   type = "kap", conf.type = "log-log")
> print(xyplot(surv ~ time, data = summary(gehan.survfit),
+      groups = gehan$treat, scales = list(y = list(at = seq(0,
+         1, by = 0.1))), ylab = "S(u)", pch = 19))
```

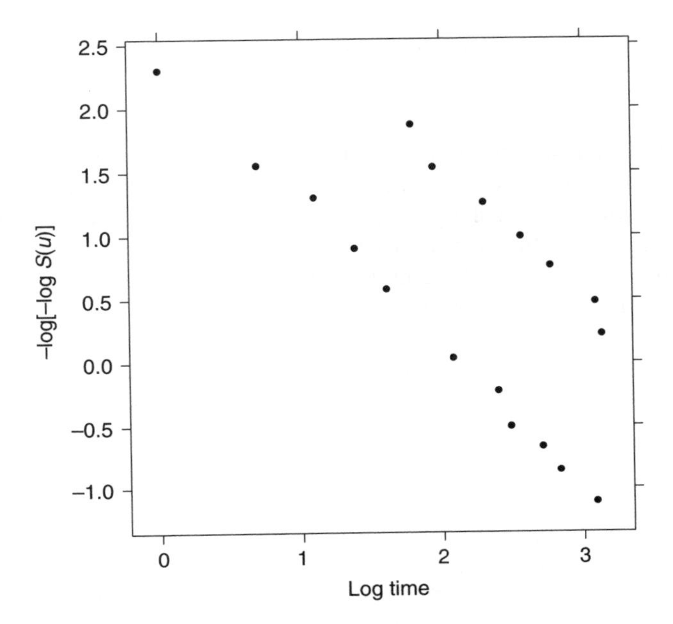

Fig. 6.11. Gehan data: log–log survivor functions

The survivor functions shown in Fig. 6.10 are very different in the two groups. The treatment has a major effect on survival.

We redraw the graph using the (negative) log–log scale for $S(t)$ and the log scale for t (Fig. 6.11).

```
> print(xyplot(-log(-log(surv)) ~ log(time),
+       data = summary(gehan.survfit),
+       groups = gehan$treat,
+       scales = list(y = list(at = pretty(-log(-log(seq(0,
+            1, by = 0.1)))))), ylab = "-log[-log S(u)]",
+       xlab = "log time", pch = 19))
```

It is striking that the two survivor functions are nearly linear and parallel on these scales with common slope nearly 1.0, consistent with an exponential distribution for survival in each group. This parallelism is a distinguishing feature of *proportional hazard models*, discussed in general in Section 6.15.

We now fit an exponential distribution with the two-group model `treat`. The fitted values are then the standardized exponential values u_i.

```
> gehan.glm <- glm(cens ~ offset(log(time)) + treat,
+       data = gehan, family = poisson)
> round(coef(summary(gehan.glm)), 3)
```

```
              Estimate Std. Error z value Pr(>|z|)
(Intercept)     -2.159      0.218  -9.896        0
treat6-MP       -1.527      0.398  -3.832        0
```

The hazard in the second group is significantly lower than in the first, with an estimated group difference of 1.527 with standard error 0.398. We now examine the survivor function for the pooled residuals. As with the `feigl` example we use the *NPL.bands* function to place 95% confidence bounds on the estimated survivor function. The simultaneous bounds for the Kaplan–Meier estimator were given by Hollander *et al.* (1997).

```
> gehan$u <- fitted(gehan.glm)
> gehan.survfit <- survfit(Surv(u, cens) ~ 1, data = gehan)
> print(xyplot((1 - lower) + (1 - upper) ~ x,
+     data = NPL.bands(summary(gehan.survfit)$time[-19]),
+     xlab = "u", ylab = "S(u)", panel = function(x,
+         y) {
+         panel.xyplot(x, y, pch = 20, col = 1)
+         panel.curve(exp(-x))
+         panel.xyplot(summary(gehan.survfit)$time[-19],
+             summary(gehan.survfit)$surv[-19])
+     }))
```

The survivor function in Fig. 6.12 appears to decrease exponentially; the bounds on the true survivor function increase in width with time, but there is little evidence of non-exponential curvature.

Transformations of the proportion scale may be used as in the `feigl` example to examine this issue further. As the fitted models are now straight lines the fine grid is not necessary.

```
> print(xyplot(log(1 - lower) + log(1 - upper) ~ x,
+     data = NPL.bands(summary(gehan.survfit)$time[-19]),
+     xlab = "u", ylab = "log S(u)", panel = function(x,
+         y) {
+         panel.xyplot(x, y, pch = 20, col = 1)
+         panel.curve(-x)
+         stmp <- summary(gehan.survfit)
+         panel.xyplot(stmp$time[-19], log(stmp$surv[-19]),
+             col = "black")
+     }))

> print(xyplot((-log(-log(1 - lower))) + (-log(-log(1 -
+     upper))) ~ log(x),
+     data = NPL.bands(summary(gehan.survfit)$time[-19]),
```

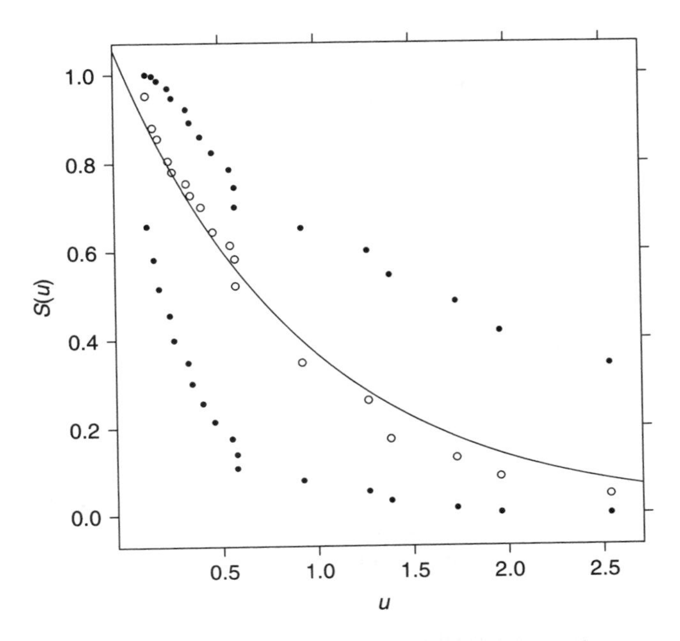

Fig. 6.12. Gehan data: survivor function and bounds

```
+       xlab = "log u", ylab = "-log[-log S(u)]", panel = function(x,
+           y) {
+           panel.xyplot(x, y, col = "black", pch = 20)
+           panel.curve(-x)
+           stmp <- summary(gehan.survfit)
+           panel.xyplot(log(stmp$time[-19]),
+           -log(-log(stmp$surv[-19])), col = "black")
+       }))
```

The points in Fig. 6.13 appear to curve away from the line for large values of scaled time, and their slope in Fig. 6.14 is slightly greater than 1 in magnitude – the fitted line is below the data points on the left, but above them on the right. However, the line is well within the bounds so it is not clear whether the exponential distribution is satisfactory. In Section 6.9 we consider the gamma, and in Section 6.11 the Weibull distribution as alternative models for the survival time.

We now consider survival distributions related to the exponential.

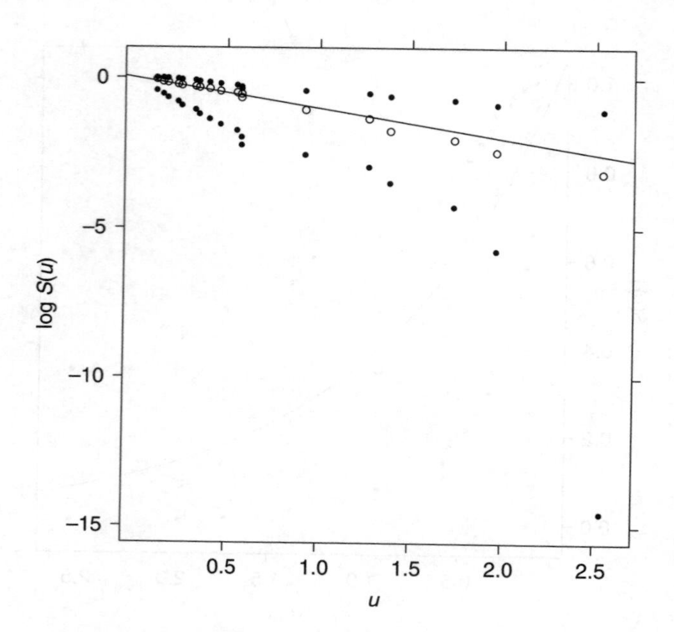

Fig. 6.13. Gehan data: log survivor function and bounds

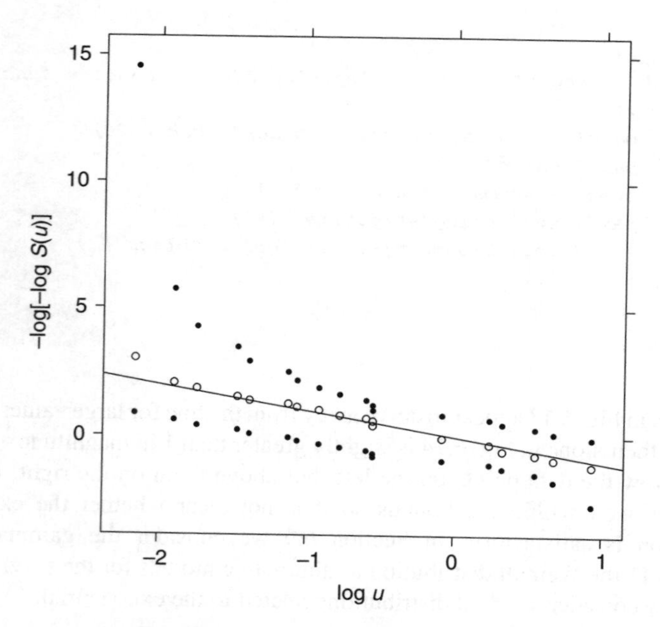

Fig. 6.14. Gehan data: log–log survivor function and bounds

6.9 The gamma distribution

Suppose that an observed lifetime is a *sum* of times for r separate stages of life, each of which is exponentially distributed with the same mean θ. Then the total lifetime has a gamma distribution with parameters r and θ:

$$f(t) = \frac{1}{\Gamma(r)\theta^r} e^{-t/\theta} t^{r-1}, \quad t, r, \theta > 0.$$

The value of r, the *shape* parameter, can more generally be any positive real number.

The mean of the distribution is $r\theta$, and we adopt the R parametrization using $\mu = r\theta$, writing

$$f(t \mid r, \mu) = \frac{r^r}{\Gamma(r)\mu^r} e^{-rt/\mu} t^{r-1}, \quad t, r, \mu > 0.$$

The variance is μ^2/r, and the exponential family *scale* parameter is $\phi = 1/r$. The *cdf* is

$$\begin{aligned} F(t \mid r, \mu) &= \int_0^t f(u \mid r, \mu)\, du \\ &= \frac{1}{\Gamma(r)} \int_0^{\frac{rt}{\mu}} e^{-z} z^{r-1}\, dz \\ &= F_r(rt/\mu), \end{aligned}$$

where $F_r(t)$ is the *cdf* of the 'standardized' gamma distribution with mean 1 and shape parameter r. This is available in R as the function `dgamma`. Graphs of the density are shown in Fig. 6.15 for $\mu = 5$ and $r = 0.5, 1, 2, 5$. Skewness decreases as r increases.

The gamma distribution does not have an analytic hazard function, and does not have a proportional hazard function except for $r = 1$, and so modelling is generally of the (log) mean rather than the (log) hazard. Graphs of the hazard function are shown in Fig. 6.16 for the same values of μ and r.

For $r < 1$ the hazard is monotone increasing. For $r = 1$ the hazard is constant, since the distribution is exponential. For $r > 1$ the hazard first increases and then decreases, with a sharper mode as r increases.

The gamma distribution can be fitted straightforwardly for a constant shape parameter r. We assume that the explanatory variables do not affect the shape parameter; Smyth (1989) has developed double modelling of both μ and r, as for the normal distribution in Chapter 3. We give details in Section 6.9.3.

We first consider uncensored observations, with a log-linear model log $\mu_i = \boldsymbol{\beta}'\mathbf{x}_i$.

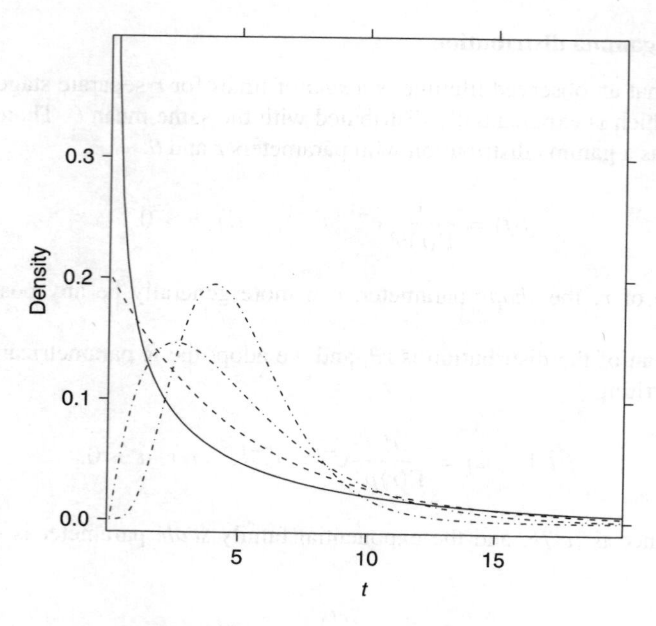

Fig. 6.15. Gamma density functions

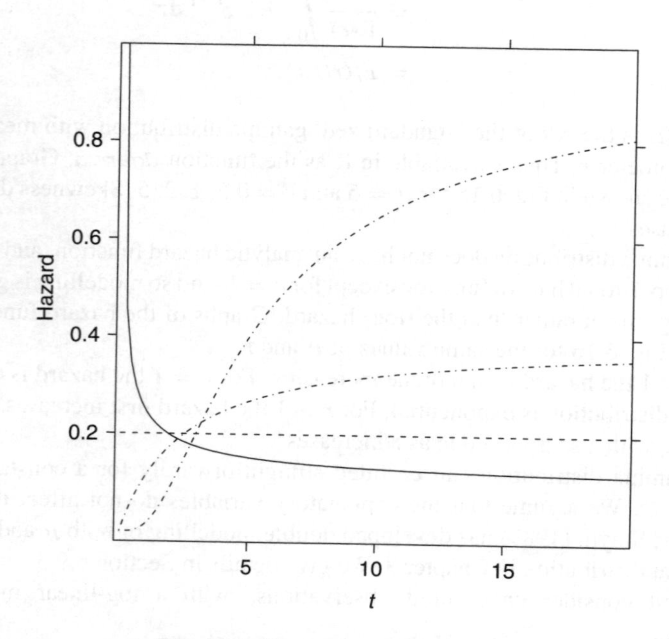

Fig. 6.16. Gamma hazard functions

6.9.1 *Maximum likelihood with uncensored data*

The log-likelihood is

$$\log L = nr \log r - n \log \Gamma(r) - r \sum \log \mu_i - r \sum (t_i/\mu_i) + (r-1) \sum \log t_i,$$

and

$$\frac{\partial \log L}{\partial \boldsymbol{\beta}} = r \sum \{(t_i/\mu_i) - 1\}\mathbf{x}_i$$
$$\frac{\partial \log L}{\partial r} = n(1 + \log r) - n\psi(r) - \sum \log \mu_i - \sum (t_i/\mu_i) + \sum \log t_i$$
$$\frac{\partial^2 \log L}{\partial \boldsymbol{\beta} \partial \boldsymbol{\beta}'} = -r \sum (t_i \mathbf{x}_i \mathbf{x}_i'/\mu_i)$$
$$\frac{\partial^2 \log L}{\partial \boldsymbol{\beta} \partial r} = \sum \{(t_i/\mu_i) - 1\}\mathbf{x}_i$$
$$\frac{\partial^2 \log L}{\partial r^2} = n/r - n\psi'(r),$$

where $\psi(r)$ and $\psi'(r)$ are the digamma and trigamma functions, respectively. Since $\mathrm{E}[T_i] = \mu_i$, the expectation of the cross-derivative is zero, and so is its sample value at the MLE $\hat{\boldsymbol{\beta}}$ (since the cross-derivative is just the score for $\boldsymbol{\beta}$ divided by r). The fact that r is estimated, rather than known, has no effect asymptotically on the precision of $\hat{\boldsymbol{\beta}}$. The MLE of $\boldsymbol{\beta}$ does not depend on r, since r cancels in the score equation for $\boldsymbol{\beta}$, but the observed and expected informations contain r as a multiplying factor. The expected information is $I(\boldsymbol{\beta}) = r \sum \mathbf{x}_i \mathbf{x}_i'$, independent of $\hat{\boldsymbol{\beta}}$.

Thus the MLE of $\boldsymbol{\beta}$ for the gamma distribution will be (in this parametrization) the same as that for the exponential distribution, and the standard errors will be $1/\sqrt{r}$ times those for the exponential. The MLE $\hat{r}$ of r is the solution of

$$n(1 + \log r) - n\psi(r) + \sum \log(t_i/\hat{\mu}_i) - \sum (t_i/\hat{\mu}_i) = 0.$$

The function *gamma.shape* from the *MASS* package calculates the MLE $\widehat{r}$ by a simple Newton–Raphson step starting from the crude estimate from the fitted model object (the mean deviance) and gives the approximate standard error based on the observed information.

A simple approximation is accurate for large r. Using Stirling's formula for the gamma function,

$$\log \Gamma(r) \doteq \frac{1}{2} \log(2\pi) - r + \left(r - \frac{1}{2}\right) \log r$$

and correspondingly

$$\psi(r) \doteq \log r - 1/(2r)$$
$$\psi'(r) \doteq 1/r + 1/(2r^2).$$

So for large r the MLE $\hat{r}$ is approximately the solution of

$$n(1 + \log r) - n\left(\log r - \frac{1}{2r}\right) - \sum(t_i/\hat{\mu}_i) + \sum \log(t_i/\hat{\mu}_i) = 0,$$

which gives

$$\hat{r} \doteq \frac{n}{2}\left(\sum(t_i/\hat{\mu}_i) - \sum \log(t_i/\hat{\mu}_i) - n\right)^{-1} = n/\text{dev},$$

where dev is the deviance, with approximate standard error $r\sqrt{2/n}$. This gives the simple approximation for $\hat{r}$ as the reciprocal of the mean deviance. However, for small r the approximation is quite inaccurate. The LR 'goodness-of-fit' test for the comparison of nested models needs care. If r were *known*, the 'saturated' model would have $\hat{\mu}_i = t_i$, with the corresponding log likelihood

$$\log L_{\text{sat}} = nr \log r - n \log \Gamma(r) - r \sum \log t_i - nr + (r-1) \sum \log t_i,$$

and the LRTS for a given model relative to the saturated model would be

$$\begin{aligned} \text{LRTS} &= -2(\log L_{\text{model}} - \log L_{\text{sat}}) \\ &= 2r\left\{\sum(t_i/\hat{\mu}_i) - n - \sum \log(t_i/\hat{\mu}_i)\right\}. \end{aligned}$$

But this test statistic can be used only when the scale factor r is known and equal under both models. When no scale parameter has been explicitly set by the user, R calculates its deviance using the above expression with r estimated using the deviance which it then uses for calculating standard errors for the parameters. (In fact R uses the deviance based on the Pearson residuals and residual degrees of freedom $n - p - 1$ rather than the sample size n.)

When the scale parameter is *known*, it can be set using the `dispersion` argument in the `summary` function (as in normal variance modelling in Chapter 3).

We illustrate with the `feigl` data of Section 6.3. We first give the standard R analysis.

```
> feigl.gamma <- glm(time ~ ag * lwbc, data = feigl,
+       family = Gamma(link = "log"))
> summary(feigl.gamma)
```

```
Coefficients:
             Estimate Std. Error t value Pr(>|t|)
(Intercept)    5.1499     0.5291   9.733 1.21e-10
ag+           -1.8704     0.8292  -2.256   0.0318
lwbc          -0.4818     0.1834  -2.627   0.0136
ag+:lwbc       0.3278     0.2820   1.162   0.2545

(Dispersion parameter for Gamma family taken to be 1.116173)

    Null deviance: 58.138  on 32  degrees of freedom
Residual deviance: 38.555  on 29  degrees of freedom
AIC: 301.74

Number of Fisher Scoring iterations: 11
```

As no argument is specified for `dispersion`, R's `summary` function estimates the dispersion parameter $\phi = 1/r$ not from the mean deviance as $38.555/29 = 1.329$, but from the deviance based on the Pearson residuals as $32.369/29 = 1.116$. The `gamma.shape` and `gamma.dispersion` functions provided in the `MASS` package calculate the correct values of $\tilde{r}$ and $\tilde{\phi}$ for the gamma distribution, 0.989 and 1.011.

For the full ML analysis, we specify the `dispersion` parameter in the `summary` function.

```
> library(MASS)
> round(coef(summary(feigl.gamma,
+       dispersion = gamma.dispersion(feigl.gamma))), 4)

             Estimate Std. Error z value Pr(>|z|)
(Intercept)    5.1499     0.5035 10.2282   0.0000
ag+           -1.8704     0.7890 -2.3705   0.0178
lwbc          -0.4818     0.1746 -2.7602   0.0058
ag+:lwbc       0.3278     0.2683  1.2216   0.2219


> round(-2 * logLik(feigl.gamma)[1], 2)

[1] 291.74
```

The MLE of α is very close to 1.0, the exponential distribution.

We do not proceed further with this example at this point, since it can be treated as exponential. However, we re-consider this example with *double modelling* of both parameters in Section 6.9.3. Note that the mean deviance estimate $\tilde{r}$ of $29/38.555 = 0.752$ differs considerably from the MLE.

We now consider the case of censored observations, again restricting our analysis to the log link.

6.9.2 *Maximum likelihood with censored data*

The likelihood constructed as in Section 6.7 requires the survivor function which is expressed in terms of the incomplete gamma function.

We solve the likelihood equations indirectly, using the EM algorithm, and considering the censored observations as incomplete. The complete data log-likelihood is linear in t_i and $\log t_i$, and so in the E-step of the algorithm the censored values of t and $\log t$ are replaced by their conditional expectations given the observed data and the current parameter estimates.

For an observation right-censored at a, the conditional distribution of T given $T > a$ is

$$f(t \mid T > a) = f(t)/S(a), \quad t > a.$$

The conditional expectation of T for a right-censored observation is therefore

$$\begin{aligned} \mathrm{E}[T \mid T > a] &= \int_a^\infty t f(t)\,\mathrm{d}t \Big/ S(a) \\ &= \mu \frac{[1 - F_{r+1}(ra/\mu)]}{[1 - F_r(ra/\mu)]}. \end{aligned}$$

To find the conditional expectation of $\log T$, we first find its conditional moment generating function. Write

$$\begin{aligned} M(\theta) &= \mathrm{E}[\mathrm{e}^{\theta \log T} \mid T > a] \\ &= \mathrm{E}[T^\theta \mid T > a] \\ &= \int_a^\infty t^\theta f(t)\,\mathrm{d}t/\mathrm{S(a)} \\ &= \frac{\mu^\theta}{r^\theta} \frac{\Gamma(r+\theta)}{\Gamma(r)} \frac{[1 - F_{r+\theta}(ra/\mu)]}{[1 - F_r(ra/\mu)]}. \end{aligned}$$

The cumulant function of $\log T$ is (omitting terms not involving θ)

$$K(\theta) = \log M(\theta) = \theta(\log \mu - \log r) + \log \Gamma(r + \theta) + \log[1 - F_{r+\theta}(ra/\mu)].$$

Differentiating w.r.t. θ and setting $\theta = 0$ gives

$$\begin{aligned} K'(0) &= \mathrm{E}[\log T \mid T > a] \\ &= \log \mu - \log r + \psi(r) + \frac{g(0 \mid ra/\mu)}{[1 - F_r(ra/\mu)]}, \end{aligned}$$

where

$$g(\theta \mid ra/\mu) = \frac{\partial}{\partial \theta}[-F_{r+\theta}(ra/\mu)].$$

We evaluate $g(0 \mid ra/\mu)$ by numerical differentiation as

$$g(0 \mid ra/\mu) = [F_{r-\delta}(ra/\mu) - F_{r+\delta}(ra/\mu)]/2\delta$$

for some small δ.

Thus in the E-step of the EM algorithm, we replace the censored values of t_i by their conditional expectations

$$t_i^* = \mu_i \frac{[1 - F_{r+1}(rt_i/\mu_i)]}{[1 - F_r(rt_i/\mu_i)]},$$

and the censored values of $\log t_i$ by their conditional expectations

$$(\log t_i)^* = \log \mu_i - \log r + \psi(r) + \frac{g(0 \mid rt_i/\mu_i)}{[1 - F_r(rt_i/\mu_i)]}.$$

In the M-step the score equations for $\boldsymbol{\beta}$ and r are solved using these 'expected data'.

The standard errors for the $\hat{\beta}_j$ are underestimated (being based on the complete data information matrix); correct standard errors can be calculated as described in Chapter 2, by equating the Wald test statistic for the parameter to the disparity change on omitting the explanatory variable. The disparity $-2\log L_{\max}$ is calculated from the summary from the censored and uncensored observations, using the MLE.

The `survreg` function in the `survival` package fits regression for a parametric survival model. We illustrate with the `gehan` example.

```
> gehan.gamma <- survreg(Surv(time, cens) ~ treat,
+       data = gehan)
> summary(gehan.gamma)


               Value Std. Error     z        p
(Intercept)   2.248       0.166 13.55 8.30e-42
treat6-MP     1.267       0.311  4.08 4.51e-05
Log(scale)   -0.312       0.147 -2.12 3.43e-02

Scale= 0.732

Weibull distribution
Loglik(model)= -106.6   Loglik(intercept only)= -116.4
    Chisq= 19.65 on 1 degrees of freedom, p= 9.3e-06
Number of Newton-Raphson Iterations: 5
n= 42

> (gehan.gamma1 <- update(gehan.gamma, . ~ 1))
```

```
Coefficients:
(Intercept)
   2.885024

Scale= 0.8765607

Loglik(model)= -116.4   Loglik(intercept only)= -116.4
n= 42

> round(exp(confint(gehan.gamma)[2, ]))

 2.5 % 97.5 %
     2      7
```

The treatment effect (on the log mean scale) is 1.267: the fitted mean survival under the treatment group is $e^{1.267} = 3.6$ times that in the control group, and the 95% confidence interval for the ratio of means is (1.9 – 6.5). The underestimated standard error of 0.240 is corrected by using the disparity change of 19.65 on omitting the treatment effect:

$$\text{S.E.} \doteq 1.267/\sqrt{19.65} = 0.286;$$

the 'complete data' S.E. underestimates by 16%. These results are qualitatively similar to those for the exponential distribution, though the treatment effect has the opposite sign, since the mean rather than the hazard is being modelled.

The disparity for the exponential model is 217.05 (see Section 6.8), so the disparity change for the gamma shape parameter is 4.17 on 1 df, moderately strong evidence for the gamma over the exponential distribution.

We now consider the modelling of both gamma parameters; we follow the approach in Section 3.19.

6.9.3 *Double modelling*

We restrict the discussion to uncensored data. We generalize the regression model to the form

$$\log \mu_i = \boldsymbol{\beta}'\mathbf{x}_i, \ \log r_i = \boldsymbol{\gamma}'z_i.$$

As for the normal distribution in Chapter 3, this ensures positivity of the scale parameter r. The variables z_i may be a subset or superset of $\mathbf{x}_i$, or may be unrelated.

The parameters $\boldsymbol{\beta}$ and $\boldsymbol{\gamma}$ are assumed to be unrelated. The log-likelihood is

$$\ell = \sum_i \{r_i \log r_i - \log \Gamma(r_i) - r_i \log \mu_i - r_i t_i/\mu_i + (r-1)\log t_i\}$$

and

$$\frac{\partial \ell}{\partial \boldsymbol{\beta}} = \sum_i r_i \mathbf{x}_i [t_i/\mu_i - 1]$$

$$\frac{\partial^2 \ell}{\partial \boldsymbol{\beta} \partial \boldsymbol{\beta}'} = -\sum_i r_i \mathbf{x}_i \mathbf{x}_i' t_i/\mu_i$$

$$\mathrm{E}\left[\frac{\partial^2 \ell}{\partial \boldsymbol{\beta} \partial \boldsymbol{\beta}'}\right] = -\sum_i r_i \mathbf{x}_i \mathbf{x}_i'$$

$$\frac{\partial \ell}{\partial \boldsymbol{\gamma}} = \sum_i r_i z_i [1 + \log r_i - \psi(r_i) + \log(t_i/\mu_i) - t_i/\mu_i]$$

$$\frac{\partial^2 \ell}{\partial \boldsymbol{\gamma} \partial \boldsymbol{\gamma}'} = \sum_i r_i z_i z_i' [1 - r_i \psi'(r_i) + 1 + \log r_i - \psi(r_i) + \log(t_i/\mu_i) - t_i/\mu_i]$$

$$\frac{\partial^2 \ell}{\partial \boldsymbol{\beta} \partial \boldsymbol{\gamma}'} = \sum_i r_i \mathbf{x}_i z_i' [t_i/\mu_i - 1].$$

For the expected information, we have immediately

$$\mathrm{E}\left[\frac{\partial^2 \ell}{\partial \boldsymbol{\beta} \partial \boldsymbol{\gamma}'}\right] = 0.$$

For the expectation of $\log(T_i)$, we have from the censored case above, letting $a \to \infty$,

$$\mathrm{E}[\log T_i] = \psi(r_i) + \log(\mu_i) - \log r_i,$$

and hence

$$\mathrm{E}\left[\frac{\partial^2 \ell}{\partial \boldsymbol{\gamma} \partial \boldsymbol{\gamma}'}\right] = \sum_i r_i z_i z_i' [1 - r_i \psi'(r_i)].$$

Since the cross-derivative has zero expectation as in the normal model, the scoring algorithm reduces to separate algorithms for $\boldsymbol{\beta}$ and $\boldsymbol{\gamma}$, conveniently applied using successive relaxation.

For given $\boldsymbol{\gamma}$, we take r_i as a prior weight in the gamma model for $\boldsymbol{\beta}$ with known scale 1, and for given $\boldsymbol{\beta}$, we formulate the scoring algorithm as an iterative weighted normal regression with adjusted dependent variate

$$z_i = \mu_i - [1 + \log r_i - \psi(r_i) + \log(t_i/\mu_i) - t_i/\mu_i]/[1 - r_i\psi'(r_i)]$$

and iterative weight variate

$$w_i = r_i[1 - r_i\psi'(r_i)].$$

The algorithm begins conveniently with the fit of the constant-r gamma model to estimate $\boldsymbol{\beta}$, followed by alternate $\boldsymbol{\gamma}$ and $\boldsymbol{\beta}$ steps until the disparity converges. At this point the standard errors for both models (based on the expected information) are correct.

This double modelling procedure is implemented in the package `dglm` using the function `dglm` (Dunn and Smyth, 2006). This package parametrizes ϕ, the dispersion parameter as a function of the covariates defining the dispersion so the parameter estimates given by this function have reflected signs of those defined here.

We illustrate with the `feigl` data. We fit interaction models to both mean and scale, and reduce first the scale model, and then the mean model, by backward elimination as in Chapter 3.

```
> library(dglm)
> feigl.double.glm <- dglm(time ~ ag * lwbc, dformula = ~ag *
+     lwbc, family = Gamma(link = "log"), data = feigl,
+     method = "ml", dlink = "log")
> print(summary(feigl.double.glm), digits = 3)

Mean Coefficients:
            Estimate Std. Error t value Pr(>|t|)
(Intercept)    4.938     0.0705  70.058 6.57e-34
ag+           -1.381     1.0780  -1.281 2.10e-01
lwbc          -0.346     0.0707  -4.899 3.36e-05
ag+:lwbc       0.076     0.5417   0.140 8.89e-01
(Dispersion Parameters for Gamma family estimated as below )

    Scaled Null Deviance: 71.3 on 32 degrees of freedom
Scaled Residual Deviance: 38.0 on 29 degrees of freedom

Dispersion Coefficients:
            Estimate Std. Error z value Pr(>|z|)
(Intercept)   -2.352      0.667   -3.52 0.000424
ag+            1.863      1.014    1.84 0.066086
lwbc           0.750      0.217    3.45 0.000562
ag+:lwbc      -0.563      0.333   -1.69 0.090940
(Dispersion parameter for Digamma family taken to be 2)
```

```
    Scaled Null Deviance: 38.1 on 32 degrees of freedom
Scaled Residual Deviance: 29.1 on 29 degrees of freedom

Minus Twice the Log-Likelihood: 283
Number of Alternating Iterations: 8
```

The interaction can be omitted from the scale model. Further simplification leads to the final model:

```
> feigl.double.glm1 <- update(feigl.double.glm, . ~
+     ag + lwbc, dformula = ~lwbc)
> print(summary(feigl.double.glm1), digits = 3)

Mean Coefficients:
            Estimate Std. Error t value Pr(>|t|)
(Intercept)    4.937     0.1013   48.73 4.02e-30
ag+           -1.289     0.4067   -3.17 3.52e-03
lwbc          -0.346     0.0781   -4.43 1.16e-04
(Dispersion Parameters for Gamma family estimated as below )

    Scaled Null Deviance: 64.3 on 32 degrees of freedom
Scaled Residual Deviance: 38 on 30 degrees of freedom

Dispersion Coefficients:
            Estimate Std. Error z value Pr(>|z|)
(Intercept)   -1.230      0.500   -2.46   0.0139
lwbc           0.424      0.165    2.57   0.0101
(Dispersion parameter for Digamma family taken to be 2 )

    Scaled Null Deviance: 41 on 32 degrees of freedom
Scaled Residual Deviance: 34.3 on 31 degrees of freedom

Minus Twice the Log-Likelihood: 287
Number of Alternating Iterations: 6
```

The disparity change from the full double interaction model to the final model is 3.51 on 3 df, and from the final model to the constant-r model is 7.23 on 1 df. There is strong evidence of a decrease in scale r, and therefore an increase in variance, with `lwbc`. The 'z'-statistic for `lwbc` in the scale model is slightly smaller than the disparity change: the squared z-statistic is $2.574^2 = 6.63$. This is to be expected in a small sample with a non-identity link function.

Over the observed range of `wbc` from 0.75 to 100, the fitted value of $\log r$ decreases from 0.72 to -1.352 , so that r decreases from 2.055 to 0.259 ; the effect on the variance of survival time, T, is a reduction by a factor of $1/2.1 = 0.49$ at the smallest value of `wbc` and an increase by a factor of 3.86 at the largest value.

An increase in variance is consistent with the appearance of the data in Fig. 6.1, where the survival times are remarkably variable in both `ag` groups at `wbc` = 100.

The variance heterogeneity does not, however, much affect the mean model parameters in this example: the estimates for `ag` and `lwbc` in the double model *`feigl.double.glm1`* are about 20% greater than those in the constant r model, though the standard errors in the former model are reduced because of the modelling of the heterogeneity. There is no doubt of the importance of both variables.

6.10 The Weibull distribution

The Weibull distribution was first used for the strength of materials corresponding to a form of 'weakest link' model: specifically, if $Z_1, \ldots, Z_r$ are independently distributed on $(0, \infty)$ and $V = \min(Z_j)$ then the distribution of V, suitably standardized, approaches the Weibull distribution as r increases.

The Weibull distribution can be expressed as a power transform of the exponential. Suppose U has an exponential distribution with mean μ and constant hazard $\lambda = 1/\mu$, and $T = U^{1/\alpha}$ where $\alpha > 0$. Then T has a Weibull distribution with density, survivor and hazard functions (expressed in terms of the exponential hazard λ)

$$f(t) = \alpha\lambda t^{\alpha-1}\mathrm{e}^{-\lambda t^{\alpha}}, \qquad t > 0$$

$$S(t) = \mathrm{e}^{-\lambda t^{\alpha}}$$

$$h(t) = \alpha\lambda t^{\alpha-1}.$$

The mean of the distribution is $\Gamma(1 + 1/\alpha)\mu^{1/\alpha}$, the median is $(\mu \log 2)^{1/\alpha}$, and the variance is $[\Gamma(1+2/\alpha) - \Gamma^2(1+1/\alpha)]\mu^{2/\alpha}$. If $\alpha = 1$, then the hazard function is constant and T has an exponential distribution; for $\alpha > 1$ the hazard function is monotone increasing with t, while for $\alpha < 1$ it is monotone decreasing. The shape of the density varies considerably with the *shape parameter* α. Graphs of the density are shown in Fig 6.17 for a mean of 5 and $\alpha = 0.5, 1, 2, 5$.

The density is generally right-skewed, but for large α relative to the mean it can be negatively skewed (as it is for mean 5 and $\alpha = 5$). Graphs of the hazard are shown in Fig. 6.18 for the same values.

If the explanatory variables **x** do not affect the shape parameter, then the Weibull distribution has the important property of a *proportional hazard function*. The effect of the explanatory variables in any regression model for λ is simply to multiply the hazard by some constant: the functional form in t remains the same. This is true regardless of the way the explanatory variables affect the parameter λ. The Weibull distribution has the further property of being an *accelerated failure*

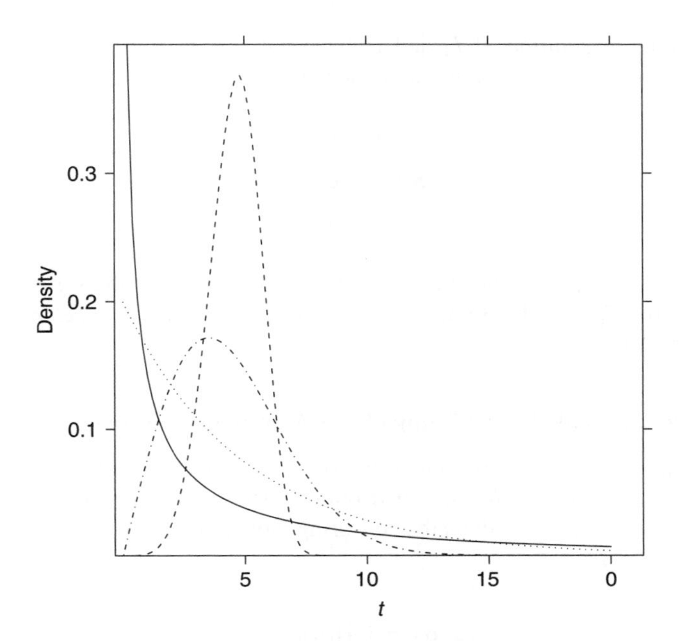

Fig. 6.17. Weibull density functions

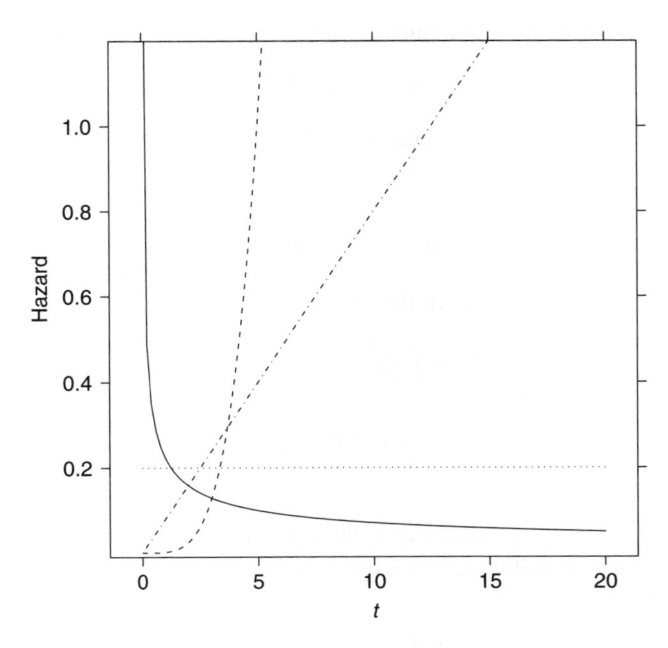

Fig. 6.18. Weibull hazard functions

(or life) *time distribution*. If T_1 and T_2 have Weibull distributions with the same shape parameter α but different parameters λ_1 and λ_2, then

$$\begin{aligned} S_1(t) &= e^{-\lambda_1 t^\alpha} \\ S_2(t) &= e^{-\lambda_2 t^\alpha} \\ &= S_1(\phi t), \end{aligned}$$

where $\phi = (\lambda_2/\lambda_1)^{1/\alpha}$. Thus the survivor function for T_2 is the same as that for T_1 if time for T_1 is scaled by the factor ϕ: death or failure is *accelerated* for T_2 relative to T_1 if $\phi > 1$.

6.11 Maximum likelihood fitting of the Weibull distribution

The Poisson likelihood representation for the exponential distribution can be easily generalized to the Weibull distribution (Aitkin and Clayton, 1980). Given observations $(t_i, w_i, \mathbf{x}_i)$ with w_i the censoring indicator, the likelihood function is, from Section 6.7,

$$L(\boldsymbol{\beta}, \alpha) = \prod_{i=1}^{n} [h(t_i)]^{w_i} S(t_i).$$

Assuming a constant shape parameter, we have for the Weibull distribution

$$\begin{aligned} h(t_i) &= \alpha \lambda_i t_i^{\alpha-1}, \\ S(t_i) &= e^{-\lambda_i t_i^\alpha}, \end{aligned}$$

and we write

$$\theta_i = \lambda_i t_i^\alpha = H(t_i),$$

where H is the integrated hazard function. Then

$$\begin{aligned} L(\boldsymbol{\beta}, \alpha) &= \prod_i [\alpha \theta_i / t_i]^{w_i} e^{-\theta_i} \\ &= \alpha^{\sum w_i} \prod_i \theta_i^{w_i} e^{-\theta_i} \Big/ \prod_i t_i^{w_i}. \end{aligned}$$

Again the term $\prod t_i^{w_i}$ is a constant, and the term $\prod \theta_i^{w_i} e^{-\theta_i}$ is a Poisson likelihood, but there is now an additional term $\alpha^{\sum w_i}$ in the full likelihood. The log of the Weibull hazard function is

$$\log h(t_i) = \log \alpha + (\alpha - 1) \log t_i + \log \lambda_i,$$

and if λ_i itself has a log-linear model:

$$\log \lambda_i = \boldsymbol{\beta}'\mathbf{x}_i,$$

then the hazard is log-linear in the explanatory variables:

$$\log h(t_i) = \log \alpha + (\alpha - 1) \log t_i + \boldsymbol{\beta}'\mathbf{x}_i.$$

As with the exponential distribution, it is possible to fit linear or reciprocal hazard models to λ by scaling the variables by t^{α} or $t^{-\alpha}$, respectively, but the fitted values are not guaranteed to be positive (as in Section 6.5) and we do not discuss this possibility further.

Since

$$\log \theta_i = \alpha \log t_i + \boldsymbol{\beta}'\mathbf{x}_i,$$

if α were known the log-linear hazard model could again be fitted using the log-linear Poisson model for w with a known offset $\alpha \log t$. On the other hand if $\boldsymbol{\beta}$ were known, α could be estimated from the likelihood equation

$$\frac{\partial \ell}{\partial \alpha} = 0.$$

Writing $n_1 = \sum w_i$, the number of uncensored observations, we have

$$\frac{\partial \ell}{\partial \alpha} = n_1/\alpha + \sum (w_i - \theta_i) \log t_i$$

$$\frac{\partial^2 \ell}{\partial \alpha \partial \boldsymbol{\beta}} = -\sum \theta_i \log t_i \mathbf{x}_i$$

and hence

$$\hat{\alpha}^{-1} = \sum (\hat{\theta}_i - w_i) \log t_i / n_1.$$

The last equation defines $\hat{\alpha}$ only recursively, and the information matrix is not block diagonal in α and $\boldsymbol{\beta}$ in any parametrization, so estimation of α affects the precision of the estimate of $\boldsymbol{\beta}$. Joint MLE of $\boldsymbol{\beta}$ and α can be carried out by the method of successive relaxation. First α is fixed at $\alpha_0 = 1$, giving an exponential distribution, and $\boldsymbol{\beta}$ is estimated from the Poisson model taking $\log t_i$ as an offset. Then α_0' is estimated from the above likelihood equation using the fitted values $\hat{\theta}_i$. A new estimate $\alpha_1 = (\alpha_0 + \alpha_0')/2$ is then used to define the offset $\alpha_1 \log t_i$ for a new fit of the Poisson model, and this process is continued until convergence. The damped estimate α_1 is used rather than α_0' because the successive estimates

α_j and α'_j oscillate about the MLE, and damping substantially accelerates the rate of convergence.

The *survreg* function from the *survival* package fits the Weibull distribution using this approach, where the scale factor estimates $1/\alpha$. The arguments are: a formula with the response variable defined as a *Surv* object and the regression formula, the dataframe and the distribution to be fitted. The formulation used by the *survival* package (Section 13.2 in Venables and Ripley, 2002) is

$$\log \theta_i = \alpha \log t_i - \boldsymbol{\beta}' \mathbf{x}_i$$

so we must reflect the signs of their estimates to match ours and their parameter estimates and their standard errors are scaled by $1/\alpha$. Standard errors for the parameter estimates $\boldsymbol{\beta}$ are slightly underestimated because the fitting algorithm treats α as known in the estimation of $\boldsymbol{\beta}$. This can be corrected by dropping each variable in turn and equating the disparity change to the squared Wald statistic. [Roger and Peacock (1982), and Roger (1985) gave direct standard error calculations using the device of extra data.] We illustrate with the gehan example.

```
> library(survival)
> gehan.weibull <- survreg(Surv(time, cens) ~ treat,
+      data = gehan, dist = "weibull")
> summary(gehan.weibull)

             Value Std. Error     z        p
(Intercept)  2.248      0.166 13.55 8.30e-42
treat6-MP    1.267      0.311  4.08 4.51e-05
Log(scale)  -0.312      0.147 -2.12 3.43e-02

Scale= 0.732

Weibull distribution
Loglik(model)= -106.6   Loglik(intercept only)= -116.4
        Chisq= 19.65 on 1 degrees of freedom, p= 9.3e-06
Number of Newton-Raphson Iterations: 5
n= 42

> (gehan.exp <- survreg(Surv(time, cens) ~ treat, data = gehan,
+     dist = "exp"))

Coefficients:
(Intercept)    treat6-MP
   2.159484     1.526614

Scale fixed at 1
```

```
Loglik(model)= -108.5   Loglik(intercept only)= -116.8
    Chisq= 16.49 on 1 degrees of freedom, p= 4.9e-05
n= 42

> (gehan.weibull1 <- update(gehan.weibull, . ~ 1))

Coefficients:
(Intercept)
   2.885024

Scale= 0.8765607

Loglik(model)= -116.4   Loglik(intercept only)= -116.4
n= 42
```

The log-likelihood reported in the summary is equal to the minus half of the disparity, which for the exponential model is 217.05, and that for the Weibull is 213.16, with an estimate $\hat{\alpha}$ of $1/0.732 = 1.366$. Hazard is increasing with time, and the hazard in the treatment group is substantially *reduced* relative to that in the control group, by the factor $e^{-1.267/0.732} = 0.177$. The change in disparity (or deviance) of 3.89 relative to the exponential is just on the 5% point of χ_1^2: there is marginal evidence to support a Weibull distribution rather than an exponential (but see Section 6.15 for further discussion of this example). Omitting the group variable gives a disparity change of 19.65, so the LRT-based standard error is

$$\text{SE} = |\beta|/\sqrt{\text{disp.change}} = (1.267/0.732)/4.433 = 0.390$$

compared with the fixed-α estimate of 0.311. There is some underestimation.

Note also that the gamma distribution (Section 6.9, *gehan.gamma*) fits equally well, with a disparity of 212.88. Comparing the two treatment parameters is simplified if we transform the Weibull parameter to the log mean scale, which requires reflecting the sign and scaling by α, giving $1.267/1.366 = 0.928$ with SE $0.311/1.366 = 0.228$, compared with the gamma value of 1.281 with SE 0.290. We cannot choose between the distributions (which have the same number of parameters) for these data.

In Section 6.5 we compared the lognormal and exponential distributions on the `feigl` data, and found that the lognormal gave a slightly better fit. Does the extra shape parameter of the Weibull improve its fit compared to the lognormal?

```
> library(survival)
> (feigl.exp <- survreg(Surv(time) ~ ag * lwbc, data = feigl,
+      dist = "exp"))
```

```
Coefficients:
(Intercept)        ag+        lwbc    ag+:lwbc
  5.1498503  -1.8704976  -0.4818293   0.3278114

Scale fixed at 1

Loglik(model)= -145.7   Loglik(intercept only)= -155.5
    Chisq= 19.58 on 3 degrees of freedom, p= 0.00021
n= 33

> feigl.weibull <- survreg(Surv(time) ~ ag * lwbc,
+      data = feigl)
> summary(feigl.weibull)

              Value Std. Error      z        p
(Intercept)  5.1536      0.527  9.787 1.28e-22
ag+         -1.8827      0.753 -2.500 1.24e-02
lwbc        -0.4868      0.186 -2.611 9.02e-03
ag+:lwbc     0.3317      0.253  1.310 1.90e-01
Log(scale)   0.0224      0.138  0.162 8.71e-01

Scale= 1.02

Weibull distribution
Loglik(model)= -145.6   Loglik(intercept only)= -153.6
    Chisq= 15.88 on 3 degrees of freedom, p= 0.0012
Number of Newton-Raphson Iterations: 6
n= 33

> feigl.lognormal <- survreg(Surv(time) ~ ag * lwbc,
+     data = feigl, dist = "lognormal")
```

The disparity for the exponential fit is 291.4, exactly that calculated in Section 6.5. The disparity for the Weibull is almost the same, and the shape parameter estimate is almost 1. Thus within the Weibull family the exponential provides almost the best fit. In Section 6.5 we found that the lognormal distribution for the same model gave a disparity of 288.37; this is 2.92 less than for the Weibull and an unimportant difference. (See Chapter 2 for interpretation of disparity differences for models with the same number of parameters.)

6.12 The extreme value distribution

If T has a Weibull distribution with parameters α and μ as defined in Section 6.10, then $Y = \log T$ has the extreme value distribution with scale parameter $\sigma = 1/\alpha$

and location parameter $\theta = \sigma \log \mu$. The density, survivor and hazard functions are

$$f(y) = \frac{1}{\sigma} \exp\left[\frac{y-\theta}{\sigma} - \exp\left(\frac{y-\theta}{\sigma}\right)\right] \quad -\infty < y < \infty,$$
$$-\infty < \theta < \infty, \quad \sigma > 0$$
$$S(y) = \exp\left[-\exp\left(\frac{y-\theta}{\sigma}\right)\right]$$
$$h(y) = \frac{1}{\sigma} \exp\left(\frac{y-\theta}{\sigma}\right).$$

The mean and variance are $\theta + \sigma\psi(1+\sigma)$ and $\sigma^2\psi'(1+\sigma)$, where ψ and ψ' are the digamma and trigamma functions, respectively. The hazard function increases exponentially, which limits the usefulness of the distribution as a model for survival data. The standard form of the density ($\theta = 0, \sigma = 1$) is plotted in Fig. 6.19. Since the extreme value is a location and scale family, all distributions have the same shape: the density has a long left tail and short right tail – it has negative skew.

The extreme value distribution can be fitted using the Poisson representation of the likelihood as for the Weibull distribution. The likelihood is, for general

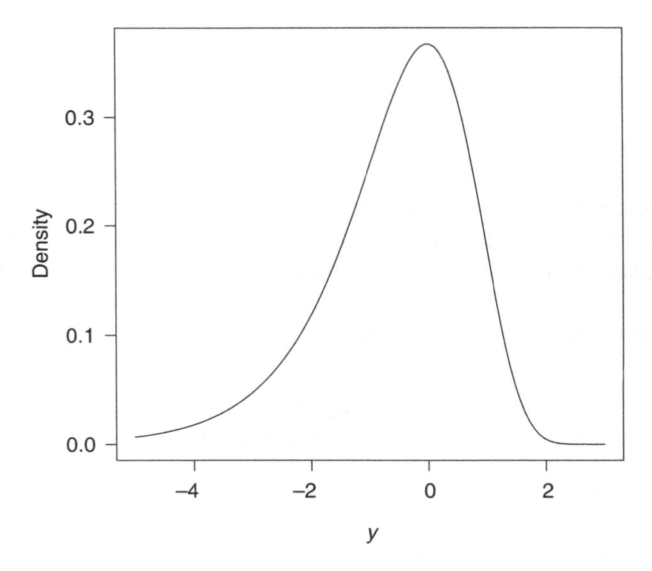

Fig. 6.19. Extreme value density function

location parameters θ_i,

$$L(\theta_1, \ldots, \theta_n, \sigma) = \prod_{i=1}^{n} [h(y_i)]^{w_i} S(y_i)$$
$$= \prod_{i=1}^{n} [H(y_i)]^{w_i} e^{-H(y_i)} \prod_{i=1}^{n} \left[\frac{h(y_i)}{H(y_i)}\right]^{w_i},$$

where $H(y_i)$ is the integrated hazard function. The first part of the above expression is a Poisson likelihood with mean

$$H(y_i) = \exp\left(\frac{y_i - \theta_i}{\sigma}\right)$$

so that

$$\log H(y_i) = y_i/\sigma - \theta_i/\sigma$$
$$= \alpha y_i - \alpha\theta_i.$$

The second term in the likelihood expression is

$$\prod_{i=1}^{n} \frac{1}{\sigma^{w_i}} = \prod_{i=1}^{n} \alpha^{w_i} = \alpha^{n_1},$$

where n_1 is the number of uncensored observations.

The likelihood can be seen to be similar to that given in Section 6.11 for the Weibull distribution, with t_i replaced by $y_i = \log(t_i)$. The parameter estimates from the Poisson fit correspond to the extreme value model (the latter including the scale parameter α) $- \alpha\theta_i = \boldsymbol{\beta}'\mathbf{x}_i$ or $\theta_i = -\boldsymbol{\beta}'\mathbf{x}_i/\alpha$, a *linear* (and proportional hazard) model for the location parameter θ. To fit the extreme value distribution in R, we again use the `survreg` function and set the `distribution` argument to `extreme`.

We illustrate with the `gehan` example.

```
> library(survival)
> summary(gehan.extreme <- survreg(Surv(time, cens) ~
+      treat, data = gehan, dist = "extreme"))

             Value Std. Error     z        p
(Intercept) 11.34       1.933  5.87 4.46e-09
treat6-MP   18.82       3.524  5.34 9.21e-08
Log(scale)   2.16       0.138 15.66 2.84e-55

Scale= 8.67
```

```
Extreme value distribution
Loglik(model)= -120   Loglik(intercept only)= -133.1
        Chisq= 26.24 on 1 degrees of freedom, p= 3e-07
Number of Newton-Raphson Iterations: 5
n= 42

> summary(gehan.extreme1 <- update(gehan.extreme, . ~
+     1))

            Value Std. Error     z        p
(Intercept) 21.13      2.193  9.63 5.73e-22
Log(scale)   2.49      0.137 18.15 1.21e-73

Scale= 12

Extreme value distribution
Loglik(model)= -133.1   Loglik(intercept only)= -133.1
Number of Newton-Raphson Iterations: 5
n= 42
```

The disparity is 240.01, much larger than the Weibull disparity of 213.16, and the estimate of α is 0.115 (corresponding to an estimate of σ of 8.67). This distribution is clearly inappropriate: the extreme value hazard increases exponentially with t ($e^{0.115t}$) compared to the Weibull power hazard ($t^{0.365}$).

For the `feigl` data, it is found similarly that the disparity for the model `ag*lwbc` is 323.7 for the extreme value distribution, much larger than the value of 291.29 for the Weibull distribution. The extreme value distribution is again inappropriate.

```
> summary(feigl.extreme <- survreg(Surv(time) ~ ag *
+     lwbc, data = feigl, dist = "extreme"))
```

6.13 The reversed extreme value distribution

The extreme value distribution is asymmetric, and is defined on the infinite range ($-\infty < y < \infty$). The distribution of $-Y = \log(1/T)$ is the *reversed extreme value distribution*, which has positive skew. The density, survivor and hazard functions are

$$f(y) = \frac{1}{\sigma}\exp\left[-\frac{y-\theta}{\sigma} - \exp\left(-\frac{y-\theta}{\sigma}\right)\right]$$

$$S(y) = 1 - \exp\left[-\exp\left(-\frac{y-\theta}{\sigma}\right)\right]$$

$$h(y) = f(y)/S(y).$$

The hazard function is monotone increasing and approaches an asymptote of $1/\sigma$. This form of the hazard function arises because for large y, $f(y)$ is essentially exponential, with constant hazard. The likelihood function for this distribution does not have the Poisson form, because the survivor function is not exponential. This distribution can be fitted by reflecting the signs of the data and fitting the extreme value distribution.

6.14 Survivor function plotting for the Weibull and extreme value distributions

In Section 6.8 we described the use of the Kaplan–Meier estimator of the survivor function to validate the exponential distribution. We can extend this procedure to assess the validity of the Weibull and extreme value distributions.

For the exponential distribution, we used the Poisson-fitted values to construct the Kaplan–Meier estimate which was then plotted against u using a suitable scale. For the Weibull distribution,

$$S(t) = \mathrm{e}^{-\lambda t^{\alpha}}$$

and the Poisson-fitted values are

$$\hat{\theta}_i = \hat{\lambda}_i t_i^{\hat{\alpha}},$$

which are again approximately (censored or uncensored) standard exponential variables. Thus the same plotting procedures are used for these values based on the Weibull distribution.

In the same way for the extreme value distribution

$$S(y) = \exp\{-\exp[(y-\theta)/\sigma]\}$$

and the Poisson-fitted values are $\exp[(y-\hat{\theta})/\hat{\sigma}]$, so again these have approximately standard exponential distributions, censored or uncensored.

We illustrate in Figs. 6.20 and 6.21 with the Weibull distribution fitted to the Gehan data, repeating the directives from the previous Kaplan–Meier fit. We omit the survivor function graph and give only the two transformed graphs. The band is unhelpfully wide here.

```
> ntimes <- (gehan$time * +
      exp(-gehan.weibull$linear.predictors))^
+ (1/gehan.weibull$scale)
> gehankm <- survfit(Surv(ntimes, gehan$cens))
> changes <- with(gehankm, c(-diff(n.risk), rev(n.event)[1]))
> surv <- rev(gehankm$surv[gehankm$n.event != 0])[-1]
```

```
> riset <- rev(gehankm$n.risk[gehankm$n.event != 0])[-1]
> r <- riset * surv
> times <- rev(gehankm$time[gehankm$n.event != 0])[-1]
> tmp <- data.frame(NPL.bands(r, riset))
> print(xyplot(log(lower) + log(upper) ~ times, data = tmp,
+     ylab = "log S(u)", xlab = "u", panel = function(x,
+         y) {
+         panel.xyplot(x, y, pch = 20)
+         panel.points(x, log(surv))
+         panel.curve(-x)
+     }))

> print(xyplot(-log(-log(lower)) + -log(-log(upper)) ~
+     log(times), data = tmp, ylab = "-log [- log S(u)]",
+     xlab = "u", panel = function(x, y) {
+         panel.xyplot(x, y, pch = 20)
+         panel.points(log(times), -log(-log(surv)))
+         panel.curve(-x)
+     }))
```

The graph in Fig. 6.20 is very close to a straight line: the curvature from the exponential model visible in Fig. 6.13 has been removed. However, in Fig. 6.21

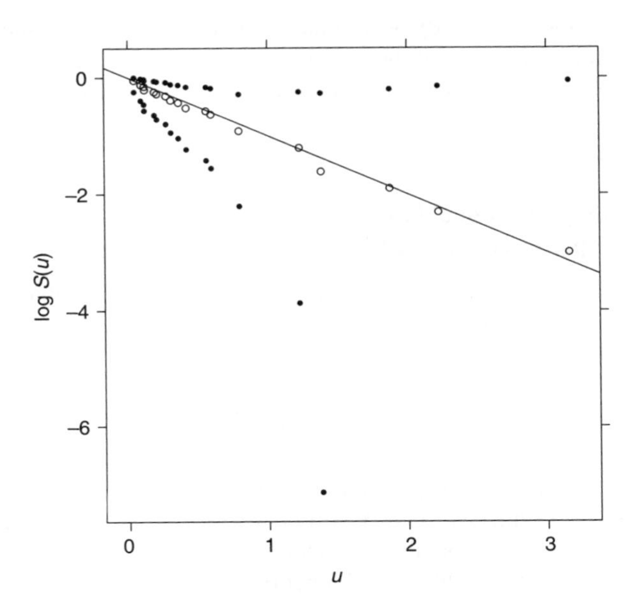

Fig. 6.20. Gehan data: log survivor function and bounds, Weibull

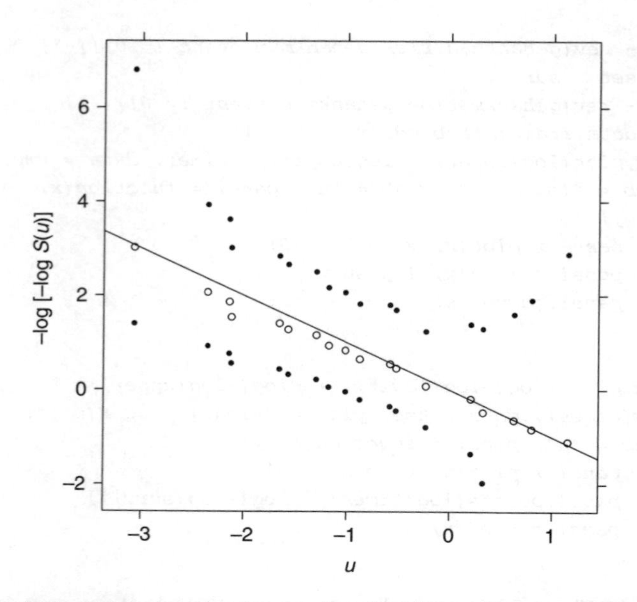

Fig. 6.21. Gehan data: log–log survivor function and bounds, Weibull

the straight line appears slightly out of place, with the magnitude of the (negative) slope overestimated because of the extreme point on the left. There is still some question of the appropriateness of the Weibull model. We consider this further in Section 6.17.

6.15 The Cox proportional hazards model and the piecewise exponential distribution

A further application of the Poisson likelihood representation enables us to fit an important model developed by Cox (1972). In this *proportional hazards model* the hazard function is

$$h(t) = \lambda_0(t)g(\mathbf{x}, \boldsymbol{\beta}),$$

where $\lambda_0(t)$ is an arbitrary, unspecified function of t – the *base-line hazard* when $\mathbf{x} = \mathbf{0}$. The effect of the covariates $\mathbf{x}$ will be assumed to be through a log-linear model, $g(\mathbf{x}, \boldsymbol{\beta}) = e^{\boldsymbol{\beta}'\mathbf{x}}$, so that the individual covariates x_j themselves affect the hazard proportionately. The likelihood function is

$$L(\boldsymbol{\beta}, \lambda_0) = \prod_i [h(t_i)]^{w_i} S(t_i)$$

$$= \prod_i [h(t_i)]^{w_i} e^{-H(t_i)}$$

$$= \prod_i [H(t_i)]^{w_i} e^{-H(t_i)} \prod_i [h(t_i)/H(t_i)]^{w_i},$$

where $H(t)$ is the integrated hazard function. Because of the proportional hazards assumption, the second product in the likelihood does not depend on $\boldsymbol{\beta}$, and the first is again a Poisson likelihood, with

$$\log H(t_i) = \log \Lambda_0(t_i) + \boldsymbol{\beta}'\mathbf{x}_i,$$

where

$$\Lambda_0(t) = \int_0^t \lambda_0(u)\, du$$

is the *integrated baseline hazard*. Since $\lambda_0(t)$ is arbitrary, it is not clear how to fit the model without parametrizing $\lambda_0(t)$ in some way.

In his ground-breaking paper, Cox (1972) used a conditioning argument to eliminate the baseline hazard in a *partial likelihood*. This partial likelihood could then be maximized without reference to the unknown baseline hazard.

We use a different semi-parametric approach based on the *piecewise exponential distribution* according to Breslow (1974); see also Whitehead (1980), and Aitkin *et al.* (1983). This approach retains the hazard but in a (semi-) parametric form which is modelled explicitly and so can be examined using the same methods as for the parametric models already discussed.

We define a set of disjoint time intervals $(a_{j-1}, a_j]$ defined by 'cut-points' $0 = a_0 < a_1 < a_2 < \cdots < a_N < a_{N+1} = \infty$. In the j-th interval $(a_{j-1}, a_j]$, we model the hazard function as *constant*, with parameter λ_j. That is, the survival time distribution $f(t)$ in $(a_{j-1}, a_j]$ is exponential with mean $1/\lambda_j$, but the mean may be different in each piece of the time axis – hence the name piecewise exponential.

The semi-parametric model has the same flexibility as the model in which $\lambda_0(t)$ is unspecified. We replace the assumption of an arbitrary hazard by the assumption of a step-function hazard with steps at $a_1, a_2, \ldots, a_N$.

We can thus write the baseline hazard as

$$\lambda_0(t) = \lambda_j = \exp(\phi_j), \quad a_{j-1} < t \le a_j, \quad j = 1, \ldots, N+1,$$

where the ϕ_j are a set of $N + 1$ time interval constants.

The proportional hazard assumption can then be written as

$$h_j = \lambda_j \exp \boldsymbol{\beta}'\mathbf{x} = \exp(\phi_j + \boldsymbol{\beta}'\mathbf{x}).$$

To construct the likelihood function, we consider as in Section 6.8 the survival experience of each individual through time, shown in the table below:

```
                         Time point
          0       a1        a2        a3        aj        aN
   1      |-------0---------0---------X
   2      |-------0---------0---------0.........0.........
   3      |-------0---------0
   :
   i      |-------0---------0---------0.........0.........X
   :
   n      |-------0---------X
Individual
```

The circles represent censoring and the crosses death. The i-th individual experiences a sequence of censorings at $a_1, a_2, \ldots$ and either final censoring or death at t_i, defined to fall in the N_i-th interval, so that $a_{N_{i-1}} < t_i < a_{N_i}$. Let h_{ij} be the hazard function for the i-th individual in the j-th interval $(a_{j-1}, a_j]$, with

$$h_{ij} = \exp(\phi_j + \boldsymbol{\beta}'\mathbf{x}_i),$$

and let w_{ij} be the censoring indicator for the i-th individual in the j-th interval. The survivor function for the i-th individual in the j-th interval is

$$\begin{aligned} S_{ij}(t) &= \exp[-H_{ij}(t)] \\ &= \exp(-h_{ij}t). \end{aligned}$$

Define e_{ij} to be the exposure time of the i-th individual in the j-th interval, so that

$$\begin{aligned} e_{ij} &= a_j - a_{j-1}, \quad j = 1, \ldots, N_i - 1 \\ &= t_i - a_{N_{i-1}}, \quad j = N_i. \end{aligned}$$

Then the contribution of the i-th individual to the likelihood is

$$L_i = \prod_{j=1}^{N_i} h_{ij}^{w_{ij}} S_{ij}(e_{ij}).$$

The full likelihood is thus

$$L(\boldsymbol{\beta}, \phi_1 \dots \phi_{N+1}) = \prod_{i=1}^{n} \prod_{j=1}^{N_i} h_{ij}^{w_{ij}} \exp(-h_{ij} e_{ij})$$
$$= \prod_i \prod_j \theta_{ij}^{w_{ij}} \exp(-\theta_{ij}) \Big/ \prod_i \prod_j e_{ij}^{w_{ij}}$$

with

$$\theta_{ij} = h_{ij} e_{ij}$$
$$= e_{ij} \exp(\phi_j + \boldsymbol{\beta}' \mathbf{x}_i).$$

The denominator in the likelihood is a constant, and the numerator is again a Poisson likelihood, with a log-linear model for θ_{ij}:

$$\log \theta_{ij} = \log e_{ij} + \phi_j + \boldsymbol{\beta}' \mathbf{x}_i.$$

6.16 Maximum likelihood fitting of the piecewise exponential distribution

We now consider the maximization of the likelihood in all the parameters $\boldsymbol{\beta}, \phi_1, \dots, \phi_{N+1}$. We have

$$\ell = \log L = \sum_{i=1}^{n} \sum_{j=1}^{N_i} (w_{ij} \log \theta_{ij} - \theta_{ij}) - \sum_{i=1}^{n} \sum_{j=1}^{N_i} w_{ij} \log e_{ij}$$
$$\frac{\partial \ell}{\partial \phi_j} = \sum_{i \in R_j} (w_{ij} - \theta_{ij}),$$

where R_j is the risk set of individuals in the j-th interval. For a maximum of the likelihood,

$$d_j = \sum_{i \in R_j} w_{ij} = \sum_{i \in R_j} \hat{\theta}_{ij} = \hat{\lambda}_j \sum_{i \in R_j} e_{ij} \exp(\hat{\boldsymbol{\beta}}' \mathbf{x}_i)$$

where d_j is the number of deaths in the j-th interval. If there are no deaths in the j-th interval, then $d_j = 0$. Since $\hat{\theta}_{ij}$ is non-negative, this equality is only possible if $\hat{\lambda}_j = 0$, whatever the model $\boldsymbol{\beta}'\mathbf{x}$. That is, the term in the maximized likelihood for this interval is 1. Thus, in comparing two models, intervals with no deaths make no contribution to the maximized log-likelihood under both models. The second term in the log-likelihood depends only on t_i and the a_j and is the same under both models.

The choice of interval cut-points determines the extent of *smoothing* of the estimated piecewise hazard function. *Minimal* smoothing is achieved by choosing the distinct death times as cut-points (Breslow, 1974), so that each interval contains just one death (or more if there are tied death times). The resulting estimate for $\hat{\boldsymbol{\beta}}$ corresponds closely to that obtained from the conditional likelihood for the Cox model (Breslow, 1974). Other choices of cut-points will in general give different estimates of $\boldsymbol{\beta}$, since the estimated hazard parameters $\hat{\phi}_j$ are not independent of $\hat{\boldsymbol{\beta}}$.

6.17 Examples

We consider first the gehan data.

The *survival* package in the *coxph* function uses the Efron approximation (Efron, 1977) by default, being more accurate in dealing with tied death times.

```
> library(survival)
> gehan.cph <- coxph(Surv(time, cens) ~ treat, data = gehan,
+      method = "breslow")
> print(gehan.cph)

              coef exp(coef) se(coef)     z       p
treat6-MP -1.51      0.221     0.410 -3.68 0.00023

Likelihood ratio test=15.2  on 1 df, p=9.61e-05  n= 42
```

The *coxph* function does not report the disparity which can be calculated, but requires the individual hazards for each of the time intervals. The *coxph.disparity* function returns the disparity from the piecewise exponential model, including the all terms in the likelihood, and is directly comparable to the disparity for the fit of other models used in this chapter. This allows the Cox proportional hazards model to be compared directly to fully parametric models. (Note that log-likelihood value stored in coxph.object is not comparable as it is based on the proportional hazards function and does not include the baseline hazard, this cancels out in the conditional probabilities that form the partial likelihood.) The individual hazards can be derived from the (cumulative) baseline hazards by differencing those reported by the *basehaz* function from the *survival* package:

```
> gz <- basehaz(gehan.cph, centered = TRUE)
> gz <- transform(gz, haz = diff(c(0, hazard)), tint = diff(c(0,
+     time)))
> ti <- apply(outer(gehan$time, gz$time, ">="), 1,
+      sum)
```

```
> gehan.cph.disparity <- -2 * sum(gehan$cens *
+      (log(gz$haz[ti]/gz$tint[ti]) +
+      gehan.cph$linear) - gz$hazard[ti] * exp(gehan.cph$linear))
```

The group effect estimate of -1.509 is very similar to that from the exponential model, and noticeably smaller in magnitude than the Weibull estimate (-1.731). The disparity is 202.11; the correct df is 23, reflecting the estimation of 1 parameter in the model and 18 hazard parameters for the 17 distinct death times. The disparity change relative to the exponential is 13.68 on 17 df, very weak evidence against the exponential.

To check the similarity of the baseline hazard to Weibull, we use as before the graph of –log(–log S_{0j}) against log t_j.

The estimated baseline survivor function is extracted from the proportional hazard model object using the function *basehaz* which returns a dataframe holding the 17 distinct event times and the estimated (integrated) baseline hazard at each time. By default, the *basehaz* computes the baseline at the covariate mean; here we ask for the baseline hazard at zero.

```
> plot(-log(hazard) ~ time, data = basehaz(gehan.cph,
+      centered = FALSE), log = "x", xlab = "time (log scale)",
+      ylab = "-log[-log(S(t))]", las = 1)
```

The log cumulative hazard function shown in Fig. 6.22 is nearly linear, and therefore close to Weibull or exponential.The group effect estimate is much closer to the exponential estimate than to the Weibull estimate.

Thus the piecewise exponential distribution provides important information about the hazard function, and may suggest a suitable parametric distribution.

We repeat the analysis with the `feigl` data.

```
> data(feigl, package = "SMIR")

> feigl <- transform(feigl, lwbc = log(wbc))
> (feigl.cph <- coxph(Surv(time) ~ ag * lwbc, data = feigl,
+      method = "breslow"))
> library(SMIR)
> round(coxph.disparity(feigl.cph), 2)

            coef exp(coef) se(coef)     z      p
ag+        2.435     11.42    0.911  2.67 0.0075
lwbc       0.642      1.90    0.211  3.04 0.0024
ag+:lwbc  -0.495      0.61    0.276 -1.79 0.0730

Likelihood ratio test=17.9  on 3 df, p=0.000454  n= 33

[1] 268.05
```

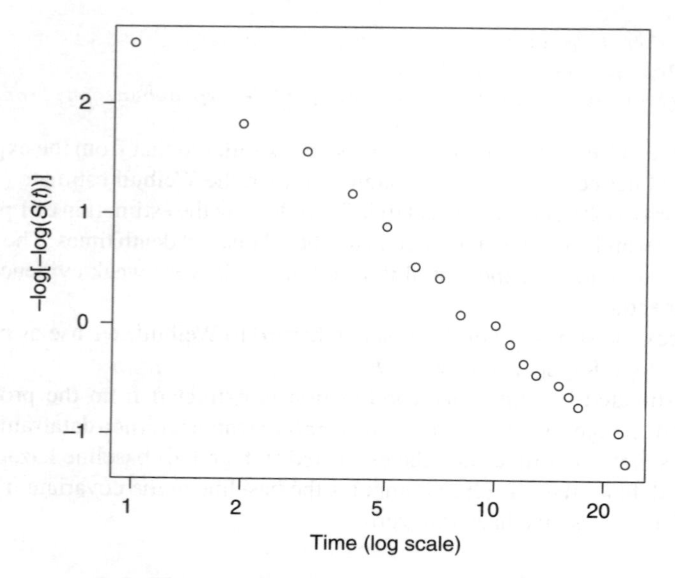

Fig. 6.22. Gehan data: baseline log cumulative hazard function estimate

The disparity is 268.05 on 8 df. We needed to write own function *coxph.disparity* to calculate the disparity for *coxph* class objects.

```
> plot(-log(hazard) ~ time, data = basehaz(feigl.cph,
+      centered = FALSE), log = "x", xlab = "time (log scale)",
+      ylab = "-log[-log(S(t))]", las = 1)
```

The hazard graph (not shown) has a decline in the middle and a steep rise at the end – the hazard does not appear consistent with the exponential distribution. The survivor function graph in Fig. 6.23 shows a corresponding double bend, and is definitely not Weibull, though the overall slope is close to -1.

The parameter estimates and standard errors from the interaction model are somewhat larger in magnitude than those for the exponential distribution (*model feigl.exp*), but less than those for the lognormal distribution (Section 6.5), and again the interaction can be omitted.

```
> print(feigl.cph1 <- update(feigl.cph, . ~ . - ag:lwbc),
+      digits = 4)

       coef exp(coef) se(coef)     z       p
ag+  1.0176     2.767   0.4235 2.403 0.01630
lwbc 0.3603     1.434   0.1355 2.659 0.00785

Likelihood ratio test=14.63  on 2 df, p=0.0006653  n=33
```

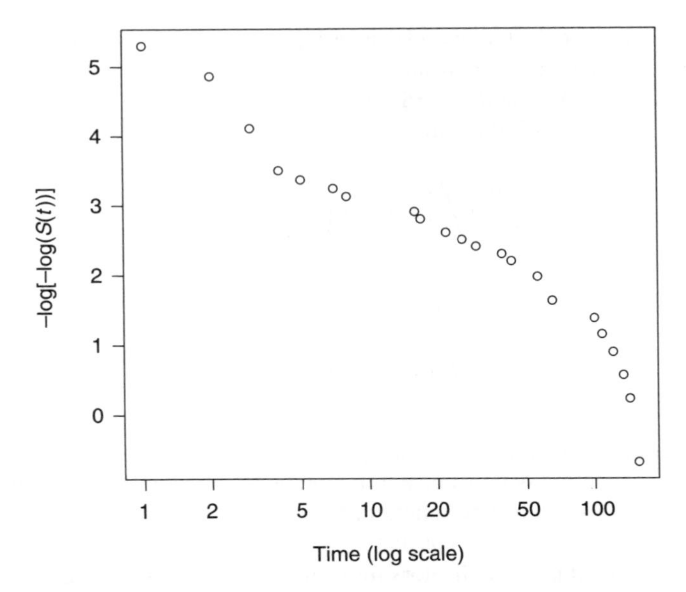

Fig. 6.23. Feigl data: baseline log–log survivor function estimate

```
> anova(feigl.cph1, feigl.cph)

Analysis of Deviance Table

Model 1: Surv(time) ~ ag + lwbc
Model 2: Surv(time) ~ ag * lwbc
  Resid. Df Resid. Dev Df Deviance
1        31    157.363
2        30    154.061  1    3.302
```

The disparity increases by 3.30. Note that there is no constant term – this is eliminated with the hazard constants.

The parameter estimates and standard errors for the main effect model are close to those from the exponential model (`feigl.glm2`), and are also closely equivalent to those reported in Cox and Oakes (1984, p. 100), obtained using the conditional likelihood, though Cox and Oakes use a different parametrization of the model, with `ag` reversed in sign and `lwbc` centred to an origin.

6.18 The logistic and log-logistic distributions

The logistic distribution is very similar to the normal but has slightly heavier tails. Its advantage in survival analysis is that both right- *and* left-censoring can be easily handled, since the likelihood can be expressed as a special form of binomial likelihood (Bennett and Whitehead, 1981).

The density, survivor and hazard functions for the general location- and scale-parameter logistic distribution are

$$f(y) = \frac{1}{\sigma}\exp\left(\frac{y-\mu}{\sigma}\right) \Big/ \left[1+\exp\left(\frac{y-\mu}{\sigma}\right)\right]^2, \qquad -\infty < y < \infty$$

$$S(y) = 1 \Big/ \left[1+\exp\left(\frac{y-\mu}{\sigma}\right)\right]$$

$$h(y) = \frac{1}{\sigma}\exp\left(\frac{y-\mu}{\sigma}\right) \Big/ \left[1+\exp\left(\frac{y-\mu}{\sigma}\right)\right];$$

the distribution mean and variance are μ and $\sigma^2\pi^2/3$, respectively.

The hazard function is logistic, and monotone increasing; the distribution does not have a proportional hazard function, but does have an ccelerated failure time, under the model $\mu_i = \boldsymbol{\beta}'\mathbf{x}_i$, σ constant.

Let c_i and b_i be dummy indicators for right- and left-censoring. Left censoring is uncommon, but may occur if individuals fail before the first measurement period occurs. For example, if survival is recorded in weeks and an individual dies within the first week, the value is left-censored at one week. (In general, recording in this way will lead to a *grouped* survival distribution, but this has little effect on inference unless the grouping is very coarse. In this case the discrete time approach of Section 6.23 should be used.)

For each individual, we have the data $(y_i, \mathbf{x}_i, b_i, c_i)$. The likelihood function is

$$L(\boldsymbol{\beta}, \sigma) = \prod_i [f(y_i)]^{1-b_i-c_i} [S(y_i)]^{c_i} [F(y_i)]^{b_i}.$$

Write

$$F(y_i) = p_i;$$

then

$$S(y_i) = 1 - p_i$$
$$f(y_i) = p_i(1-p_i)/\sigma$$

and

$$L(\boldsymbol{\beta}, \sigma) = \frac{1}{\sigma^m}\prod_i p_i^{1-c_i}(1-p_i)^{1-b_i},$$

where m is the number of uncensored observations. The product has the form of the likelihood from n binomial observations with $r_i = (1 - c_i)$ successes in $n_i = (2 - b_i - c_i)$ trials, and the success probability is

$$p_i = \exp\left(\frac{y_i - \mu_i}{\sigma}\right) \bigg/ \left[1 + \exp\left(\frac{y_i - \mu_i}{\sigma}\right)\right]$$

so that

$$\text{logit } p_i = (y_i - \mu_i)/\sigma = y_i/\sigma - \boldsymbol{\beta}'\mathbf{x}_i/\sigma .$$

(This parametrization is different from that in Bennett and Whitehead who absorb σ into β and use $1/\sigma$ as the scale parameter.)

Maximization of the likelihood can be achieved by a relaxation method using the binomial error model to estimate $\boldsymbol{\beta}$, and an additional step for the estimation of σ. For fixed σ, the logit model is fitted to the number of successes $1 - c_i$ with binomial denominator $2 - b_i - c_i$, using an offset of y_i/σ. The quantities $(y_i - \hat{\mu}_i)$ are obtained by multiplying the linear predictor by σ. A new estimate of σ is then obtained by solving the likelihood equation $\partial \log L/\partial\sigma = 0$, which reduces to

$$m\hat{\sigma} = \sum_i (n_i \hat{p}_i - r_i)(y_i - \hat{\mu}_i).$$

Alternate estimation of $\boldsymbol{\beta}$ and σ continues until convergence, with damping of successive estimates of σ. As with the Weibull macro, the standard errors of the estimated parameters in $\hat{\boldsymbol{\beta}}$ are slightly underestimated because σ is taken as known in each iteration. This can be corrected using the approach of Roger and Peacock (1982), or by dropping each variable in turn as in Chapter 2.

Before considering an example, we describe the log-logistic distribution. If Y is logistic, then $T = e^Y$ is log-logistic. The density, survivor and hazard functions are

$$f(t) = \frac{\alpha}{\theta}\frac{(t/\theta)^{\alpha-1}}{[1 + (t/\theta)^{\alpha}]^2}, \qquad t, \theta, \alpha > 0$$

$$S(t) = \frac{1}{1 + (t/\theta)^{\alpha}}$$

$$h(t) = \frac{\alpha}{\theta}\frac{(t/\theta)^{\alpha-1}}{1 + (t/\theta)^{\alpha}},$$

where $\theta = e^{\mu}$, $\alpha = 1/\sigma$.

For $\alpha \leq 1$ the hazard is monotone decreasing, while for $\alpha > 1$ it has a single maximum. The log-logistic distribution is a convenient approximation to the lognormal since it possesses similar positive skew. The parameter estimates again differ from those in Bennett and Whitehead by the factor σ, and the disparity is different from Bennett and Whitehead's deviance as they omit the term $2\sum(1-b_i-c_i)\log t_i$ required to make the logistic and log-logistic comparable.

We illustrate with the gehan data.

```
> library(survival)
> (gehan.logistic <- survreg(Surv(time, cens) ~ treat,
+      data = gehan, dist = "logistic"))

Coefficients:
(Intercept)   treat6-MP
   8.289423   13.980749

Scale= 5.597992

Loglik(model)= -118.5   Loglik(intercept only)= -126.5
        Chisq= 16.1 on 1 degrees of freedom, p= 6e-05
n= 42

> round(gehan.logistic$loglik[2], 2)

[1] -118.46
```

The disparity is $-2 \times -118.46 = 236.92$ with an estimated σ of 5.60.

```
> gehan.loglogistic <- update(gehan.logistic,
+     dist = "loglogistic")
> summary(gehan.loglogistic)

              Value Std. Error     z        p
(Intercept)   1.893      0.208  9.12 7.78e-20
treat6-MP     1.265      0.326  3.89 1.02e-04
Log(scale)   -0.604      0.150 -4.02 5.72e-05

Scale= 0.547

Log logistic distribution
Loglik(model)= -107.7   Loglik(intercept only)= -115.4
        Chisq= 15.38 on 1 degrees of freedom, p= 8.8e-05
Number of Newton-Raphson Iterations: 4
n= 42

> round(gehan.loglogistic$loglik[2], 2)

[1] -107.66

> (gehan.loglogistic.null <- update(gehan.loglogistic,
+     . ~ 1))
```

```
Coefficients:
(Intercept)
   2.470496

Scale= 0.6378985

Loglik(model)= -115.4   Loglik(intercept only)= -115.4
n= 42
```

The disparity is $2\times 107.66 = 215.32$, and the estimated group difference is 1.265, with (underestimated) standard error 0.326; the estimated σ is 0.547. Fitting the null model gives a disparity of $2 \times 115.35 = 230.70$, and a corrected standard error for the group difference parameter of $1.265/\sqrt{15.38} = 0.323$. The disparity is greater than that for the Weibull (`gehan.weibull`, 213.16) in Section 6.11, and much lower than that for the logistic, as we should expect from the skewness of the survival times.

6.19 The normal and lognormal distributions

The density, survivor, and hazard functions for the normal (μ, σ) distribution for Y are

$$f(y) = \frac{1}{\sqrt{2\pi}\sigma} \exp\left\{-\frac{1}{2}\frac{(y-\mu)^2}{\sigma^2}\right\}$$
$$S(y) = 1 - \Phi\left(\frac{y-\mu}{\sigma}\right)$$
$$h(y) = f(y)/S(y).$$

The hazard function is monotone increasing, and $h(y) \to (y-\mu)/\sigma$ for large y.

The density, survivor and hazard functions for the lognormal distribution for T (a *positive* random variable) with parameters (μ, σ) are

$$f(t) = \frac{1}{\sqrt{2\pi}\sigma t} \exp\left\{-\frac{1}{2}\frac{(\log t-\mu)^2}{\sigma^2}\right\}$$
$$S(t) = 1 - \Phi\left(\frac{\log t-\mu}{\sigma}\right)$$
$$h(t) = f(t)/S(t)$$

The hazard function first increases and then decreases to zero.

In Section 6.5 we compared the exponential and lognormal distributions for the `feigl` data, where there were no censored observations. As with the gamma distribution, censoring complicates the fitting of the normal model because the survivor function does not have a simple analytic form. However, it is possible to fit the normal model with both left- and right-censoring; for simplicity we consider only right censoring (see Wolynetz, 1979 for the general case). For right censoring with censoring indicator w_i we have

$$L(\boldsymbol{\beta}, \sigma) = \prod_i [f(y_i)]^{w_i} [S(y_i)]^{1-w_i},$$

where

$$f(y_i) = \frac{1}{\sqrt{2\pi}\sigma} \exp\left[-\frac{1}{2\sigma^2}(y_i - \mu_i)^2\right]$$

$$\mu_i = \boldsymbol{\beta}'\mathbf{x}_i.$$

Maximization of the log-likelihood can be achieved using an EM algorithm (Dempster *et al.*, 1977; Wolynetz, 1979; Aitkin, 1981).

In the normal model the log-likelihood is linear in y_i and y_i^2. In the E-step, the values of y_i and y_i^2 for the censored observations are replaced by their conditional expectations given the current parameter estimates. In the M-step, the likelihood equations are solved to give new parameter estimates using the conditional expectations obtained in the E-step as real data.

For an observation right censored at a, the conditional density is

$$f(y \mid Y > a) = f(y) \Big/ S\left(\frac{a-\mu}{\sigma}\right), \quad y > a.$$

The conditional expectation of Y is

$$\begin{aligned} \mathrm{E}[Y \mid Y > a] &= \int_a^\infty y f(y)\,\mathrm{d}y \Big/ S\left(\frac{a-\mu}{\sigma}\right) \\ &= \mu + \sigma h\left(\frac{a-\mu}{\sigma}\right) \end{aligned}$$

and that of Y^2 is

$$\begin{aligned} \mathrm{E}[Y^2 \mid Y > a] &= \int_a^\infty y^2 f(y)\,\mathrm{d}y \Big/ S\left(\frac{a-\mu}{\sigma}\right) \\ &= \mu^2 + \sigma^2 + \sigma(\mu + a) h\left(\frac{a-\mu}{\sigma}\right), \end{aligned}$$

where $h(y)$ is the hazard function.

Thus in the E-step, we construct from the current parameter estimates the expected observations

$$\tilde{y}_i = w_i y_i + (1 - w_i) \left[\mu_i + \sigma h \left(\frac{y_i - \mu_i}{\sigma} \right) \right]$$

$$\widetilde{y^2}_i = w_i y_i^2 + (1 - w_i) \left[\mu_i^2 + \sigma^2 + \sigma(\mu_i + y_i) h \left(\frac{y_i - \mu_i}{\sigma} \right) \right].$$

In the M-step, we estimate the parameters $\boldsymbol{\beta}$ and σ. Thus for the identity link, we have the estimates

$$\hat{\boldsymbol{\beta}} = \left(\sum_i \mathbf{x}_i \mathbf{x}_i' \right)^{-1} \left(\sum \mathbf{x}_i \tilde{y}_i \right)$$

$$\hat{\sigma}^2 = \sum_i \left(\widetilde{y^2}_i - 2\mu_i \tilde{y}_i + \hat{\mu}_i^2 \right) / n,$$

where $\hat{\mu}_i = \hat{\boldsymbol{\beta}}' \mathbf{x}_i$.

The normal and lognormal distributions can be fitted using the *survreg* function and specifying the *dist* argument as *gaussian* and *lognormal*, respectively. The survival time and censoring indicator variables are first converted to a *Surv* object which becomes the response variable in the *survreg* regression formula.

The standard errors based on $\hat{\sigma}^2 (\sum \mathbf{x}_i \mathbf{x}_i')^{-1}$ may be in error. Correct standard errors can be obtained by dropping each variable in turn from the model.

We illustrate with the `gehan` data.

```
> library(survival)
> (gehan.normal <- survreg(Surv(time, cens) ~ treat,
+     data = gehan, dist = "gaussian"))

Coefficients:
(Intercept)   treat6-MP
   8.666667   13.862632

Scale= 9.558973

Loglik(model)= -118   Loglik(intercept only)= -126.5
        Chisq= 17.02 on 1 degrees of freedom, p= 3.7e-05
n= 42

> round(gehan.normal$loglik[2], 2)

[1] -117.97
```

```
> gehan.lognormal <- update(gehan.normal, dist = "lognormal")
> summary(gehan.lognormal)

               Value Std. Error      z        p
(Intercept)   1.8251      0.202  9.053 1.39e-19
treat6-MP     1.3468      0.316  4.255 2.09e-05
Log(scale)   -0.0792      0.132 -0.598 5.50e-01

Scale= 0.924

Log Normal distribution
Loglik(model)= -106.7   Loglik(intercept only)= -115.4
        Chisq= 17.38 on 1 degrees of freedom, p= 3.1e-05
Number of Newton-Raphson Iterations: 4
n= 42

> round(gehan.lognormal$loglik[2], 2)

[1] -106.7
```

The disparity for the lognormal (213.4) is slightly greater than that for the Weibull (213.2, `gehan.weibull`) in Section 6.11, slightly less than for the log-logistic (215.3), and much lower than that for the normal (235.9), as we might expect.

```
> gehan.lognormal1 <- update(gehan.lognormal, . ~ 1)
> print(anova(gehan.lognormal1, gehan.lognormal), digits = 4)

  Terms Resid. Df -2*LL Test Df Deviance P(>|Chi|)
1     1         40 230.8        NA       NA        NA
2 treat         39 213.4         1    17.38 3.066e-05
```

The disparity increases by 17.38, giving a standard error of $1.347/\sqrt{17.38} = 0.323$. The standard error of `treat` is underestimated by 2%.

Note that the variance estimates for the lognormal and log-logistic distributions of T are very similar. The estimated variance of the log-logistic is $\hat{\sigma}^2\pi^2/3$, that is, 0.983 with $\hat{\sigma} = 0.547$, while that of the lognormal is 0.854. It is very difficult to distinguish between these distributions in small samples.

The empirical survivor function of the residuals from the normal or lognormal model can be constructed as in Section 6.8.

6.20 Evaluating the proportional hazard assumption

We illustrate the importance of the proportional hazard assumption with a complex example of medical survival data. The data are adapted from Prentice (1973); Prentice's data are reproduced in Kalbfleisch and Prentice (1980; pp. 223–4), and are held in the dataset `prentice`. They consist of survival times t in days of 137 lung cancer patients from a Veteran's Administration Lung Cancer trial, together with explanatory variables: `status`, a measure of general medical status on a continuous scale 1–9.9, with 1–3 completely hospitalized, 4–6 partial confinement to hospital, 7–9.9 able to care for self; age in years `age`; time in months from diagnosis `mfd` to starting on the study; a factor prior therapy `prior` (1 no, 2 yes); a factor treatment `treat` (1 standard, 2 test) and a factor tumour type `type` (1 squamous, 2 small, 3 adeno, 4 large). There are three censored observations; the censoring indicator is `censor`.

We first examine the survivor functions for the four different cell types. This is achieved by selecting the appropriate subset using the *subset* argument and using the *survfit* function The survival probabilities stored in the *survfit* objects are then log-log transformed as described in Section 6.8 and plotted on the same graph. If the other explanatory variables have little effect and the hazard functions are proportional, the transformed survivor functions should be approximately parallel curves; if they are linear as well, a Weibull distribution with a common shape parameter is well supported.

```
> data(prentice, package = "SMIR")
> library(survival)

> prentice.survfit <- survfit(Surv(time, censor) ~
+       Type, data = prentice)
> print(xyplot(-log(-log(surv)) ~ log(time), groups = strata,
+       summary(prentice.survfit),
+       ylab = expression("-log[-log S(t)]"),
+       xlab = "log t"))
```

The survivor functions in Fig. 6.24 show an interesting pattern, which is more clearly visible in colour than in the monotone graph printed here. Three of the cell types show nearly linear, and closely parallel, log-log survivor functions. The fourth (circles – the squamous cell type) is also linear but appears to have a different slope. Thus all four appear to follow Weibull survival, but the squamous cell type has a shape parameter different from the other three.

This complicates the analysis, since our Weibull analysis assumes the shape parameter is unaffected by the explanatory variables. We deal with this by separating the squamous cell type data from the others, and analysing them separately. We begin with the squamous cell type, fitting first the piecewise exponential distribution with the full two-way interaction model and examining the hazard. A graph of survival time against `mfd` (not shown) shows considerable skew in this variable, which is removed by a log transformation `lmfd`. We use this transformed variable in subsequent modelling. We first extract the explanatory variable values for this cell type.

```
> prentice <- transform(prentice, lmfd = log(mfd),
+       Prior = factor(prior), Treat = factor(treat),
```

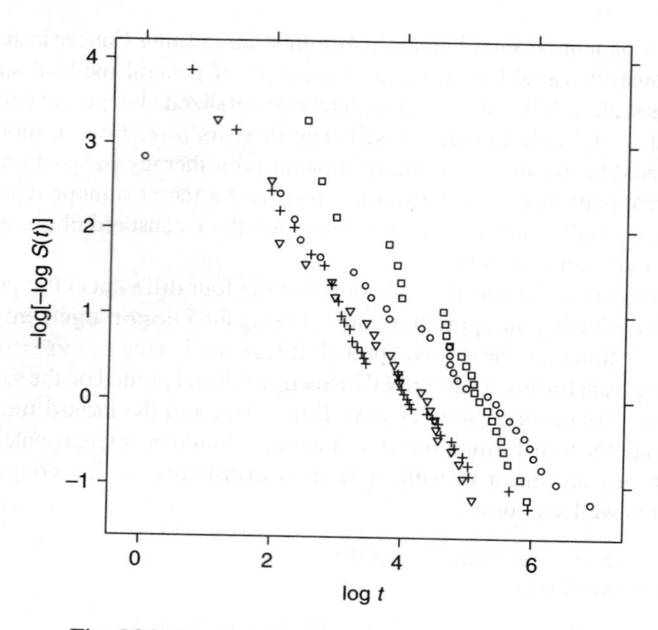

Fig. 6.24. Prentice data: cell type survivor functions

```
+       Type = factor(type))
> library(survival)
> prentice.type1 <- subset(prentice, Type == "1")
> prentice.cph1 <- coxph(Surv(time, censor) ~ status +
+     lmfd + age + Prior + Treat, data = prentice.type1,
+     method = "breslow")
> print(summary(prentice.cph1)$coef, digits = 3)

           coef exp(coef) se(coef)      z      p
status  -0.3329     0.717   0.1120 -2.974 0.0029
lmfd     0.5168     1.677   0.2481  2.083 0.0370
age      0.0352     1.036   0.0251  1.404 0.1600
Prior10 -0.3710     0.690   0.4279 -0.867 0.3900
Treat2  -0.1342     0.874   0.4096 -0.328 0.7400

> plot(-log(hazard) ~ time, data = basehaz(prentice.cph1,
+     centered = FALSE), log = "x", xlab = "time (log scale)",
+     ylab = "-log(-log S)")
```

The log-log survivor function in Fig. 6.25 is roughly linear in $\log t$ but with the two extreme points well out of line. Hazard increases with (log) months from diagnosis and decreases with status, but treatment shows no effect for this cell type.

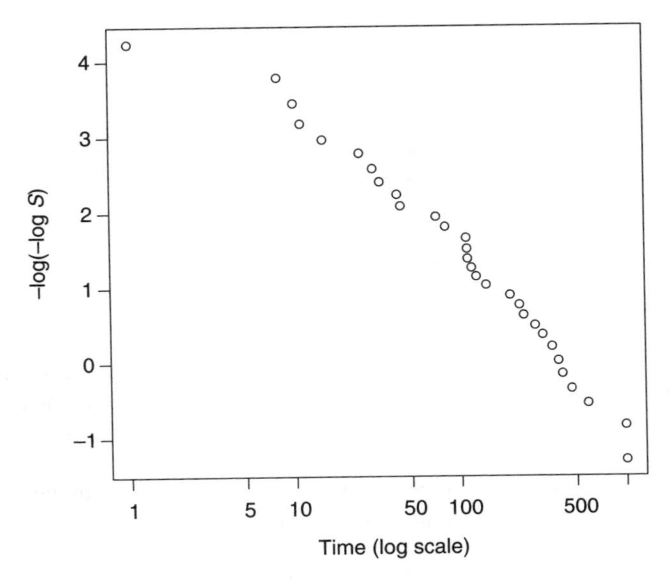

Fig. 6.25. Prentice data: type1 survivor function

We extend the model to include all two-way interactions. Backward elimination gives a final model with the interaction of `prior` and `status` as well as `lmfd`:

```
> prentice.cph2 <- update(prentice.cph1, . ~ .^2)
> prentice.cph3 <- update(prentice.cph2, . ~ status +
+       lmfd + Prior + Prior:status)
```

Hazard declines with increasing status much more rapidly for those with prior therapy, but is higher in the prior therapy group for patients with low status, and lower for patients with high status, than in the group without prior therapy. The cross-over occurs at `status` $= 0.4942/0.5427 = 0.9$. A Weibull distribution gives a similar model:

```
> prentice.weibull <- survreg(Surv(time, censor) ~
+       lmfd + status * Prior, data = prentice.type1)
> print(summary(prentice.weibull), digits = 3)

                    Value Std. Error       z        p
(Intercept)       4.58154      0.993  4.6150 3.93e-06
lmfd             -0.45419      0.225 -2.0208 4.33e-02
status            0.22340      0.128  1.7405 8.18e-02
Prior10          -2.47045      1.382 -1.7877 7.38e-02
status:Prior10    0.44339      0.224  1.9830 4.74e-02
Log(scale)       -0.00170      0.139 -0.0123 9.90e-01
```

```
Scale= 0.998

Weibull distribution
Loglik(model)= -188   Loglik(intercept only)= -197.2
        Chisq= 18.29 on 4 degrees of freedom, p= 0.0011
Number of Newton-Raphson Iterations: 5
n= 35

> round(prentice.weibull$loglik[2], digits = 2)

[1] -188.03
```

The disparity is increased by 32.38, with 29 df. The Weibull hazard is exponential, with a shape parameter estimate of 1.00. The actual hazard however is not exponential, though this does not affect the final model, nor much affect the parameter estimates.

Now we analyse the other three cell types. We begin again with the main effect model, and then extend it with all two-way interactions.

```
> (prentice.t234.cph <- coxph(Surv(time, censor) ~
+      status + lmfd + age + Prior + Type + Treat, data = prentice,
+      subset = Type != "1", method = "breslow"))

            coef exp(coef) se(coef)     z       p
status   -0.3398     0.712   0.0645 -5.27 1.4e-07
lmfd     -0.2655     0.767   0.1573 -1.69 9.1e-02
age      -0.0191     0.981   0.0102 -1.87 6.2e-02
Prior10   0.4214     1.524   0.3133  1.34 1.8e-01
Type2     0.6572     1.929   0.2765  2.38 1.7e-02
Type3     0.8735     2.395   0.3112  2.81 5.0e-03
Type4         NA        NA   0.0000    NA      NA
Treat2    0.5898     1.804   0.2461  2.40 1.7e-02

Likelihood ratio test=48  on 7 df, p=3.55e-08  n= 102
```

Hazard decreases with `status`, but unexpectedly *increases* with the new treatment. We extend the model to include all two-way interactions. Backward elimination gives the final model:

```
> (prentice.t234.cph2 <- update(prentice.t234.cph,
+      . ~ status * Prior + Treat + I(Type == "4")))

                       coef exp(coef) se(coef)     z       p
status               -0.232     0.793   0.0655 -3.54 0.00039
Prior10               1.714     5.551   0.7709  2.22 0.02600
Treat2                0.608     1.837   0.2233  2.72 0.00650
I(Type == "4")TRUE   -0.636     0.530   0.2544 -2.50 0.01200
status:Prior10       -0.286     0.751   0.1337 -2.14 0.03300
```

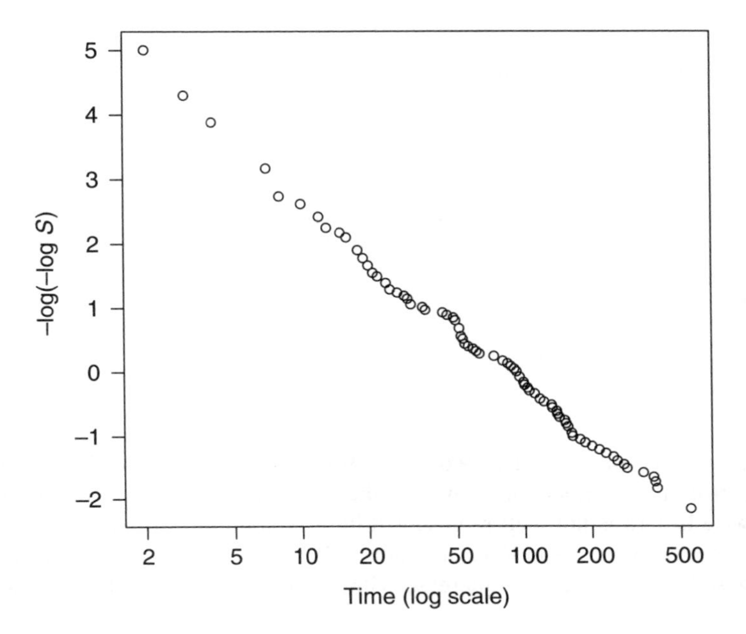

Fig. 6.26. Prentice data: types 2–3–4 survivor function

```
Likelihood ratio test=46  on 5 df, p=9.07e-09  n= 102

> round(coxph.disparity(prentice.t234.cph2), 2)

[1] 988.06
```

The `status:Prior` interaction is similar to that for Type 1, but there is no `lmfd` effect and the adverse treatment effect remains. The hazard for Type 4 cells is lower than for Types 2 and 3 (`I(Type == "4")` is a dummy for this cell type) (Fig. 6.26).

We fit the same model using the Weibull distribution.

```
> prentice.t234.weibull <- survreg(Surv(time, censor) ~
+      status * Prior + Treat + I(Type == "4"), data = prentice,
+      dist = "weibull", subset = Type != "1")
> summary(prentice.t234.weibull)

                     Value Std. Error     z        p
(Intercept)          3.495     0.3278 10.66 1.52e-26
status               0.188     0.0524  3.59 3.30e-04
Prior10             -1.496     0.6220 -2.41 1.62e-02
Treat2              -0.491     0.1708 -2.88 4.03e-03
I(Type == "4")TRUE   0.535     0.2028  2.64 8.38e-03
```

```
status:Prior10       0.246     0.1078  2.29 2.23e-02
Log(scale)          -0.188     0.0759 -2.48 1.31e-02

Scale= 0.828

Weibull distribution
Loglik(model)= -518.9   Loglik(intercept only)= -543.2
        Chisq= 48.64 on 5 degrees of freedom, p= 2.6e-09
Number of Newton-Raphson Iterations: 5
n= 102

> round(prentice.t234.weibull$loglik[2], 2)

[1] -518.87
```

The deviance increases by 49.67, for 69 fewer degrees of freedom. It would appear from this comparison alone that the Weibull should be a good fit, but the piecewise hazard is not well represented by a Weibull hazard, and the deviance change between the models is also not well represented by χ^2_{69}, because the piecewise hazard parameters increase with n and so asymptotic theory does not apply.

The Weibull parameter estimates are similar to those above, except that for the `status` variable, which is very small and positive for those with no prior therapy, instead of significantly large and negative.

The evaluation of the proportional hazards assumption can clearly have an important bearing on the interpretation of the data: the conclusions from this re-analysis are importantly different from those which treated all types as having the same Weibull shape parameter, and which consequently mis-stated the treatment effect, the form of the hazard and the importance of prior therapy and months from diagnosis.

6.21 Competing risks

The survival time modelling of this chapter can be extended to an important class of processes in which failure may be from one of several causes. In the example considered below of heart transplantation patients, death of the patient may occur by rejection of the heart or from other causes not related to rejection (e.g. from infections due to lowered resistance caused by immuno-suppressant drugs). If treatment or patient background variables affect the hazard differently for different causes of death, then an analysis which does not distinguish the different causes may misrepresent both the importance of the explanatory variables and the nature of the hazard function.

Consider in general the case of n individuals on whom we observe $(t_i, w_i, j_i, \mathbf{x}_i)$ for $i = 1, \ldots, n$. Here t_i, w_i and $\mathbf{x}_i$ are the survival time, censoring indicator and explanatory variables as in previous sections, and j_i is the cause of failure, taking one of the values $1, 2, \ldots, k$ for the k possible causes of failure. The associated random variable will be denoted by J.

We define the *cause-specific hazard function* or *sub-hazard function* (Crowder, 2001) $h_j(t)$ by

$$h_j(t)\mathrm{d}t = \Pr(t < T \leq t + \mathrm{d}t,\, J = j \mid T > t)\,.$$

That is, $h_j(t)$ is the instantaneous failure rate for failure from the j-th cause at time t, given survival to time t. The overall hazard function is then

$$h(t) = \sum_{j=1}^{k} h_j(t)$$

since failure must be from one of the k given causes. The survivor function is then

$$S(t) = \mathrm{e}^{-H(t)}$$

with

$$H(t) = \int_0^t h(u)\,\mathrm{d}u = \sum_{j=1}^{k} H_j(t),$$

where $H_j(t)$ is the cause-specific integrated hazard function.

The *cause-specific density function* $f_j(t)$ for survival time for the j-th cause is then given by

$$\begin{aligned} f_j(t)\mathrm{d}t &= \Pr(t < T \leq t + \mathrm{d}t,\, J = j) \\ &= \Pr(T > t)\Pr(t < T < t + \mathrm{d}t,\, J = j \mid T > t) \\ &= S(t)h_j(t)\,\mathrm{d}t \end{aligned}$$

so that

$$f_j(t) = h_j(t)S(t), \qquad j = 1, \ldots, k.$$

To construct the likelihood function, define a set of k dummy indicators d_{ij} for the i-th individual by

$$d_{ij} = \begin{cases} 1 & \text{if failure for the } i\text{-th individual is from cause } j, \\ 0 & \text{otherwise,} \qquad (j = 1, \ldots, k). \end{cases}$$

Then $\sum_j d_{ij} = w_i$.

The likelihood function can be written as

$$
\begin{aligned}
L &= \prod_{i=1}^{n} \left[f_{j_i}(t_i) \right]^{w_i} [S(t_i)]^{1-w_i} \\
&= \prod_{i} \left[h_{j_i}(t_i) \right]^{w_i} S(t_i) \\
&= \prod_{i=1}^{n} \prod_{j=1}^{k} \left[h_j(t_i) \right]^{d_{ij}} \mathrm{e}^{-H_j(t_i)}.
\end{aligned}
$$

Interchanging the products, we see that the likelihood is a product of k factors, the j-th being the likelihood obtained by treating death from cause j as the outcome, and deaths from any other cause as censoring. Since no assumptions have been made about $h_j(t)$, we can fit completely unrelated models to each cause of death very simply. We illustrate with a much-analysed set of data from the Stanford Heart Transplantation programme (Crowley and Hu, 1977). For a discussion of the data and a detailed analysis see Aitkin *et al.* (1983); slightly different data from the same study were presented and analysed in Kalbfleisch and Prentice (1980) and Cox and Oakes (1984). A more extensive data set is given in Miller and Halpern (1982) but without cause of death.

The data can be found as the data set `stan` in the `SMIR` package and contains the data on 65 transplanted patients, consisting of the patient's age at transplantation `age`, prior open-heart surgery `surg` (1 = yes, 0 = no), a censoring indicator `died` (1 = yes, 0 = no), the survival time in days after transplant `surv`, a score `mm` representing the mismatch between the patient's and the donor's tissue type (values range from 0.00 to 3.05), and an indicator `rej` for death by rejection (1 = yes, 0 = no). One zero survival time is recoded to 0.5. There are 41 deaths and 24 censored survivals, with 39 distinct death times. We begin by fitting the piecewise exponential distribution, without distinguishing the causes of death.

```
> data(stan, package = "SMIR")

> library(survival)
> (stan.cph <- coxph(Surv(surv, died) ~ age + surg +
+       mm, data = stan, method = "breslow"))

          coef exp(coef) se(coef)      z     p
age     0.0555     1.057   0.0223   2.49 0.013
surg   -0.8504     0.427   0.4839  -1.76 0.079
mm      0.4426     1.557   0.2918   1.52 0.130

Likelihood ratio test=14.7  on 3 df, p=0.00213  n= 65
```

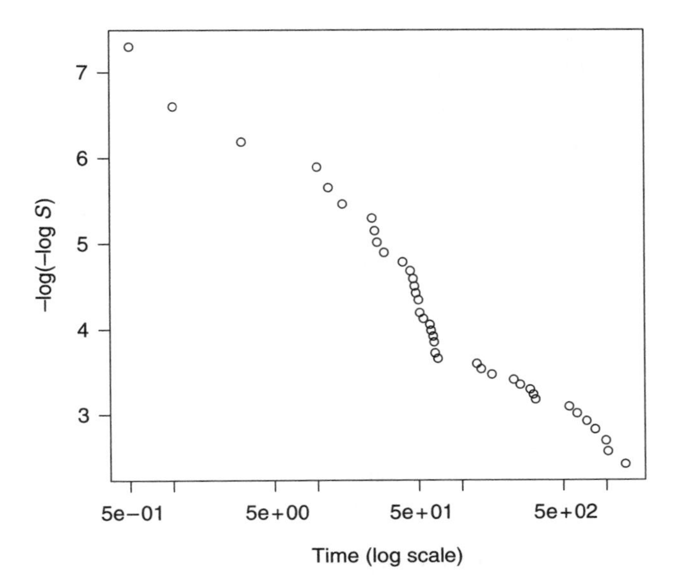

Fig. 6.27. Transplant survival log hazard, all causes

Age is clearly important: risk increases with age as would be expected. The importance of surgery and mismatch score is not clearly established (we obtain the same conclusions by looking at disparity changes due to omitting each variable in turn from the model). We now graph the log hazard and log–log survivor functions against log survival time.

```
> plot(-log(hazard) ~ time, basehaz(stan.cph, center = F),
+      log = "x", las = 1, xlab = "time (log scale)",
+      ylab = "-log(-log S)")
```

Figure 6.27 (of the log–log survivor function) shows a peculiar feature. There is a sudden fall in the survivor function around 55 days followed by another change in slope.

These changes can be identified in the ordered death times, where there are 13 deaths between 44 and 68 days, and then a gap to 127 days. Of these 13 deaths, 11 are by rejection, compared with 29 out of 41 overall. Could this peculiar behaviour be due to different hazards for death by rejection and other causes?

We first consider death by rejection:

```
> stan <- transform(stan, drej = rej * died)
> (stan.reject.cph <- coxph(Surv(surv, drej) ~ age +
+      surg + mm, data = stan, method = "breslow"))
```

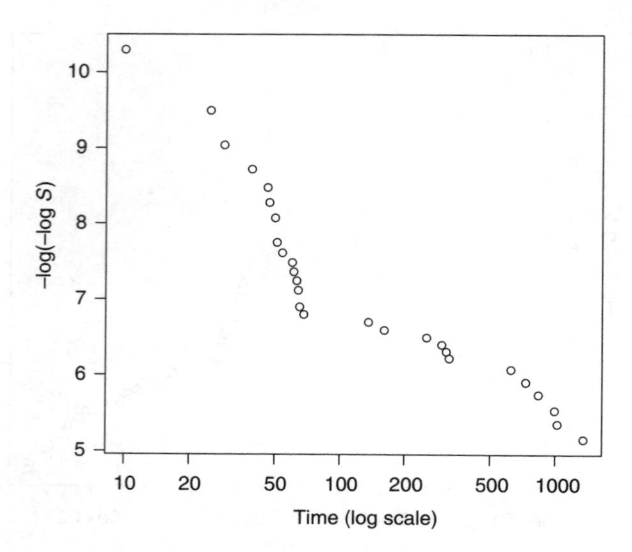

Fig. 6.28. Transplant survival, rejection

```
        coef exp(coef) se(coef)     z      p
age   0.0988     1.104   0.0314  3.15 0.0016
surg -1.0342     0.356   0.6202 -1.67 0.0950
mm    0.9528     2.593   0.3618  2.63 0.0085

Likelihood ratio test=25.9  on 3 df, p=9.9e-06  n= 65
```

The standard errors have increased because the effective sample size – the number of deaths – has decreased. However, all three variables have increased in importance, and mismatch is now clearly significant, though surgery appears not, with a Wald test statistic of -1.67. However, if we drop surgery from the model, the deviance increases by 3.55, so the importance of surgery is still equivocal.

```
> plot(-log(hazard) ~ time, basehaz(stan.reject.cph,
+    centered = FALSE), log = "x", las = 1,
+     xlab = "time (log scale)", ylab = "-log(-log S)")
```

Figure 6.28 shows a precipitous fall in the log–log survivor function from 10 to 65 days, followed by a very slow decline from 130 days onward. This behaviour does not correspond to any standard survival distribution. (The lognormal distribution does not provide an adequate fit, with a disparity of 407.67 compared with 359.42 for the piecewise exponential, a change of 48.24 for 26 df.)

We now analyse the deaths from other causes.

```
> stan <- transform(stan, nrej = (1 - rej) * died)
> (stan.nrej.cph <- coxph(Surv(surv, nrej) ~ age +
+     surg + mm, data = stan, method = "breslow"))
```

```
          coef exp(coef) se(coef)      z    p
age   -0.00736     0.993   0.0319 -0.231 0.82
surg -0.40457      0.667   0.7884 -0.513 0.61
mm   -0.43006      0.650   0.5404 -0.796 0.43

Likelihood ratio test=0.96  on 3 df, p=0.812  n= 65
```

The standard errors have increased further because of the small number of deaths, and *none* of the variables appears important. We examine the hazard and survivor function.

```
> plot(-log(hazard) ~ time, data = basehaz(stan.nrej.cph,
+      centered = FALSE), log = "x", las = 1,
+      xlab = "time (log scale)")
```

The log–log survivor function in Fig. 6.29 decreases nearly linearly with log time, suggesting a Weibull distribution for survival time for deaths from other causes.

We fit the null model.

```
> (stan.nrej.null <- coxph(Surv(surv, nrej) ~ 1, data = stan,
+      method = "breslow"))
```

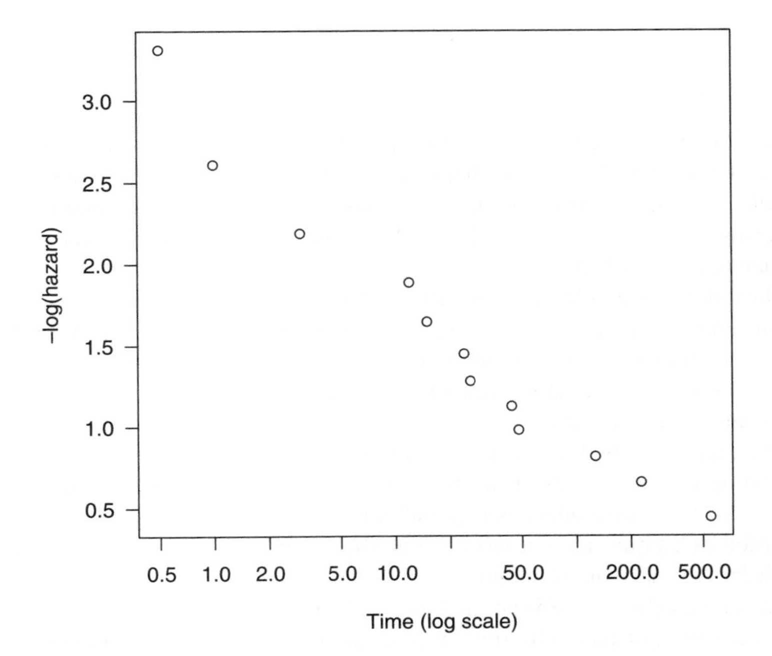

Fig. 6.29. Transplant survival, other causes

```
Null model
  log likelihood= -46.40238
  n= 65

> round(coxph.disparity(stan.nrej.null), 2)

[1] 165.87
```

The disparity increases by 0.96 on 3 df. We fit the Weibull model.

```
> stan.weibull <- survreg(Surv(surv, nrej) ~ 1, data = stan,
+     subset = surv > 0, dist = "weibull")
> summary(stan.weibull)

             Value Std. Error     z        p
(Intercept) 10.11       1.368 7.39 1.49e-13
Log(scale)   1.03       0.258 3.99 6.54e-05

Scale= 2.80

Weibull distribution
Loglik(model)= -91   Loglik(intercept only)= -91
Number of Newton-Raphson Iterations: 7
n= 65

> round(-2 * stan.weibull$loglik[2], 2)

[1] 182.03
```

The Weibull disparity is an increase of 16.16 on 11 df compared to the piecewise exponential. The MLE of the shape parameter is $1/2.80 = 0.357$: risk declines rapidly with time. Fitting the previous model with age, surgery and mismatch confirms their irrelevance for the Weibull distribution, with the same deviance change of 0.96 on 3 df.

Thus in this example the competing risk framework reveals the importance of mismatch and the very rapid increase in hazard in the first 60 days for deaths by rejection, and the rapidly declining hazard for deaths from other causes which is unrelated to the explanatory variables, but reflects the normal recovery process following major surgery.

The cause-specific hazard analysis combines deaths from causes other than the one being examined with 'true' censoring from survival into a single censored group. An alternative analysis is possible in which the censored observations are regarded as coming from a *mixture* distribution (Chapter 7), since the cause of death for the survivors is unknown.

Larson and Dinse (1985) applied this analysis to the Stanford Heart Transplant data; their mixture model for the survivors used a logistic regression model for the probability of death by rejection, with the same covariates used in the logistic model as in the cause-specific piecewise exponential hazard models. The covariates were mismatch and age (both scaled to have mean zero and variance 1), and waiting time for transplant, but not prior surgery.

Waiting time was irrelevant in all the models, and the logistic model could be reduced to the null model, giving a constant probability of death by rejection.

The practical conclusions are the same as in the analysis above: age and mismatch are important for death by rejection, but not for death by other causes.

6.22 Time-dependent explanatory variables

All of the analysis in this chapter has been based on the assumption that the explanatory variables $\mathbf{x}$ are *constant over time*, that is that they do not change during the lifetime of the individual. Many medical variables are of this type, but others may change their values during the individual's lifetime. For example, measures of physical or physiological status may be available during the course of treatment; these may be important predictors of survival time, and more relevant than the same measures taken before treatment begins. Variables of this type are called *time-dependent*.

Such variables can be incorporated into a proportional hazards model using the piecewise exponential distribution. We need first two notational changes. The time-varying explanatory variables may be changing their values continuously in time, but in practice they are measured only at follow-up or interview times, which will generally not correspond to the times a_j at which the hazard function changes. We assume that these variables can be taken as constant over the time intervals between measurements, and we extend the set of cut-points a_j to include the measurement times for each individual.

We now write $\mathbf{x}_{ij}$ for the values of the explanatory variables, measured at a_{j-1}, for the i-th individual in the j-th time interval $a_{j-1} < t \leq a_j$. In practice many, perhaps most, components of $\mathbf{x}_{ij}$ will be constant over time intervals.

The piecewise exponential model of Section 6.15 now applies with

$$h_{ij} = \exp\left(\phi_j + \boldsymbol{\beta}'\mathbf{x}_{ij}\right).$$

We do not illustrate the use of time-varying variables here as these are usually associated with *repeated event* times, for example with several recurrences of a disease. The analysis of repeated events requires *variance component* or *frailty* models, which are discussed in Chapter 9.

6.23 Discrete time models

The discussion so far in this chapter has assumed that time is measured on a continuous scale, or at least that the discreteness of the recording of time is small compared to the range of possible time values.

Time may, however, be grouped into quite broad categories; for example, when deaths of laboratory animals are recorded at relatively long intervals, or when human studies use follow-up periods which are long in comparison to the progress of the disease.

In such cases the likelihood construction of Section 6.15 is not appropriate, and a model for discrete time is needed. We follow the development in Prentice and Gloeckler (1978) (see also Kalbfleisch and Prentice 1980, p. 98 and Thompson 1981).

Suppose the continuous distribution with density $f(t)$, survivor function $S(t)$ and hazard $h(t)$ is grouped into s intervals $A_j = (t_{j-1}, t_j]$ $j = 1, \ldots, s$, with $t_0 = 0, t_s = \infty$. Write f_j, s_j and h_j for the probability mass, survivor and hazard functions for the resulting discrete distribution.
Then

$$\begin{aligned} f_j &= \Pr(T \in A_j) \\ &= S\left(t_{j-1}\right) - S\left(t_j\right) \\ &= s_j - s_{j+1}, \\ h_j &= \Pr\left(T \in A_j \mid T > t_{j-1}\right) \\ &= f_j / \left(f_j + f_{j+1} + \cdots + f_s\right) \\ &= f_j / s_j \end{aligned}$$

so that

$$s_j = \prod_{k=1}^{j-1} (1 - h_k)$$

as in Section 6.8. Consider the survival experience of each individual through time as in Section 6.15. The i-th individual experiences a sequence of censorings at $t_1, t_2, \ldots$ and either dies or is finally censored in the j_i-th interval. Let w_{ij} be the censoring indicator and h_{ij} the hazard for the i-th individual in the j-th interval; the i-th contribution to the likelihood is then

$$\begin{aligned} L_i &= \prod_{j=1}^{j_i} h_{ij}^{w_{ij}} s_{ij} \\ &= \prod_{j=1}^{j_i} h_{ij}^{w_{ij}} \left(1 - h_{ij}\right) \\ &= \prod_{j=1}^{j_i} h_{ij}^{w_{ij}} \left(1 - h_{ij}\right)^{1-w_{ij}}. \end{aligned}$$

For the proportional hazards model,

$$h(t, \mathbf{x}) = \lambda_0(t)\, \mathrm{e}^{\boldsymbol{\beta}'\mathbf{x}}$$
$$S(t, \mathbf{x}) = \exp\Bigl(-\Lambda_0(t)\, \mathrm{e}^{\boldsymbol{\beta}'\mathbf{x}}\Bigr).$$

Thus

$$s_j = \exp\Bigl(-\Lambda_0(t_{j-1})\, \mathrm{e}^{\boldsymbol{\beta}'\mathbf{x}}\Bigr)$$
$$h_j = 1 - s_{j+1}/s_j$$
$$= 1 - \exp\Bigl[-\mathrm{e}^{\boldsymbol{\beta}'\mathbf{x}} \left\{\Lambda_0(t_j) - \Lambda_0(t_{j-1})\right\}\Bigr]$$

so that

$$h_{ij} = 1 - \exp\Bigl(-\mathrm{e}^{\boldsymbol{\beta}'\mathbf{x}_i + \psi_j}\Bigr)$$

where

$$\psi_j = \log\bigl\{\Lambda_0(t_j) - \Lambda_0(t_{j-1})\bigr\}.$$

Since nothing is assumed about $\lambda_0(t)$, the ψ_j are unrelated, unknown constants as in the proportional hazards model. The likelihood function over all individuals is then

$$L = \prod_{i=1}^{n} \prod_{j=1}^{j_i} h_{ij}^{w_{ij}} \left(1 - h_{ij}\right)^{1-w_{ij}} = \prod_{j=1}^{s} \prod_{i \in R_j} h_{ij}^{w_{ij}} \left(1 - h_{ij}\right)^{1-w_{ij}},$$

where R_j is the set of individuals at risk in the k-th time interval. This is a product of $\sum j_i$ Bernoulli likelihoods, one for each individual in each interval. In each interval the individuals currently at risk either die or survive to the next interval. The death probability in the j-th interval for the i-th individual is h_{ij}, and

$$\log\bigl\{-\log\left(1 - h_{ij}\right)\bigr\} = \boldsymbol{\beta}'\mathbf{x}_i + \psi_j.$$

Thus the discrete time proportional hazard model can be fitted by treating the observations in each time interval as independent across intervals, with the censoring indicator as the response variable with a Bernoulli distribution, and a complementary log–log link function (Section 4.2); the regression model contains the explanatory variables and the interval parameters ψ_j. If the explanatory variables are themselves categorical, the data have the form of a contingency table, and the response variable is the number of deaths in the interval, which has a binomial distribution with the number at risk in the interval as the binomial denominator.

We illustrate with the gehan example. Suppose that remission times are recorded in months (units of 4 weeks) instead of weeks. The grouped data are presented below as a contingency table, with the number at risk and the number dying in each month for each treatment group. Censoring within an interval is treated as censoring at the end of the previous interval.

		Time					
		1-4	5-8	9-12	13-16	17-20	21-24
control	d	7	6	4	1	1	2
	r	21	14	8	4	3	2
6-MP	d	0	4	1	2	0	2
	r	21	20	13	12	7	7

The analysis is straightforward.

```
> gehand <- data.frame(d = c(7, 6, 4, 1, 1, 2, 0, 4,
+     1, 2, 0, 2), r = c(21, 14, 8, 4, 3, 2, 21, 20,
+     13, 12, 7, 7), month = gl(6, 1), treat = gl(2,
+     6, labels = c("control", "6-MP")))
> library(gnm)
> gehand.gnm <- gnm(cbind(d, (r - d)) ~ 1, eliminate = month,
+     data = gehand, family = binomial(link = "cloglog"),
+     verbose = F)
> (gehand.gnm1 <- update(gehand.gnm, . ~ . + treat))

Coefficients of interest:
treat6-MP
   -1.698

Deviance:            6.801872
Pearson chi-squared: 5.278622
Residual df:         5

> anova(gehand.gnm, gehand.gnm1)

Analysis of Deviance Table

Model 1: cbind(d, (r - d)) ~ 1
Model 2: cbind(d, (r - d)) ~ treat
  Resid. Df Resid. Dev Df Deviance
1        6    24.8841
2        5     6.8019  1  18.0823
```

The estimate of `treat6-MP` is reasonably close to that (-1.509) from the piecewise exponential model (*gehan.cph*, Section 6.17), and is between those for the exponential (1.527) and Weibull distributions (1.267).

An alternative to fitting the grouped time variable as a factor in the model is to construct a conditional likelihood which depends only on the common group difference. We condition on the total number at risk in each time interval, in the same way as for the 2×2 table in Section 4.6. Various tests based on the resulting conditional hypergeometric distribution are possible; the best known is the *log rank test* (Mantel, 1966).

It is not necessary to have categorical explanatory variables to fit the discrete time model. As for the proportional hazards model in continuous time with time dependent covariates, the data vector w_{ij} is expanded to length $\sum N_i$, as are the explanatory variables $\mathbf{x}_i$. N_i (defined in Section 6.15) is the number of intervals through which person i passes before death or final censoring.

We illustrate with the `feigl` data, using the distinct death times as discrete cut-points.

```
> Nn <- sum(feigl$time)
> I <- nrow(feigl)
> Feigl <- feigl[rep(1:I, feigl$time), ]
> vP <- NULL
> for (i in 1:I) vP <- c(vP, seq(1, feigl$time[i],
+      by = 1))
> Feigl <- transform(Feigl, y = ifelse(time == vP,
+      1, 0), lwbc = log(wbc), Period = factor(vP))
> feigl.discrete.glm <- glm(y ~ Period + ag * lwbc -
+      1, data = Feigl, family = binomial(link = "cloglog"))
> print(summary(feigl.discrete.glm)$coef[-c(1:156),
+      ], digits = 3)

          Estimate Std. Error z value Pr(>|z|)
ag+          2.873      0.938    3.06 0.002188
lwbc         0.747      0.222    3.36 0.000766
ag+:lwbc    -0.598      0.281   -2.13 0.033535
```

The hazard estimates are not $\log h_{ij}$ but $\log[-\log(1-h_{ij})]$, which is approximately $\log h_{ij}$ if h_{ij} is small (see below). The estimates (not shown here) are initially fairly constant but show a steady increase with cutpoint number from cutpoint 14 on. The behaviour of these estimates does not however in general provide information about the form of the hazard function, since it depends on both the spacing of the cut-points and on $\lambda(t)$. The interaction appears significant compared to its standard error. We try omitting it from the model.

```
> feigl.discrete.glm1 <- update(feigl.discrete.glm,
+      . ~ . - ag:lwbc)
```

```
> print(summary(feigl.discrete.glm1)$coef[-c(1:156),
+    ], digits = 4)

     Estimate Std. Error z value Pr(>|z|)
ag+    1.1257     0.4289   2.625 0.008672
lwbc   0.3922     0.1371   2.861 0.004227

> anova(feigl.discrete.glm1, feigl.discrete.glm, test = "Chisq")

Analysis of Deviance Table

Model 1: y ~ Period + ag + lwbc - 1
Model 2: y ~ Period + ag * lwbc - 1
  Resid. Df Resid. Dev  Df Deviance P(>|Chi|)
1      1191    178.064
2      1190    173.396   1    4.668     0.031
```

The disparity change, 4.67 is significant, and the main-effect parameter estimates are similar to those from the piecewise exponential model; the observed survival times in weeks are not heavily grouped relative to the variation in individual survival.

It is easily seen that if $t_j - t_{j-1}$ is small,

$$\begin{aligned}\psi_j &= \log\left[\Lambda_0(t_j) - \Lambda_0(t_{j-1})\right]\\ &\approx \log\left[(t_j - t_{j-1})\lambda_0(t_j)\right]\\ &= \log(e_j) + \phi_j\end{aligned}$$

from the piecewise exponential. As h_{ij} will also be very small,

$$\log\left\{-\log\left(1 - h_{ij}\right)\right\} \approx \log h_{ij},$$

so we obtain the previous piecewise exponential model as the Poisson limit of the binomial model for precise measurement in continuous time.

7
Finite mixture models

7.1 Introduction

The exponential family of distributions and its extensions discussed in earlier chapters are extremely useful in statistical analysis, but they cannot represent all types of data of scientific interest.

In this chapter, we consider a general extension of the exponential family to *mixtures* of distributions from this family. The need for this extension will become clear in subsequent chapters in which we discuss *random effect models* and their fitting by maximum likelihood.

Mixtures are extensively discussed in the books by Everitt and Hand (1981); Maritz and Lwin (1989); Titterington *et al.* (1985); McLachlan (1988); Lindsay (1995); Böhning (1999); and McLachlan and Peel (2000). Our treatment is much more limited.

We restrict our discussion to the *mixed* exponential family distribution

$$m(y \mid \theta, \phi) = \int f(y \mid z, \phi) g(z \mid \theta)\, dz,$$

where $f(y|z, \phi)$ is a 'kernel' exponential family density or mass function depending on a parameter ϕ and a random variable Z, and $g(z \mid \theta)$ is the 'mixing' distribution of Z, depending on the parameter θ. The value of z is unobservable, and we can observe only y.

Two different classes of problems arise:

1. We know or specify the form of the mixing distribution $g(z \mid \theta)$ up to the unknown value of θ, and want to draw likelihood inferences about θ and ϕ.
2. The distribution $g(z \mid \theta)$ is unknown, and we want to draw inferences about ϕ, without making any assumption about $g(z)$; we may also want to draw inferences about $g(z)$.

We deal with the first class of problems in subsequent chapters, as no new problems arise from them. In this chapter we consider in detail the second class, in the framework of *finite mixtures*.

We first illustrate the need for mixture distributions with the girl birthweight example in Chapter 2.

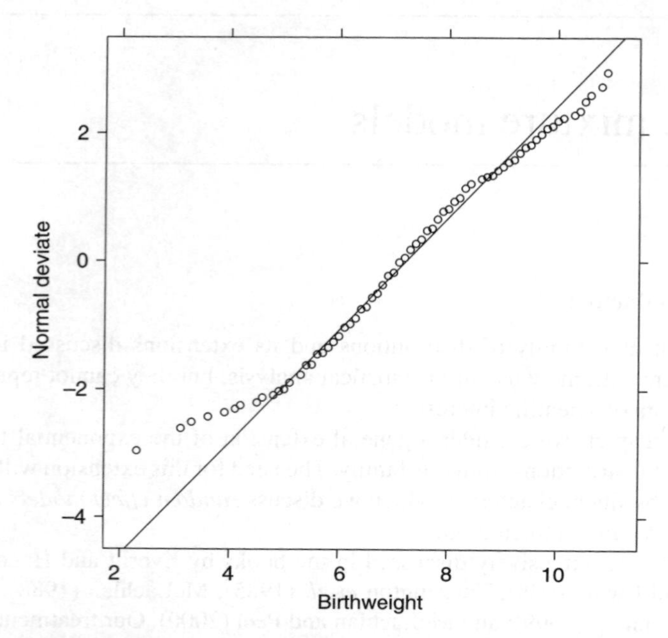

Fig. 7.1. Birthweight: normal deviate of cumulative proportion, girls

7.2 Example – girl birthweights

We saw in Chapter 2 that the *cdf* plot for the girl birthweights was not normal, with two notable bends in the Q–Q plot (reproduced in Fig. 7.1).
Such changes in slope of the Q–Q plot are a good indication of the presence of a mixture; the observations from the separate components, if we could identify them, would each have a straight-line plot if the component distributions were normal, but the Q–Q plot of the mixture of distributions has to bend to accommodate the different means (and possibly variances) of the component distributions.

7.3 Finite mixtures of distributions

If the distribution of Z is *discrete*, on the set of values (support points) $z_1, \ldots, z_K$ with probabilities $\pi_1, \ldots, \pi_K$, then the distribution of Y is a *finite mixture*. In this case, we write the *mixed* probability distribution for the observed Y in the form

$$m(y \mid z_1, \ldots, z_K, \pi_1, \ldots, \pi_{K-1}, \phi) = \sum_{k=1}^{K} \pi_k f_k,$$

where $f_k = f(y \mid z_k, \phi),\ k = 1, \dots, K$ is an exponential family density or mass function depending on a parameter z_k unique to f_k, and a parameter ϕ common to all the K component densities. The proportions π_k of each component in the mixture are non-negative and sum to 1, so there are only $K - 1$ distinct proportion parameters. (To remove unnecessary generality, we will in fact require the π_k to be *positive*, so that the mixture has K non-empty components.)

An appealing interpretation of mixtures is that they arise because of the omission or suppression of a K-category group identifier G which, if observed, would allow us to fit a one-way classification model. We make use of this interpretation implicitly in the maximum likelihood fitting of the model.

7.4 Maximum likelihood in finite mixtures

Given a random sample $y_1, \dots, y_n$ from the mixture distribution, the likelihood is

$$L = L(z_1, \dots, z_K, \pi_1, \dots, \pi_{K-1}, \phi) = \prod_{i=1}^{n} m_i = \prod_i \left(\sum_k \pi_k f_{ik} \right),$$

where

$$m_i = m(y_i \mid z_1, \dots, z_K, \pi_1, \dots, \pi_{K-1}, \phi),\ f_{ik} = f(y_i \mid z_k, \phi).$$

The log-likelihood is

$$\ell = \ell(z_1, \dots, z_K, \pi_1, \dots \pi_{K-1}, \phi) = \sum \log m_i = \sum_i \log \left(\sum_k \pi_k f_{ik} \right)$$

and the score for z_k is

$$\begin{aligned} \frac{\partial \ell}{\partial z_k} &= \sum_i \frac{\pi_k}{m_i} \frac{\partial f_{ik}}{\partial z_k} \\ &= \sum_i \frac{\pi_k f_{ik}}{m_i} \frac{\partial \log f_{ik}}{\partial z_k}. \end{aligned}$$

Write

$$w_{ik} = \frac{\pi_k f_{ik}}{m_i};$$

then

$$\frac{\partial \ell}{\partial z_k} = \sum_i w_{ik} s_{ik}(z_k),$$

where

$$s_{ik}(z_k) = \frac{\partial \log f_{ik}}{\partial z_k}$$

is the score for z_k in the k-th component distribution. Similarly,

$$\frac{\partial \ell}{\partial \phi} = \sum_i \sum_k w_{ik} s_{ik}(\phi).$$

Thus the likelihood (score) equations

$$\frac{\partial \ell}{\partial z_k} = 0, \qquad \frac{\partial \ell}{\partial \phi} = 0$$

are *weighted* forms of the single-distribution score equations for these parameters, with weights w_{ik}. We now apply Bayes' theorem to the hypotheses that the observation y_i comes from component k. The prior probabilities are π_k from the original mixture formulation, and

$$\Pr(y_i \mid G = k) = f_{ik},$$

so the posterior probability that y_i comes from component k is

$$\Pr(G = k \mid y_i) = \frac{\pi_k f_{ik}}{\sum_\ell \pi_\ell f_{i\ell}} = w_{ik},$$

and the weights w_{ik} are just these posterior probabilities.

Now consider the mixture probabilities π_k. Differentiating the log-likelihood subject to the constraint $\sum \pi_k = 1$ using a Lagrange multiplier λ, we have

$$\begin{aligned}
\ell^* &= \ell - \lambda\left(\sum \pi_k - 1\right) \\
\frac{\partial \ell^*}{\partial \pi_k} &= \sum_i \frac{f_{ik}}{m_i} - \lambda \\
&= \sum_i \frac{w_{ik}}{\pi_k} - \lambda.
\end{aligned}$$

Solving the score equations gives immediately

$$\hat{\pi}_k = \sum_i w_{ik}/\lambda,$$

and multiplying by λ and summing over k gives

$$\hat{\lambda} = \sum_k \sum_i w_{ik} = \sum_i \sum_k w_{ik} = n,$$

so

$$\hat{\pi}_k = \sum_i w_{ik}/n.$$

MLE of the model parameters can now proceed iteratively by an EM algorithm Day, 1969; Dempster *et al.*, 1977; Aitkin, 1980; McLachlan and Krishnan, 1997) In the M-step, the z_k and ϕ are estimated by solving the weighted score equations with given weights w_{ik}, while in the E-step, the weights w_{ik} are updated using the new parameter estimates in the previous M-step.

The weights w_{ik} can be interpreted as *conditional expectations* of the missing or unobserved K-category group identifier G described in Section 7.3. Define $G_{i1}, \ldots, G_{iK}$ to be the binary indicators for membership of the i-th observation in groups $1, 2, \ldots, K$. If these were observed, the 'complete data' likelihood would be

$$L^* = \prod_i \prod_k (\pi_k f_{ik})^{G_{ik}},$$

and the conditional expectation of the log-likelihood used in the E-step would replace G_{ik} by $\mathrm{E}[G_{ik} \mid y_i] = \Pr(G_{ik} = 1|y_i) = w_{ik}$.

Multiple maxima are a common feature of mixture likelihoods and different starting values for the EM algorithm need to be used to locate these maxima. Finch *et al.* (1989) discussed strategies for this.

We use the function `alldist` included in the package npmlreg (Einbeck *et al.*, 2006). This function uses the EM algorithm, and initial estimates of the mass-point locations are obtained from the *Gaussian quadrature* mass-points used for the normal distribution (see Section 8.3 for details). Searching for multiple maxima is assisted through the argument `tol` which scales in or out these initial estimates of the mass-point locations. This is not a completely general method for locating multiple maxima. We give an example in Section 7.8.

An important point is that the above results for ML estimation are not restricted to exponential family models: they apply quite generally. Further the densities $f_k(y)$ do not even have to be of the same form – we could for example have a mixture of a gamma and a lognormal distribution, though we will not in fact consider such mixtures.

7.5 Standard errors

A general feature of the EM algorithm is that it gives MLEs , but not their standard errors. These require additional computation beyond EM itself. The most direct method to obtain standard errors is via computation of the observed information matrix. We write the density or mass function of the observed data as $m(y \mid \psi)$, depending on the parameter ψ in the joint distribution of Y and Z. Then

$$m(y \mid \psi) = \int f(y \mid z, \psi) g(z \mid \psi)\,\mathrm{d}z.$$

Writing $m(y_i \mid \psi) = m_i$, $f(y_i \mid z, \psi) = f_i$ and $g(z_i \mid \psi) = g_i$, the contribution of a single observation y_i to the log-likelihood and its derivatives for a sample are:

$$\begin{aligned}
\ell_i = \ell_i(\psi) &= \log m(y_i \mid \psi) \\
&= \log \int f_i g_i \,\mathrm{d}z_i \\
s_i(\psi) = \frac{\partial \ell_i}{\partial \psi} &= \frac{\frac{\partial}{\partial \psi} \int f_i g_i \,\mathrm{d}z_i}{m_i} \\
&= \frac{\int \left[\frac{\partial f_i}{\partial \psi} g_i + f_i \frac{\partial g_i}{\partial \psi} \right] \mathrm{d}z_i}{m_i} \\
&= \int \frac{f_i g_i}{m_i} \left[\frac{\partial \log f_i}{\partial \psi} + \frac{\partial \log g_i}{\partial \psi} \right] \mathrm{d}z_i \\
&= \int h_i \left[\frac{\partial \log f_i}{\partial \psi} + \frac{\partial \log g_i}{\partial \psi} \right] \mathrm{d}z_i.
\end{aligned}$$

Here

$$h_i = \frac{f_i g_i}{m_i}$$

is the conditional distribution of Z_i given y_i. The term $(\partial \log f_i / \partial \psi) + (\partial \log g_i / \partial \psi)$ is called the 'complete data score', since it corresponds to observing the 'complete data' Y *and* Z. Then

$$\begin{aligned}
s_i(\psi) &= \mathrm{E}_{Z_i \mid y_i} \left[\frac{\partial \log f_i}{\partial \psi} + \frac{\partial \log g_i}{\partial \psi} \right] \\
&= \mathrm{E}_{Z_i \mid y_i} [s_{ci}(\psi)],
\end{aligned}$$

where the subscript c indicates 'complete data'. Here $s_{ci}(\psi)$ is the i-th component of the complete-data score, and we assume sufficient regularity in the density f to allow the interchange of differentiation and integration. Then, summing over all n observations, *the observed-data score is equal to the conditional expectation of the complete-data score.*

Before taking the second derivative, we note that

$$\log h_i = \log f_i + \log g_i - \log m_i$$

and hence

$$\begin{aligned}\frac{\partial \log h_i}{\partial \psi} &= \frac{\partial \log f_i}{\partial \psi} + \frac{\partial \log g_i}{\partial \psi} - \frac{\partial \log m_i}{\partial \psi}\\ &= s_{ci}(\psi) - s_i(\psi),\end{aligned}$$

the difference between the complete-data and the observed-data score. Then the observed data Hessian is

$$\begin{aligned}H(\psi) = \sum_i H_i(\psi) &= \sum_i \frac{\partial^2 \ell_i}{\partial \psi \partial \psi'}\\ &= \sum_i \frac{\partial}{\partial \psi'} \int h_i s_{ci}(\psi)\, \mathrm{d}z_i\\ &= \sum_i \int \left[h_i H_{ci}(\psi) + s_{ci}(\psi) \frac{\partial h_i}{\partial \psi'} \right] \mathrm{d}z_i\\ &= \sum_i \int [H_{ci}(\psi) + s_{ci}(\psi)[s_{ci}(\psi) - s_i(\psi)]']h_i\, \mathrm{d}z_i.\end{aligned}$$

Interchanging the order of summation and integration, the second term in the integral, the matrix $\sum_i s_{ci}(\psi)[s_{ci}(\psi) - s_i(\psi)]'$, is the *sample covariance matrix* of the complete-data score. The first term in the integral is the complete data Hessian, and so (Louis, 1982; Oakes, 1999) *the observed-data information matrix is equal to the conditional expectation of the complete-data observed information matrix minus the conditional covariance matrix of the complete-data score*. In many incomplete-data problems in which the EM algorithm is used with a package which analyses complete data, standard errors reported by the package are those for the complete-data information matrix. These are always too small, since they correspond to treating the conditional expectations of the unobserved data as though they were known, but can be corrected by computing the conditional covariance matrix of the complete-data score. Although we could compute this using the matrix functions available in R we use an alternative approach, decribed in Chapter 2 which is often useful (Aitkin, 1994; Dietz and Böhning, 1995).

In the applications considered in the next chapters, the common parameter ϕ is frequently a regression coefficient vector $\boldsymbol{\beta}$ and is the parameter of principal interest. Standard errors for an individual coefficent β_j can be obtained by omitting the corresponding explanatory variable x_j from the regression model, recording the disparity change $\Delta\mathrm{disp}_j$, and equating the Wald test statistic for the hypothesis

$\beta_j = 0$ to the LRTS, giving

$$\left[\frac{\hat{\beta}_j}{\mathrm{SE}(\hat{\beta}_j)}\right]^2 = \Delta\mathrm{disp}_j$$

whence

$$\mathrm{SE}(\hat{\beta}_j) \doteq \frac{|\hat{\beta}_j|}{\sqrt{\Delta\mathrm{disp}_j}}.$$

In small samples from non-linear models the likelihood may be far from normal and there may be poor agreement between the LRT and Wald test. In this case any standard error is less useful, as likelihood-based confidence intervals for the parameter will not be symmetric around the MLE. Nevertheless the approximate standard error is still useful in this case as it gives the correct impression of the importance of the variable in the model; the usual standard error is misleading if interpreted in a Wald significance test sense.

7.6 Testing for the number of components

We have assumed above that the number of components K in the mixture is specified. However, even with well-separated components it may be unclear how many components are needed to provide an adequate fit to the data. It might be expected that this question could be resolved straightforwardly by increasing the number of components until the decrease in disparity becomes non-significant. Two theoretical properties of maximum likelihood in finite mixtures, however prevent us from using the standard asymptotic properties of the LRT:

1. The mixture model is non-regular, in the sense that the $(K-1)$-component mixture is not a simple restriction of the K-component mixture: for example, restricting the component-specific parameters so that $z_{K-1} = z_K$ leaves these two components indistinguishable, with π_{K-1} and π_K not separately identifiable. Equivalently, if π_K is set to zero, z_K is not identifiable.
2. In single-parameter models like the binomial and Poisson, the disparity does not continue to decrease indefinitely as K increases. Surprisingly, at some value K_0 the disparity stabilizes, and increasing K further gives the same disparity, the fitted $(K_0 + 1)$-component mixture degenerating to a K_0-component mixture, and similarly for larger numbers of components.

This unusual result can be expressed quite generally, following Kiefer and Wolfowitz (1956); Laird (1978), and Lindsay (1983). Under certain conditions (discussed at length in, e.g. Maritz and Lwin, 1989), the mixing distribution can be consistently estimated, that is, the estimate $\hat{g}(z)$ converges to the true $g(z)$ as

$n \to \infty$. This was established by Kiefer and Wolfowitz (1956). The form of the MLE $\hat{g}(z)$ of $g(z)$ was established by Laird (1978) and Lindsay (1983), and is a *discrete* distribution on $\hat{K}$ points of support, with probability masses $\hat{\pi}_1, \ldots, \hat{\pi}_{\hat{K}}$ at mass-points $\hat{z}_1, \ldots, \hat{z}_{\hat{K}}$. The number, location and masses of these mass-points have to be determined computationally.

The one-parameter exponential family distributions satisfy the conditions for consistent estimation of $g(z)$, except for the Bernoulli distribution. It is easy to see why the latter distribution fails: if $Y \mid Z$ is Bernoulli $b(1, Z)$ and Z has some distribution $g(z)$ on (0,1) with mean μ, then the distribution of Y is

$$p(y) = \int z^y (1-z)^{1-y} g(z) - dz.$$

So

$$p(1) = \int z g(z) dz = \mu,$$

and $p(0) = 1 - \mu$. Thus Y again has a Bernoulli distribution with success probability μ. The distribution of Y is the same for all distributions $g(z)$ with the same mean, and so $g(z)$ cannot be consistently estimated.

For binomial distributions $b(n, Z)$ with $n > 1$, the marginal probabilities from the compound binomial distribution are not binomial: for $n = 2$, for example, they are $\mu^2 + \sigma^2$ for $r = 2$, $2\mu(1-\mu) - 2\sigma^2$ for $r = 1$ and $(1-\mu)^2 + \sigma^2$ for $r = 0$, where μ and σ^2 are the mean and variance of Z. These probabilities are binomial only if $\text{Var}[Z] = 0$. Thus the distribution of Z is identified, but only up to its first two moments. For general n the first n moments of Z are identified; this does not define uniquely the distribution of Z.

The form of the general MLE of $g(z)$ means that the estimated distribution of Y is a *finite mixture of exponential families*:

$$\hat{m}(y) = \sum_{k=1}^{\hat{K}} f(y \mid \hat{z}_k) \hat{\pi}_k.$$

The number $\hat{K}$, locations $\hat{z}_k$ and masses $\hat{\pi}_k$ have to be determined computationally to maximize the likelihood. This brings us back to the finite mixture problem discussed at the beginning of the chapter: we have a straightforward EM algorithm for maximizing finite mixture likelihoods, and in a more general form than for the simple example in Section 7.2. The ML estimate $\hat{K}$ of the number of components K in the one-parameter family is found by increasing K until the maximized likelihood stabilizes. We make extensive use of this approach in subsequent chapters. In the two-parameter family there is no MLE $\hat{K}$. The maximizing $\hat{g}(z)$, where it exists, is called the *non-parametric maximum likelihood* (*NPML*) *estimate* of $g(z)$ – non-parametric because no parametric model is assumed for $g(z)$.

(In models with additional regression parameters, this estimate of $g(z)$ is often called *semi-parametric*, because of the additional model parameters. However, these additional parameters do not affect the nature of the estimate, so we retain the term *non-parametric* even when there are additional model parameters.)

Since no general distributional results are available for the distribution of the LRTS in the general mixture problem (though we give some results for special cases in subsequent chapters), *bootstrap* methods are the only general methods available, though they are not very satisfactory either (McLachlan and Peel, 2000, p. 198). For the special case of testing a two-component normal mixture against a single normal distribution, Thode *et al.* (1988) gave tables of simulated percentage points of the LRTS.

A common alternative approach is through the Akaike or the Bayesian Information Criteria (AIC or BIC), in which the disparity for the K-component mixture is penalized by a measure of model complexity depending on the number of model parameters. The penalty functions for these criteria are derived by assuming the large-sample normality of the likelihood function around the parameter MLEs. Since this assumption fails for mixture likelihoods with poorly defined components, the applicability of these criteria to mixture likelihoods is questionable. A more detailed discussion of criteria for assessing the number of components is given in McLachlan and Peel (2000, Chapter 6).

In the bootstrap approach, to test the null hypothesis of a K-component mixture against the alternative of a $(K + 1)$-component mixture, we simulate a sample of n observations from the fitted K-component mixture, and fit the K- and $(K + 1)$-component mixture models to the simulated sample. The value of the LRTS for the null hypothesis is then computed for this sample. This sampling procedure and model fitting are replicated to give R independent samples, and the value of the LRTS for the observed sample is compared with the R simulation values.

If the null hypothesis is true, the probability that the observed value is larger than any of the R simulation values is approximately $1/(R + 1)$. So for a 5% level test, we need 19 simulation values, and the hypothesis is rejected if the observed sample value is larger than all 19 simulation values. For a 1% level test, we need 99 simulated values.

The test size is only approximate because we are treating the MLEs of the parameters from the given sample data as the true values in the simulations. McLachlan and Krishnan (1997) showed that the P-values from the test tend to be non-conservative, giving overstatements of significance.

The bootstrap test procedure was proposed by Hope (1968); an illustration of its use in mixture models was given in Aitkin *et al.* (1981). McLachlan (1987) provided FORTRAN code for the simulations assuming a multivariate normal mixture.

We illustrate with the girl birthweight example.

7.6.1 *Example*

We first fit a mixture of normals with different means μ_k but a common variance σ^2. Applying the above approach gives the score equations

$$\frac{\partial \ell}{\partial \mu_k} = \sum_i w_{ik}(y_i - \mu_k) = 0$$

$$\frac{\partial \ell}{\partial \sigma} = \sum_i \sum_k w_{ik} \left[-\frac{1}{\sigma} + \frac{1}{\sigma^3}(y_i - \mu_k)^2 \right] = 0$$

giving

$$\hat{\mu}_k = \sum_i w_{ik} y_i / \sum_i w_{ik}$$

$$\hat{\sigma}^2 = \sum_i \sum_k w_{ik}(y_i - \mu_k)^2 / n$$

and

$$\hat{\pi}_k = \sum_i w_{ik}/n.$$

The EM algorithm is extremely simple, alternating weighted mean and sums of squares calculations with recalculation of the weights using

$$w_{ik} = \frac{\pi_k f_{ik}}{\sum_\ell \pi_\ell f_{i\ell}}$$

with

$$f_{ik} = \frac{1}{\sqrt{2\pi}\sigma} \exp\left\{ -\frac{1}{2\sigma^2}(y_i - \mu_k)^2 \right\}.$$

The algorithm is implemented in the R function `alldist` available in the package `npmlreg`. The function can be used in a similar manner to the R function `glm`. For simple mixtures we specify the 'fixed' and 'random' formulas with a simple intercept term. The number of components is specified by the k argument with $k = 1, 2, \ldots, 10, 21$ mass-points (components) implemented. The normal distribution is the default.

```
> data(statlab, package = "SMIR")
> girls <- subset(statlab, sex == "girl")
> library(npmlreg)
> girls.k1 <- alldist(c.b.wgt ~ 1, data = girls, k = 1)
> summary(girls.k1)
```

```
Coefficients:
      Estimate Std. Error  t value
MASS1 7.223       0.045    160.467

Mixture proportions:
MASS1
    1


Component distribution - MLE of sigma:      1.146
Random effect distribution - standard deviation:        0

-2 log L:        2014.4     Convergence at iteration  0

> round(girls.k1$disparity, 2)

[1] 2014.39
```

With only one component, there is no mixture, and we obtain as parameter estimates the mean birthweight of 7.22 pounds (the mean MASS1 of the 'first' component) and the MLE $\hat{\sigma} = 1.146$ pounds. We now increase the number of components.

```
> girls.k2 <- update(girls.k1, k = 2, plot.opt = 0,
+      verbose = FALSE)
> girls.k3 <- update(girls.k2, k = 3)
> girls.k4 <- update(girls.k3, k = 4)
> girls.k5 <- update(girls.k4, k = 5)
> girls.k6 <- update(girls.k5, k = 6)
> summary(girls.k2)

Coefficients:
      Estimate Std. Error  t value
MASS1 7.100     0.041      171.585
MASS2 7.395     0.049      151.406

Mixture proportions:
   MASS1     MASS2
   0.582     0.418

Component distribution - MLE of sigma:      1.136
Random effect distribution - standard deviation:        0.145

-2 log L:        2014.43    Convergence at iteration  86
> round(girls.k2$disparity, 2)

[1] 2014.43

> summary(girls.k3)
```

```
Coefficients:
      Estimate Std. Error   t value
MASS1 3.892     0.150        25.949
MASS2 7.078     0.021       339.395
MASS3 9.132     0.063       145.256

Mixture proportions:
     MASS1          MASS2          MASS3
     0.017          0.885          0.097

Component distribution - MLE of sigma:      0.864
Random effect distribution - standard deviation:        0.750

-2 log L:           1972.05     Convergence at iteration  41

> round(girls.k3$disparity, 2)

[1] 1972.05

> summary(girls.k4)

Coefficients:
      Estimate Std. Error   t value
MASS1 3.567     0.122        29.175
MASS2 6.181     0.031       199.079
MASS3 7.308     0.017       433.968
MASS4 9.266     0.045       207.612

Mixture proportions:
     MASS1          MASS2          MASS3          MASS4
     0.013          0.202          0.687          0.098

Component distribution - MLE of sigma:      0.710
Random effect distribution - standard deviation:        0.898

-2 log L:           1968.82     Convergence at iteration  262
```

The disparity for the two-component mixture is equal (to 1 dp) to that for the one-component model. The two components differ in mean by 0.29, but this is only 0.26 (estimated) standard deviations, and a two-component mixture with such a small separation between the components is indistinguishable from a single normal distribution even in this large sample.

For the three-component mixture the disparity decreases by 42.37, a value so large that a formal test is hardly needed (though we give one below). The four-component model reduces the disparity by 3.23, and the five-component model (not shown) increases the disparity by only 0.07; increasing the number of components to six leaves the disparity essentially unchanged. It is clear that the second and third components in the four-component model are simply a partition of the second component in the three-component model, with means differing by 1.13, which is 1.6 standard deviations.

The three-component model is made up of a large sub-population of 88% with mean 7.1 pounds, a 'high birthweight' sub-population of about 10% with mean 9.1 pounds, and a 'low birthweight' sub-population of about 2% with mean 3.9 pounds; the common standard deviation is 0.86 pounds.

We finally consider the LRT for the existence of a four-component mixture over a three-component mixture using the bootstrap approach discussed earlier.

We generate $M = 19$ independent samples of size 648 from the three-component normal distribution with means 3.892, 7.078, and 9.132, standard deviation of 0.8645, and mixing proportions 0.017, 0.885, and 0.097, and fit the three- and four-component mixture distributions, saving the disparity changes. Then we compare the disparity change in the real data with that from the simulations.

```
> lrts <- function(mu, sd, p, N, k1) {
+      k <- length(p)
+      sigma <- ifelse(length(sd) == 1, rep(sd, k),
+          sd)
+      n1 <- round(N * p[-k])
+      nk <- N - sum(n1)
+      n <- c(n1, nk)
+      y <- NULL
+      for (i in 1:k) y <- c(y, rnorm(n[i], mu[i], sigma[i]))
+      data <- data.frame(yy = y)
+      fit1 <- alldist(yy ~ 1, k = k, data = data, plot.opt = 0,
+          verbose = FALSE)
+      fit2 <- update(fit1, k = k1)
+      fit1$disparity - fit2$disparity
+ }
> lrts3vs4 <- NULL
> for (R in 1:19) {
+      lrts3vs4 <- c(lrts3vs4, lrts(mu = c(3.892, 7.078,
+          9.132), sd = 0.8645, p = c(0.017, 0.885,
+          0.0974), N = 648, k1 = 4))
+ }

> round(sort(lrts3vs4), 4)

 [1] -0.0250 -0.0061 -0.0049 -0.0001  0.8404  1.3206  1.3493
 [8]  1.7608  1.8038  2.0960  3.3243  3.3899  4.6468  4.9129
[15]  4.9152  6.4939  6.7793  7.7377 10.6105
```

The observed data test statistic of 3.23 lies between the 9th and 10th of the 19 simulated values, giving an approximate p-value greater than 0.5 for the bootstrap test of the null hypothesis. There is no support for a four-component mixture over a three-component mixture.

The bootstrap procedure can be repeated to compare the three- with the one-component mixture. The hypothesis of a single normal is firmly rejected in favour of the three-component mixture since the observed LRT value of 43.59 is much larger than all 19 bootstrap values.

```
> sort(round(lrts1vs3, 2))

 [1]  1.19  1.49  1.60  2.06  2.34  3.30  3.35  3.59  3.69  4.04
[11]  4.50  4.53  5.02  5.04  5.11  5.77  6.48  7.47 10.32
```

In Chapter 2, we raised the possibility that the variation was caused by other variables, like mother's weight and age. We examine this possibility, by fitting these variables explicitly in a regression model and refitting the mixture model.

```
> girl.regk1 <- alldist(c.b.wgt ~ m.b.ag + m.b.wgt,
+      data = girls, k = 1, plot.opt = 0, verbose = FALSE)
> summary(girl.regk1)

Coefficients:
         Estimate  Std. Error    t value
MASS1    5.223      0.284        18.363
m.b.ag   0.011      0.007         1.533
m.b.wgt  0.013      0.002         6.567

Mixture proportions:
MASS1
    1

Component distribution - MLE of sigma:      1.101
Random effect distribution - standard deviation:        0

-2 log L:             1961.12    Convergence at iteration  0
```

Mother's weight is very important, with a z-statistic (here reported as a t-value) of 6.57, but mother's age is not. We drop mother's age and refit the model with one and three components.

```
> girl.regk1a <- update(girl.regk1, . ~ . - m.b.ag)
> summary(girl.regk1a)

Coefficients:
         Estimate  Std. Error    t value
MASS1    5.431      0.250        21.691
m.b.wgt  0.014      0.002         7.267

Mixture proportions:
MASS1
    1
```

```
Component distribution - MLE of sigma:      1.103
Random effect distribution - standard deviation:      0

-2 log L:          1963.47    Convergence at iteration  0

> girl.regk3a <- update(girl.regk1a, k = 3)
> summary(girl.regk3a)

Coefficients:
         Estimate  Std. Error   t value
m.b.wgt  0.013      0.001        15.904
MASS1    2.161      0.183        11.780
MASS2    5.366      0.111        48.553
MASS3    7.241      0.127        56.842
Mixture proportions:
     MASS1        MASS2        MASS3
     0.018        0.887        0.096

Component distribution - MLE of sigma:      0.842
Random effect distribution - standard deviation:      0.709

-2 log L:          1919.88    Convergence at iteration  34
```

The three-component model appears clearly again, with a disparity change of 43.59. The mixture cannot be attributed to mother's weight. Though highly significant, the effect of mother's weight is not very large: a 10-pound increase in mother's weight is associated with an increase in the mean baby's weight of only 0.133 pounds: even a 50-pound increase in mother's weight corresponds to an increase of only 0.66 pounds, less than the standard deviation of 0.841 pounds. The 'unexplained' variability about the regression remains large.

Note that the R standard error for mother's weight is too small, being based on the complete data conditional expected information. The indirect estimate of the standard error by equating the disparity change and the squared Wald statistic is

$$\mathrm{SE} = \frac{|\hat{\beta}|}{\sqrt{\Delta \mathrm{disp}}} = \frac{0.01328}{\sqrt{52.2}} = 0.00188,$$

almost unchanged from the (1-component) normal regression model. The 'complete data' value is a very substantial underestimate.

7.7 Likelihood 'spikes'

A computational difficulty occurs in normal mixtures with different variances. If one observation is remote from the others, it may be identified as a mixture component with one observation, with a variance which goes to zero. This will invalidate the computation of the likelihood, since the 'spike' density for the degenerate component will go to infinity, which is obviously invalid.

This difficulty arises from the failure to represent the measurement precision (discussed in Chapter 2) properly in this case. Consider a *single* observation y drawn from a population with a normal distribution model $N(\mu, \sigma^2)$ for Y. Let the measurement precision be δ. The likelihood for the single observation is then

$$\begin{aligned} L(\mu, \sigma) &= \Pr(y - \delta/2 < Y < y + \delta/2 \mid \mu, \sigma) \\ &= \Phi([y + \delta/2 - \mu]/\sigma) - \Phi([y - \delta/2 - \mu]/\sigma), \end{aligned}$$

where $\Phi(z)$ is the standard normal *cdf*. If the measurement precision δ is small compared with σ, then as in Chapter 2 we may approximate the likelihood by the normal density at y multiplied by δ. But if σ is small, as we are considering if it may tend to zero, this approximation cannot be used, and the exact expression above must be retained. The MLE of μ may be shown by differentiation to be $\hat{\mu} = y$. Substituting in the likelihood gives the profile likelihood in σ:

$$P(\sigma) = \Phi(\delta/[2\sigma]) - \Phi(-\delta/[2\sigma]) = 2\Phi(\delta/[2\sigma]) - 1.$$

This does not depend on the data value y at all, only the measurement precision δ. Thus the single observation is uninformative about σ, unless we have independent information about μ. This is not surprising – to obtain information about variability from data, we need to have observations which *show* variability!

It can be verified that the profile likelihood in σ is maximized at $\sigma = 0$, when $P(\sigma) = 1$. (If the invalid normal density approximation is used for the likelihood, again the MLE of σ is zero, but at this value the likelihood appears to be infinite, a sure sign that the calculation is invalid: since the likelihood is a probability, how can it be infinite?) This result is often presented as a criticism of direct likelihood inference – it is interpreted to mean that the single observation somehow provides misleading information that σ is very small – indeed, if $\sigma = 0$, then the observation y is *certain* to occur! In response, we note that if we are only ever allowed a single observation from a population, then any model for it will be irrelevant. But if, given the model, only one observation is to be taken, and in the absence of any other information about the population, what is more natural than to suppose that the observation we draw is certain to occur? We find this unreasonable because we have very strong prior information that degenerate populations with zero variances do not occur. It is this information, not that provided by the data, that leads us to the conclusion.

However, we have to deal with this difficulty in the computation of the likelihood for the mixture model. The usual solution is to bound the component standard deviations below by some small σ_0. This is easy to implement computationally, but the maximized likelihood resulting will depend on the value of σ_0.

So long as we do *not* have single observations belonging to mixture components, the problem does not arise, and the likelihood can be computed in its usual mixed-density form. Suppose now that one and only one observation belongs to component k uniquely, and no other observation has non-negligible probability of belonging to it. Then the likelihood for this observation should be maximized at 1, not at ∞. This can be achieved simply by removing this observation from the likelihood computation by assigning it weight zero. The maximized likelihood will then be correct for this mixture model, allowing an extra degenerate component for the excluded observation.

However, if we consider further the general model with different variances, for n sample observations it is possible to fit a mixture model with n components with different means and variances, with all the variances equal to zero. The same model with *equal* variances would also fit the data with n components with zero variance. These 'mixture' models in fact give the *empirical mass function* – the non-parametric estimate of the density when no assumption is made about it.

The empirical mass function is not a mixture in any real sense – any two-parameter distribution would give the same result as the normal. However, the possibility of 'pushing the mixture to the limit' means that the nonparametric MLE of the mixing distribution is well-defined only for one-parameter kernel distributions whose variances cannot be zero. For normal and other two-parameter distributions we need to rely on the bootstrap LRT to determine reasonable numbers of components.

7.8 Galaxy data

We discuss a second example: the well-known 'galaxy' data (Roeder, 1990; Richardson and Green, 1997; Aitkin, 2001). The data are the recession velocities, in units of 10^3 km/s, of 82 galaxies receding from our own, sampled from six well-separated conic sections of space. The astronomers Postman *et al.* (1986) gave the full data by region. They are part of the R package *MASS* as data set `galaxies`, except that the 78th data point has been entered incorrectly and should be 26960 and not 26690.

One question of scientific interest is whether these galaxies form distinct super-clusters surrounded by voids in space. If so, some clustering of velocities would be expected, with the overall distribution of velocity being multimodal. Following all authors, including the astronomers, we do not analyse separately the data from the six regions: the individual regions have very small data sets, from which not much can be learned about clustering among or within regions.

The data are shown below in increasing order scaled by 0.001.

```
     Recession velocities of 82 galaxies

 9.172  9.350  9.483  9.558  9.775 10.227 10.406 16.084 16.170
18.419 18.552 18.600 18.927 19.052 19.070 19.330 19.343 19.343
19.440 19.473 19.529 19.541 19.547 19.663 19.846 19.856 19.863
```

```
19.914 19.918 19.973 19.989 20.166 20.175 20.179 20.196 20.215
20.221 20.415 20.629 20.795 20.821 20.846 20.875 20.986 21.137
21.492 21.701 21.814 21.921 21.960 22.185 22.209 22.242 22.249
22.314 22.374 22.495 22.746 22.747 22.888 22.914 23.206 23.241
23.263 23.484 23.538 23.542 23.666 23.706 23.711 24.129 24.285
24.289 24.366 24.717 24.990 25.633 26.960 26.995 32.065 32.789
34.279
```

There is a gap, or jump, between the seven smallest observations around 10 and the large central body of observations between 16 and 26, and another gap between 27 and 32, for the three largest observations. So we might expect to find at least three components in the data.

Figure 7.2 shows the 95% simultaneous confidence band for the population *cdf* and the superimposed fitted normal *cdf*. The R code is given below:

```
> data(galaxies, data = "SMIR")

> galaxies[78] <- 26960
> velocity <- galaxies/1000
> print(xyplot(lower + upper ~ x, data = NPL.bands(velocity),
+     pch = 20, col = 1, xlab = "velocity", ylab = "probability",
+     panel = function(...) {
+         panel.xyplot(...)
```

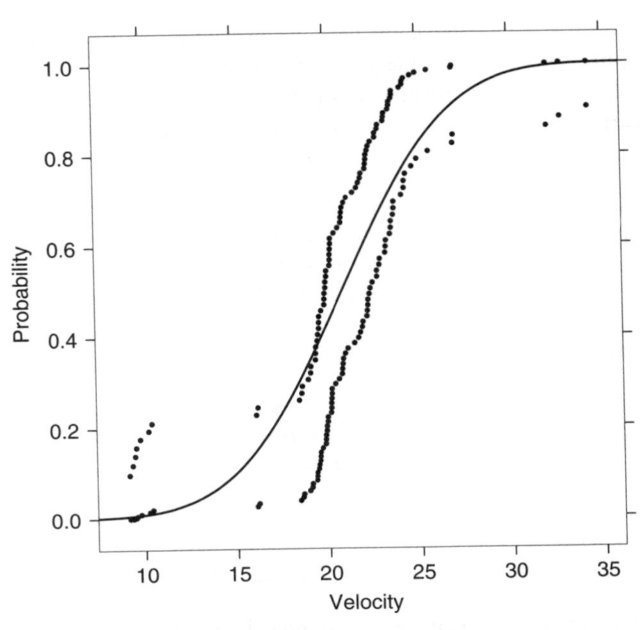

Fig. 7.2. Galaxy data *cdf* bounds and fitted normal

```
+           mgalaxy <- mean(velocity)
+           sdgalaxy <- sd(velocity)
+           panel.curve(pnorm(x, mgalaxy, sdgalaxy),
+               lwd = 2)
+       }))
```

It is clear that a single normal distribution is not appropriate.

An immediate question with all mixture analyses is – what kernel density should be used? We follow previous discussions of these data and fit mixtures of normals with both equal and unequal variances; the latter are fitted

by the same approach, but the common standard deviation σ is replaced by component-specific standard deviations σ_k. The R implementation of fitting mixtures, using the package *npmlreg*'s `alldist` function, allows unequal standard deviations σ_k, $k = 1, \ldots, K$ to vary over the components and incorporates a smoothing kernel defined by *lambda*. The default setting *lambda=0* is automatically mapped to *lambda=1/k* and corresponds to the 'maximal smoothing' case homogenous variance case, while *lambda=1* means 'no smoothing', that is, unequal variances. Smoothing in the `alldist` function is acheived by means of the discrete kernel

$$W(x, y|\lambda) = \begin{cases} \lambda & \text{if } y = x \\ (1-\lambda)/(K-1) & \text{if } y \neq x \end{cases}$$

(Aitchinson and Aitken, 1976). In the R implementation, Einbeck *et al.* (2006) use a 'damping' procedure in the initial cycles of the algorithm which reduces the sensitivity of the EM algorithm to the optimal choice of `tol` for exponential family densities possessing a dispersion parameter, see Einbeck and Hinde (2005).

```
> library(npmlreg)
> galaxies.df <- data.frame(velocity = velocity)
> galaxy.k1 <- alldist(velocity ~ 1, data = galaxies.df,
+     k = 1, plot.opt = 0, lambda = 0, verbose = FALSE)
> galaxy.k2 <- update(galaxy.k1, k = 2)

:

> galaxy.k8 <- update(galaxy.k1, k = 8)
> galaxy.k2u <- update(galaxy.k2, lambda = 1, spike.protect = 1)
> galaxy.k3u <- update(galaxy.k2u, k = 3)

:

> galaxy.k8u <- update(galaxy.k2u, k = 8, lambda = 0.99)
```

Table 7.1 gives the component means, standard deviations and proportions, and the disparity, for up to eight components.

Table 7.1. Mixture estimates

K	k	Equal variances				Unequal variances			
		Mean	Prop	SD	Disparity	Mean	Prop	SD	Disparity
1	1	20·83	1	4.57	480.84	20·83	1	4.57	480.84
2	1	21·88	0.913	3.03	461.00	21·35	0.740	1.87	440.72
	2	9·87	0.087			19·36	0.260	8.14	
3	1	32·94	0.037	2.08	425.36	33·04	0.037	1.13	409.05
	2	21·40	0.877			21·40	0.878	2.20	
	3	9·75	0.086			9·71	0.085	0.64	
4	1	33·04	0.037	1.32	416.50	33·04	0.037	0.98	405.42
	2	23·50	0.352			23·11	0.409	1.68	
	3	20·00	0.526			19·91	0.469	1.36	
	4	9·71	0.085			9·71	0.085	0.50	
5	1	33·04	0.037	1.10	410.69	33·04	0.037	0.95	387.97
	2	23·60	0.345			22·81	0.505	1.70	
	3	20·17	0.508			19·73	0.349	3.05	
	4	16·21	0.025			16·13	0.024	0.40	
	5	9·71	0.085			9·71	0.085	0.47	
6	1	33·04	0.037	0.81	394.58	33·04	0.037	0.92	379.12
	2	26·24	0.044			26·98	0.024	0.26	
	3	23·05	0.357			22·92	0.425	1.20	
	4	19·93	0.453			19·79	0.404	0.68	
	5	16·14	0.025			16·13	0.024	0.26	
	6	9·71	0.085			9·71	0.085	0.44	
7	1	33·04	0.037	0.66	388.86	33·04	0.037	0.92	359.94
	2	26·60	0.033			26·98	0.024	0.02	
	3	23·87	0.172			22·92	0.426	1.21	
	4	12·30	0.219			19·79	0.403	0.67	
	5	19·83	0.427			16·13	0.024	0.04	
	6	16·13	0.024			10·32	0.024	0.09	
	7	9·71	0.085			9·47	0.061	0.20	
8	1	33·04	0.037	0.59	388.18	34·28	0.012	0.10	359.49
	2	26·57	0.035			32·43	0.024	0.37	
	3	23·90	0.170			26·98	0.113	0.07	
	4	22·34	0.213			22·92	0.354	1.20	
	5	20·19	0.253			19·79	0.375	0.67	
	6	19·40	0.183			16·13	0.036	0.08	
	7	16·13	0.024			10·32	0.051	0.11	
	8	9·71	0.085			9·47	0.035	0.21	

For the unequal variance case, two of the component standard deviations are small at $K = 6$ (corresponding to two nearly identical observations with velocities of 16.084 and 16.170 and 26.960 and 26.995), and the same 'spikes' appear for larger K. As we increase the number of components beyond six, these components, and the extreme components with means of around 9 and 33, remain stable for $K = 7$, while the main body of observations is split into further closely spaced components with very little change in ML. Beyond $K = 7$ the highest velocity group is split further into components containing a single observations and a group of two. For the unequal variance model, at most six components (with 17 parameters) appear meaningful.

For the equal variance case, for $K > 6$ similar mean values for the components appear, though with slightly different masses. The six-component equal variance model has 12 parameters and fits slightly worse than the unequal variance model with 14 parameters. For all models with $K > 2$, the smallest 7 and largest 3 observations form stable components (note that $7/82 = 0.085$, $3/82 = 0.037$). For $K > 3$ the central group of observations is split into successively smaller groups, with the largest subgroup stable at a mean of around 19.8. The evidence for at least three components looks quite strong.

The equal- and unequal-variance models can be compared by the LRT to assess the variance heterogeneity. Although the mixture models are non-regular in their mean and proportion parameters, the hypothesis of equal variances is interior to the parameter space and does not possess any singularity (though by analogy with standard tests of variance heterogeneity, the test may be affected badly by departures from the specified null mixture distribution). Comparing the disparities for the two models for each K, the unequal-variance model appears to be needed: the homogeneity of variance test gives a disparity difference of 16.31 on 2 df for $K = 3$, 11.07 on 3 df for $K = 4$, 22.72 on 4 df for $K = 5$ and 15.46 on 5 df.

However, bootstrapping this test gives a quite different picture. Generating 99 samples from the fitted six-component equal-variance model and then fitting the six-component equal- and unequal-variance models gives the following ordered disparity differences:

```
 [1] -2.04  1.41  2.42  2.66  3.22  3.55  3.74  3.99  4.29  4.42
[11]  4.43  4.70  4.84  4.90  5.56  5.81  6.03  6.11  6.58  6.87
[21]  6.99  7.06  7.15  7.29  7.29  7.30  7.39  7.58  7.60  8.48
[31]  8.61  8.63  8.68  8.89  8.94  9.24 10.38 10.47 11.31 11.43
[41] 12.28 12.52 12.53 12.72 13.05 13.06 13.50 13.59 13.63 14.63
[51] 15.15 15.16 15.57 15.61 15.77 16.30 16.52 16.96 17.58 17.99
[61] 18.50 19.06 19.85 20.05 20.25 20.61 20.81 21.32 21.73 22.33
[71] 22.36 22.61 22.74 25.24 25.48 25.90 26.62 26.75 27.17 27.62
[81] 27.91 28.80 28.84 30.80 30.80 32.74 32.94 34.91 35.76 36.13
[91] 36.44 37.90 38.00 40.63 40.92 45.61 48.98 50.13 54.58
```

The negative value reflects the inability of the unequal-variance algorithm to find the point in the subspace of equal variances with much higher likelihood from the Gaussian quadrature-based starting values. This could be corrected by starting the EM algorithm for the unequal variance model from the estimates from the equal variance model.

Nevertheless, it is clear that large values of the LRTS for variance homogeneity are quite likely: the observed value of 15.46 is the 53rd largest in the above table and so has a P-value of 0.48. This does not provide any evidence of variance heterogeneity for six components.

To assess the number of components needed, we again use the bootstrap test. We first compare the single normal model with the three-component unequal- variance model. The 99 simulated LRTSs are computed for $K = 1$ to 3, using the estimated parameters in Table 7.1.

```
 [1]  0.1339  0.4934  0.9884  1.0216  1.0735  1.1149  1.1225
 [8]  1.1266  1.1510  1.1882  1.2592  1.3118  1.4538  1.4951
[15]  1.5259  1.6132  1.6514  1.7633  1.8330  1.8513  1.8708
[22]  1.9812  2.1133  2.2398  2.2888  2.3692  2.3806  2.3851
[29]  2.4087  2.4310  2.5714  2.5922  2.5943  2.8723  2.8806
[36]  2.9119  2.9290  3.0586  3.2304  3.2784  3.3272  3.3816
[43]  3.4297  3.4346  3.5424  3.5927  3.6037  3.7432  3.9733
[50]  4.1762  4.2156  4.3410  4.5057  4.5178  4.5364  4.5453
[57]  4.9485  4.9720  5.1984  5.2339  5.4780  5.5430  5.5586
[64]  5.6162  5.6983  5.7205  5.8220  5.9257  5.9484  6.1462
[71]  6.1522  6.2348  6.4741  6.5714  6.5753  6.8553  6.9909
[78]  7.2287  7.3458  7.3901  7.4605  7.6624  7.8227  7.9006
[85]  8.2375  8.2515  8.8614  9.2677  9.3977  9.4672  9.7461
[92] 10.7831 11.0101 11.5033 12.1782 12.4712 15.9012 17.0818
[99] 21.6641
```

The observed LRTS is 55.48, far beyond the largest value. We reject the hypothesis $K = 1$ in favour of $K = 3$. We repeat the simulations using the model parameters for $K = 3$ and fit the three- and four-component equal variance models. The observed LRTS of 8.87 is the 95th largest in the full set of 100, so it has a P-value of 0.05.

Extending the bootstrapping further, we find the P-values (based on 99 bootstrap samples) for the 4–5, 5–6, and 6–7 comparisons are 0.11, 0.02, and 0.21. For a direct 3–6 comparison, the P-value of the observed LRTS of 30.78 is 0.01. For the equal-variance model, the bootstrap LRT supports six components; we note however the results of McLachlan and Peel (1997) concerning the overstatement of significance for larger numbers of components.

We also generate bootstrap comparisons for the unequal variance cases. For unequal variances, which allow the same component mean separation but smaller standard deviations in one component and hence a higher likelihood, the bootstrap LRT was applied to the galaxy data by McLachlan and Peel in the discussion of Richardson and Green (1997); they reported P-values for 1–2, 2–3, 3–4, 4–5, 5–6, and 6–7 components of 0.01, 0.01, 0.01, 0.04, 0.02, and 0.22. Our bootstrapped tests for models with unequal variances support three components.

Table 7.2. Bootstrap generated P-values for testing an additional component for mixtures with equal and unequal dispersions and testing for heterogeneity for each value of K.

$K \to K+1$	1	2	3	4	5	6
Equal variances	0.01	0.01	0.05	0.11	0.02	0.21
Unequal variances	0.02	0.04	0.45	0.14	0.34	0.08
$\sigma \to \sigma_k$		0.01	0.01	0.1	0.02	0.48

The P-values comparing models for the number of components in both the equal and unequal models and for unequal versus equal variances for various K are given in Table 7.2. The models best supported define a choice between a six-component equal-variance mixture, with a disparity of 394.6 and an unequal-variance three-component mixture having a disparity of 409.1.

We now directly compare the three-component unequal-variance model to the six-component equal-variance model. The P-value of the observed LRTS of 14.47 is 0.11 showing that the six component mixture model is no better than the three-component with unequal variances.

The fitted six-component equal-variance model (solid curve) and the unequal-variance three-component model (dotted curve) *cdfs* are shown in Fig. 7.3 with the 95% simultaneous confidence band, while Fig. 7.4 gives the corresponding densities.

```
> print(xyplot(upper + lower ~ x,
+     data = NPL.bands(galaxies.df$velocity),
+     xlab = "velocity", ylab = "probability",
+     panel = function(...) {
+         panel.xyplot(..., pch = 20)
+         panel.curve(0.085 * pnorm(x, 9.71, 0.81) +
+             0.025 * pnorm(x, 16.14, 0.81) + 0.453 *
+             pnorm(x, 19.93, 0.81) + 0.357 * pnorm(x,
+             23.05, 0.81) + 0.044 * pnorm(x, 26.24,
+             0.81) + 0.037 * pnorm(x, 33.04, 0.81),
+             lwd = 2)
+         panel.curve(0.08536 * pnorm(x, 9.71, 0.64) +
+             0.878 * pnorm(x, 21.4, 2.2038) + 0.037 *
+             pnorm(x, 33.04, 1.13), lty = 2, lwd = 2)
+     }))
```

The fitted *cdfs* are very similar, though the densities are not, with the unequal variance model fitting the small velocities slightly better. Increasing the number of components beyond four does not materially affect the fitted model: the additional complexity of the model is used only to interpolate, more and more precisely, the sample fluctuations in the empirical *cdf* (Aitkin, 2001).

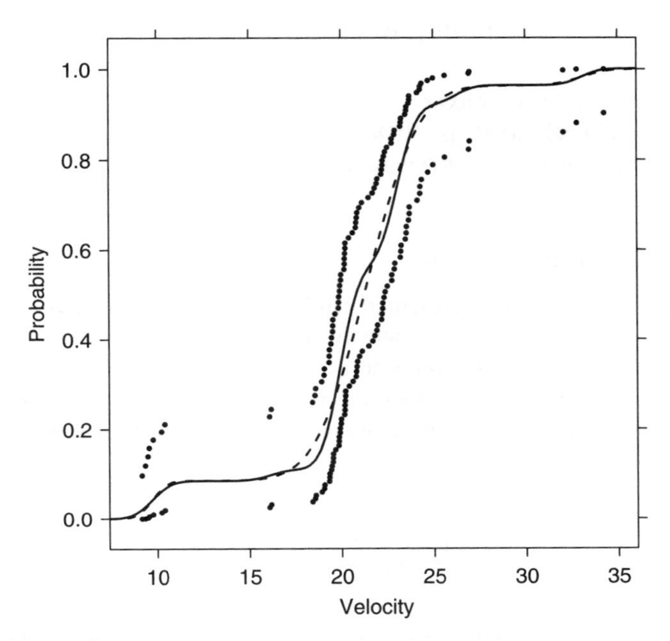

Fig. 7.3. Galaxy data *cdf* bounds and fitted three-component unequal variance and six-component equal variance normal *cdf*s.

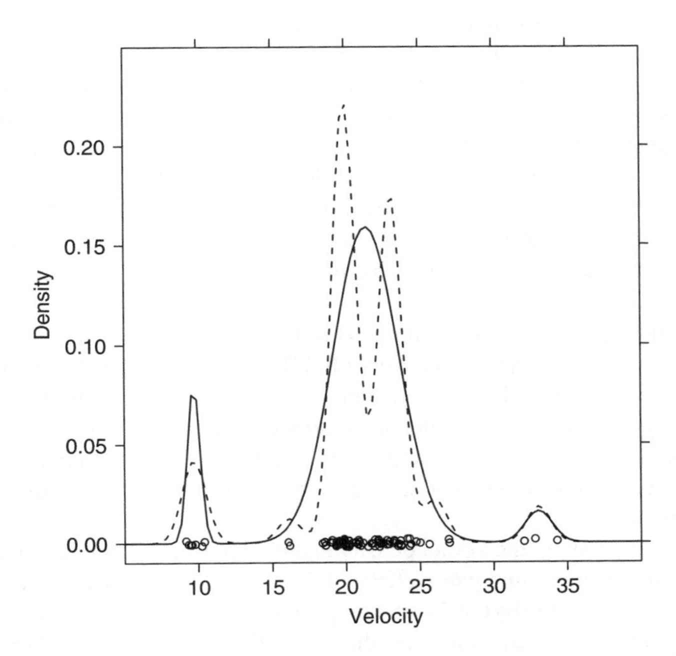

Fig. 7.4. Galaxy data fitted three-component unequal and six-component equal variance normal densities.

Further discussion of mixture fitting for the galaxy data can be found in McLachlan and Peel (2000, pp. 194–196) and Aitkin (2001) who also compared several Bayesian analyses of the galaxy data.

7.9 Kernel density estimates

The normal mixture we have presented in this chapter is an alternative to the widely used *kernel density estimate* of a density, using a normal (Gaussian) kernel (see, e.g. Wand and Jones, 1994). Suppose we have observations $y_1, \ldots, y_n$ from an unknown continuous density (model) $f(y)$. We want to give a smooth estimate of the density, without a strong model assumption about it. The kernel density estimate $\tilde{f}(y)$ of f is defined by

$$\tilde{f}(y) = \frac{1}{nh} \sum_{i=1}^{n} K\left(\frac{y - y_i}{h}\right),$$

where $K(x)$ is the *kernel* density function. This is frequently taken as the normal $N(0, 1)$ density, though other kernels are also used; the choice of kernel is much less critical than the choice of the scale parameter h. We will consider only the normal kernel. The scale parameter (standard deviation) h is generally called the 'bandwidth' parameter or 'tuning constant'. From our discussion of normal mixtures, the kernel density estimate can be immediately recognized as an equally weighted n-component normal mixture, with component means equal to the observed values y_i, and common standard deviation h. It can be viewed alternatively as a *smoothing* of the NPML estimate of $f(y)$, which is just the *empirical mass function* with mass $1/n$ at y_i. The empirical mass function is smoothed out by assigning to y_i a normal density with positive standard deviation scaled by mass $1/n$ instead of a degenerate spike mass.

The difficulty in using this estimate is in the choice of the bandwidth h. From our discussion of the normal mixture with different variances, we can see that the same problem arises with the kernel density estimate. If we try to estimate h by ML, the maximized likelihood increases monotonically as $h \to 0$, and in the limit we recover the empirical mass function again. Considerable research has therefore been devoted to other ways of choosing the bandwidth, and this remains an active area of research.

In Fig. 7.5 we show the kernel density estimate for the galaxy data, for a range of values of the smoothing parameter h: 0.5, 1, 2, 4, together with the data.

These were produced by the `lattice` package function `panel.densityplot` which requires a specification of the bandwidth parameter, either a value

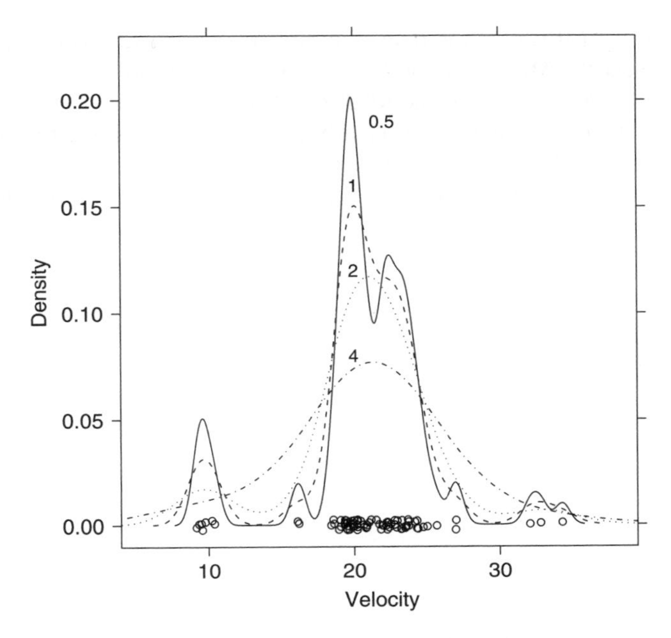

Fig. 7.5. Kernel densities for galaxy data

representing the standard deviation of the smoothing kernel or a string defining a choice of rules.

The data are plotted at the zero value of the density.

```
> library(lattice)
> print(densityplot(~velocity, data = galaxies.df,
+     xlab = "velocity", ylim = c(-0.01, 0.23),
+     panel = function(x, ...) {
+         panel.densityplot(x, darg = list(bw = 0.5))
+         panel.densityplot(x, darg = list(bw = 1),
+             plot.points = FALSE, lty = 2)
+         panel.densityplot(x, darg = list(bw = 2),
+             plot.points = FALSE, lty = 3)
+         panel.densityplot(x, darg = list(bw = 4),
+             plot.points = FALSE, lty = 4)
+         panel.text(c(22, 20, 20, 20), c(0.19, 0.16,
+             0.12, 0.08), c(0.5, 1, 2, 4))
+     }))
```

The density estimates become progressively more 'peaked' as the bandwidth h decreases. At $h = 4$, near the MLE of σ for $K = 1$, the density is unimodal. At

$h = 2$, near the MLE of σ for $K = 3$, a three-mode structure is clearly visible. At $h = 1$, near the MLE of σ for $K = 5$, the density is still trimodal, but with 'shoulders'. At $h = 0.5$, near the MLE of σ for $K = 7$, the density has seven modes. The choice of bandwidth is clearly critical if the density estimate is to be used to indicate multi-modality consistent with a mixture.

In the next chapter we consider the application of mixtures in *random effect models*.

8
Random effect models

8.1 Overdispersion

In Chapter 5 we found overdispersion in the fabric fault data; the Poisson GLM did not fit or represent the data adequately. The failure of a generalized linear model to fit may be due to several causes. The distribution of Y may not be the specified exponential family member, or the regression model fitted may be mis-specified. A fruitful way of expressing the problem, which unifies these two possibilities, is through *omitted variables* from the regression model: the full model we have fitted is missing one or more important variables. If these have been recorded then they can be added to the model and fitted in the usual way (e.g. interactions). The common reality however, especially in survey or observational studies, is that we do not know what variables *should* have been recorded or measured to construct an adequate model.

We now make the (model) assumption that there is another set (vector) of unobserved variables $\mathbf{z}$, in addition to the observed set $\mathbf{x}$ in our model, and that the 'correct' linear predictor is $\boldsymbol{\beta}'\mathbf{x} + \boldsymbol{\gamma}'\mathbf{z}$, where $\boldsymbol{\gamma}$ is the vector of regression coefficients for the unobservable variables. Since $\mathbf{z}$ was not measured, we know nothing about it – it is varying over the dataset in an unknown way, and is therefore a *random* vector as far as we are concerned. Since $\boldsymbol{\gamma}$ is also an unknown vector, the term $\boldsymbol{\gamma}'\mathbf{z}$ is in fact a *single* (scalar) unobserved random variable or *random effect* Z, and so we may write the linear predictor without any loss of generality as $\boldsymbol{\beta}'\mathbf{x} + Z$, where the scalar variable Z is random with an unknown distribution. This model including Z is called a *random effect model*. It should be noted that we *have* imposed a model restriction, namely that Z appears *additively* in the model on the scale of the linear predictor, that is that Z does not interact with $\mathbf{x}$. We will relax this restriction later. We call the model for Y an *overdispersed* GLM.

We have also assumed that the distribution of Z does not depend on $\mathbf{x}$, for example, through a regression of Z on $\mathbf{x}$. Under the assumption of linear dependence, however, we can still proceed. Suppose that the distribution $g(z)$ of Z does in fact depend on $\mathbf{x}$ through a location parameter, and so can be written as $g(z - \boldsymbol{\delta}'\mathbf{x})$. Then we may write

$$m(y) = \int f(y|z)g(z - \boldsymbol{\delta}'\mathbf{x})\,\mathrm{d}z = \int f(y|z^*)g(z^*)\,\mathrm{d}z^*,$$

where the distribution of $Z^* = Z - \boldsymbol{\delta}'\mathbf{x}$ does not depend on $\mathbf{x}$, and the dependence of Y on Z^* is through the linear predictor

$$\eta = \boldsymbol{\beta}'\mathbf{x} + Z = \boldsymbol{\beta}'\mathbf{x} + Z^* + \boldsymbol{\delta}'\mathbf{x} = \boldsymbol{\beta}^{*\prime}\mathbf{x} + Z^*,$$

where $\boldsymbol{\beta}^* = \boldsymbol{\beta} + \boldsymbol{\delta}$. So if the distribution of Z does have a regression on $\mathbf{x}$ of this form, we can still fit the model assuming that Z^* is independent of $\mathbf{x}$, but the regression coefficient $\boldsymbol{\beta}$ will be affected by the dependence. A substantial change in the estimate of the regression coefficient $\boldsymbol{\beta}$ when we allow for overdispersion is a warning that the omitted variable(s) represented by Z may be correlated with the explanatory variables $\mathbf{x}$, and so the regression of Y on $\mathbf{x}$ integrated over Z may be quite different from the conditional regression of Y on $\mathbf{x}$ given Z.

We now restrict consideration to the case of 'homogeneous' Z, with distribution independent of $\mathbf{x}$. Since Z is unobserved, we can observe only the distribution of Y, which is now a *compound* or *mixed* exponential family distribution. We have lost the original distribution by compounding. Formally,

$$m(y) = \int f(y|z)g(z)\,\mathrm{d}z,$$

where $g(z)$ is the density or mass function of Z. Consequently,

$$\begin{aligned}
\mathrm{E}[Y] &= \mathrm{E}[\mathrm{E}[Y|Z]] \\
\mathrm{Var}[Y] &= \mathrm{E}[\mathrm{Var}[Y|Z]] + \mathrm{Var}[\mathrm{E}[Y|Z]].
\end{aligned}$$

Write $\mu(Z)$ and $V(Z)$ for the mean and variance of Y given Z; then

$$\begin{aligned}
\mathrm{E}[Y] &= \mathrm{E}[\mu(Z)] \\
\mathrm{Var}[Y] &= \mathrm{E}[V(Z)] + \mathrm{Var}[\mu(Z)].
\end{aligned}$$

The variance of Y is always *inflated* by the compounding with Z, and the mean/variance relationship of the original ('kernel') distribution of Y is lost by the compounding. For example, if $Y|Z$ has a Poisson distribution with $V(Z) = \mu(Z)$ and log-linear model $\log \mu(Z) = \boldsymbol{\beta}'\mathbf{x} + Z = \eta + Z$, then

$$\begin{aligned}
\mathrm{E}[Y] &= \mathrm{E}[\mathrm{e}^{\eta+Z}] = \mathrm{e}^{\eta}M(1) \\
\mathrm{Var}[Y] &= \mathrm{E}[\mathrm{e}^{\eta+Z}] + \mathrm{Var}[\mathrm{e}^{\eta+Z}] \\
&= \mathrm{e}^{\eta}M(1) + \mathrm{e}^{2\eta}[M(2) - M^2(1)],
\end{aligned}$$

where $M(t)$ is the moment-generating function of Z. Thus

$$\mathrm{Var}[Y] = \mathrm{E}[Y] + \phi \mathrm{E}^2[Y],$$

where $\phi = \left[M(2)/M^2(1)\right] - 1$, for any distribution of Z. These results can be expressed equivalently, and very simply, in terms of the distribution of $W = \mathrm{e}^Z$: $M(1)$ is the mean, and $M(2) - M^2(1)$ is the variance of W, so ϕ is the *squared coefficient of variation* of W. (We use $g(z)$ for the pdf of the random effect distribution and $g(\mu)$ for the link function. Since these rarely appear in the same expression we hope the reader will not be confused.)

In general, for non-linear link functions g, the mean of the marginal distribution of Y no longer has the same link to the explanatory variables. Writing h for the inverse link function, using a second-order Taylor expansion we have for the random effect model

$$\mu(Z) = h(\eta + Z) \approx h(\eta) + Zh'(\eta) + \frac{1}{2}Z^2 h''(\eta).$$

So for the compound model

$$\mathrm{E}[Y] = \mathrm{E}[\mu(Z)] \approx h(\eta) + \frac{1}{2}\sigma^2 h''(\eta),$$

where σ^2 is the variance of Z and the mean of Z is taken as zero without any loss of generality (since a non-zero mean could be included with the intercept term in the regression model). Then $g\,(\mathrm{E}[Y]) = h^{-1}(\mathrm{E}[Y]) \neq \eta$ in general. However, if σ is small relative to the variation in η the link will still hold approximately, and for two special links the original link holds exactly: for the identity link we have immediately

$$\mathrm{E}[Y] = \eta + \mathrm{E}[Z] = \eta,$$

and for the log link

$$\mathrm{E}[Y] = \mathrm{E}[\mathrm{e}^{\eta+Z}] = \mathrm{e}^{\eta} M(1) = \mathrm{e}^{\eta+\delta},$$

where $\delta = \log M(1)$. In this case the intercept β_0 of the regression model is affected but the regression coefficients for the explanatory variables are not: the mean of Y still has a log-linear model in these variables.

Without a specific distributional assumption for Z, we cannot proceed directly to fit the model by ML, and are limited to *quasi-likelihood* (LQ) analyses (Wedderburn, 1974; McCullagh and Nelder, 1989) which use only the mean and variance of Y. We do not, discuss, this approach further in this book, since we emphasize the use of specific models for Z. We note, however, that the QL approach is not sensitive to variations in the distribution of Z: for example in the Poisson model above, the normal and the log-gamma distributions for Z, scaled to have the same mean and variance, will have the same value of ϕ, though the marginal distributions of Y and therefore the likelihoods will be different.

8.1.1 *Testing for overdispersion*

In the general framework of random effect modelling in this chapter, we use the LRT discussed below for model simplification, including testing for the presence of overdispersion. Earlier treatments of this model generally used the score test, since as we noted in Chapter 2 this does not require the fitting of the overdispersed model, only that of the null hypothesis model without overdispersion. We illustrate the score test with a brief example of heterogeneity in a simple Poisson model.

8.1.1.1 *Poisson heterogeneity*

Consider the heterogeneous Poisson model $Y_i \sim P(\mu_i)$, where there is no specific model for the heterogeneity in terms of explanatory variables. We want a test for heterogeneity, against the null homogeneous model $Y_i \sim P(\mu)$. We write $\log \mu_i = \theta + \delta_i$; the null hypothesis is $H_0 : \delta_i = 0$ for all i, and the alternative hypothesis is H_1 : not all $\delta_i = 0$. Here θ is the 'nuisance parameter' of Chapter 2 and the δ_i are the parameters of interest. This model is over-parametrized with $n + 1$ parameters for n observations. We set $\delta_1 = 0$ for identifiability.

The log-likelihood and its derivatives are

$$\ell = \sum_i [-\mu_i + y_i \log \mu_i]$$

$$\frac{\partial \ell}{\partial \theta} = \sum_i [-\mu_i + y_i]$$

$$\frac{\partial \ell}{\partial \delta_i} = -\mu_i + y_i$$

$$\frac{\partial^2 \ell}{\partial \theta^2} = -\sum_i \mu_i$$

$$\frac{\partial^2 \ell}{\partial \theta \partial \delta_i} = -\mu_i$$

$$\frac{\partial^2 \ell}{\partial \delta_i^2} = -\mu_i.$$

The score components evaluated at the MLE $\hat{\mu}_i = e^{\hat{\theta}_0} = \bar{y}$ under H_0 give

$$\mathbf{s}_0' = (0, y_2 - \bar{y}, \ldots, y_n - \bar{y}),$$

and the expected information is

$$\mathcal{I} = \begin{bmatrix} \mu_+ & \mu_2 & \mu_3 & \dots & \mu_n \\ \mu_2 & \mu_2 & 0 & \dots & 0 \\ \mu_3 & 0 & \mu_3 & \dots & 0 \\ \dots & \dots & \dots & \dots & \dots \\ \mu_n & 0 & 0 & \dots & \mu_n \end{bmatrix},$$

where $\mu_+ = \sum_1^n \mu_i$. Evaluating this under H_0 gives

$$\mathcal{I}_0 = \begin{bmatrix} n\bar{y} & \bar{y} & \bar{y} & \dots & \bar{y} \\ \bar{y} & \bar{y} & 0 & \dots & 0 \\ \bar{y} & 0 & \bar{y} & \dots & 0 \\ \bar{y} & 0 & 0 & \dots & \bar{y} \end{bmatrix}$$

with inverse matrix

$$\mathcal{I}_0^{-1} = \frac{1}{\bar{y}} \begin{bmatrix} 1 & -n\mathbf{1}' \\ -n\mathbf{1} & \mathrm{I} - \mathbf{1}\mathbf{1}' \end{bmatrix},$$

where $\mathbf{1}$ is the unit vector of length $n-1$, and the score test statistic is

$$\mathbf{s}_0' \mathcal{I}_0^{-1} \mathbf{s}_0 = \frac{\sum_{i=1}^n (y_i - \bar{y})^2}{\bar{y}}.$$

Under H_0 the score statistic has an asymptotic χ^2_{n-1} distribution.

Since we concentrate on the formulation and fitting of the alternative model, we use the LRT where it is available. While the score test is sometimes *locally* most powerful (i.e. in the near neighbourhood of the null hypothesis), when departures from the null hypothesis are large, the score test may have substantially lower power than the LRT, and in any case we need to use the alternative model.

The normal model, and any other *specific* parametric model for Z, allows the use of the LRT for testing the null hypothesis $\sigma = 0$ with a *known* asymptotic distribution, given originally by Chernoff (1954) and frequently re-derived in various forms. For the general class of overdispersed models with a single overdispersion parameter, the LRTS is distributed asymptotically as an equal mixture of a degenerate distribution with mass 1 at 0, and a χ^2_1 distribution. The degenerate distribution arises because if the null exponential family model is correct but the overdispersed model is fitted, in large samples the estimate of σ will frequently be zero, on the boundary of the parameter space, and so the disparities for the null and overdispersed models will be the same.

Thus a size α test can be obtained (asymptotically) as follows: if the LRTS is zero, accept H_0. If it is non-zero, compare it with the $100(1 - 2\alpha)\%$ point of the

χ_1^2 distribution. A comprehensive discussion of the likelihood ratio and score tests for variance component models can be found in Verbeke and Molenberghs (2003).

We consider now several parametric distributions for the random effect Z.

8.2 Conjugate random effects

Each of the exponential family distributions has its own *conjugate* distribution for the canonical parameter θ of the distribution. The conjugate distribution for Z has the same analytic form in Z as the likelihood in Z from the distribution of $Y|Z$, giving a simple form for the marginal distribution of Y. In some cases the conjugate distribution for Z can be used to model the extra variation.

8.2.1 *Normal kernel: the t-distribution*

For the random effect model for the normal mean, the conjugate distribution is also normal, but the resulting model is *unidentifiable*: if $Y|Z \sim N(\theta + Z, \sigma^2)$ conditionally and $Z \sim N(\mu, \phi^2)$ marginally, then $Y \sim N(\theta+\mu, \sigma^2+\phi^2)$ marginally. If all the parameters are unknown, then μ is aliased with θ and ϕ^2 with σ^2: we simply have another unknown normal distribution marginally. So the normal distribution cannot be extended by conjugate compounding on the mean.

However, it *can* be extended by conjugate compounding on the *variance*. Let $Y|Z \sim N(\mu, \sigma^2/Z)$ with Z having a conjugate gamma $(r/2, 2/r)$ distribution with mean 1 as in Section 6.9. The marginal distribution of Y is then

$$
\begin{aligned}
m(y) &= \frac{(r/2)^{r/2}}{\sqrt{2\pi}\sigma\Gamma(r/2)} \int z^{\frac{r-1}{2}} \exp\left\{-z\left[\frac{r}{2} + \frac{(y-\mu)^2}{2\sigma^2}\right]\right\} dz \\
&= \frac{(r/2)^{r/2}\Gamma(\frac{r+1}{2})}{\sqrt{2\pi}\sigma\Gamma(r/2)} \left[\frac{r}{2} + \frac{(y-\mu)^2}{2\sigma^2}\right]^{-\frac{(r+1)}{2}} \\
&= \frac{\Gamma(\frac{r+1}{2})}{\sqrt{\pi r}\sigma\Gamma(r/2)} \left[1 + \frac{(y-\mu)^2}{r\sigma^2}\right]^{-\frac{(r+1)}{2}}
\end{aligned}
$$

which is a t-distribution of $(y-\mu)/\sigma$ with r degrees of freedom. Here Y has mean μ (provided $r > 1$) and variance $r\sigma^2/(r-2)$ (provided $r > 2$). The *Cauchy* distribution is the special case with $r = 1$: it is a *heavy-tailed* distribution with infinite mean and variance.

Thus, the t-distribution arises naturally as a conjugate extension of the normal – it is an *inverse gamma mixture of normals*, mixed on the variance. It is a candidate for modelling in the presence of *outliers*. We will illustrate this with an example, but first we show that the t-distribution can be fitted as a GLM with an additional iterative component to the algorithm. ML for a linear model $\mu_i = \boldsymbol{\beta}'\mathbf{x}_i$ in the t-distribution is readily achieved (Lange *et al.*, 1989). To simplify the log-likelihood derivatives, we first re-parametrize the distribution to $(\boldsymbol{\beta}, \phi, r)$ where $\phi = r\sigma^2$, and write

$$\mathrm{e}_i = y_i - \mu_i, \qquad w_i = \left[1 + \frac{(y_i - \mu_i)^2}{\phi}\right]^{-1}.$$

The log-likelihood for a regression model with n observations is

$$\ell = n\left[\log\Gamma\left(\frac{r+1}{2}\right) - \frac{1}{2}\log\pi - \log\Gamma\left(\frac{r}{2}\right) - \frac{1}{2}\log\phi\right] + \frac{r+1}{2}\sum_{i=1}^{n}\log w_i.$$

The derivatives involve the expressions

$$\begin{aligned}\frac{\partial w_i}{\partial\boldsymbol{\beta}} &= \frac{2}{\phi}w_i^2 e_i \mathbf{x}_i\\ \frac{\partial w_i}{\partial\phi} &= \frac{1}{\phi^2}w_i^2 e_i^2,\end{aligned}$$

while w_i is not a function of r. Then

$$\begin{aligned}\frac{\partial\ell}{\partial\boldsymbol{\beta}} &= \frac{r+1}{\phi}\sum_i w_i e_i \mathbf{x}_i\\ \frac{\partial\ell}{\partial\phi} &= -\frac{n}{2\phi} + \frac{r+1}{2\phi^2}\sum_i w_i e_i^2\\ \frac{\partial\ell}{\partial r} &= \frac{n}{2}\left[\psi\left(\frac{r+1}{2}\right) - \psi\left(\frac{r}{2}\right)\right] + \frac{1}{2}\sum_i \log w_i,\end{aligned}$$

where ψ is the di-gamma function.

The first two score equations can be expressed as

$$\begin{aligned}\hat{\boldsymbol{\beta}} &= \left(\sum_i w_i \mathbf{x}_i \mathbf{x}_i'\right)^{-1}\sum_i w_i \mathbf{x}_i y_i\\ &= (X'WX)^{-1}X'W\mathbf{y}\end{aligned}$$

and

$$\hat{\phi} = (\hat{r}+1)\sum_i \frac{w_i e_i^2}{n},$$

giving

$$\hat{\sigma}^2 = \left(\frac{\hat{r}+1}{\hat{r}}\right) \sum_i \frac{w_i(y_i - \hat{\boldsymbol{\beta}}'\mathbf{x}_i)^2}{n}$$
$$= \left(\frac{\hat{r}+1}{\hat{r}}\right) \frac{\text{RSS}}{n},$$

where RSS is the *weighted* residual sum of squares from the fitted model. The MLE of $\boldsymbol{\beta}$ is a weighted least squares estimate with weights w_i, though since the weights are themselves functions of $\boldsymbol{\beta}$ the weighting has to be done iteratively. The MLE of σ^2 depends on $\hat{r}$, which is the solution of

$$\psi\left(\frac{r+1}{2}\right) - \psi\left(\frac{r}{2}\right) = -\sum_i \frac{\log w_i}{n}.$$

The effect of the weights w_i is to downweight the observations with large residuals e_i from the fitted model, so reducing the effect of outliers: the t-distribution provides a parametric model for *robust regression*.

By direct calculation of the second derivatives and their expected values, Lange *et al.* (1989) showed that the expected information matrix is block diagonal, with a zero $\boldsymbol{\beta}$, (ϕ, r) off-diagonal block. Thus in large samples the estimation of ϕ and r does not affect the precision of $\hat{\boldsymbol{\beta}}$, and the iterative procedure gives the correct standard errors for $\hat{\boldsymbol{\beta}}$.

Fitting the t-distribution with known r is particularly simple; we iterate until convergence between fitting the model with weights w_i and recomputing the weights from the current estimates of $\boldsymbol{\beta}$ and ϕ. This is implemented in the function `treg`. For unknown r, we can compute the profile likelihood over a grid of values and hence obtain the joint MLEs of $\boldsymbol{\beta}$, τ and r. Alternatively, we can introduce an additional level to our procedure, iterating between the estimation of $\boldsymbol{\beta}$ and τ for known r, and solving the score equation in r directly.

The iteratively weighted least squares algorithm described above is closely related to the EM algorithm: if we regard the random effect Z as unobserved data, we can proceed by EM, maximizing the conditional expected complete-data log-likelihood. It is straightforward to show that the EMEs of $\boldsymbol{\beta}$ and σ^2 are also weighted least squares estimates, with weights $(1 + 1/r)w_i$, which are a constant multiple of the w_i. This multiple does not affect the estimates.

8.2.1.1 *Example – the Brownlee stack-loss data*

We reproduce here the analysis of Brownlee's (1965) stack-loss data from Lange *et al.* (1989). The stack-loss data have been analysed many times for outliers: the observations numbered 1, 3, 4, and 21 have frequently been identified as outliers. A detailed study of the data, with a full list of previous analyses, can be found in Dodge (1999). The data consist of 21 observations on stack-loss `y` (the loss of acid

Table 8.1. Stack-loss data

Obs.	y	x_1	x_2	x_3	Obs.	y	x_1	x_2	x_3
1	42	80	27	89	12	13	58	17	88
2	37	80	27	88	13	11	58	18	82
3	37	75	25	90	14	12	58	19	93
4	28	62	24	87	15	8	50	18	89
5	18	62	22	87	16	7	50	18	86
6	18	62	23	87	17	8	50	19	72
7	19	62	24	93	18	8	50	19	79
8	20	62	24	93	19	9	50	20	80
9	15	58	23	87	20	15	56	20	82
10	14	58	18	80	21	15	70	20	91
11	14	58	18	89					

through the stack) in a chemical plant for the conversion of ammonia to nitric acid, with three explanatory variables: air flow x_1, cooling water inlet temperature x_2 and acid concentration x_3. The (coded) data are listed in Table 8.1, and are held in the dataset `stackloss`.

```
> data(stackloss, package = "SMIR")

> stackloss.lm <- lm(y ~ x1 + x2 + x3, data = stackloss)
> print(summary(stackloss.lm)$coeff, 4)

            Estimate Std. Error t value  Pr(>|t|)
(Intercept) -39.9197    11.8960 -3.3557 3.750e-03
x1            0.7156     0.1349  5.3066 5.799e-05
x2            1.2953     0.3680  3.5196 2.630e-03
x3           -0.1521     0.1563 -0.9733 3.440e-01
```

The disparity is

$$n\{1 + \log(2\pi) + \log(\text{RSS}/n)\} = 104.58.$$

The variables x_1 and x_2 appear important but x_3 does not. Increasing levels of the important variables increases the stack-loss. We check the leverage of the individual observations.

```
> round(influence(stackloss.lm)$hat, 3)

    1     2     3     4     5     6     7     8     9    10
0.302 0.318 0.175 0.129 0.052 0.077 0.219 0.219 0.140 0.200
   11    12    13    14    15    16    17    18    19    20
0.155 0.217 0.158 0.206 0.190 0.131 0.412 0.161 0.175 0.080
   21
0.285
```

The average leverage, $(p+1)/n$, is 0.19, and twice this is 0.38. Only the 17th observation exceeds this value.

We now fit the t model with fixed r, over a grid of r using the function `treg` from the `SMIR` package.

```
> library(SMIR)

> rvalues <- c(30, 20, 10, 8, 6, 5, 4, 3, 2, 1.1, 1,
+     0.5, 0.4, 0.3, 0.2)
> stackloss.treg <- NULL
> for (s in seq(along = rvalues)) {
+     stackloss.treg <- c(stackloss.treg, list(treg(stackloss.lm,
+         r = rvalues[s], verbose = FALSE)))
+ }
```

The estimates and disparities are shown in Table 8.2. The likelihood is fairly flat for large r but more peaked near $\hat{r} = 1.1$. The change in disparity between the normal model at $r = \infty$ and that at $\hat{r}$ is 5.44. The disparity is not monotone for $r < 1$ – it increases as r decreases from $r = 1.1$ to $r = 0.4$, and then decreases again with further decrease in r. With decreasing r, x_1 becomes more important and x_2 and x_3 less important. We now examine the weights and leverage values for $r = 1.1$.

```
> (stackloss.treg1.1 <- treg(stackloss.lm, r = 1.1,
+     verbose = FALSE))
```

Table 8.2. Stackloss estimates

r	1	x_1	x_2	x_3	σ	Disparity
∞	−39.9	0.716	1.30	−0.152	3.24	104.58
30	−40.2	0.740	1.21	−0.145	2.81	104.50
20	−40.3	0.753	1.17	−0.142	2.75	104.43
10	−40.6	0.793	1.03	−0.133	2.56	104.12
8	−40.7	0.811	0.968	−0.129	2.46	103.92
6	−40.7	0.835	0.877	−0.124	2.30	103.56
5	−40.5	0.847	0.817	−0.120	2.19	103.27
4	−40.1	0.857	0.746	−0.115	2.03	102.85
3	−39.1	0.854	0.657	−0.104	1.76	102.14
2	−38.1	0.848	0.557	−0.089	1.34	100.63
1.1	−38.4	0.852	0.491	−0.071	0.93	99.14
1	−38.6	0.852	0.489	−0.068	0.88	99.16
0.5	−40.8	0.840	0.536	−0.044	0.38	101.09
0.4	−40.1	0.834	0.562	−0.055	0.18	101.91
0.3	−39.9	0.833	0.567	−0.058	0.043	100.53
0.2	−39.7	0.834	0.566	−0.060	0.000	72.61

```
Coefficients:
(Intercept)          x1          x2          x3
   -39.9197      0.7156      1.2953     -0.1521

> round(stackloss.treg1.1$weights, 3)

    1     2     3     4     5     6     7     8     9    10
0.032 0.915 0.029 0.015 0.495 0.304 0.767 0.817 0.455 0.987
   11    12    13    14    15    16    17    18    19    20
0.771 0.997 0.097 0.253 0.344 0.982 0.882 0.979 0.646 0.236
   21
0.010

> round(influence(stackloss.treg1.1)$hat, 3)

    1     2     3     4     5     6     7     8     9    10
0.302 0.318 0.175 0.129 0.052 0.077 0.219 0.219 0.140 0.200
   11    12    13    14    15    16    17    18    19    20
0.155 0.217 0.158 0.206 0.190 0.131 0.412 0.161 0.175 0.080
   21
0.285
```

The weights on observations 1, 3, 4, and 21 are very low, below 0.033, and that on observation 13 is 0.096. As small weights correspond to large residuals, these points are identified as outliers, downweighted and effectively excluded from this robust regression model fit. An issue of concern is that the total weight may be substantially less than n – here it is 11.01. Nearly half the data have been 'lost' in the downweighting, and the parameter estimates become more strongly dependent on the remaining observations with high weights and leverages, particularly 2 and 17.

For smaller values of r these effects become extreme. The weights and leverage values for $r = 0.5$ are

```
    1     2     3     4     5     6     7     8     9    10
0.003 0.954 0.002 0.001 0.047 0.023 0.066 1.000 0.036 1.000
   11    12    13    14    15    16    17    18    19    20
0.310 0.846 0.008 0.018 0.054 0.995 0.703 0.791 0.145 0.024
   21
0.001

    1     2     3     4     5     6     7     8     9    10
0.302 0.318 0.175 0.129 0.052 0.077 0.219 0.219 0.140 0.200
   11    12    13    14    15    16    17    18    19    20
0.155 0.217 0.158 0.206 0.190 0.131 0.412 0.161 0.175 0.080
   21
0.285
```

The regression is now being determined only by observations 2, 8, 10, 12, 16, 17, and 18 – the total weight is only 7.03. At $r = 0.2$ only four weights are 1.0, on observations 2, 8, 12, and 18, and all the other weights are zero. The four-parameter model fits almost exactly the total observation weight of four, with a variance estimate of almost zero and an almost undefined disparity.

We consider the stack-loss data further below (Section 8.7.4) with a different random effect model, and come to a quite different conclusion.

8.2.2 *Poisson kernel: the negative binomial distribution*

For the Poisson distribution of $Y|Z$, with $\log \mu(Z) = \eta + Z$, the conjugate distribution of Z is the log-gamma. It is simpler to work with $W = \mathrm{e}^Z$, so that $\log \mu(W) = \eta + \log W$. W has a gamma distribution, which we take in the standard form (Section 6.9)

$$g(w) = \frac{r^r}{\Gamma(r)} \mathrm{e}^{-rw} w^{r-1}, \quad w, r > 0,$$

so that W has mean 1, variance $1/r$, and coefficient of variation r. The mean of the corresponding log-gamma distribution of Z is $\psi(r) - \log r$ which is approximately $-1/2r$, and the variance is $\psi'(r)$ which is approximately $1/r + 1/2r^2$.

The marginal distribution of Y is then negative binomial, a *gamma mixture of Poissons*:

$$\begin{aligned} m(y) &= \frac{r^r}{\Gamma(r)y!} \int_0^\infty \mathrm{e}^{-w\mathrm{e}^\eta} (w\mathrm{e}^\eta)^y \mathrm{e}^{-rw} w^{r-1}\, \mathrm{d}w \\ &= \frac{r^r \mathrm{e}^{\eta y}}{\Gamma(r)y!} \int_0^\infty \mathrm{e}^{-w(\mathrm{e}^\eta + r)} w^{y+r-1}\, \mathrm{d}w \\ &= \frac{r^r \mathrm{e}^{\eta y}}{\Gamma(r)y!} \frac{\Gamma(y+r)}{(\mathrm{e}^\eta + r)^{y+r}} \\ &= \frac{\Gamma(y+r)}{\Gamma(r)y!} \left(\frac{\mathrm{e}^\eta}{\mathrm{e}^\eta + r}\right)^y \left(\frac{r}{\mathrm{e}^\eta + r}\right)^r, \end{aligned}$$

with mean $\mu = \mathrm{e}^\eta$ and variance $\mu + \mu^2/r$. To fit the negative binomial as a GLM we write

$$m(y) = \frac{\Gamma(y+r)}{\Gamma(r)y!} p^y (1-p)^r$$

with

$$p = \frac{\mathrm{e}^\eta}{\mathrm{e}^\eta + r}$$

so that

$$\text{logit } p = \eta - \log r.$$

The intercept is altered in this formulation, but the other regression coefficients are unaffected.

The likelihood for a sample $(y_1, \mathbf{x}_1), \ldots, (y_n, \mathbf{x}_n)$ is then

$$L(\boldsymbol{\beta}, r) = \prod_{i=1}^{n} \frac{\Gamma(y_i + r)}{\Gamma(r) y_i!} p_i^{y_i} (1 - p_i)^r$$

with

$$\text{logit } p_i = \boldsymbol{\beta}' \mathbf{x}_i - \log r.$$

To fit the model, if r were known $\boldsymbol{\beta}$ could be estimated directly from a binomial logit model for p_i, treating y_i as the number of successes in $y_i + r$ Bernoulli trials. The intercept would then be an estimate of $\beta_0 - \log r$. The correct intercept is achieved by using an offset of $-\log(r)$ in the model. For the estimation of r, we first consider the information matrix for $\boldsymbol{\beta}$ and r. From the log-likelihood

$$\ell(\boldsymbol{\beta}, r) = \sum \{\log \Gamma(y_i + r) - \log \Gamma(r) - \log y_i! + y_i \eta_i + r \log r - (y_i + r) \log(e^{\eta_i} + r)\}$$

we have

$$\frac{\partial \ell}{\partial r} = \sum \left\{ \psi(y_i + r) - \psi(r) + 1 + \log r - \left(\frac{y_i + r}{e^{\eta_i} + r} + \log(e^{\eta_i} + r) \right) \right\}$$

$$\frac{\partial^2 \ell}{\partial r^2} = \sum \left\{ \psi'(y_i + r) - \psi'(r) + \frac{1}{r} - \left(\frac{e^{\eta_i} - y_i}{(e^{\eta_i} + r)^2} + \frac{1}{e^{\eta_i} + r} \right) \right\}$$

$$\frac{\partial^2 \ell}{\partial \boldsymbol{\beta} \partial r} = -\sum \left\{ -\frac{y_i + r}{(e^{\eta_i} + r)^2} e^{\eta_i} + \frac{e^{\eta_i}}{e^{\eta_i} + r} \right\} \mathbf{x}_i.$$

To evaluate the expected information, we note that

$$\mathrm{E}[Y_i] = \mathrm{E}[\mathrm{E}[Y_i | W_i]] = \mathrm{E}[\mu_i(Z_i)] = \mathrm{E}[W_i e^{\eta_i}] = e^{\eta_i}$$

and hence

$$\mathrm{E}\left[\frac{\partial^2 \ell}{\partial \boldsymbol{\beta} \partial r} \right] = 0$$

$$\mathrm{E}\left[\frac{\partial^2 \ell}{\partial r^2} \right] = \sum \left\{ \mathrm{E}[\psi'(Y_i + r)] - \psi'(r) + \frac{1}{r} - \frac{1}{e^{\eta_i} + r} \right\}.$$

Since the expected information is diagonal, full ML estimation is easily attained by successive relaxation, alternating between estimating $\boldsymbol{\beta}$ for fixed r and r for fixed $\boldsymbol{\beta}$, the latter by a simple Newton algorithm as for the gamma distribution in Chapter 6. The expectation of the digamma function in the second equation has no standard form, and it is therefore simpler to work with the observed information in this parameter. This is implemented in the `glm.nb` function. Note also that the estimation of r does not affect the precision of $\hat{\boldsymbol{\beta}}$ asymptotically, because of the diagonal form of the information matrix.

8.2.2.1 *Example: the fabric fault data*

In Chapter 5 we considered the data on fabric faults and modelled them by the Poisson distribution. We saw that the residual deviance from the model was quite large (64.54 on 30 df), suggesting that other relevant factors were varying over the data. The disparity for this model is 187.84. We now fit the Poisson, and the negative binomial distribution using the `glm.nb` function made available in the *MASS* package, and save the fitted values and the fitted two-standard deviation bands around the fitted values, for both distributions.

The *summary* method prints out a deviance, an estimate of r (labelled as `theta`) and its standard error as well as the disparity, and can thus be used to compare the negative binomial to a Poisson model.

The standard deviation printed out is the approximate standard deviation of Z, that is, $\sqrt{1/r + 1/2r^2}$.

```
> data(faults)

> faults <- transform(faults, ll = log(l))
> faults.glm <- glm(n ~ ll, data = faults, family = poisson)
> round(summary(faults.glm)$coeff[, 1:2], 4)

             Estimate Std. Error
(Intercept)   -4.1730     1.1352
ll             0.9969     0.1759

> library(MASS)
> faults.negbin.model <- glm.nb(n ~ ll, data = faults)
> summary(faults.negbin.model, corr = F)

Coefficients:
             Estimate Std. Error z value Pr(>|z|)
(Intercept)   -3.795      1.458   -2.603   0.009
ll             0.938      0.229    4.114 >0.000

(Dispersion parameter for Negative Binomial(8.6674)
    family taken to be 1)

    Null deviance: 50.28  on 31  degrees of freedom
```

```
Residual deviance: 30.67  on 30  degrees of freedom
AIC: 181.39

Number of Fisher Scoring iterations: 1

              Theta:  8.67
          Std. Err.:  4.17

 2 x log-likelihood:  -175.39

> round(-2 * as.numeric(logLik(faults.negbin.model)),
+     2)

[1] 175.39

> faults.newdata <- data.frame(l = seq(min(faults$l),
+     max(faults$l), length = 101))
> faults.newdata$ll <- log(faults.newdata$l)
> faults.newdata$pfv <- predict(faults.glm,
+     new = faults.newdata, type = "response")
> faults.newdata$nbfv <- predict(faults.negbin.model,
+     new = faults.newdata, type = "response")
> faults.newdata <- transform(faults.newdata,
+     nbsd = sqrt(nbfv + nbfv^2/faults.negbin.model$theta))
> faults.newdata <- transform(faults.newdata, pll = pfv -
+     2 * sqrt(pfv), pul = pfv + 2 * sqrt(pfv), nbll = nbfv -
+     2 * nbsd, nbul = nbfv + 2 * nbsd)
> summary(faults.negbin.null.model <- update(faults.negbin.model,
+     . ~ 1))%

Coefficients:
            Estimate Std. Error z value Pr(>|z|)
(Intercept)   2.183      0.106    20.61   <2e-16

(Dispersion parameter for Negative Binomial(4.0615)
    family taken to be 1)

    Null deviance: 32.49  on 31  degrees of freedom
Residual deviance: 32.49  on 31  degrees of freedom
AIC: 195.06

Number of Fisher Scoring iterations: 1
              Theta:  4.06
          Std. Err.:  1.46

 2 x log-likelihood:  -191.062
```

The disparity for the negative binomial model is 175.39, a reduction of 12.45 compared with the Poisson model. This is very large compared to $\chi^2_{1,0.90} = 2.71$. There is no doubt about the presence of overdispersion. The parameter estimates are slightly affected by the overdispersion modelling, with a lower slope and a higher intercept. The standard error of $\hat{\beta}_1$ is increased by 30%. (Note that we have two standard error estimates here: dropping the roll-length variable and fitting the null negative binomial model gives a disparity change of $191.06 - 175.39 = 15.67$, and the corresponding standard error of $\hat{\beta}_1$ is $0.938/\sqrt{15.67} = 0.237$, slightly larger than the information-based standard error of 0.228. The profile likelihood in β_1 is slightly skewed.)

In Fig. 8.1, we show the fitted Poisson and negative binomial models, together with 2 SD limits around the fitted values. We restrict the vertical scale to non-negative values; the negative values within the variability bounds reflect the asymmetry of the Poisson and negative binomial distributions, and the unsuitability of 2 SD limits for these distributions for small means.

The fitted values agree closely, despite the differences in the parameter estimates, but the variability bounds for the negative binomial are substantially wider than those for the Poisson.

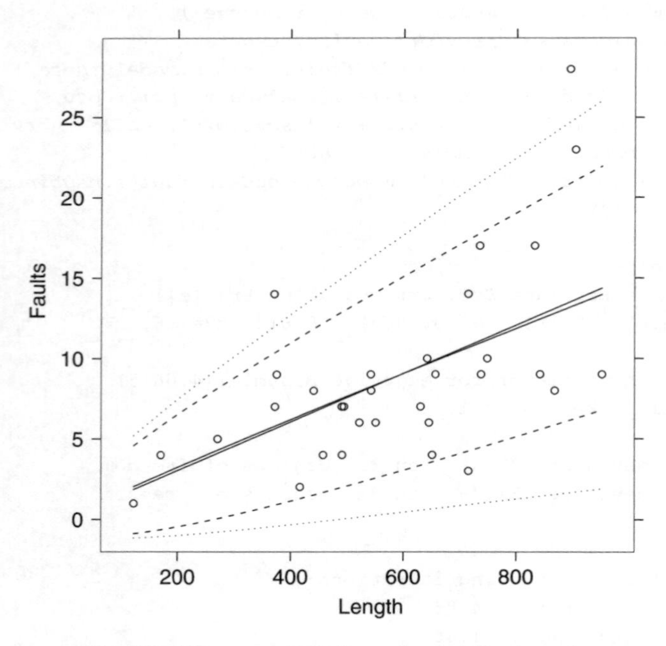

Fig. 8.1. Poisson and NB fits and variability bounds

```
> print(xyplot(pfv + pul + pll + nbfv + nbll + nbul ~
+     1, type = "l", data = faults.newdata, ylim = c(-2,
+     30), lwd = 1.5, xlab = "length", ylab = "faults",
+     lty = c(1, 2, 2, 1, 3, 3), panel = function(...) {
+         panel.xyplot(...)
+         panel.points(faults$l, faults$n)
+     }))
```

8.2.3 *Binomial kernel: beta-binomial distribution*

Consider first the natural parametrization of the binomial distribution in terms of the logit parameter $\theta = \log[p/(1-p)]$. The probability of y successes in n trials is

$$
\begin{aligned}
P(y|n, p) &= \binom{n}{y} p^y (1-p)^{n-y} \\
&= \binom{n}{y} \frac{e^{\theta y}}{(1+e^{\theta})^n}.
\end{aligned}
$$

For a simple null random effect model $\theta = Z$ the conjugate distribution for Z is the logistic, equivalent to the beta distribution on the p scale, giving the marginal *beta-binomial* distribution for Y as

$$
\begin{aligned}
m(y) &= \binom{n}{y} \frac{1}{B(a,b)} \int_{-\infty}^{\infty} \frac{e^{z(y+a)}}{(1+e^z)^{n+a+b}}\, dz \\
&= \binom{n}{y} \frac{B(y+a, n-y+b)}{B(a,b)},
\end{aligned}
$$

a hypergeometric distribution. However for any non-null model $\theta = \eta + Z$, with η the linear predictor and Z having the above logistic distribution, we have

$$
m(y) = \binom{n}{y} \frac{1}{B(a,b)} \int_{-\infty}^{\infty} \frac{e^{(\eta+z)y}}{(1+e^{\eta+z})^n} \cdot \frac{e^{az}}{(1+e^z)^{a+b}}\, dz
$$

and the integral cannot be evaluated explicitly. Thus conjugate random effect modelling is limited to the trivial null model in this distribution.

This difficulty can be circumvented by using a different model for the extra variation (Crowder, 1978). Suppose that $Y_i|P_i \sim b(n_i, P_i)$ conditionally, while $P_i \sim \text{beta}(a_i, b_i)$ marginally, with mean $\mu_i = a_i/(a_i + b_i)$ and variance $\phi_i\mu_i(1-\mu_i)$, where $\phi_i = (a_i + b_i + 1)^{-1}$. Then marginally Y_i has the beta-binomial (hypergeometric) distribution as before, with mean $E[Y_i] = n_i\mu_i$ and variance

$$\begin{aligned} V_i &= n_i^2 \mathrm{Var}[P_i] + \mathrm{E}[n_i P_i (1 - P_i)] \\ &= n_i^2 \phi_i \mu_i (1 - \mu_i) + n_i (1 - \phi_i) \mu_i (1 - \mu_i) \\ &= n_i \mu_i (1 - \mu_i)(1 + (n_i - 1)\phi_i). \end{aligned}$$

Without some restriction on the a_i and b_i, this model is unidentifiable. We therefore set the ϕ_i equal to ϕ. The model with logistic link for μ_i can now be fitted by QL through the mean–variance relationship provided that ϕ can be estimated in some way. ML fitting of the model is however very difficult since the beta distribution parameters appearing in the likelihood have a complex dependence on $\boldsymbol{\beta}$ and ϕ. We do not consider this approach further.

8.2.4 *Gamma kernel*

For a gamma distribution we have

$$f(y|\theta, r) = \frac{\theta^r}{\Gamma(r)} \mathrm{e}^{-\theta y} y^{r-1},$$

where the mean of the distribution is $\mu = r/\theta$. Assuming the random variation is in the mean of the gamma distribution, with the parameter r constant, the conjugate distribution for the canonical parameter θ is again the gamma distribution. For a simple null random effect model $\theta = Z$, we have for the conjugate distribution

$$g(z) = \frac{\phi^s}{\Gamma(s)} \mathrm{e}^{-\phi z} z^{s-1}$$

and the marginal distribution of Y is

$$\begin{aligned} m(y|r, s, \phi) &= \frac{y^{r-1}}{\Gamma(r)} \frac{\phi^s}{\Gamma(s)} \int_0^\infty \mathrm{e}^{-z(\phi+y)} z^{r+s-1} \, \mathrm{d}z \\ &= \frac{1}{y} \frac{\Gamma(r+s)}{\Gamma(r)\Gamma(s)} \left(\frac{y}{\phi + y} \right)^r \left(\frac{\phi}{\phi + y} \right)^s \end{aligned}$$

which is essentially a beta distribution of $y/(\phi + y)$. We do not consider this further.

8.2.5 *Difficulties with the conjugate approach*

The conjugate random effect approach is commonly used with the Poisson distribution: the negative binomial is a popular model in many application areas, for example, consumer purchasing (Chatfield and Goodhardt, 1970). The t-distribution is also used increasingly as a long-tailed distributional model. However, the conjugate approach does not work in binomial models and other approaches are necessary. Because of this lack of generality of the conjugate approach, we turn in the next section to using the normal distribution for the random effect. The resulting normal random effect models are widely used.

8.3 Normal random effects

The non-conjugate normal distribution for Z means that the likelihood of $Y|Z$ cannot be integrated analytically (we exclude the normal likelihood, as discussed in Section 8.2.1). The need for numerical integration of the likelihood by Gaussian quadrature has in the past discouraged the use of normal random effect models, but Bock and Aitkin (1981) and Hinde (1982) showed that the numerical integration can be expressed as a simple finite mixture of the type considered in Section 7.1, and so these models can be fitted straightforwardly using the EM algorithm.

As before, we have an unobserved random effect Z_i for the i-th observation, but now the Z_i are independently normally distributed $N(0, 1)$, and conditional on Z_i, Y_i has a GLM with linear predictor $\eta_i = \boldsymbol{\beta}' x_i + \sigma Z_i$. The likelihood is then

$$L(\boldsymbol{\beta}, \sigma) = \prod_{i=1}^{n} \int f(y_i|\boldsymbol{\beta}, \sigma, z_i)\phi(z_i)\ \mathrm{d}z_i,$$

where $\phi(z)$ is the standard normal density function.

Since the integral is not analytic except for Y normal, we approximate it by Gaussian quadrature: we replace the integral over the normal Z by a finite sum over K Gaussian quadrature mass-points z_k with masses π_k; the z_k and π_k are given in standard references (Abramowitz and Stegun, 1970; Stroud and Sechrest, 1966). The likelihood is then

$$L(\boldsymbol{\beta}, \sigma) \approx \prod_{i=1}^{n} \sum_{k=1}^{K} \pi_k f(y_i|\boldsymbol{\beta}, \sigma, z_k).$$

The likelihood is thus (approximately) the likelihood of a finite mixture of exponential family densities with *known* mixture proportions π_k at *known* mass-points z_k, with the linear predictor for the i-th observation in the k-th mixture component being

$$\eta_{ik} = \boldsymbol{\beta}' x_i + \sigma z_k.$$

We may also regard this as the exact likelihood for this discrete mixing distribution for Z. This is inherently of interest because the NPML estimate of the mixing distribution, when it exists (Chapter 7) is a discrete distribution on a finite number of mass-points. In Section 8.6 we consider the joint estimation of $\boldsymbol{\beta}$, the π_k and the mass-points z_k, but for the moment consider the latter quantities as fixed.

ML in this model follows immediately from the results of Section 7.1, with some simplifications. We identify the ϕ parameter there with $\boldsymbol{\beta}$ and σ, and there are no unknown component-specific parameters θ_k. The mixture proportions π_k are known and do not have to be estimated.

The score component for $\boldsymbol{\beta}$ is

$$\frac{\partial \ell}{\partial \boldsymbol{\beta}} = \sum_i \frac{\sum_k \pi_k f_{ik} \frac{\partial \log f_{ik}}{\partial \boldsymbol{\beta}}}{\sum_\ell \pi_\ell f_{i\ell}} = \sum_i \sum_k w_{ik} \mathbf{s}_{ik}(\boldsymbol{\beta}),$$

where w_{ik} is the posterior probability that observation y_i comes from component k:

$$w_{ik} = \frac{\pi_k f_{ik}}{\sum_\ell \pi_\ell f_{i\ell}}$$

and $\mathbf{s}_{ik}(\boldsymbol{\beta})$ is the $\boldsymbol{\beta}$-component of the score for observation i in component k:

$$\mathbf{s}_{ik}(\boldsymbol{\beta}) = \frac{(y_i - \mu_{ik})}{V_{ik} g'_{ik}} \mathbf{x}_i$$

(Chapter 2). Similarly for σ,

$$s_{ik}(\sigma) = \frac{y_i - \mu_{ik}}{V_{ik} g'_{ik}} z_k.$$

Equating the score to zero gives likelihood equations which are simple weighted sums of those for an ordinary GLM with weights w_{ik}; alternately solving these equations for given weights w_{ik}, and updating these weights from the current parameter estimates, is an EM algorithm. The double summation over i and k is conveniently (if inefficiently) handled in R by expanding the data vectors to length $K \times n$ by replicating y and $\mathbf{x}$ K times, with each block of values corresponding to a specific quadrature point, and in this expanded version each quadrature value is repeated n times (Hinde, 1982; Anderson and Aitkin, 1985). Model fitting is then identical to that of a single sample of size Kn with prior weight vector w. Initial estimates for the first E-step for $\boldsymbol{\beta}$ are conveniently obtained from the ordinary GLM fit, and for σ by arbitrary specification other than zero (e.g. $\sigma = 1$).

This approach was discussed at length in Aitkin (1995), and is implemented in the function `alldist`, mentioned in Chapter 7 for normal mixtures. We give examples of its use below. We note one peculiarity of Gaussian quadrature: the disparity does not reduce monotonically as we increase the number of mixture components K. This is because all the models have the same number of unknown parameters, since the mass-point locations and masses are known. Variations in the disparity are due to the better or worse representation of the extra variation by the discrete 'normal' distributions with different numbers of components. As the number of mass-points increases these variations decrease, but it is not easy to give a value of the disparity for the 'normal' random effect distribution, without using a very large number of mass-points. Crouch and Speigelman (1990) found that for the logistic model with a normal random effect, even 20-point Gaussian quadrature may not give the integral accurately if the variance of Z is large. Lesaffre and Spiessens (2001) found variation in the integral with even larger numbers of mass-points, and recommended adaptive quadrature and routine investigation of this variation with the number of mass-points.

We address this problem below.

8.3.1 *Predicting from the normal random effect model*

While the fitted random effect model correctly assesses the effects of the explanatory variables, it cannot be used *predictively*, to predict the mean response for an individual given the values of the explanatory variables, since the random effect for the individual is unobservable. For this purpose we need the fitted *marginal* model, integrating the *conditional* (on the random effect) model over the random effect.

For the Poisson distribution the mean and variance can be expressed analytically, since

$$\mathrm{E}[Y] = \mathrm{E}[\mathrm{E}[Y \mid Z]] = \mathrm{E}[\mathrm{e}^{\beta'\mathbf{x}+\sigma Z}] = \mathrm{e}^{\beta'\mathbf{x}+\frac{1}{2}\sigma^2},$$

$$\mathrm{Var}[Y] = \mathrm{E}[Y] + (\mathrm{e}^{\sigma^2} - 1)\mathrm{E}^2[Y],$$

but for binomial models the logit and other transformations do not give an analytic mean or variance function.

The fitted mean values can be computed numerically from the conditional model (assuming a normal random effect) and are the π_k-weighted sums of the means of the component distributions. These can be obtained directly from the output of `alldist`.

The variable `ebp` holds the empirical Bayes predictions for each observation – the posterior means of the random effects Z_i. This vector can be graphed against the explanatory variables, but the linearity of the original linear predictor is lost. The variability of the prediction is much harder to assess, except for simple models like the Poissson, and we do not give results here.

We now give several examples of Gaussian quadrature for normal random effect models.

8.4 Gaussian quadrature examples

8.4.1 *Overdispersion model fitting*

To fit an overdispersion model, we use the `alldist` function from the `npmlreg` package used in Chapter 7. We specify the mean model which contains the regression variables to be fitted, and a random model formula which for simple overdispersion models contains just the intercept (we consider more general models in Section 8.8). The number of components K is defined using the argument `k`.

Gaussian quadrature is carried out using the function `alldist` and defining the argument `random.distribution` equal to the string `gq`.

8.4.2 *Poisson – the fabric fault data*

We have already seen the negative binomial analysis of the fabric fault data. We extend this analysis with the Poisson/normal model (Hinde, 1982; Aitkin, 1996*b*). The results are summarized in Table 8.3. Standard errors for $\hat{\beta}_1$ are given beside the estimates. They are calculated as described in Chapter 7, by omitting the roll-length variable, refitting the model and equating the disparity change to the squared Wald statistic.

```
> library(npmlreg)
> faults.z10 <- alldist(n ~ ll, random = ~1, data = faults,
+      family = poisson, random.distribution = "gq",
+      k = 10, verbose = FALSE, plot.opt = 0)
> faults.z20 <- update(faults.z10, k = 20)
```

Table 8.3 also gives the negative binomial estimates, and Nelder's (1985) QL estimates and standard errors for these parameters, based on the variance/mean relation $V(\mu) = \mu + \phi\mu^2$, where $\phi = \sigma^2$ is estimated from the mean residual deviance. (Note that for the normal random effect model, ϕ, the coefficient of variation of $W = e^Z$, is $e^{\sigma^2} - 1 = 0.124$ for 20-point quadrature compared to $\phi = 0.36^2 = 0.13$ for the QL method.)

The normal mass-point models show reasonable consistency in the regression slope estimates, apart from that for $K = 3$, which is somewhat smaller than for

Table 8.3. Estimates and disparities for fabric fault data

K	β_0	β_1	SE(β_1)	σ	Disparity
1	−4.173	0.997	0.176	NA	187.84
2	−4.413	1.033	0.206	0.339	175.58
3	−3.309	0.849	0.234	0.357	174.26
4	−3.769	0.924	0.238	0.322	175.21
5	−3.833	0.935	0.239	0.351	175.12
6	−3.748	0.921	0.236	0.344	174.91
10	−3.771	0.925	0.241	0.341	175.04
20	−3.765	0.924	0.240	0.342	175.02
NB	−3.795	0.938	0.228	0.340	175.39
QL	−3.86	0.94	0.23	0.36	

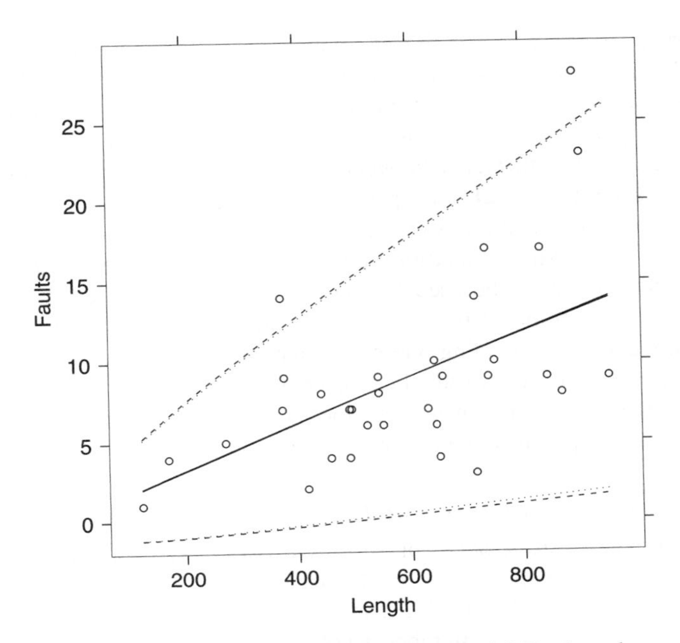

Fig. 8.2. Poisson/normal and NB fits and variability bounds

other values of K. They appear to be converging at $K = 20$. The disparity is not monotone in K (as we noted above, it need not be). The disparities for the normal and negative binomial models are very close in this small sample: we cannot distinguish here the normal distribution from the log-gamma for the unobserved variation.

In Fig. 8.2, we graph the fitted values and 2 SD limits for the negative binomial and Poisson/normal ($K = 20$) compound models. They are almost identical.

```
> faults.newdata$pnfv <- predict(faults.z20, new = faults.newdata,
+     type = "response")
> faults.newdata <- transform(faults.newdata, pnsd = sqrt(pnfv +
+     (exp(0.342^2) - 1) * pnfv^2))
> faults.newdata <- transform(faults.newdata, pnul = pnfv +
+     2 * pnsd, pnll = pnfv - 2 * pnsd)
> print(xyplot(pnfv + pnul + pnll + nbfv + nbul + nbll ~
+     l, type = "l", data = faults.newdata, ylim = c(-2,
+     30), lwd = 1.5, xlab = "length", ylab = "faults",
+     lty = c(1, 2, 2, 1, 3, 3), panel = function(...) {
+         panel.xyplot(...)
+         panel.points(faults$l, faults$n)
+     }))
```

The residual deviance for the compound normal model (compared to the saturated Poisson model) is 51.72, substantially above the 'degrees of freedom' of 29. This result has often caused confusion, and has been interpreted to mean that the compound model also does not 'fit' the data. This misunderstanding arises because of the difference between one- and two-parameter families. In the former the 'saturated model' has a parameter for every observation, and provides a 'baseline' against which we assess the current model. In two-parameter families, however (with an extra variability parameter), the model with a parameter for each observation cannot be used to reproduce the data, as the estimate of the scale parameter becomes zero and the model degenerates. So in two-parameter families obtained by extending one-parameter families by overdispersion, there is no interpretation of the 'residual deviance' relative to the saturated model as a 'goodness-of-fit' test of the overdispersion model – only the change in deviance is of inferential importance, representing the importance of the extra variation.

8.4.3 *Binomial – the toxoplasmosis data*

We consider the overdispersed logistic regression model for incidence of the parasitic condition toxoplasmosis in 34 cities of El Salvador, discussed by Efron (1986) and in the GLIM4 manual (Francis *et al.*, 1993), and re-analysed in Aitkin (1996*b*). Table 8.4 shows the number (n) of men tested and the number (r) with a positive test for toxoplasmosis in each city, together with the annual rainfall (x) in metres in the city. These data are in the dataset `toxoplas`.

We fit polynomial models in rainfall to the logit of the proportion with toxoplasmosis. We suppress the output.

```
> data(toxoplas, package = "SMIR")

> toxoplas <- transform(toxoplas, x2 = x^2, x3 = x^3)
```

Table 8.4. Toxoplasmosis data

r	n	x	r	n	x	r	n	x
2	4	1.735	3	10	1.936	1	5	2.000
2	2	1.750	3	5	1.800	2	8	1.750
7	19	2.077	3	6	1.920	8	10	1.800
7	24	2.050	0	1	1.830	15	30	1.650
4	22	2.200	0	1	2.000	6	11	1.770
0	1	1.920	33	54	1.770	4	9	2.240
5	18	1.620	0	1	1.650	8	11	2.250
41	77	1.796	24	51	1.890	7	16	1.871
46	82	2.063	9	13	2.100	23	43	1.918
53	75	1.834	8	13	1.780	3	10	1.900
1	6	1.976	23	37	2.292			

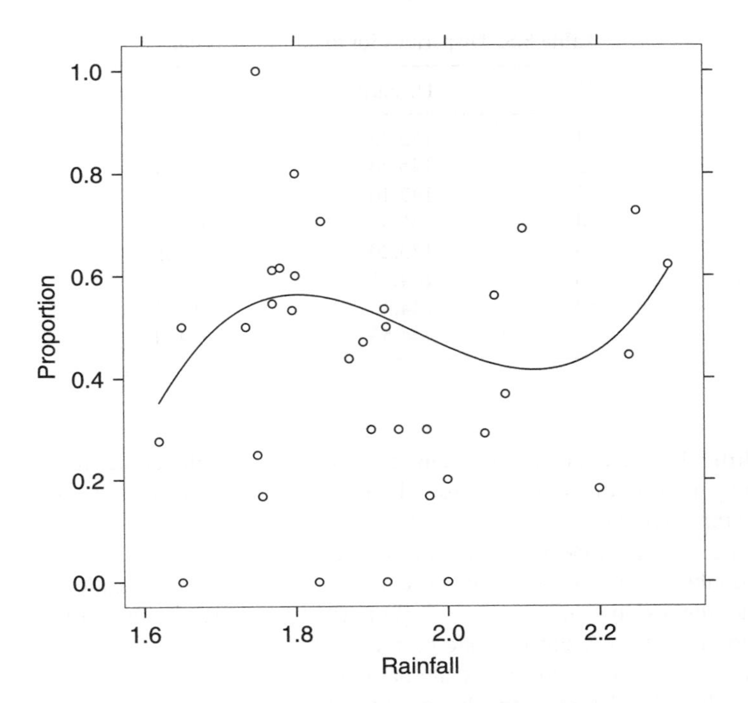

Fig. 8.3. Observed proportions and fitted cubic

```
> toxoplas.glm <- glm(cbind(y, (n - y)) ~ x + x2 +
+     x3, data = toxoplas, family = binomial)
```

The regression shows a strong cubic trend, with disparities of 164.90, 164.78, 164.78, and 153.33 for the null, linear, quadratic, and cubic models, respectively. We graph the fitted cubic and the data on the proportion scale in Fig. 8.3.

```
> toxoplas.newdata <- data.frame(x = seq(min(toxoplas$x),
+     max(toxoplas$x), len = 101))
> toxoplas.newdata <- transform(toxoplas.newdata, x2 = x^2,
+     x3 = x^3)
> toxoplas.newdata$fp <- predict(toxoplas.glm,
+     new = toxoplas.newdata, type = "response")
> print(xyplot(fp ~ x, data = toxoplas.newdata, type = "l",
+     ylim = c(-0.05, 1.05), xlab = "rainfall",
+     ylab = "proportion", panel = function(...) {
+         panel.xyplot(...)
+         op <- toxoplas$y/toxoplas$n
+         panel.points(toxoplas$x, op)
+     }))
```

Table 8.5. Disparities for toxoplasmosis data

K	Disparity	σ
1	153.33	NA
2	145.74	0.361
3	142.16	0.479
4	145.84	0.339
5	143.20	0.509
6	144.81	0.386
10	144.49	0.427
20	144.53	0.424

It is difficult to understand the scientific significance of the cubic model. We do not show the variability bounds around the fitted values as these vary considerably with sample size and give a very complex picture.

Overdispersion appears substantial; the residual deviance of 62.64 with 30 df casts doubt on the existence of the cubic trend. (This interpretation implicitly assumes the residual deviance has a χ^2_{30} distribution, which is not correct, as noted in Section 2.10.4.) A simple scale correction for the overdispersion (McCullagh and Nelder, 1989) would set a scale parameter ϕ equal to the mean deviance of 2.09, scaling the change in the deviance for the cubic term to 5.48.

We give in Table 8.5 the disparities for two to six and 10 and 20 mass-points for the cubic model using Gaussian quadrature. The disparities show substantial changes between odd and even numbers of mass-points when K is small. In all cases the reduction in disparity relative to the binomial model is substantial, between 9 and 12. This greatly exceeds the 10% point of χ^2_1: there is real extra variation. Refitting successively the null, linear and quadratic models with overdispersion using 20-point quadrature gives disparities of 150.45, 150.45, and 150.28.

The disparity change for the cubic term relative to the quadratic is still rather large (5.74 on 1 df), but the disparity change for the cubic relative to the null model is almost the same (5.92 on 3 df). A plausible interpretation is that the cubic model is unnecessary and that there is no strong evidence of *any* trend with rainfall, but strong evidence of variation with other unmeasured factors.

Efron (1986) used the *double exponential family* model to analyse these data; this introduced an extra 'variability' parameter into the generalized binomial distribution, to unlink the mean and variance of the distribution. However, the normalizing constant in the resulting distribution is not available analytically, and this affects the parameter estimation. A discussion of this problem in the Poisson distribution was given in Aitkin (1995).

8.5 Other specified random effect distributions

We may use the Gaussian quadrature approach to fit other specified random effect distributions with very small changes in the procedure. Suppose we wish to use a specific distribution $g(z)$ for Z, for example, a t_r-distribution for specified r. One simple approach is still to use the Gaussian mass-point locations z_k (rather than the mass-point locations for the t_r distribution) and to scale the masses π_k by the density ratio $t_r(z_k)/\phi(z_k)$. This is closely related to *importance sampling* (Tanner, 1996). The resulting masses are not, however, *optimal* for the t_r-distribution because the mass-points z_k are optimized for Gaussian quadrature assuming the normal distribution.

8.6 Arbitrary random effects

We now consider the most general form of random effect model, in which we make *no* assumption about the distribution of the random effects. It may seem impossible to proceed if we know nothing about this distribution, but as we signalled in Section 7.5, the identification of general mixtures provides a very general framework for random effect models (Clayton and Kaldor, 1987).

Consider again the general overdispersion model of Section 8.1, in which a random effect Z_i with a distribution $g(z)$ is added to the linear predictor $\boldsymbol{\beta}'\mathbf{x}_i$ for the i-th observation. We observe n observations $y_1, \ldots, y_n$ from the marginal or compound distribution

$$m(y) = \int f(y|z)g(z)\ \mathrm{d}z.$$

Now we make no assumption about the form of $g(z)$, but instead regard it as another unknown, which we wish to estimate together with $\boldsymbol{\beta}$. The likelihood is

$$L(\boldsymbol{\beta}, g(z)) = \prod_{i=1}^{n} \int f(y_i|z_i, \boldsymbol{\beta})g(z_i)\ \mathrm{d}z_i,$$

where the linear predictor is

$$\eta_i = \boldsymbol{\beta}'\mathbf{x}_i + z_i.$$

As discussed in Chapter 7, the NPML estimate of $g(z)$, when it exists, is known (Laird, 1978; Lindsay, 1983, 1995) to be a discrete distribution on a finite number $\hat{K}$ of mass-points $\hat{z}_k$, with masses $\hat{\pi}_k$. Thus the *profile* likelihood in $\boldsymbol{\beta}$, maximized over $g(z)$, is the $\hat{K}$-component finite mixture likelihood

$$P(\boldsymbol{\beta}) = \prod_{i=1}^{n} \left\{ \sum_{k=1}^{\hat{K}} f(y_i | \hat{z}_k, \boldsymbol{\beta}) \hat{\pi}_k \right\},$$

where $\hat{K}$, $\hat{z}_k$, and $\hat{\pi}_k$ are functions of $\boldsymbol{\beta}$.

In order to maximize this profile likelihood, we can reformulate the problem as the maximization of the *joint* likelihood

$$L(\boldsymbol{\beta}, K, \pi_1, \ldots, \pi_{K-1}, z_1, \ldots z_K) = \prod_{i=1}^{n} \left\{ \sum_{k=1}^{K} f(y_i | z_k, \boldsymbol{\beta}) \pi_k \right\}$$

over $\boldsymbol{\beta}$ and all of the parameters of the mixture. The number of components K is unknown, so the likelihood has to be maximized over this as well. This maximization problem is exactly that for the general finite mixture, discussed in Chapter 7. So we have a very general solution to the problem of modelling overdispersion, without any parametric model assumption being required for the random effects.

The estimated components in the mixture may have a direct interpretation. A simple example is the existence of *outliers*: an observation remote from the main body may be split off, and identified as a single component with mass $\hat{\pi}_k = 1/n$. Less commonly, the components may correspond to actual *latent classes* of observations, corresponding to an omitted factor. The posterior probabilities of component membership are frequently informative: outliers can be immediately identified as described above, and latent classes can be identified by a high proportion of 1s and 0s. In general, however, no substantive meaning can be attached to the components – they merely represent in a discrete form the variation which we usually think of as continuous.

Testing for overdispersion in the more general mixture model is greatly complicated by the estimation of the mixing distribution. As we noted in Chapter 7, there is no satisfactory general theory for the distribution of the LRTS in this model, and for semi-formal tests we need to resort to bootstrapping. It is clear that the distribution of the LRTS must be stochastically larger than it is in the normal random effect model case, and so if the disparity change is not significant by this test, it will certainly not be so by a formal test. This result is, however, not very helpful in assessing the reality of larger disparity changes.

We now re-examine the previous examples for which we made normal random effect model assumptions, using the NPML approach. For this purpose we use the function `alldist`, previously used for normal random effect modelling.

The posterior probabilities are held in the variable `post.prob` inside the `glmmGQ` object generated by `alldist`.

8.7 Examples

8.7.1 *The fabric fault data*

We fit the model with one, two and three mass-points:

```
> library(npmlreg)
> (faults.np1 <- alldist(n ~ ll, data = faults, family = poisson,
+      random.distribution = "np", k = 1, plot.opt = 0,
+      verbose = FALSE))

Coefficients:
 MASS1       ll
-4.173    0.997

Random effect distribution - standard deviation:    0

Mixture proportions:
MASS1
    1
-2 log L:        187.8

> (faults.np2 <- update(faults.np1, k = 2))

Coefficients:
     ll     MASS1     MASS2
 0.805    -3.165    -2.402

Random effect distribution - standard deviation:    0.308

Mixture proportions:
    MASS1       MASS2
    0.7940      0.2060
-2 log L:        172.7

> (faults.np3 <- update(faults.np1, k = 3))

Coefficients:
    ll    MASS1    MASS2    MASS3
 0.798   -3.154   -3.114   -2.353

Random effect distribution - standard deviation:    0.308

Mixture proportions:
    MASS1       MASS2       MASS3
    0.1319      0.6667      0.2014
-2 log L:        172.7

> faults.gq20 <- update(faults.np1, k = 20,
+     random.distribution = "gq")
```

Further increases in K leave the disparity essentially unaltered, with replication of the mass-points and split masses. The NPML estimate is a two-point distribution, a somewhat surprising result. The disparity is 172.66 compared to 175.02 for 20-point Gaussian quadrature. For comparability with the results from Gaussian quadrature we rescale the mass-point locations to give the mixing distribution mean zero; the resulting mass-points are -0.157 with mass 0.794 and 0.606 with mass 0.206. The standard deviation of the estimated mixing distribution is 0.309, slightly less than that for the normal models. With the rescaling of the mixture distribution the estimate of the intercept parameter is $\hat{\beta}_0 = -2.402 - 0.606 = -3.008$. The slope estimate, $\hat{\beta}_1 = 0.804$, is substantially below those for the simple Poisson and Poisson/normal models, and the negative binomial estimate. As noted earlier we can obtain a standard error for $\hat{\beta}_1$ by omitting the term and equating the disparity change $(188.46 - 172.66)$ to the squared Wald statistic, giving a value of 0.202. We note that the slope is still quite consistent with a true slope of 1. Fitted values and 2 SD bounds for the Poisson/non-parametric mixture may be calculated as for the Poisson/normal, with the slight change that the value of ϕ has to be computed directly from the moment generating function of the two-point mixing distribution (from Section 8.1 $\phi = [M(2)/M^2(1)] - 1$):

```
> M1 <- 0.794 * exp(-0.157) + 0.206 * exp(0.606)
> M2 <- 0.794 * exp(2 * -0.157) + 0.206 * exp(2 * 0.606)
> phi <- M2/M1^2 - 1
> faults.newdata <- transform(faults.newdata, npfv = exp(0.80446 *
+       ll + 0.794 * (-3.1645) + 0.206 * (-2.4017)))
> faults.newdata <- transform(faults.newdata, pnpsd = sqrt(npfv +
+       phi * npfv^2))
> faults.newdata <- transform(faults.newdata, pnpul = npfv +
+       2 * pnpsd, pnpll = npfv - 2 * pnpsd)
> print(xyplot(npfv + pnpul + pnpll + nbfv + nbll +
+       nbul ~ ll, type = "l", data = faults.newdata,
+       ylim = c(-2, 30), lwd = 1.5, xlab = "length",
+       ylab = "faults", lty = c(1, 2, 2, 1, 3, 3),
+       panel = function(...) {
+           panel.xyplot(...)
+           panel.points(faults$l, faults$n)
+       }))
```

The fitted values and 2 SD bounds are shown in Fig. 8.4 for the two-point mixture model and the negative binomial model.

Despite the substantial differences in the estimates, the fitted values are not very different, but the two-component mixture bounds (dot-dashed curves) are slightly narrower than those for the negative binomial (dotted curves), reflecting the smaller standard deviation of the two-point distribution.

In Fig. 8.5, the posterior probabilities of membership in component 2 are plotted as a function of the standardized residuals from the Poisson fit.

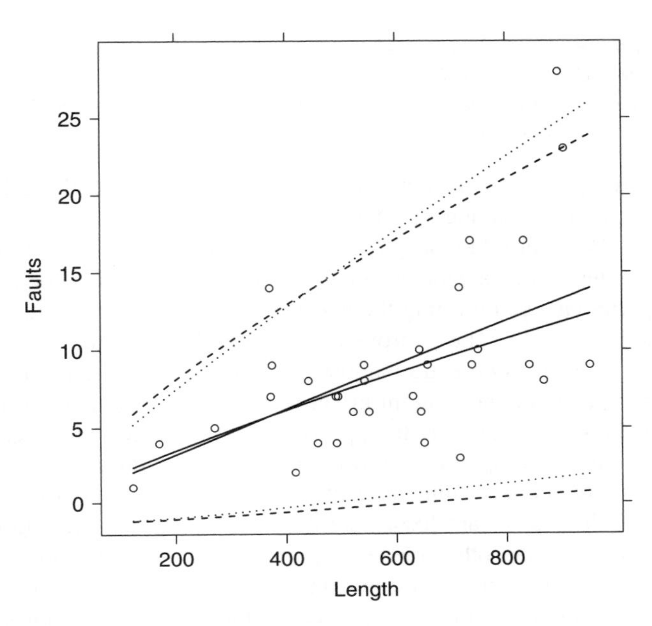

Fig. 8.4. Poisson/non-parametric mixture and NB fits and variability bounds

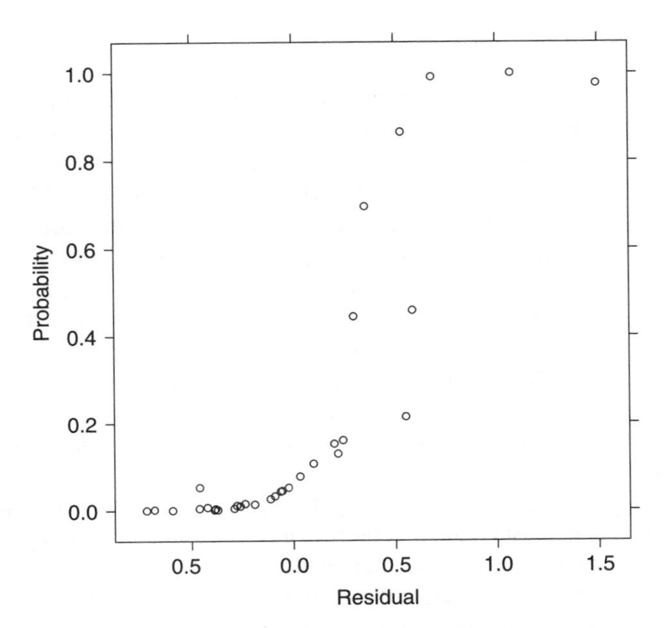

Fig. 8.5. Posterior probability against residual

```
> fabric.pp <- faults.np2$post.prob[, 2]
> fabric.resid <- resid(faults.np1)
> print(xyplot(fabric.pp ~ fabric.resid, xlab = "residual",
+      ylab = "probability"))
```

The posterior probability increases almost monotonically with the residual, with only five observations having a posterior probability of more than 0.5 of being in component 2. We referred to this possibility in Section 8.1: if Z is correlated with $\mathbf{x}$, then fitting the overdispersion model will change the regression coefficient on $\mathbf{x}$.

The disparity change on fitting the general mixture compared with the simple Poisson model is 15.17. The improvement in disparity over the normal random effect model (2.36 for two extra parameters) is quite small, suggesting that the normal assumption for the random effect is reasonable. Comparing the NPML fit with the Gaussian quadrature fit appears to raise a difficult model comparison problem, as we are comparing a normal model for Z with an unspecified one. However, we are actually comparing two discrete distributions for the random effect; in one the masses are fixed, and the mass-points are fixed up to a scale parameter, while in the other the mass-points and masses are both estimated. Since the parameter estimates are not on a boundary, we can invoke standard asymptotic theory to compare the disparities for the two models by the usual LRT. The difference of 2.36 is far from significant compared with χ^2_2, and we conclude that the normal model (*or* the negative binomial) is quite satisfactory.

8.7.2 *The toxoplasmosis data*

Refitting the cubic model with a non-parametric model, the NPML estimate is a three-point distribution which is close to the mass-points and masses for three-point Gaussian quadrature. Mass-points (centered at 0) for the three-point estimate are 0.796, -0.022, and -1.106 with masses of 0.109, 0.720, and 0.171, respectively. Disparities for two and three mass-point models are 141.84 and 141.84 (compared with 142.16 for three-point Gaussian quadrature). There is again a substantial reduction in the significance of the cubic term, with the disparities for the null, linear, quadratic and cubic models 146.87, 146.62, 146.16, and 141.84, respectively. A plot (not shown) of the posterior probabilities of component membership against the residuals from the simple logit model shows that the large positive residuals are in the first component, the large negative residuals in the third, and the intermediate residuals (between -1.5 and $+1$) are predominantly in the second.

As for the normal random effect analysis, a plausible interpretation is that the cubic model is unnecessary and that there is no strong evidence of *any* trend with rainfall, but strong evidence of variation with other unmeasured factors.

The disparity improvement from three-point Gaussian quadrature is very small: the estimated mixing distribution is almost exactly the same as that for the quadrature, apart from the scale factor.

8.7.3 *Leukaemia remission data*

We consider briefly the exponential survival example of Feigl and Zelen (1965) discussed in Chapter 6. Recall that the exponential model `ag + lwbc` with a log link gave a deviance of 40.32 on 30 df. Weibull and gamma models gave deviances and parameter estimates very close to the exponential values.

For any number of mass-points greater than 2, Gaussian quadrature gives the same disparity of 293.04. The three-point NPML estimate gives a disparity of 287.56.

8.7.4 *The Brownlee stack-loss data*

We conclude with a re-analysis of the stack-loss data. As noted by many critics of outlier analyses of these data, the data do not come from a randomized experiment and so conventional modelling approaches are suspect, but we present here for interest the results of the finite mixture estimation. Aitkin and Wilson (1980) presented a detailed investigation of normal mixture models for outliers, with suspected outliers assigned to specific mixture components. These components had their own means, unrelated to the regression structure modelled for the 'good' observations. Our approach here (following Aitkin, 1996*b*) is different since the 'outliers' still contribute to the estimation of the regression model parameters.

For Gaussian and gamma mixtures the R implementation (Einbeck *et al.*, 2006) uses by default a damping procedure in the first cycles of the EM algorithm (Einbeck and Hinde, 2006), which stabilizes the algorithm and makes it less sensitive to the optimal choice of `tol`.

For the two-component model the best solution gives a disparity of 100.39; parameter estimates are shown for this model, and for three- and four-component models, in Table 8.6. We abbreviate `MASS` to `M`. The mixture models are fitted without intercepts.

Posterior probabilities of component membership are given for the same three models in Table 8.7. The two-component model splits the observations into two groups, with each observation having posterior probability of at least 0.907 of falling in one component. The three-component model isolates observations 1, 3, and 4 in the third component, and observation 21 in the first. (Observations 1, 3, 4, and 21 are those repeatedly identified as outliers in previous analyses.)

Table 8.6. Mixture fits for stack-loss data

K	x_1	x_2	x_3	σ	M_(1)	M_(2)	M_(3)	M_(4)	Disparity
1	0.716	1.295	−0.152	3.24	−39.9				105.01
2	0.605	1.829	−0.280	1.38	−36.2	−30.5			100.39
3	0.765	0.619	−0.060	1.03	−45.5	−37.2	−30.0		85.72
4	0.874	0.912	0.033	0.51	−67.4	−59.5	−55.5	−50.9	72.96

Table 8.7. Posterior probabilities of component membership

Obs.	K								
	2		3			4			
	1	2	1	2	3	1	2	3	4
1	0.001	0.999	0	0	1	0	0	1	0
2	1.000	0.000	0	1	0	0	1	0	0
3	0.000	1.000	0	0	1	0	0	1	0
4	0.000	1.000	0	0	1	0	0	0	1
5	0.998	0.002	0	1	0	0	1	0	0
6	1.000	0.000	0	1	0	0	1	0	0
7	1.000	0.000	0	1	0	0	1	0	0
8	0.998	0.002	0	1	0	0	1	0	0
9	1.000	0.000	0	1	0	0	1	0	0
10	0.007	0.993	0	1	0	0	0	1	0
11	0.000	1.000	0	1	0	0	0	1	0
12	0.000	1.000	0	1	0	0	0	1	0
13	0.907	0.093	0	1	0	0	1	0	0
14	0.012	0.988	0	1	0	0	1	0	0
15	0.000	1.000	0	1	0	0	0	1	0
16	0.028	0.972	0	1	0	0	0	1	0
17	1.000	0.000	0	1	0	0	0	1	0
18	0.991	0.009	0	1	0	0	0	1	0
19	0.998	0.002	0	1	0	0	0	1	0
20	0.089	0.911	0	1	0	0	0	1	0
21	1.000	0.000	1	0	0	1	0	0	0

For the four-component model the MLE of the residual SD is 0.5, half a measurement unit. Aitkin and Wilson (1980) noted a similar phenomenon in their mixture modelling: with four components the residual standard deviation was so small that the normal density representation of the likelihood breaks down unless measurements can be recorded to another decimal place. This model splits the observations into four groups, each observation having posterior probability at least 0.999 of falling in one component. Observation 4 falls in the fourth component, and 21 in the first. The other observations are split in several long sequences between the second and third components. This result strongly suggests a real latent class model, with a 'missing factor' from the regression model. The appearance of the successive observations 5–9 in the second component and observations 15–20 in the third component suggests a five-day week effect.

We define a `week` factor with five levels: week 1 contains observations 1–4, week 2 observations 5–9, week 3 observations 10–14, week 4 observations 15–19, and week 5 observations 20 and 21. Adding `week` to the three-variable normal model with no mixture structure gives a disparity of 76.28 which is close to that for the four-component mixture model.

```
> stackloss <- transform(stackloss, week = factor(cut(1:21,
+     breaks = c(0, 4, 9, 14, 19, 21))))
> library(npmlreg)
> stackloss.week.glm <- glm(y ~ x1 + x2 + x3 + week,
+     data = stackloss)
> summary(stackloss.week.glm)

Coefficients:
              Estimate Std. Error t value Pr(>|t|)
(Intercept) -12.006      15.340    -0.783 0.4480
x1            0.371       0.128     2.892 0.0126
x2            0.884       0.592     1.493 0.1592
x3           -0.026       0.102    -0.256 0.8022
week(4,9]   -10.881       2.010    -5.414 0.0001
week(9,14]  -10.375       4.095    -2.533 0.0250
week(14,19] -13.050       3.806    -3.429 0.0045
week(19,21] -11.795       3.335    -3.536 0.0037

(Dispersion parameter for gaussian family taken to be 3.575)

    Null deviance: 2069.238  on 20  degrees of freedom
Residual deviance:   46.479  on 13  degrees of freedom
AIC: 94.28

Number of Fisher Scoring iterations: 2

> stackloss.glm <- glm(y ~ x1 + x2 + x3, data = stackloss)
> round(-2 * as.numeric(logLik(stackloss.glm)), 2)

[1] 104.58

> round(-2 * as.numeric(logLik(stackloss.week.glm)),
+     2)

[1] 76.28
```

The disparity change between the normal regression model and that including `week` is 28.30 on 4 df. Equivalently, the F-statistic for the contribution of `week` is 9.25 for $F_{4,13}$; these are both significant beyond the 0.1% level. There is no question of the importance of this factor.

The importance of x_2 has decreased in this model and it appears that airflow x_1 is the only important variable when the weekly time sequence of the data is taken into account. The weeks after week 1 (which contain the 'outlier' observations 1,

3, and 4) show a dramatically lower stack-loss, with week 4 lower than weeks 2, 3, and 5.

'Week' however is not an explanatory variable in any real sense: the need for this factor in the model is a strong indication that there are other very important omitted process variables which apparently changed their values in successive operating weeks, as did the variables x_1 and x_2: airflow was reduced in the second week, reduced again, with water temperature, in the second, further reduced in the fourth, and increased in the fifth.

This analysis was repeated with six- and seven-day weeks, and for five-day weeks starting with different observations in the sequence, with totally negative results: there is no evidence at all for a week effect longer than five days, or of the sequence starting at a different day.

We conclude that the persistent identification of outliers in these data may be a result of model mis-specification: the data are consistent with consecutive daily observations of a five-day working week process, in which changes in the process were made at the ends of some working weeks. These changes were apparently in addition to the substantial changes in the level of airflow and water temperature between weeks. When the week is included as a factor in the model, the evidence for a mixture disappears.

The detailed discussion of the data in Dodge (1999) appears to bear out the near-daily time sequencing of the observations, though no full list of the dates of the observations appears to exist.

We now consider more general versions of random effect models, in which the model regression coefficients are random, as well as the intercept.

8.8 Random coefficient regression models

The overdispersion model can be interpreted as a *random intercept* model. The random effect Z_i can be combined with the model intercept term β_0 to give a random intercept B_{0i}, with mean β_0 and variance σ^2, the variance of Z.

This formulation of the random effect model can be extended to more general *random coefficient models*. Consider a simple example with a single variable x_{1i} whose coefficient β_1 varies across the data, in addition to the intercept. We represent it by $B_{1i} = \beta_1 + U_i$, where U_i corresponds to variation in the regression coefficient about a 'mean' β_1. The remaining regression coefficients $\boldsymbol{\beta}_2$ are fixed. Then conditional on U_i *and* Z_i, the regression model is

$$\eta_i = \beta_0 + \beta_1 x_{1i} + \boldsymbol{\beta}_2' \mathbf{x}_{2i} + Z_i + U_i x_{1i},$$

while Z_i and U_i have some joint distribution $g(z, u)$. The likelihood is then

$$L(\boldsymbol{\beta}) = \prod_i \int \int f(y_i \mid z, u) g(z, u) \, dz \, du.$$

If the distribution $g(z, u)$ is assumed Gaussian with unknown covariance matrix, we need to numerically integrate over both parameters. This can be achieved by two independent single parameter numerical integrations by using parameter orthogonalization. We rewrite the model as

$$\eta_i = \beta_0 + \beta_1 x_{1i} + \boldsymbol{\beta}_2' \mathbf{x}_{2i} + \sigma_Z Z_i + \sigma_U U_i x_{1i},$$

where Z_i and U_i are bivariate normal with means zero, variances 1 and $\text{corr}(Z_i, U_i) = \rho$. The distribution of (U, Z) can be re-expressed as the independent $N(0, 1)$ distributions of U and Z^*, the residual from the regression of Z on U:

$$Z^* = \frac{Z - \rho U}{\sqrt{1 - \rho^2}}.$$

The model above can then be re-expressed as

$$\begin{aligned} \eta_i &= \beta_0 + \beta_1 x_{1i} + \boldsymbol{\beta}_2' \mathbf{x}_{2i} + \sigma_Z \left(\rho U_i + \sqrt{1 - \rho^2} Z_i^* \right) + \sigma_U U_i x_{1i} \\ &= \beta_0 + \beta_1 x_{1i} + \boldsymbol{\beta}_2' \mathbf{x}_{2i} + \lambda_0 Z_i^* + (\lambda_1 + \lambda_2 x_{1i}) U_i. \end{aligned}$$

Approximating the double integral by two finite sums over the independent Z_i^* and U_i, we have

$$L(\boldsymbol{\beta}) \approx \prod_i \sum_k \sum_\ell f(y_i \mid z_k, u_\ell) \pi_k \pi_\ell$$

with linear predictor

$$\eta_{ik\ell} = \beta_0 + \beta_1 x_{1i} + \boldsymbol{\beta}_2' \mathbf{x}_{2i} + \lambda_0 z_k + (\lambda_1 + \lambda_2 x_{1i}) u_\ell.$$

The unknown variances and covariance of Z and U are replaced by the three regression coefficients of z_k, u_ℓ and the interaction $x_{1i} u_\ell$.

The double integration still requires K^2 terms in the summation, and with R random parameters we would need K^R terms, making this approach impractical for a large number of random parameters. We have not implemented Gaussian quadrature for random coefficient models, only for random intercept models.

However, when we estimate the joint distribution of both Z_i and U_i non-parametrically, we again obtain the NPML estimate of $g(z, u)$ as a discrete distribution on a finite number $\hat{K}$ of points in the (z, u) plane, with an estimated mass $\hat{\pi}_k$ and estimated mass-points $(\hat{z}_k, \hat{u}_k)$ for the k-th component. The likelihood maximized over $g(z, u)$ can again be expressed as a finite mixture, of regressions with different intercept and slope for x_1 in each component of the mixture, and common slopes for $\mathbf{x}_2$.

In the R implementation of this model, the components are indexed by the 'component factor'. The random coefficients can be handled by including in the regression model, both the main effect of the factor z_k, and the interaction of this factor with the explanatory variable x_1. Again the number of mass-points has to be determined by sequential increase from 1, and, in general, more mass-points may be required than for the case of fixed β_1. This process is quite general, and is invoked simply by specifying in the `random` formula the terms which are to have random regression coefficients (the intercept term is always included). We stress that this approach is valid only for NPML estimation (i.e. with `random.distribution="np"`): if Gaussian quadrature is used instead with a random coefficient model, the resulting analysis is *not* a full Gaussian quadrature over all the random parameters.

We now illustrate this extension with an example.

8.8.1 *Example – the fabric fault data*

We extend the model to a random linear regression on log length:

```
> library(npmlreg)
> (faults.rc2 <- alldist(n ~ ll, data = faults, family = poisson,
+     random = ~ll, random.distribution = "np", plot.opt = 0,
+     verbose = FALSE, k = 2))

Coefficients:
     ll      MASS1      MASS2  MASS1:ll
  1.126     -2.108     -4.571    -0.491

Random effect distribution - standard deviation:   1.094609

Mixture proportions:
    MASS1      MASS2
0.7291639  0.2708361
-2 log L:     171.6

> (faults.rc3 <- update(faults.rc2, k = 3, tol = 1))
Coefficients:
     ll     MASS1      MASS2      MASS3   MASS1:ll   MASS2:ll
  2.196    -0.829    -13.554    -10.377     -1.755      0.262

Random effect distribution - standard deviation:   5.9551

Mixture proportions:
     MASS1       MASS2       MASS3
    0.6551      0.3188      0.0260
-2 log L:       169.4
```

```
> (faults.rc4 <- update(faults.rc2, k = 4, tol = 0.3))

Coefficients:
       ll        MASS1        MASS2        MASS3        MASS4
   0.531       -1.279       -1.199      -13.771       -0.609
 MASS1:ll    MASS2:ll     MASS3:ll
 -0.0216      -0.033        1.961

Random effect distribution - standard deviation:    5.664

Mixture proportions:
     MASS1        MASS2        MASS3        MASS4
     0.0456       0.6219       0.2800       0.0524
-2 log L:        168.7
```

Identification of the NPML estimate for different K requires some searching over values of `tol`. For $K = 2$ or 3 the disparity is 171.58 with the default setting of `tol`$= 0.5$, while for $K = 3$ it is 169.37 with `tol` $= 1$ and for $K = 4$ it is 168.73 with `tol` $= 0.3$, though the first and second components are essentially the same. For $K > 4$ the disparity stabilizes at 168.73. With this small sample and complex model, several local maxima appear.

The change in disparity of 3.30 relative to the simple overdispersion model is quite small; if this change were for a *fixed* distribution of (Z, U) it would barely be significant compared to $\chi^2_{1,0.90}$. There is no real evidence of non-constant regression slope.

8.9 Algorithms for mixture fitting

We use the EM algorithm for fitting mixture models. It has the great virtues of relatively easy programming and guaranteed convergence to at least a local maximum, but locating the global maximum may require extensive searching over starting points for the algorithm. Other algorithms are available which are able to locate where additional mass-points should be placed to increase the likelihood, given the current mass-point configuration. These are *vertex* algorithms based on optimal design theory, and are discussed in Lindsay (1995); Böhning (1999), among others.

These approaches do not necessarily locate the same maximum, as illustrated in the following example.

8.9.1 *The trypanosome data*

Follman and Lambert (1989) gave an example of a logistic regression with a varying intercept term. The data consist of numbers `y` of trypanosomes killed out of `n` treated at a treatment dose `x`. The data are shown in Table 8.8, and are graphed in Fig. 8.6. The logistic regression on log `x` is a very poor fit, with deviance 24.66

Table 8.8. Trypanosome data

x	n	y
4.7	55	0
4.8	49	8
4.9	60	18
5.0	55	18
5.1	53	22
5.2	53	37
5.3	51	47
5.4	50	50

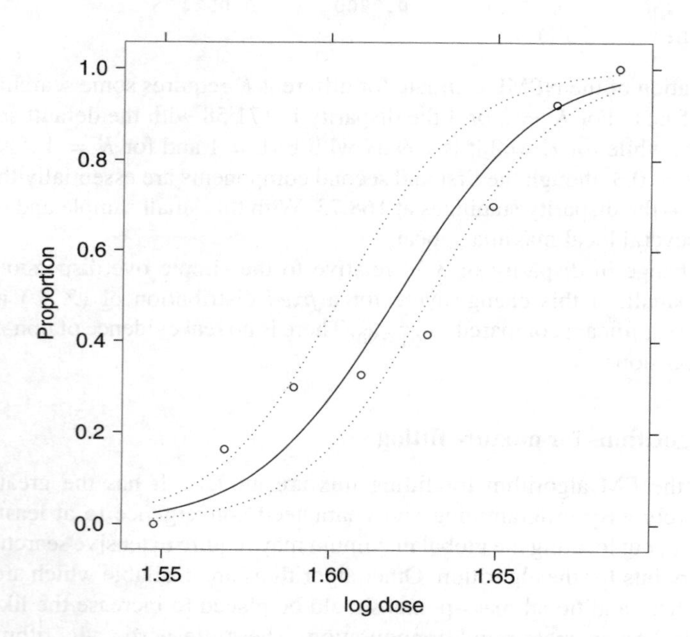

Fig. 8.6. Observed proportions and EM mixture model

with 6 df. (A CLL link gives a much better fit, with a deviance of 15.83, but we do not pursue this here.)

For comparison with Follman and Lambert we fit the random intercept model using log dose, although a slightly better fit is obtained with dose x.

```
> data(trypanos, package = "SMIR")

> library(npmlreg)
> trypanos.np1 <- alldist(cbind(y, (n - y)) ~ log(x),
```

```
+       random = ~1, data = trypanos, family = binomial,
+       plot.opt = 0, verbose = FALSE, k = 1)
> trypanos.np2 <- update(trypanos.np1, k = 2)
```

The NPML estimate of the mixing distribution found by this approach has two mass-points with location $(-0.546, 0.894)$ and masses $(0.621, 0.379)$ and disparity 40.74. Parameter estimates are $\hat{\beta}_0 = -87.78$, $\hat{\beta}_1 = 53.854$. The fitted proportions from the marginal mixed logistic model are obtained from the fitted values and are shown in Fig. 8.6.

```
> tryp.newdata <- data.frame(lx = log(seq(min(trypanos$x),
+       max(trypanos$x), len = 101)))
> tryp.newdata <- transform(tryp.newdata, p1 = 1/(1 +
+       exp(-(-87.78 + 53.85 * lx))), p2 = 1/(1 + exp(-(-86.34 +
+       53.85 * lx))))
> tryp.newdata <- transform(tryp.newdata, mfp = 0.621 *
+       p1 + 0.379 * p2)
> print(xyplot(p1 + p2 + mfp ~ lx, xlab = "log dose",
+       ylab = "proportion", data = tryp.newdata, panel = function(x,
+           y, ...) {
+           panel.xyplot(x, y, type = "l", lty = c(3,
+               3, 1), lwd = 1.5, ...)
+           lpoints(log(trypanos$x), trypanos$y/trypanos$n)
+       }))
```

For the EM estimates, the posterior probabilities of membership in the upper component are (0.075, 0.997, 1.000, 0.001, 0.000, 0.000, 0.179, 0.784). Thus the EM estimates separate off the second and third observations from the others, with the last observation having high weight for this component as well. The dashed curves in the figure are the fitted component models.

Using an optimal design algorithm, Follman and Lambert found a two-point distribution for the intercept with location $(6.27, -3.23)$, masses $(0.34, 0.66)$ and a deviance of 3.38. Parameter estimates are $\hat{\beta}_0 = -202.47$, $\hat{\beta}_1 = 124.8$. Tabulation over a grid of β_0 and β_1 shows that the deviance evaluated at these parameter values is indeed the minimum of 3.38, but the EM algorithm as implemented here cannot find this point, despite searching over starting values using `tol_`. Figure 8.7 shows the two logistic components and the fitted model.

```
> print(xyplot(y/n ~ log(x), data = trypanos, xlab = "log dose",
+       ylab = "proportion", panel = function(x, y, ...) {
+           panel.xyplot(x, y, ...)
+           panel.curve(1/(1 + exp(-(-196.2 + 124.8 *
+               x))), lty = 3, lwd = 1.5)
+           panel.curve(1/(1 + exp(-(-205.7 + 124.8 *
```

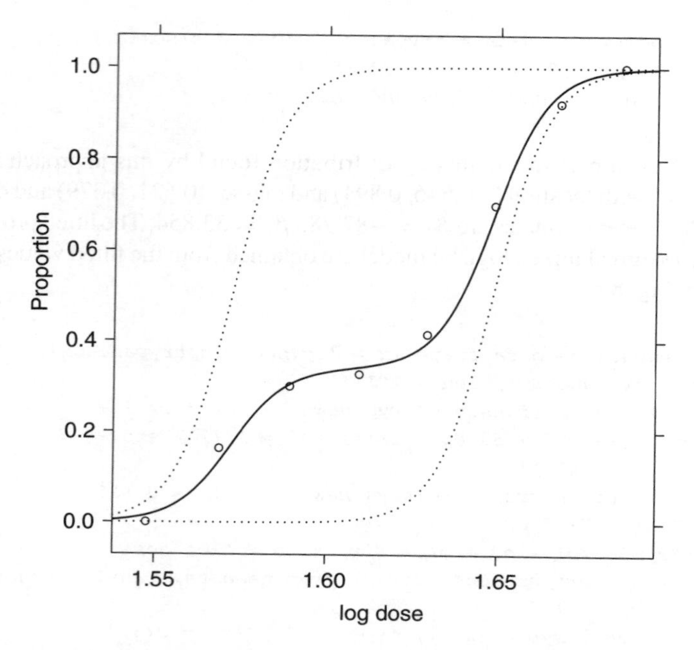

Fig. 8.7. Observed proportions and FL mixture model

```
+            x))), lty = 3, lwd = 1.5)
+        panel.curve(0.34 * (1/(1 + exp(-(-196.2 +
+            124.8 * x)))) + 0.66 * (1/(1 + exp(-(-205.7 +
+            124.8 * x)))), lwd = 1.5)
+    }))
```

The fitted probabilities nearly interpolate the observed proportions, with a very low deviance. The posterior probabilities are 1 for all observations except the fifth – the first four fall in the first component, and the last three in the second.

For the second mixture, the standard deviation of the estimated mixing distribution is 4.50, compared with 0.70 for the first. The two components are completely separated in the second mixture, with a 'low-dose' sub-population of 34% which is killed almost completely by a log-dose of 1.62; the remaining 'high-dose' sub-population of 66% requires higher doses for mortality.

A peculiarity of the EM algorithm here is that if it is started from the Follman/Lambert estimates, it jumps in one iteration to the NPML estimates above. Further study of this example, and others, would be rewarding. Aitkin (2000*b*) discussed a second example.

8.10 Modelling the mixing probabilities

We have assumed that the mixing distribution of the random effect Z does not depend on the explanatory variable $\mathbf{x}$ except through a location shift, which does not affect the model-fitting, though it may change its interpretation.

An alternative approach, widely used in computer science, is to model Z as discrete, with component proportions π_k modelled as multinomial logit functions of the explanatory variables, in addition to the model for the kernel parameter. These *double* models are known as *mixtures of experts* models in the computer science literature (see, e.g. Jacobs *et al.*, 1991) an 'expert' being a regression model within a single component of the mixture. The usual form of the model allows a different regression model for the kernel density parameter within each mixture component. The double model can be fitted straightforwardly by successive relaxation, by alternating the mixture fitting of the kernel models and the fitting of the multinomial logit model. This is similar to the double modelling in Chapter 3 of the normal mean and variance parameters. The discrete mixture model likelihood of Section 8.6:

$$L(\boldsymbol{\beta}, K, \pi_1, \ldots, \pi_{K-1}, z_1, \ldots z_K) = \prod_{i=1}^{n} \left\{ \sum_{k=1}^{K} f(y_i | z_k, \boldsymbol{\beta}) \pi_k \right\}$$

is generalized using the multinomial model

$$\theta_{ik} = \log(\pi_{ik}/\pi_{i1}) = \boldsymbol{\delta}_k' \mathbf{x}_i$$

and the component-specific regression parameter vector $\boldsymbol{\beta}_k$, to give the mixture of experts likelihood

$$L(\boldsymbol{\beta}_1, \ldots, \boldsymbol{\beta}_K, \boldsymbol{\delta}_1, \boldsymbol{\delta}_2, \ldots, \boldsymbol{\delta}_K, K,) = \prod_{i=1}^{n} \left\{ \sum_{k=1}^{K} f(y_i | \boldsymbol{\beta}_K) \pi_{ik} \right\},$$

where $\boldsymbol{\delta}_1 = 0$, with log-likelihood

$$\ell = \log L = \sum_i \log \left\{ \sum_k \pi_{ik} f_{ik} \right\} = \sum_i \log f_i.$$

The derivatives of the log-likelihood with respect to β_k follow from Section 8.3:

$$\frac{\partial \ell}{\partial \boldsymbol{\beta}_k} = \sum_i w_{ik} \mathbf{s}_{ik}(\boldsymbol{\beta}_k),$$

and those with respect to $\delta_k (k > 1)$ are

$$\frac{\partial \ell}{\partial \delta_k} = \sum_i \frac{1}{f_i} \left\{ \frac{\partial \pi_{ik}}{\partial \delta_k} f_{ik} - \frac{\partial \pi_{iK}}{\partial \delta_K} f_{iK} \right\}$$

$$= \sum_i \left\{ \frac{\pi_{ik} f_{ik}}{f_i} \mathbf{x}_i - \frac{\pi_{iK} f_{iK}}{f_i} \mathbf{x}_i \right\}$$

$$= \sum_i (w_{ik} - w_{iK}) \mathbf{x}_i.$$

We do not give details of algorithm implementation.

The model is heavily parametrized, and as for the double normal model there may be identifiability difficulties in assessing the effects of variables in the two sets of models. The expected information matrix may not be a reliable indicator of effect importance.

An important special case of the mixture of experts model is one in which the component-specific regression models are the same: only the logistic regression models are different for each component. This special case was proposed by Aitkin and Foxall (2003) as an alternative to the *feed-forward neural network* or *multi-layer perceptron.*

8.11 Mixtures of mixtures

An interesting question is whether mixtures of exponential family distributions can be generalized further by mixing. Can we develop and fit mixtures of mixtures?

Two issues arise immediately. The first is that for exponential family distributions, mixing with a conjugate random effect distribution gives in some cases closed-form distributions with unlinked mean and variance parameters, giving greater generality than the original exponential family distribution. However, the resulting compound distribution may not have a conjugate mixing distribution, and numerical integration will then be necessary, as it is for the normal mixing distribution with Poisson and binomial distributions.

The second is that the possible improvement in likelihood, by mixing the mixed distribution, is limited by the ML for the NPML estimate of the mixing distribution for the original (one-parameter) exponential family distribution.

We illustrate these limitations with mixtures of negative binomials for the fabric fault data. The Poisson model using the log length variable had a deviance (relative to the saturated model) of 64.54 on 30 df, and fitting the negative binomial model, a gamma mixture of Poissons, reduced this to 52.09. If we now mix the negative binomial by introducing a new random effect w with marginal distribution $h(w \mid \xi)$ into its linear predictor, there is no conjugate form for this, so we will be limited to normal or other non-conjugate mixing requiring numerical integration. The marginal distribution of the negative binomial mixture is

$$m(y) = \int nb(y \mid \eta, r, w)h(w \mid \xi)\, dw,$$

where

$$nb(y \mid \eta, r, w) = \frac{\Gamma(y+r)}{\Gamma(r)y!}\left(\frac{e^{\eta+w}}{e^{\eta+w}+r}\right)^y\left(\frac{r}{e^{\eta+w}+r}\right)^r.$$

But since the negative binomial is itself the gamma mixture of Poissons, we have

$$m(y) = \int h(w \mid \xi)\left[\int P(y \mid \eta, z, w)g(z \mid r)dz\right]dw$$

where

$$P(y \mid \eta, z, w) = \frac{e^{-e^{\eta+z+w}}e^{(\eta+z+w)y}}{y!}.$$

Changing variables to $u = z + w$ and z, with Jacobian 1, we have

$$\begin{aligned} m(y) &= \int P(y \mid \eta, u)\left[\int h(u-z \mid \xi)g(z \mid r)dz\right]du \\ &= \int P(y \mid \eta, u)k(u \mid \xi, r)\, du. \end{aligned}$$

So the mixed negative binomial distribution is a mixed Poisson distribution with respect to a different mixing distribution k, the convolution of h and g. But for the single-parameter Poisson distribution, the minimized value of the deviance across *all* mixing distributions cannot be less than 49.37, the deviance for the NPML-estimated mixing distribution. So any compound negative binomial distribution for the fault data could only reduce the negative binomial deviance by at most 2.72 (from 52.09 to 49.37).

Such results greatly limit the possibilities for iterating mixing for any exponential family distribution – the NPML estimate will be the same for double mixing as for single mixing. An example is the *hierarchical mixture of experts* model (Jordan and Jacobs, 1994) in which the mixture of experts model is again mixed: each of the K mixture components is itself regarded as a mixture of L components with mixing probabilities π^*_{ikl} depending logistically on a parameter

$\boldsymbol{\delta}^*_{kl}$. The marginal distribution of the response Y_i in such a model is then

$$\begin{aligned} m(y_i \mid \mathbf{x}_i, \delta 2, \ldots, \boldsymbol{\delta}_K, \boldsymbol{\delta}^*_{22}, \ldots \boldsymbol{\delta}^*_{2L}, \ldots, \boldsymbol{\delta}^*_{K2}, \ldots \boldsymbol{\delta}^*_{KL}) \\ = \sum_{k=1}^{K} \pi_{ik}(\boldsymbol{\delta}_k) \sum_{l=1}^{L} \pi^*_{ikl}(\boldsymbol{\delta}^*_{kl}) f(y_i \mid \theta_{ikl}) \\ = \sum_{k=1}^{K} \sum_{l=1}^{L} \pi_{ik}(\boldsymbol{\delta}_k) \pi^*_{ikl}(\boldsymbol{\delta}^*_{kl}) f(y_i \mid \theta_{ikl}), \end{aligned}$$

where

$$\sum_{k=1}^{K} \sum_{l=1}^{L} \pi_{ik}(\boldsymbol{\delta}_k) \pi^*_{ikl}(\boldsymbol{\delta}^*_{kl}) = 1.$$

Write

$$\pi_{ik}(\boldsymbol{\delta}_k) \pi^*_{ikl}(\boldsymbol{\delta}^*_{kl}) = \pi^{**}_{im}(\boldsymbol{\gamma}_m),$$

where m runs over the range $1, \ldots, M = KL$, with $\sum_{m=1}^{M} \pi^{**}_{im}(\boldsymbol{\gamma}_m) = 1$, and $\boldsymbol{\gamma}$ is a parametric function of the $\boldsymbol{\delta}$ and $\boldsymbol{\delta}^*$ to be determined. Now

$$\begin{aligned} \pi_{ik}(\boldsymbol{\delta}_k) \pi^*_{ikl}(\boldsymbol{\delta}^*_{kl}) &= \frac{e^{\boldsymbol{\delta}_k \mathbf{x}_i}}{\sum_{k=1}^{K} e^{\boldsymbol{\delta}_k \mathbf{x}_i}} \cdot \frac{e^{\boldsymbol{\delta}^*_{kl} \mathbf{x}_i}}{\sum_{l=1}^{L} e^{\boldsymbol{\delta}^*_{kl} \mathbf{x}_i}} \\ &= \frac{e^{(\boldsymbol{\delta}_k + \boldsymbol{\delta}^*_{kl}) \mathbf{x}_i}}{\sum_{k=1}^{K} \sum_{l=1}^{L} e^{(\boldsymbol{\delta}_k + \boldsymbol{\delta}^*_{kl}) \mathbf{x}_i}} \\ &= \frac{e^{\boldsymbol{\gamma}_m \mathbf{x}_i}}{\sum_{m=1}^{M} e^{\boldsymbol{\gamma}_m \mathbf{x}_i}}, \end{aligned}$$

where

$$\boldsymbol{\gamma}_m = \boldsymbol{\gamma}_{kl} = \boldsymbol{\delta}_k + \boldsymbol{\delta}^*_{kl}.$$

Thus the marginal distribution of Y_i can be expressed as

$$m(y_i \mid \mathbf{x}_i, \boldsymbol{\gamma}_2, \ldots, \boldsymbol{\gamma}_M) = \sum_{m=1}^{M} \pi^{**}_{im}(\boldsymbol{\gamma}_m) f(y_i \mid \theta_{im}),$$

which is an M-component mixture with multinomial logit parameters regressing on $\mathbf{x}$ with coefficients $\boldsymbol{\gamma}_m$.

Thus, the 'hierarchical' mixture is equivalent to an M-component mixture of the original type, though with apparently many more components. Since there is an identifiable limit to the number of components, hierarchical mixing will not identify more components than simple mixing, and does not give greater generality.

9
Variance component models

9.1 Models with shared random effects

In this chapter we consider GLMs with *shared random effects* arising through variance component or repeated measures structure, for example, in two-stage sample designs, or longitudinal data. As we will see, the analysis of these models parallels closely that for overdispersion models, with slightly greater complexity in the EM algorithm. The range of possible models in this category is surprising.

We begin for simplicity of exposition with the two-level variance component model, for a two-stage random sample with upper- or second-level or *primary* sampling units (PSUs) indexed by $i = 1, \ldots, r$, and lower- or first-level or *secondary* sampling units (SSUs) indexed by j, sampled within each upper-level unit, where $j = 1, \ldots, n_i$. On each first-level unit we measure or record a response y_{ij}, and we have explanatory variables $\mathbf{x}$ which may be measured at both upper ($\mathbf{x}_{1i}$) and lower ($\mathbf{x}_{2ij}$) levels. We represent the distribution of the response Y by an exponential family member, with a link function and linear predictor involving the explanatory variables at both levels, and perhaps their cross-level interactions.

The common membership of the responses y_{ij} and $y_{ij'}$ in the same PSU implies, in most sampled populations, a greater homogeneity of these responses than of those in different PSUs. This homogeneity should be reflected in the model.

A natural way of representing this 'common variation' is by adding a common unobserved random effect to the linear predictor for each lower-level unit in the same upper-level unit. Thus the common variation is modelled as an extra unobserved variable Z_i on the same scale as the linear predictor, but this variable is now shared between lower-level units in the same upper-level unit, rather than being unique to each observation as in the overdispersion model.

9.2 The normal/normal model

For the normal/normal model, we take the distribution of both Y and Z to be normal, with identity link: conditional on Z_i, the Y_{ij} are independently normal

$N(\mu_{ij}, \sigma^2)$, with

$$\mu_{ij} = \eta_{ij} = \boldsymbol{\beta}'\mathbf{x}_{ij} + \sigma_A Z_i,$$
$$\boldsymbol{\beta}'\mathbf{x}_{ij} = \boldsymbol{\beta}_1'\mathbf{x}_{1ij} + \boldsymbol{\beta}_2'\mathbf{x}_{2i},$$
$$Z_i \sim N(0, 1),$$

where σ_A^2 is the 'among PSU' variance component, and σ^2 is the 'within PSU' variance component. It follows immediately that marginally,

$$Y_{ij} \sim N(\boldsymbol{\beta}_1'\mathbf{x}_{1ij} + \boldsymbol{\beta}_2'\mathbf{x}_{2i},\ \sigma^2 + \sigma_A^2),$$

$\text{Cov}[Y_{ij}, Y_{ij'}] = \sigma_A^2$ for $j \neq j'$, and $\text{Cov}[Y_{ij}, Y_{i'j'}] = 0$ for $i \neq i'$, so that correspondingly

$$\text{corr}[Y_{ij}, Y_{ij'}] = \rho = \frac{\sigma_A^2}{\sigma^2 + \sigma_A^2},$$

for $j \neq j'$, and $\text{corr}[Y_{ij}, Y_{i'j'}] = 0$ for $i \neq i'$.

The random effect induces an *intra-class* correlation between the lower-level responses on the same upper-level unit, and so *does* represent the greater homogeneity of the responses in the same upper-level unit. An equivalent form of the model is more familiar from *multi-level modelling* (Goldstein, 2003):

$$Y_{ij} \mid Z_i \sim N(\boldsymbol{\beta}_1'\mathbf{x}_{1ij} + Z_i, \sigma^2), \quad j = 1, \ldots, n_i,$$
$$Z_i \sim N(\boldsymbol{\beta}_2'\mathbf{x}_{2i}, \sigma_A^2), \quad i = 1, \ldots, r.$$

In this form the role of the upper-level variables is clearer, in reducing or 'explaining' the upper-level variance σ_A^2. It is an important aspect of these models that the parameter σ_A can appear either as a variance in the second form, or as a regression coefficient in the first. In fitting the model using an EM algorithm, the rate of convergence of the algorithm is affected by the size of σ_A in the second parametrization, but not in the first (Aitkin *et al.*, 1981; Meng and van Dyk, 1997). In the normal model, the EM algorithm is not the most efficient algorithm for ML estimation, the scoring algorithm being the fastest (Longford, 1993), but in non-normal models the EM algorithm is very effective and relatively simple to program.

We now give the EM algorithm for the normal model using the first parametrization above. We maximize in the E-step the expected complete data log-likelihood, treating the Z_i as missing data. The complete data likelihood can be written as

$$L^* = \prod_{i=1}^{r}\left[\prod_{j=1}^{n_i}\frac{1}{\sqrt{2\pi}\sigma}\exp\left\{-\frac{1}{2\sigma^2}(y_{ij} - \boldsymbol{\beta}'\mathbf{x}_{ij} - \sigma_A Z_i)^2\right\}\right]\exp\left\{-\frac{1}{2}Z_i^2\right\}.$$

The last term is the normal density of the Z_i, which does not involve any parameters, and so we may drop it from the likelihood. In the complete data model, Z_i is an additional explanatory variable, and the complete data log-likelihood involves the missing data only through Z_i and Z_i^2; in the E-step these values are replaced by their conditional expectations given the observed data y_{ij} (and $\mathbf{x}_{ij}$): write

$$\begin{aligned} \mathrm{E}[Z_i \mid y_{ij}] &= \tilde{z}_i, \\ \mathrm{E}[Z_i^2 \mid y_{ij}] &= \mathrm{Var}[Z_i \mid y_{ij}] + E^2[Z_i \mid y_{ij}] \\ &= V_i + \tilde{z}_i^2. \end{aligned}$$

For the M-step, the normal equations for $\boldsymbol{\beta}$ and σ_A involve the cross-products of $\mathbf{x}$ with Z; in these equations Z_i is replaced by $\tilde{z}_i$, and the sum of squares $\sum Z_i^2$ is replaced by $\sum \tilde{z}_i^2 + \sum V_i$. The score equation for σ^2 is adjusted similarly.

For the E-step, it is easily seen that, given $\bar{Y}_{i\cdot}$, Z_i is independent of $(Y_{ij} - \bar{Y}_{i\cdot})$, where $\bar{Y}_{i\cdot} = \sum_j Y_{ij}/n_i$. Since

$$\bar{Y}_{i\cdot} \mid Z_i \sim N(\boldsymbol{\beta}'\bar{\mathbf{x}}_{i\cdot} + \sigma_A Z_i, \sigma^2/n_i)$$

and $Z_i \sim N(0, 1)$, it follows easily that

$$Z_i \mid \bar{Y}_{i\cdot} \sim N\left(\frac{\sigma_A}{\sigma_A^2 + \sigma^2/n_i}(\bar{Y}_{i\cdot} - \boldsymbol{\beta}'\bar{\mathbf{x}}_{i\cdot}), 1 - \frac{\sigma_A^2}{\sigma_A^2 + \sigma^2/n_i}\right).$$

The EM algorithm alternates between fitting the linear model to $\mathbf{x}$ and z, with z_i replaced by $\tilde{z}_i$ in the cross-products, and a diagonal loading on the SSP matrix of $\sum V_i$ corresponding to σ_A, and computing $\tilde{z}_i$ and V_i from the current parameter estimates.

The difficulty with other exponential family distributions is that the normal random effect model is not conjugate, and so numerical integration or other approximations are necessary. We give in Section 9.3 the same treatments of these models by Gaussian quadrature and NPML as for the overdispersion models of Chapter 8. Other approximate approaches to the variance component model have been extensively discussed, and we consider these in Section 9.4.

9.3 Exponential family two-level models

Restating the general model, we have a two-stage random sample y_{ij} with $i=1,\ldots,r,\ j=1,\ldots,n_i$, and $\sum n_i = n$, from an exponential family distribution $f(y \mid \theta)$ with canonical parameter θ and mean μ, and explanatory variables $X = (\mathbf{x}_{ij})$, related to μ through a link function $\eta_{ij} = g(\mu_{ij})$ with linear predictor $\eta_{ij} = \boldsymbol{\beta}'\mathbf{x}_{ij}$. Here the X matrix is understood to include both upper- and lower-level explanatory variables (and their interactions if any), with $\mathbf{x}_{ij} = \mathbf{x}_i$ for all j for an upper-level variable. Thus the upper-level variable $\mathbf{x}_i$ is replicated n_i times for the n_i lower-level units in the i-th upper-level unit.

The random effect model has an unobserved common random effect Z_i for each lower-level unit in the i-th upper-level unit, the Z_i being initially assumed independently normally distributed $Z_i \sim N(0, 1)$, and conditionally on the Z_i, the Y_{ij} have independent GLMs with linear predictor $\eta_{ij} = \boldsymbol{\beta}'\mathbf{x}_{ij} + \sigma_A Z_i$. The random effect is modelled as acting on the same scale as the linear predictor; we again denote the standard deviation of the random effect by σ_A as in Section 9.2.

The likelihood is then

$$L(\boldsymbol{\beta}, \sigma_A) = \prod_{i=1}^{r} \left\{ \int \left[\prod_{j=1}^{n_i} f(y_{ij} \mid \boldsymbol{\beta}, \sigma_A, z_i) \right] \phi(z_i)\, \mathrm{d}z_i \right\},$$

where $\phi(z)$ is the standard normal density function.

The integral in the likelihood does not have a closed form except for Y normal, and so for other response models we approximate it by Gaussian quadrature: we replace the integral over the normal Z_i by the finite sum over K Gaussian quadrature mass-points z_k with masses π_k. The likelihood is then

$$L(\boldsymbol{\beta}, \sigma_A) \doteq \prod_{i=1}^{r} \left\{ \sum_{k=1}^{K} \left[\prod_{j=1}^{n_i} f(y_{ij} \mid \boldsymbol{\beta}, \sigma_A, z_k) \right] \pi_k \right\}.$$

The likelihood is thus (approximately) the likelihood of a finite mixture of exponential family densities with known mixture proportions π_k at known mass-points z_k, with the linear predictor for the ij-th observation in the k-th mixture component being

$$\eta_{ijk} = \boldsymbol{\beta}'\mathbf{x}_{ij} + \sigma_A z_k.$$

Thus z_k becomes another observable variable in the regression, with regression coefficient σ_A.

The log-likelihood is

$$\ell(\boldsymbol{\beta}, \sigma_A) = \sum_i \log \left\{ \sum_k \pi_k m_{ik} \right\},$$

where for compactness we write

$$m_{ik} = \prod_j f_{ijk},$$

$$f_{ijk} = f(y_{ij} \mid \boldsymbol{\beta}, \sigma_A, z_k)$$

$$= \exp\{\theta_{ijk} y_{ij} - b(\theta_{ijk}) + c(y_{ij})\}$$

with

$$\mu_{ijk} = b'(\theta_{ijk}), \quad V_{ijk} = b''(\theta_{ijk}),$$

$$\eta_{ijk} = g(\mu_{ijk}) = \boldsymbol{\beta}'\mathbf{x}_{ij} + \sigma_A z_k,$$

$$g'_{ijk} = g'(\mu_{ijk}).$$

Then

$$\frac{\partial \ell}{\partial \boldsymbol{\beta}} = \sum_i \frac{\sum_k \pi_k m_{ik} \frac{\partial \log m_{ik}}{\partial \boldsymbol{\beta}}}{\sum_k \pi_k m_{ik}} = \sum_i \sum_j \sum_k w_{ik} \mathbf{s}_{ijk}(\boldsymbol{\beta}),$$

where w_{ik} is the posterior probability that observation y_{ij} comes from component k:

$$w_{ik} = \frac{\pi_k m_{ik}}{\sum_\ell \pi_\ell m_{i\ell}}$$

and $\mathbf{s}_{ijk}(\boldsymbol{\beta})$ is the $\boldsymbol{\beta}$-component of the score for observation (ij) in component k:

$$\mathbf{s}_{ijk}(\boldsymbol{\beta}) = (y_{ij} - \mu_{ijk})\mathbf{x}_{ij} / V_{ijk} g'_{ijk}.$$

Similarly

$$s_{ijk}(\sigma_A) = (y_{ij} - \mu_{ijk}) z_k / V_{ijk} g'_{ijk}.$$

Equating the score to zero gives likelihood equations which are simple weighted sums of those for an ordinary GLM with weights w_{ik}; alternately solving these equations for given weights w_{ik}, and updating these weights from the current parameter estimates, is an EM algorithm. The triple summation over (ij) and k is conveniently (though inefficiently) handled as in the simpler overdispersion case by expanding the data vectors to length Kn by replicating y and X K times, and the Gaussian quadrature variable z n times. Model fitting is then identical to that of a single sample of Kn observations with prior weight vector w. Initial estimates for the first E-step for $\boldsymbol{\beta}$ are conveniently obtained from the ordinary GLM fit, and for σ_A by arbitrary specification other than zero (e.g. $\sigma_A = 1$). A distinctive feature of the weights compared to those in the overdispersion model is that they are calculated for each upper-level unit in the E-step but applied to all lower-level units in this upper-level unit in the M-step.

9.4 Other approaches

A number of approaches have been suggested to avoid the full likelihood analysis above. First, the log-likelihood function can be approximated by a quadratic in the random effect, and standard computational methods for the normal variance component model can then be used, giving approximate ML or REML estimation (Laird and Ware, 1986). This approach has been implemented generally in slightly different forms by Breslow and Clayton (1993); McGilchrist (1994); Goldstein (2003), and Longford (1993), giving penalized quasi-likelihood (PQL) analyses. The success of the approximation depends on the closeness to normality (in the random effect) of the observed data likelihood, and may fail badly, for example for binary response data (Rodriguez and Goldman, 1995).

Second, the integrals required in the E-step of the EM algorithm can be avoided by Laplace approximations (Steele, 1996) or by Monte Carlo integration (Walker, 1996; McCulloch, 1997).

Third, the problem can be circumvented by the generalized estimating equation (GEE) approach (Liang and Zeger, 1986; Diggle *et al.*, 2002). Here the *marginal* distribution of Y is assumed to be exponential family, and the repeated measures structure is represented by a covariance matrix model whose parameters are estimated by a form of QL (marginal quasi-likelihood MQL) which does not require a full parametric specification for the random effect distribution. This approach is widely used. It is formally inconsistent with the random effect model, since this model is always overdispersed because of the variance inflation from the random effect. However in Bernoulli models in which overdispersion cannot be identified, the analyses may be very similar (see Section 9.7.2 for an example).

Fourth, a fully Bayes approach can be followed, with the additional structure of a prior distribution on all the model parameters, and Markov Chain Monte Carlo methods can be used to obtain marginal posterior distributions of the parameters. Gelman *et al.* (1995) gave a detailed exposition of this approach, which is becoming increasingly popular with the widespread dissemination of Bayesian software.

A disadvantage of any approach using a specified parametric form for the 'mixing' distribution of the unobserved random effects is the possible sensitivity of the conclusions to this specification. The influential paper by Heckman and Singer (1984) showed substantial changes in parameter estimates with quite small changes in mixing distribution specification; Davies (1987) showed similar effects. This difficulty can be avoided by NPML estimation of the mixing distribution in one-parameter exponential family distributions, concurrently with the structural model parameters. Clayton and Kaldor (1987) gave an example of this approach, in the simpler framework of a single-level overdispersion model.

9.5 NPML estimation of the masses and mass-points

As in Chapter 8, we may drop completely the parametric distributional assumption for the random effects, and estimate this distribution nonparametrically in single-parameter exponential family distributions. The analysis parallels that in Chapter 8: we treat the masses and mass-points as unknown parameters; the number K of mass-points is also unknown but is treated as fixed, and sequentially increased until the likelihood is maximized. Since the variance of the mixing distribution is a function of the unknown parameters, we absorb the scale parameter σ_A into the mass-point parameters z_k, with linear predictor

$$\eta_{ijk} = \boldsymbol{\beta}'\mathbf{x}_{ij} + z_k.$$

Thus z_k acts as an intercept parameter for the k-th component: it is estimated as in the overdispersion model by including a 'component factor' in the model with K levels. One of the z_k parameters will be aliased with the intercept term β_0; alternatively the intercept can be removed from the model.

Differentiating the log-likelihood with respect to π_k, and using $\pi_K = 1 - \sum_1^{K-1} \pi_k$, we have directly

$$\frac{\partial \ell}{\partial \pi_k} = \sum_i \frac{m_{ik} - m_{iK}}{\sum_\ell \pi_\ell m_{i\ell}} = \sum_i \left\{ \frac{w_{ik}}{\pi_k} - \frac{w_{iK}}{\pi_K} \right\}.$$

Equating this to zero gives simply

$$\hat{\pi}_k = \sum_i w_{ik}/n,$$

the standard mixture ML result. The same EM algorithm applies with the additional calculation in each M-step of the estimate of π_k from the weights. Initial estimates of the z_k can be taken as the Gaussian quadrature values scaled by a factor depending on the standard deviation of Y. An R implementation of the EM algorithm for this model is provided in the `npmlreg` package by the `allvc` function. Hypothesis testing for regression model parameters (nested model comparisons) may be carried out via the LRT using differences of disparities in the usual way. The theoretical justification for this application of standard theory is given in Murphy and van der Vaart (2000).

9.6 Random coefficient models

The analysis can be extended to a class of random coefficient models as in Chapter 8. A frequent case of interest is when lower-level variables have slopes which vary across upper-level units. Consider a simple example with a single lower-level variable x_{1ij} whose coefficient β_1 varies across the upper-level units. We index it by $\beta_{1i} = \beta_1 + U_i$, where U_i represents variation about a 'mean' β_1. The remaining regression coefficients $\boldsymbol{\beta}_2$ are fixed. Then conditional on U_i *and* Z_i, the regression model is

$$\eta_{ij} = \beta_1 x_{1ij} + \boldsymbol{\beta}_2' \mathbf{x}_{2ij} + Z_i + U_i x_{1ij},$$

while marginally Z_i and U_i have an unknown joint distribution $g(z, u)$. The likelihood is then

$$L(\boldsymbol{\beta}) = \prod_i \left\{ \int \left[\prod_j f(y_{ij} \mid z_i, u_i) \right] g(z_i, u_i) \, \mathrm{d}z_i \, \mathrm{d}u_i \right\}.$$

By estimating the joint distribution of z_i and u_i nonparametrically, we again obtain the NPML estimate (in the one-parameter exponential family) as a discrete distribution on a finite number of mass-points $(\hat{z}_k, \hat{u}_k)$ in the (z, u) plane, with an estimated mass $\hat{\pi}_k$ for the k-th component. As in Chapter 8, the likelihood is again that of a finite mixture, of regressions on x_1 with a different slope and intercept in each component of the mixture, and on $\mathbf{x}_2$ with the same regression coefficient in each component. The approach can be generalized to any number of explanatory variables, but the number of parameters in the model increases rapidly.

The fitting of these models parallels exactly that in Chapter 8: the random coefficients are handled in the computational implementation by including in the regression model, in addition to the main effect of the component factor, the interaction of this factor with the explanatory variable x_{1ij}. As in Chapter 8 the user defines a formula involving the fixed-effect terms in the model, and a `random` model involving (lower-level) variables with random regression coefficients (the intercept term is always included).

Upper-level variable slopes may also be allowed to vary over upper-level units by including them in interactions with the random factor in the same way. This represents a form of generalized overdispersion at the upper level. However, incorporating both *lower-level* overdispersion *and* variance component structure requires two sets of random effects which are *both* modelled non-parametrically. This is beyond the scope of the present discussion. However, it is important to note that since compound Bernoulli models are unidentifiable, one set of random effects is sufficient for two-level Bernoulli models.

9.7 Variance component model fitting

To fit a variance component model, we use the function *`allvc`* from the *`npmlreg`* package with the same fixed formula as used for the overdispersion modelling and specify the number of mass-points and mixing distribution as before. We specify the two-level structure via the *random* formula in a similar way to that used in the *nlme* package (see Pinheiro and Bates, 2000). Gaussian quadrature is achieved by setting the *random.distribution* argument to '*gq*', and to '*np*' for NPML.

If the data are binomial, as with the *glm* function, the response is defined by a two column matrix with the first column holding the number of 'successes' and the other column, the number of 'failures'. Alternatively a vector of proportions can be set as the response with the number of trials (successes plus failures) defined as *prior.weights*. Binary responses can be specified as a 'factor' or a binary (0, 1) vector or a vector of the logical constants (*FALSE*, *TRUE*) (when the first level denotes failure and the second success).

We now consider several examples.

9.7.1 *Children's height development*

The first example is from Harrison and Brush (1990), also analysed in Goldstein (2003, pp. 132–133). The heights of 26 boys in Oxford were measured on nine occasions over two years, initially at age 12; the occasions were almost equally spaced in time through the school year. The data, `Oxboys` provided in the *SMIR* package, are the height in cm, age expressed in months relative to an origin of 13 years, and months relative to an annual time origin of January 1. The heights at each actual age are shown in Fig. 9.1 for each boy. The points are joined by lines to identify the individual boys.

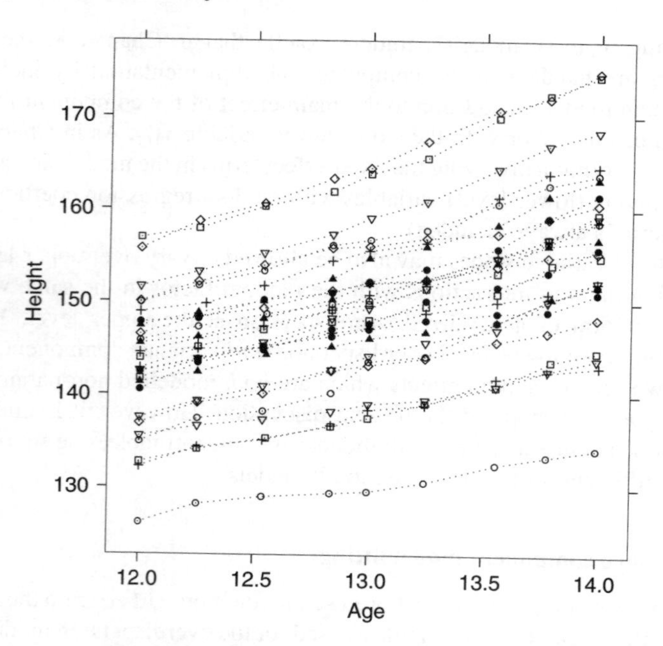

Fig. 9.1. Heights of 26 boys over two years

```
> data(Oxboys, package = "SMIR")

> library(lattice)
> print(xyplot(height ~ I(age + 13), xlab = "age",
+      group = Subject, data = Oxboys, type = "b", lty = 3))
```

The growth patterns are very similar though there is naturally variation in initial height, which increases with age. It also appears if the lines diverge suggesting that the growth rate for the smaller boys is less than for the larger boys.

We begin with a normal random effect model, fitted by 20-point Gaussian quadrature.

```
> library(npmlreg)
> data(Oxboys, package = "nlme")
> oxboys.gq20 <- allvc(height ~ age, random = ~1 |
+      Subject, data = Oxboys, random.distribution = "gq",
+      k = 20, verbose = FALSE, plot.opt = 0)
> summary(oxboys.gq20)

Coefficients:
              Estimate Std. Error   t value
(Intercept) 148.958     0.030      4977.920
age           6.524     0.046       141.241
z             4.769     0.018       264.649

Component distribution - MLE of sigma:      1.506
Random effect distribution - standard deviation:    4.769

-2 log L:            991.82     Convergence at iteration  15
```

The linear regression on age is very strong and the intra-class correlation is very high: $\sigma_A^2/(\sigma_A^2 + \sigma^2) = 0.909$!

The disparity of 991.82 is substantially larger than the value of 940.0 reported by Pinheiro and Bates (2000) using the analytic likelihood. To obtain a similar value to that of Pinheiro and Bates using the `allvc` function we need to increase the number of quadrature points nominally to 500, although this produces only 218 ‘real’ mass-points since those with negligible weight are dropped.

```
> oxboys.gq500 <- update(oxboys.gq20, k = 500)

> summary(oxboys.gq500)
```

```
Coefficients:
              Estimate Std. Error   t value
(Intercept) 149.797    0.015       9985.081
age           6.524    0.023        281.749
z             7.387    0.014        530.365

Component distribution - MLE of sigma:      1.299
Random effect distribution - standard deviation:   7.387

-2 log L:            940.1      Convergence at iteration  260
```

The intraclass correlation is even higher: 0.972.

At this *growth spurt* period, is growth linear, or is it accelerating? We fit the quadratic term.

```
> Oxboys <- transform(Oxboys, a2 = age^2)
> oxboys.gq500quad <- update(oxboys.gq500, . ~ . +
+      I(age^2))

> summary(oxboys.gq500quad)

Coefficients:
              Estimate Std. Error    t value
(Intercept) 149.454      0.021      6980.217
age           6.515      0.022       297.171
I(age^2)      0.741      0.038        19.342
z             6.758      0.012       560.261

Component distribution - MLE of sigma:      1.272
Random effect distribution - standard deviation:   6.758

-2 log L:            931.3      Convergence at iteration  197
```

The disparity change is 8.80 and the standard error of the quadratic term is $0.741/\text{sqrt}(8.80) = 0.2498$. There is strong evidence of curvature.

Is the assumption of a normal random effect distribution reasonable? With only 26 boys we do not have much information about it, but we try the discrete random effect model with quadratic age, with increasing numbers of mass-points.

```
> oxboys.np2 <- update(oxboys.gq500quad,
+      random.distribution = "np", k = 2)
> oxboys.np3 <- update(oxboys.np2, k = 3)
```

The successive disparities and values of $\hat{\sigma}$ and K are shown in the table below along with the values from the 20- and 500-point Gaussian quadrature models:

K	Disparity	$\hat{\sigma}$
2	1466.10	5.16
3	1319.54	3.62
4	1210.41	2.79
5	1151.49	2.40
6	1049.44	1.90
7	1011.59	1.74
8	922.64	1.41
9	1011.59	1.74
10	922.64	1.41
GQ20	983.87	1.51
GQ500	931.33	1.30

Beyond $K = 8$ some mass-points are duplicated with the same total mass as in lower-dimension models, though the remaining mass is assigned differently over the other (different) mass-points and the disparity increases.

```
> summary(oxboys.np8)

Coefficients:
            Estimate Std. Error     t value
age            6.515      0.055     117.455
I(age^2)       0.741      0.097       7.647
MASS1        129.890      0.187     693.564
MASS2        138.106      0.113    1220.744
MASS3        143.072      0.113    1265.220
MASS4        147.039      0.091    1610.010
MASS5        150.957      0.080    1884.156
MASS6        155.480      0.010    1555.529
MASS7        159.212      0.187     850.129
MASS8        164.574      0.135    1214.611

Mixture proportions:
    MASS1       MASS2       MASS3       MASS4       MASS5
    0.0385      0.1154      0.1154      0.1923      0.2692
    MASS6       MASS7       MASS8
    0.1539      0.0385      0.0769

Component distribution - MLE of sigma:     1.406
```

```
Random effect distribution - standard deviation:   7.918

-2 log L:           922.6     Convergence at iteration  10
```

This suggests that a quadratic-age, 8-point discrete model is sufficiently complex. The disparity change from the 500-point Gaussian model is 8.69 which is not large. The within-boy standard deviation increases from 1.272 in the Gaussian model to 1.406 for the 8 mass-point quadratic-age discrete model.

We consider the interpretation of this model below, but first consider a random slope model. Is the linear growth trend consistent for all boys? We extend the 8-point model by allowing the quadratic trend to vary across boys.

```
> oxboys.np8lin <- update(oxboys.np8, height ~ age,
+      random = ~1 | Subject)
> oxboys.np8linlin <- update(oxboys.np8lin, random = ~age |
+      Subject)
> oxboys.np8quad <- update(oxboys.np8, height ~ age +
+      I(age^2), random = ~1 | Subject)
> oxboys.np8quadlin <- update(oxboys.np8quad, random = ~age |
+      Subject)
> oxboys.np8quadquad <- update(oxboys.np8, random = ~(age +
+      I(age^2)) | Subject)
```

The disparity decreases by 101.97 on 7 df, showing clear heterogeneity in slope. Examination of the slope differences shows that the age slope decreases smoothly with increasing mass-points, that is, with decreasing height at age 12. Smaller boys increase their heights less rapidly than taller boys. Before examining the fitted model, we check again for an acceleration term.

```
> summary(oxboys.np8quadlin)

Coefficients:
            Estimate Std. Error     t value
age            9.206      0.175      52.651
I(age^2)       0.746      0.085       8.799
MASS1        129.950      0.164     793.285
MASS2        138.135      0.099    1395.983
MASS3        143.059      0.099    1446.443
MASS4        147.063      0.080    1841.203
MASS5        150.953      0.070    2154.443
MASS6        155.466      0.087    1778.325
MASS7        159.162      0.164     971.597
MASS8        164.512      0.119    1388.139
MASS1:age     -5.490      0.303     -18.129
MASS2:age     -4.009      0.226     -17.775
MASS3:age     -2.154      0.226      -9.544
MASS4:age     -3.786      0.207     -18.311
```

```
MASS5:age  -2.566      0.198        -12.941
MASS6:age  -2.126      0.214         -9.923
MASS7:age  -0.542      0.303         -1.790

Mixture proportions:
     MASS1         MASS2         MASS3         MASS4         MASS5
     0.0385        0.1154        0.1154        0.1923        0.2692
     MASS6         MASS7         MASS8
     0.1538        0.0385        0.0769

Component distribution - MLE of sigma:      1.152
Random effect distribution - standard deviation:    7.893692

-2 log L:           829.4     Convergence at iteration  10
```

Could this be a higher-order curvature? Adding the cubic term in age to the fixed model gives a disparity change of only 1.27.

Is the quadratic curvature itself varying across boys? Adding the quadratic term in age to the `random` term gives a disparity change of 3.65 on 6 df. We present a summary of the disparities from the models involving the linear, quadratic and cubic trends in Table 9.1.

Is there a seasonal variation in growth? We define the four seasons represented by the four measurement times in each year (they occur in June/July, September, January, and April/May) by a four-level factor, and add this to the fixed part of the model.

```
> Oxboys$Season <- Oxboys$Occasion
> levels(Oxboys$Season) <- levels(Oxboys$Season)[c(1,
+     2, 3, 4, 1, 2, 3, 4, 1)]
> oxboys.np8seasonrclin <- update(oxboys.np8quadlin,
+     . ~ . + Season, tol = 0.4)
> round(coef(oxboys.np8seasonrclin), 3)
```

Table 9.1. Disparities for random intercept and random coefficient models relating boys' height to age

Fixed formula	Random formula	Disparity
height ~ age	~1 | Subject	931.38
height ~ age	~age | Subject	842.44
height ~ age + I(age^2)	~1 | Subject	922.64
height ~ age + I(age^2)	~age | Subject	829.41
height ~ age + I(age^2)	~(age + I(age^2 | Subject	825.75
height ~ age + I(age^2) + I(age^3)	~age | Subject	828.13

```
      age  I(age^2)  Season.L  Season.Q  Season.C     MASS1
    9.185     0.740     0.066     0.112     0.061   129.953
    MASS2     MASS3     MASS4     MASS5     MASS6     MASS7
  138.138   143.062   147.066   150.956   155.469   159.165
    MASS8 MASS1:age MASS2:age MASS3:age MASS4:age MASS5:age
  164.515    -5.490    -4.009    -2.154    -3.786    -2.566
MASS6:age MASS7:age
   -2.126    -0.542
```

```
> round(oxboys.np8seasonrclin$disparity, 2)
```

```
[1] 828.63
```

The disparity increases by 0.77 with three extra parameters!

We conclude that the random slope but fixed quadratic model is appropriate. We examine the posterior probabilities of component membership for this model. We suppress probabilities smaller than 0.01 and group boys by component.

```
> oxboys.post.prob <- post(oxboys.np8quadrclin,
+     level = "upper")$prob
> boy.cluster <- which(round(oxboys.post.prob) == 1,
+     arr.ind = TRUE)[, 2]
> tabn <- apply(round(oxboys.post.prob), 2, sum)
> ord <- names(sort(tabn))
> oxboys.post.prob <- oxboys.post.prob[names(boy.cluster),
+     ]
> dimnames(oxboys.post.prob) <- list("boy   " = names(boy.cluster),
+     "        K" = unique(boy.cluster))
> print.pp <- function(x) print.noquote(ifelse(x <
+     0.01, "    ", fc0(x)))
> print.pp(oxboys.post.prob)
```

```
           K
boy    1   2    3    4    5    6    7    8
   10 1
   9       1
   25      1
   26      1
   2            1
   15           1
   17           1
   1                 1
   6                 1
   7                 1
   8                 1
   16                1
   5                      1
   11                     1
```

```
18                   1
20                   1
21                   1
23                   1
24                   1
3                         1
12                        1
13                        1
22                        1
14                             1
4                                   1
19                                  1
```

The boys are assigned to a single component with estimated probability 1.0. The random intercept and slope model classifies the boys into eight homogeneous groups with two boys allocated their own groups and one group with intercept 151 cm containing 7 boys. The fitted models for each component are shown in Fig. 9.2, superimposed on the previous graph.

```
> Oxboys$fitted <- predict(oxboys.np8quadrclin)
> Oxboys$cluster <- factor(boy.cluster[Oxboys$Subject %in%
```

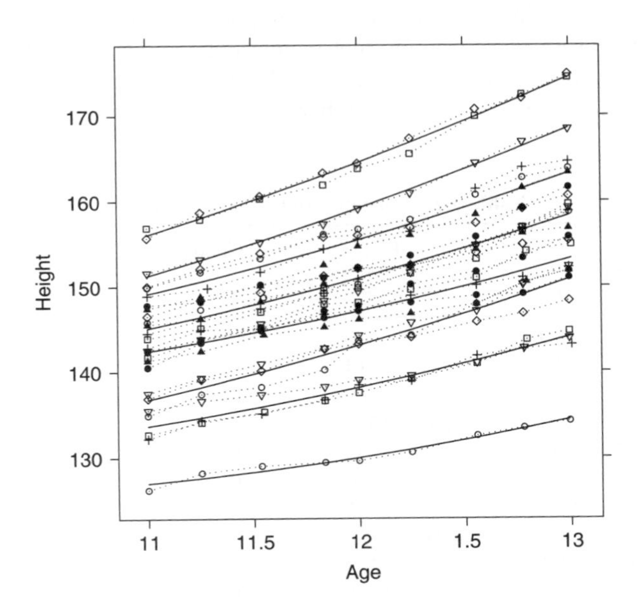

Fig. 9.2. Height trends of 26 boys over two years

```
+       names(boy.cluster)])
> print(xyplot(height ~ age, group = Subject, data = Oxboys,
+       subscripts = TRUE, scales = list(x = list(at = c(-1,
+           -0.5, 0, 0.5, 1), labels = c(11, 11.5, 12,
+           1.5, 13))), panel = function(x, y, subscripts,
+           ...) {
+           panel.superpose(x, y, subscripts, type = "b",
+               lty = 3, ...)
+           panel.superpose(x, Oxboys$fitted, lty = 1,
+               subscripts, type = "l", lwd = 1.2, ,
+               ...)
+       }))
```

The increasing variability of heights with increasing age is now seen to be a combination of a common acceleration and different slopes for the different groups.

9.7.2 *Multi-centre trial of beta-blockers*

The second example is the 22-centre clinical trial of beta-blockers for reducing mortality after myocardial infarction, described by Yusuf *et al.* (1984) and analysed in detail in Gelman *et al.* (1995) and Aitkin (1999). The data are shown in Table 9.2, and are in the dataset `betablok`. In each centre *r* patients died out of *n* treated, under either the treatment T or control C condition. Studies involving multiple centres are common in drug trials, where an important issue is the *generalizability* of the treatment effect across different patient populations. The analysis of the multiple tables, or of multiple data sets of other kinds, is frequently called *meta-analysis* – *meta* in the sense of 'at a higher level' than that of a single trial.

Table 9.2. 22-centre trial of beta-blockers

Centre	C		T		Centre	C		T	
	r	*n*	*r*	*n*		*r*	*n*	*r*	*n*
1	3	39	3	38	12	47	266	45	263
2	14	116	7	114	13	16	293	9	291
3	11	93	5	69	14	45	883	57	858
4	127	1520	102	1533	15	31	147	25	154
5	27	365	28	355	16	38	213	33	207
6	6	52	4	59	17	12	122	28	251
7	152	939	98	945	18	6	154	8	151
8	48	471	60	632	19	3	134	6	174
9	37	282	25	278	20	40	218	32	209
10	188	1921	138	1916	21	43	364	27	391
11	52	583	64	873	22	39	674	22	680

The outcomes are represented by a two-level model, with centres as the upper level and patients as the lower level. There is only one explanatory variable, the treatment assignment at the lower level.

We input the data, and fit both normal and nonparametric models for Z.

```
> data(betablok, package = "SMIR")

> library(npmlreg)
> (betablok.gq1 <- allvc(cbind(r, (n - r)) ~ treat,
+     data = betablok, random = ~1 | centre, family = binomial,
+     random.distribution = "gq", k = 1, verbose = FALSE,
+     plot.opt = 0))

Coefficients:
(Intercept)        treatT
     -2.197        -0.257
Random effect distribution - standard deviation:   0

-2 log L:        523.2
```

The model fitted with one mass-point is identical to the fixed-effect model ignoring among-centre variation. We increase successively the number of mass-points, from 2 to 10 and 20. The parameter estimates, standard errors (expected complete data information-based) and the disparities are shown in Table 9.3.

```
> betablok.gq2 <- update(betablok.gq1, k = 2)
> betablok.gq3 <- update(betablok.gq1, k = 3)

                    ⋮   ⋮

> betablok.gq20 <- update(betablok.gq1, k = 20)
```

Table 9.3. Gaussian quadrature for beta-blockers

K	α	SE	β	SE	σ_A	Disparity
1	−2.197	0.034	−0.257	0.049	NA	523.19
2	−2.034	0.035	−0.257	0.050	0.366	365.49
3	−2.239	0.034	−0.258	0.031	0.360	320.98
4	−2.092	0.033	−0.262	0.031	0.295	331.03
5	−2.238	0.034	−0.258	0.030	0.455	319.76
6	−2.091	0.034	−0.262	0.031	0.359	322.47
7	−2.238	0.034	−0.259	0.031	0.528	321.03
8	−2.088	0.034	−0.262	0.031	0.410	319.36
9	−2.238	0.034	−0.259	0.031	0.588	322.77
10	−2.238	0.034	−0.262	0.030	0.454	318.16
20	−2.180	0.034	−0.261	0.030	0.432	316.72

The disparities vary considerably with small numbers of mass-points. The standard deviation σ_A also varies, but the treatment effect estimate $\hat{\beta}$ and its standard error are almost unchanged from the independence model, despite the very large change in disparity. To check the standard error calculation, we drop the treatment variable:

```
> (betablok.gq20.1 <- update(betablok.gq20, . ~ 1))

Coefficients:
(Intercept)             z
    -2.307         0.431
Random effect distribution - standard deviation:    0.431

-2 log L:             344.2
```

The disparity change is 27.50, so the LRTS-based standard error is $0.2609/\sqrt{27.50} = 0.04974$, compared with 0.04977 from the summary output. There is almost no underestimation. The treatment produces a significant, though small, reduction in death risk compared to the control: the odds of death under the betablocker treatment are reduced by 23% relative to those under the control treatment. Increasing the number of quadrature points to 100 results in little change in disparity.

For the nonparametric estimation, we redefine the random distribution to be 'np'.

```
> betablok.np1 <- update(betablok.gq1, random.distribution = "np")
> betablok.np2 <- update(betablok.gq2, random.distribution = "np")
> betablok.np3 <- update(betablok.np2, k = 3)
> betablok.np4 <- update(betablok.np2, k = 4)

> betablok.np5 <- update(betablok.np2, k = 5)
> betablok.np6 <- update(betablok.np2, k = 6)
> betablok.np7 <- update(betablok.np2, k = 7)
```

The mass-points, mixture proportions, parameter estimate, standard error and the disparities are shown in Table 9.4.

The decrease in the disparities becomes small after $K = 5$. The disparity is now 310.39, less than that for the 20-point Gaussian quadrature analysis. For the six mass-point model only five mass-points are needed – the last mass-point has no mass and is identical to the fifth mass. A further increase in K to $K = 7$, shows little change in the disparity, so the five mass-points appear to define the NPML estimate of the mixing distribution.

```
> summary(betablok.np5)

Coefficients:
       Estimate Std. Error    t value
treatT -0.258     0.0498       -5.182
```

Table 9.4. Non-parametric mass-points and mixture proportions (row below, to 2 dp) and treatment estimate and its standard error for beta-blockers

K	1	2	3	4	5	6	7	α	SE	Disparity
1	−2.20							−0.257	0.049	523.19
2	−2.39	−1.65						−0.255	0.050	362.66
	0.70	0.30								
3	−2.83	−2.25	−1.61					−0.258	0.050	318.72
	0.24	0.51	0.25							
4	−2.83	−2.26	−1.79	−1.44				−0.258	0.050	311.51
	0.24	0.48	0.10	0.18						
5	−2.98	−2.69	−2.26	−1.79	−1.44			−0.258	0.050	310.39
	0.16	0.08	0.48	0.10	0.18					
6	−2.98	−2.69	−2.26	−1.79	−1.44	−1.44		−0.258	0.050	310.39
	0.16	0.08	0.48	0.10	0.18	0.00				
7	−2.97	−2.97	−2.68	−2.20	−1.77	−1.44	−2.35	−0.259	0.050	309.47
	0.00	0.16	0.07	0.37	0.08	0.18	0.13			

```
MASS1   -2.975        0.1047        -28.410
MASS2   -2.687        0.0989        -27.176
MASS3   -2.258        0.0404        -55.947
MASS4   -1.787        0.0657        -27.196
MASS5   -1.440        0.0690        -20.859

Mixture proportions:
      MASS1         MASS2         MASS3         MASS4         MASS5
      0.1631        0.0766        0.4812        0.0994        0.1797

Random effect distribution - standard deviation:    0.488

-2 log L:            310.4     Convergence at iteration   41
```

The masses are not increasing and decreasing smoothly, but are fairly symmetrical. This non-monotone behaviour makes the NPML estimate difficult to find from the Gaussian quadrature starting points with smaller K.

The treatment effect estimate $(\widehat{\beta})$ is almost identical to that in the Gaussian quadrature analyses, but the standard deviation of the mass-point distribution is 0.488, larger than for the normal models, partly because of the relatively large masses at the extremes of the distribution.

Is there evidence of non-normal variation in the intercepts? The disparity change between the five-point Gaussian quadrature and the NPML analyses is 9.37 for the eight additional parameters. This is not at all persuasive (by the usual χ_8^2 test).

We finally check on the constancy of the treatment effect, by fitting a random treatment model.

```
> betablok.np5rc <- update(betablok.np5, random = ~treat |
+     centre)
> summary(betablok.np5rc)

Coefficients:
             Estimate Std. Error    t value
treatT         -0.159      0.130     -1.225
MASS1          -3.675      0.971     -3.786
MASS2          -2.905      0.098    -29.565
MASS3          -2.254      0.045    -50.148
MASS4          -1.683      0.083    -20.346
MASS5          -1.486      0.089    -16.680
MASS1:treatT  0.543      1.217      0.446
MASS2:treatT  0.074      0.192      0.384
MASS3:treatT -0.100      0.145     -0.691
MASS4:treatT -0.328      0.182     -1.798

Mixture proportions:
    MASS1       MASS2       MASS3       MASS4       MASS5
    0.0148      0.2247      0.4929      0.0860      0.1816

Random effect distribution - standard deviation:   0.510

-2 log L:      306.2     Convergence at iteration  10

> round(betablok.np5rc$disparity, 2)

[1] 306.21
```

The disparity reduces to 306.21. This is a reduction in disparity of 4.18 for the four additional parameters. Since these interaction parameters do not affect the mixture structure, their zero null hypothesis values are internal to the parameter space, and the sample sizes are relatively large, the usual asymptotic χ^2 theory applies to the disparity difference. The *P*-value of 4.18 is 0.382 for χ^2_4 – there is no evidence for treatment heterogeneity.

We examine the posterior probabilities of component membership for each centre. We table these probabilities below, rounded to 2 dp, omitting probabilities less than 0.01. We adopt the convention that a centre is identified with a component if the posterior probability of the centre belonging to the component is at least 0.90, and present the centers ordered by cluster which are themselves ordered by their mass.

```
> betablok.post.prob <- apply(betablok.np5$post.prob,
+     2, tapply, betablok$centre, mean)
> betablok.post.prob <- betablok.post.prob[,
+     rev(order(betablok.np5$masses))]
```

```
> center.cluster <- sort(apply(betablok.post.prob,
+       1, which.max))
> betablok.post.prob <-
+       betablok.post.prob[names(center.cluster), ]
> dimnames(betablok.post.prob) <- list("center   " = names(center
  .cluster),
+       "           K" = rev(order(betablok.np5$masses)))
> print.pp2(betablok.post.prob)
```

```
                K
center    3      5      1      4      2
      1   0.74   0.01   0.1    0.06   0.09
      2   0.94                 0.04   0.02
      3   0.87                 0.09   0.02
      4   1
      5   0.98                        0.02
      6   0.83   0.01   0.03   0.08   0.05
      8   0.99                 0.01
      9   0.55                 0.45
      10  1
      11  1
      17  0.63                 0.37
      21  1
      12         1
      15         0.99          0.01
      16         0.96          0.04
      20         0.96          0.04
      13                0.87          0.13
      18  0.06          0.7           0.24
      19                0.92          0.08
      22                0.93          0.07
      7                        1
      14                0.03          0.97
```

Component 3, which contains 48% of the mass, is identified by the seven centres 2, 4, 5, 8, 10, 11, and 21.

Component 5, containing 18% of the mass is identified by centres 12, 15, 16, and 20.

Centres 1, 3, 6, 9, 13, 17, and 18 are not identified with a single component. These centres have small or moderate size trials, which do not estimate the treatment effect very precisely. The large trials identify precisely single components; the small trials have posterior probabilities which are spread over these components.

Centre 14 stands out as an 'outlier' being the only centre associated with component 2.

The analysis of this study by Aitkin (1999) used a three mass-point estimate for the intercept and slope variation which was incorrectly thought to be the NPML estimate. This analysis consequently found no significant evidence of variation in the treatment effect. Gelman *et al.* (1995) found a posterior median for the treatment effect of −0.25 with a posterior standard deviation of about 0.07. The estimated treatment effect is −0.258 and standard deviation 0.045 which are very similar, but our analysis identifies centre 14 as an outlier.

9.7.3 *Longitudinal study of obesity*

The third example is of data from the Muscatine, Iowa study of child development. A component of the Muscatine study was a longitudinal study of childhood obesity, discussed in Woolson and Clarke (1984). Part of the data from Woolson and Clarke were reanalysed by Fitzmaurice *et al.* (1994) and Aitkin (1999). These data were binary indicators of obesity on 1,014 children who were 7–9 years old in 1977, and were followed up in 1979 and 1981. Children were classified as obese if their weights were more than 210% of the population median weight for their gender and height; about 20% of children were classified as obese. The repeated binary response of interest `y` is whether the child is obese (1) or not (0) at each occasion. The `sex` of the child is also recorded as binary: male 0, female 1. Data on many children are incomplete, and only 460 children had complete data from all three occasions. Table 9.5 gives the child's obesity status at all three occasions for the 1,014 children with complete or incomplete data; the data are in the dataset `Obesity`.

Fitzmaurice *et al.* (1994) modelled the *marginal* probability of response by a logistic model with linear and quadratic age terms and their interactions with sex, and saturated the covariance matrix between occasions. This is similar to the GEE approach of Liang and Zeger (1986) but the parameters are estimated by full maximum likelihood. They analysed both the subset of children with complete data and the full sample with both complete and incomplete data, and found substantial changes in the conclusions, demonstrating the need for inclusion of all the data in the analysis.

We repeat the analysis of the full sample with the random-effect model, with the two-level structure of children and occasions within child. The intra-class correlation structure on the logit scale is simple, though on the probability scale this corresponds to a general correlation structure because of the non-linear transformation.

```
> data(Obesity, package = "SMIR")

> library(npmlreg)
> obesity.np1 <- allvc(cbind(status, (1 - status)) ~
+     sex * (age1 + age2), data = Obesity, tol = 0.2,
```

Table 9.5. Obesity status on three occasions

1977	1979	1981	Males	Females
0	0	0	150	154
0	0	1	15	14
0	1	0	8	13
0	1	1	8	19
1	0	0	8	2
1	0	1	9	6
1	1	0	7	6
1	1	1	20	21
—	1	1	13	8
—	1	0	3	1
—	0	1	2	4
—	0	0	42	47
1	—	1	3	4
1	—	0	1	0
0	—	1	6	16
0	—	0	16	3
1	1	—	11	11
1	0	—	1	1
0	1	—	3	3
0	0	—	38	25
—	—	1	14	13
—	—	0	55	39
—	1	—	4	5
—	0	—	33	23
1	—	—	7	7
0	—	—	45	47

```
+      family = binomial, random = ~1 | id2, k = 1,
+      plot.opt = 0, verbose = FALSE)

> summary(obesity.np1)

Coefficients:
             Estimate Std. Error     t value
MASS1     -1.328       0.074       -17.950
sex        0.034       0.105         0.322
age1       0.146       0.091         1.607
age2       0.012       0.052         0.238
sex:age1   0.113       0.130         0.865
sex:age2  -0.082       0.073        -1.124
```

```
Mixture proportions:
MASS1
    1

Random effect distribution - standard deviation:   0

-2 log L:          2271     Convergence at iteration  0

> fc2(obesity.np1$disparity)

[1] 2271.01

> obesity.np2 <- update(obesity.np1, k = 2)

> summary(obesity.np2)

Coefficients:
             Estimate Std. Error      t value
sex           0.031     0.159         0.195
age1          0.339     0.140         2.421
age2          0.031     0.080         0.390
sex:age1      0.341     0.195         1.748
sex:age2     -0.190     0.113        -1.685
MASS1        -3.309     0.155       -21.330
MASS2         1.152     0.130         8.860

Mixture proportions:
    MASS1       MASS2
    0.7598      0.2402

Random effect distribution - standard deviation:   1.906

-2 log L:          1892.06    Convergence at iteration  58

> obesity.np3 <- update(obesity.np1, k = 3)

> summary(obesity.np3)

Coefficients:
             Estimate Std. Error      t value
sex           0.026     0.159         0.160
age1          0.339     0.140         2.423
age2          0.032     0.080         0.399
sex:age1      0.344     0.195         1.759
sex:age2     -0.192     0.113        -1.695
MASS1        -5.228     0.617        -8.475
MASS2        -2.963     0.155       -19.153
MASS3         1.185     0.132         9.004
```

```
Mixture proportions:
    MASS1       MASS2       MASS3
    0.2068      0.5567      0.2365

Random effect distribution - standard deviation:    2.206

-2 log L:       1892.06    Convergence at iteration  29

> obesity.np4 <- update(obesity.np1, k = 4)

> summary(obesity.np4)

Coefficients:
            Estimate Std. Error    t value
sex          0.184     0.160       1.149
age1         0.328     0.137       2.396
age2         0.032     0.079       0.404
sex:age1     0.380     0.198       1.922
sex:age2    -0.197     0.113      -1.739
MASS1       -8.076     4.053      -1.993
MASS2       -6.494     0.678      -9.579
MASS3       -0.707     0.123      -5.750
MASS4        1.745     0.176       9.898

Mixture proportions:
    MASS1        MASS2        MASS3        MASS4
    0.0728       0.5479       0.2301       0.1492

Random effect distribution - standard deviation:    3.469

-2 log L:       1891.29    Convergence at iteration  40
```

The disparity changes slightly from two to four mass-points, as the large negative mass-points are driven further down the scale. This does not change the meaning of the model, and changes the other model parameters only slightly, apart from the sex effect. The large number of children who are not obese on all three occasions implies large masses at large negative values on the logit scale, but their precise locations have little effect on the fitted model. We proceed to simplify the model with two mass-points; we suppress the full output.

```
> obesity.np2a <- update(obesity.np2, . ~ . - sex:age2)
> obesity.np2b <- update(obesity.np2a, . ~ . - sex:age1)
> obesity.np2c <- update(obesity.np2b, . ~ . - age2)
> obesity.np2d <- update(obesity.np2c, . ~ . - sex)
> obesity.np2e <- update(obesity.np2d, . ~ . - age1)
```

Successive disparity changes for the omitted effects are 2.58 (`sex:age2`), 2.60 (`sex:age1`), 1.22 (`age2`), 0.00 (`sex`), and 23.07 (`age1`). There is a linear age effect, the same for both sexes and no sex difference. Parameter estimates for the `age1` model are given below to 3dp.

```
> print(obesity.np2d, digits = 3)

Coefficients:
  age1    MASS1    MASS2
 0.504   -3.301    1.129

Random effect distribution - standard deviation:    1.898056

Mixture proportions:
    MASS1         MASS2
    0.7578        0.2422
-2 log L:          1898.5
```

The LRTS-based standard error for `age1` can be computed as $0.504/\sqrt{23.07} = 0.105$, only 4% larger than the *glm* figure.

The two-component mixture is identified by the 'consistently obese' and 'consistently non-obese' children, not surprisingly. The fitted probability of obesity is shown below with the response on each occasion (1 = obese).

Pattern	Obesity Probability	Pattern	Obesity Probability
0 0 0	.005	1 - 0	.621
- 0 0	.014	0 1 -	.706
0 - 0	.019	1 0 -	.706
0 0 -	.028	- - 1	.822
0 - -	.095	- 1 -	.871
- 0 -	.120	1 - -	.905
- - 0	.178	0 1 1	.972
1 0 0	.293	1 0 1	.972
0 1 0	.293	1 1 0	.972
0 0 1	.293	- 1 1	.990
- 0 1	.538	1 - 1	.993
- 1 0	.538	1 1 -	.995
0 - 1	.621	1 1 1	1.0

The fitted conditional (broken curves) and population-average (solid curve) models are shown in Fig. 9.3.

```
> inv.logit <- function(eta) exp(eta)/(1 + exp(eta))
> fval <- inv.logit(as.numeric(names(table(
+     obesity.np2d$linear.predictors))))
```

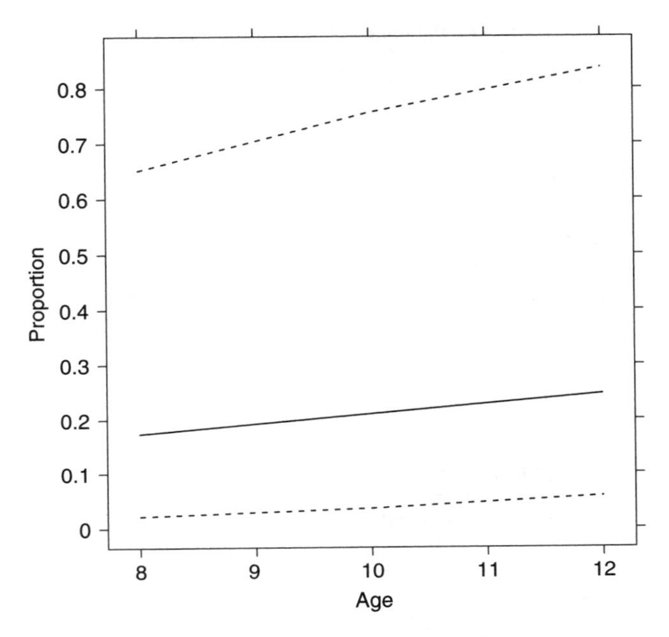

Fig. 9.3. Conditional and population-average obesity proportions

```
> age <- rep(c(8, 10, 12), 2)
> group <- gl(2, 3)
> prob <- rep(obesity.np2d$masses, c(3, 3))
> paop <- as.vector(apply(matrix(prob * fval, nrow = 3,
+     ncol = 2), 1, sum))
> print(xyplot(fval ~ age, groups = group, ylab = "proportion",
+     scales = list(y = list(at = seq(0, 1, by = 0.1),
+         labels = seq(0, 1, by = 0.1))), panel = function(x,
+         y, ...) {
+         panel.superpose.2(x, y, type = "l", lty = 2,
+             lwd = 1.3, ...)
+         panel.lines(c(8, 10, 12), paop, lwd = 1.3,
+             ...)
+     }))
```

Table 9.6 gives the parameter estimates and standard errors for the full model, from both the Fitzmaurice *et al.* and the random effect model approaches, and Table 9.7 gives those for the final model (Fitzmaurice *et al.* did not give the estimates for the final linear model).

It is of interest that both ML analyses lead to the same models with nearly proportional parameter estimates and standard errors, though the models being

Table 9.6. Parameter estimates for full model

Parameter	Marginal ML estimate	SE	NPML estimate	SE
Intercept	−1.356	0.098	1.194	
GENDER	0.043	0.138	0.031	0.159
AGE(L)	0.142	0.063	0.339	0.140
AGE(Q)	0.014	0.035	0.031	0.080
GENDER.AGE(L)	0.162	0.096	0.341	0.195
GENDER.AGE(Q)	−0.089	0.049	−0.190	0.113
Disparity			1892.1	

Table 9.7. Parameter estimates for final model

Parameter	Marginal ML estimate	SE	NPML estimate	SE
Intercept	−1.321	0.094	1.151	
AGE(L)	0.220	0.059	0.506	0.101
AGE(Q)	−0.033	0.032	−0.065	0.056
Deviance			1897.3	
Intercept			1.165	
AGE(L)			0.504	0.101
Deviance			1898.5	

fitted are different: a marginal logit model in the first case and a conditional model in the second. The inflation of estimates and standard errors for the NPML analysis is a consequence of its much greater variance on the logit scale: $\pi^2/3+\sigma_A^2 = 9.29$ compared with $\pi^2/3 = 3.29$ for the marginal logit model. The corresponding intraclass correlation is 0.65. Neuhaus and Jewell (1990) reported similar inflation in a paired study, and Neuhaus (1992) gave a general discussion of the relation between parameter estimates by these and other approaches.

Further discussion of these and other examples, and extensive references, can be found in Aitkin (1999). The data given in Woolson and Clarke (1984) are much more extensive, with five cohorts of children aged 5–7, 7–9, 9–11, 11–13, and 13–15 followed for four years, with obesity assessment every two years. The Muscatine study was even larger, with longer follow-up.

The data are in `obesityfull`, and the cohorts are identified by the weight variables `wt6`, `wt8`, `wt10`, `wt12`, `wt14`. We leave further analysis as an exercise for the reader.

9.8 Autoregressive random effect models

A limitation of the simple random effect model in Examples 2 and 3 is that it forces the correlation between any two repeated measures to be the same (at least on the underlying latent variable scale). It is a common experience, however, that such correlations 'decay' with increasing distance in time. For this reason models with other forms of correlation structure are widely used in longitudinal studies; a common one is the *autoregressive* model, in which the response on occasion j depends explicitly on responses at previous times. Autoregressive models are quite straightforward to fit, at least with complete longitudinal data, and provide a valuable generalization of the simple random effect model, but care is needed in the model specification when both random effect and autoregressive structure are modelled, because of a well-known difficulty called the *initial conditions problem* (see, e.g. Blundell and Bond, 1998; Aitkin and Alfó, 2003).

A simple example is the AR1 model, where the subscript t indexes time, for the response repeatedly measured within individual. Suppose for simplicity that each individual i is measured on T occasions, with an exponential family distribution for the response Y_{it}:

$$Y_{i,t} \mid y_{i,t-1}, z_i \sim ef(\eta_{i,t}), \quad t = 2, \dots, T \text{ independently},$$

where

$$\eta_{i,t} = \boldsymbol{\beta}'\mathbf{x}_{it} + \gamma y_{i,t-1} + z_i,$$

and as before $z_i \sim \pi(z)$ for some random effect distribution π. Here γ is the autoregressive parameter. A difficulty with this model is that it does not specify the distribution of $Y_{i,1}$ at the first outcome. The difficulty could be avoided by conditioning on this outcome; it might appear that

$$f\left(y_{i,2}, \dots, y_{i,T} \mid y_{i,1}\right) = \int \prod_{t=2}^{T} f\left(y_{i,t} \mid y_{i,t-1}, z_i\right) \pi(z_i)\, \mathrm{d}z_i.$$

But individual i will have the same 'propensity' random effect on the first measurement occasion as on the other occasions, and so $Y_{i,1}$ cannot be assumed independent of Z_i. In fact using Bayes' theorem

$$\pi(z_i \mid y_{i,1}) = f(y_{i,1} \mid z_i)\pi(z_i)/f(y_{i,1}),$$

and we could restate the model in the following way:

$$f(y_{i,1}, \ldots, y_{i,T}) = \int \prod_{t=2}^{T} f(y_{i,t} \mid y_{i,t-1}, z_i) f(y_{i,1} \mid z_i) \pi(z_i) \, dz_i$$

$$= \int \prod_{t=2}^{T} f(y_{i,t} \mid y_{i,t-1}, z_i) f(y_{i,1}) \pi(z_i \mid y_{i,1}) \, dz_i$$

$$= f(y_{i,1}) \int \prod_{t=2}^{T} f(y_{i,t} \mid y_{i,t-1}, z_i) \pi(z_i \mid y_{i,1}) \, dz_i$$

or equivalently,

$$f(y_{i,2}, \ldots, y_{i,T} \mid y_{i,1}) = \int \prod_{t=2}^{T} f(y_{i,t} \mid y_{i,t-1}, z_i) \pi(z_i \mid y_{i,1}) \, dz_i.$$

The result is a random effect model with the $Y_{i,t}$ $(t = 2, \ldots, T)$ conditioned on the initial outcome, but the distribution of Z_i has changed due to the conditioning, and depends on the initial outcome for each observation; it can no longer be considered the previous homogeneous distribution $\pi(z)$. This difficulty invalidates in general the simple approach of conditioning and using the standard random effect model; however, for the conjugate normal/normal model for Y *and* the random effect Z, the conditioning on the first outcome still leaves a normal variance component model with a normal random effect, and this model can be re-parametrized to transfer the (conditional) mean of the normal random effect to the linear predictor for the response model, allowing the standard random effect analysis to be applied in the conditional model given the first outcome.

Aitkin and Alfó (2003) discussed alternative approaches to the conditioned model; a different solution is to complete the likelihood by specifying the initial outcome. The absence of a previous outcome means that the regression coefficients on the explanatory variables cannot in general be the same, nor can it be assumed that the variability of the random effect is unaffected by conditioning. We might therefore specify

$$Y_{i,1} \mid z_i \sim ef(\eta_{i,1}),$$

where

$$\eta_{i,1} = \boldsymbol{\delta}' \mathbf{x}_{i,1} + \lambda z_i,$$

where λ is a scale factor.

The full random effect AR1 model can be fitted by re-structuring the data so that each outcome is an explanatory variable for the subsequent one; the first outcome is given a dummy zero variable as its 'previous outcome'. We define a dummy variable `t=0` for the first outcome and 1 for subsequent outcomes; the `fixed` formula argument then includes the previous outcome, the `t` variable and its interactions with the explanatory variables. The `random` formula includes the `t` variable, to allow for different variances of the random effect in the first unconditional model and the subsequent conditional models.

We illustrate with the subset of complete data from the Woolson and Clarke study (the missingness in the full data set greatly complicates this analysis, without providing much information about the autoregressive parameter). We read in the dataset `woolson` and extend the data structure as shown below:

```
> data(woolson, package = "SMIR")

> woolson <- transform(woolson, a1 = (age - 10)/2,
+      a2 = 3 * ((age - 10)/2)^2 - 2)
> woolson <- transform(woolson, sa1 = sex * a1, sa2 = sex *
+      s2, t = ifelse(age > 8, 1, 0))
> woolson <- transform(woolson, tsex = t * sex, ta1 = t *
+      a1, tsa1 = t * sex * a1)
> woolson <- woolson[order(woolson$age, 1:48), ]
> wt <- c(150, 15, 8, 8, 8, 9, 7, 20, 154, 14, 13,
+      19, 2, 6, 6, 21)
> Wt <- rep(wt, 3)
> Woolson <- data.frame(lapply(woolson, function(x) rep(x,
+      Wt)))
> Woolson$Child <- rep(1:460, 3)

> library(npmlreg)
> obesity.allvc1 <- allvc(y ~ sex + t + x + tsex +
+      ta1 + tsa1, random = ~t | Child, random.distribution = "np",
+      k = 1, data = Woolson, family = binomial, verbose = FALSE,
+      plot.opt = 0)

> summary(obesity.allvc1)

Coefficients:
         Estimate Std. Error   t value
MASS1 -1.414       0.168      -8.414
sex    -0.329      0.249      -1.322
t      -0.895      0.273      -3.282
x       2.739      0.199      13.780
tsex    1.006      0.364       2.762
ta1     0.352      0.272       1.290
tsa1   -0.748      0.374      -2.002
```

```
Mixture proportions:
MASS1
   1

Random effect distribution - standard deviation:   0

-2 log L:          1197.4     Convergence at iteration  0
```

The one-mass-point model has no random effect, but the autoregression coefficient is very large. We increase the number of mass-points.

```
> obesity.allvc2 <- update(obesity.allvc1, k = 2)

> summary(obesity.allvc2)

Coefficients:
            Estimate Std. Error    t value
sex       -0.542      0.345      -1.572
t         -0.661      0.369      -1.792
x          0.595      0.257       2.318
tsex       1.323      0.465       2.847
ta1        0.462      0.311       1.486
tsa1      -0.517      0.438      -1.179
MASS1    -11.414     18.905      -0.604
MASS2      0.540      0.248       2.173
MASS1:t    8.399     18.907       0.444

Mixture proportions:
    MASS1      MASS2
   0.6965     0.3035

Random effect distribution - standard deviation:   5.496

-2 log L:          1156.2     Convergence at iteration  52

> obesity.allvc3 <- update(obesity.allvc1, k = 3)

> summary(obesity.allvc3)

Coefficients:
            Estimate Std. Error    t value
sex       -0.614      0.351      -1.751
t         -0.586      0.388      -1.511
x          0.449      0.269       1.670
tsex       1.358      0.472       2.875
ta1        0.466      0.317       1.469
tsa1      -0.496      0.445      -1.116
MASS1    -10.916     25.036      -0.436
```

```
MASS2     -4.134       0.628      -6.587
MASS3      0.704       0.262       2.690
MASS1:t    4.581      25.092       0.183
MASS2:t    1.664       0.699       2.381

Mixture proportions:
    MASS1       MASS2       MASS3
    0.2395      0.4814      0.2791

Random effect distribution - standard deviation:   4.181

-2 log L:        1156.2     Convergence at iteration  83
```

The disparity change when a random effect is introduced is 41.17, but little changes when the number of mass-points is increased from two to three; even with autoregression on the previous outcome, there is a strong random effect dependence. The dependence on the previous outcome is greatly reduced, and there is little sign of different random effect variances. We simplify the random model.

```
> obesity.allvc3a <- update(obesity.allvc3, random = ~1 |
+     Child)

> summary(obesity.allvc3a)

Coefficients:
        Estimate Std. Error      t value
sex    -0.769         0.346      -2.221
t       0.035         0.375       0.094
x      -0.195         0.300      -0.648
tsex    1.360         0.484       2.813
tal     0.515         0.340       1.513
tsal   -0.433         0.471      -0.918
MASS1  -7.413         1.890      -3.922
MASS2  -2.888         0.296      -9.752
MASS3   0.933         0.262       3.564

Mixture proportions:
    MASS1       MASS2       MASS3
    0.2528      0.5037      0.2435

Random effect distribution - standard deviation:   2.947

-2 log L:        1157.2     Convergence at iteration  32
```

The disparity increases by only 1.05. We omit the `t` effect, followed by the dependence on the previous outcome.

```
> obesity.allvc3b <- update(obesity.allvc3a, . ~ . -
+     t)

> summary(obesity.allvc3b)

Coefficients:
        Estimate Std. Error     t value
sex    -0.807      0.313      -2.579
x      -0.202      0.279      -0.726
tsex    1.402      0.363       3.858
ta1     0.535      0.300       1.781
tsa1   -0.452      0.437      -1.036
MASS1  -6.598      1.320      -4.998
MASS2  -2.903      0.219     -13.273
MASS3   0.976      0.223       4.376

Mixture proportions:
    MASS1      MASS2      MASS3
    0.2346     0.5228     0.2426

Random effect distribution - standard deviation:   2.617

-2 log L:        1157.2     Convergence at iteration  29

> obesity.allvc3c <- update(obesity.allvc3b, . ~ . -
+     x)

> summary(obesity.allvc3c)

Coefficients:
        Estimate Std. Error     t value
sex    -0.681      0.292      -2.336
tsex    1.286      0.334       3.846
ta1     0.484      0.293       1.654
tsa1   -0.433      0.435      -0.995
MASS1  -5.622      0.873      -6.440
MASS2  -3.036      0.221     -13.759
MASS3   0.790      0.182       4.339

Mixture proportions:
    MASS1      MASS2      MASS3
    0.2088     0.5401     0.2510

Random effect distribution - standard deviation:   2.212

-2 log L:        1157.4     Convergence at iteration  25
```

The omission of `t` gives a change in disparity of 0.02, and of the previous outcome gives a disparity change of 0.15. Further reductions can be made in this model, but we conclude that the random effect model, rather than the autoregressive model, is the appropriate representation: given the random effect, no more complex dependence structure is necessary.

9.9 Latent variable models

Variance component models are a special case of more general *latent variable models* in which the covariance structure of a set of observed variables is expressed in terms of their joint *conditional independence* given a set of latent variables.

We begin with *factor models.*

9.9.1 *The normal factor model*

Given a vector of continuous observable response variables $\mathbf{Y}' = (Y_1, \ldots, Y_r)$, and a vector of explanatory variables $\mathbf{x}' = (x_1, \ldots, x_p)$, we model the joint dependence of the responses, given the explanatory variables, by a set of *conditionally independent* regressions of the Y_j on a vector of *latent* or *factor* variables $\mathbf{Z}' = (Z_1, \ldots, Z_s)$:

$$\mathbf{Y} \mid \mathbf{Z} \sim N(\boldsymbol{\mu} + \Lambda\mathbf{Z}, \Psi),$$
$$\mathbf{Z} \sim N(\mathbf{0}, I),$$

where $\boldsymbol{\mu}$ is an r-vector of intercepts, Λ is an $r \times s$ matrix of *factor loadings* (regression coefficients) and Ψ is an $r \times r$ diagonal matrix of *specific variances.* The marginal distribution of the observable responses $\mathbf{Y}$ is then

$$\mathbf{Y} \sim N(\boldsymbol{\mu}, \Lambda\Lambda' + \Psi).$$

The zero mean and diagonal covariance matrix of $\mathbf{Z}$ appear restrictive, but a more general mean and covariance matrix are not identifiable. The independent factors can be *rotated* to any specific covariance structure, but this rotation is absorbed into the loading matrix, leaving the overall marginal distribution invariant.

The factor model can be fitted either by directly maximizing the observed data log-likelihood, or by using an EM algorithm (Rubin and Thayer, 1982) in which the latent variables Z_t are the unobserved data. We follow the latter approach.

In the complete data log-likelihood, the sufficient statistics are the sums involving the Z_t and the sums of squares and cross-products of the Z_t for the independent multiple regressions of each Y_j on $\mathbf{Z}$, and the residual sum of squares. These terms are replaced by their conditional expectations in the E-step, and the regression models are fitted separately in the M-step with appropriate adjustments to the SSP matrices, as in the variance component model.

The classical factor model does not contain any explanatory variables; these can be incorporated by a simple extension to the latent variable model:

$$Z_t \mid \mathbf{x} \sim N(\boldsymbol{\beta}_t'\mathbf{x}, 1), \quad t = 1, \ldots, s.$$

The general model can be fitted using an EM algorithm in the same way. In the complete data log-likelihood, the sufficient statistics now include the sums of squares and cross-products of the Z_t with the variables in $\mathbf{x}$. Here the representation of the model with independent factors allows the factor-explanatory variable regressions to be fitted independently as well.

We do not consider normal factor models further, though as shown below the closely related *item response models* which do not have a free variance parameter *can* be fitted directly using the variance component formulae.

9.10 IRT models

A popular model for examinees' responses to binary test items is based on *item response theory* (IRT), first developed by Lord (1952), Rasch (1960) and Birnbaum (Lord and Novick, 1968) and extensively developed since by many psychometricians and statisticians (see van der Linden and Hambleton 1997 for a review). We consider here only binary items, though ordered categorical response items are also possible.

The basic structure of such test data is a two-way array of binary responses Y_{ij}, indexed by person (examinee) i and test item j.

9.10.1 *The Rasch model*

The simplest IRT model (the 'one-parameter' model) is the *Rasch* model (Rasch, 1960), developed and popularized later by Andersen (1973, 1980). The Rasch model is a main-effect logistic regression model with parameters for person i and item j:

$$Y_{ij} \sim b(1, p_{ij}) \text{ independently}, \quad i = 1, \ldots, n, \quad j = 1, \ldots, r,$$
$$\text{logit } p_{ij} = \alpha_j + \theta_i,$$

where p_{ij} is the probability of a correct answer ($Y_{ij} = 1$) by person i on item j. The parameter α_j is the 'easiness' of the j-th item, and θ_i is the 'ability' of the i-th person.

If both sets of parameters are regarded as fixed effects, the model may be fitted directly as a Bernoulli GLM with n- and r-level factors for persons and items. However, the ability estimates $\hat{\theta}_i$ are not consistent as $n \to \infty$ for fixed r, as their number $\to \infty$ as well.

This problem can be circumvented by conditioning: because of the simple model structure, as in the logistic model for 2×2 tables in Chapter 4, the marginal totals

$Y_{+j} = \sum_i Y_{ij}$ – the item totals (or proportions correct) – are sufficient for the item parameters α_j, and the $Y_{i+} = \sum_j Y_{ij}$ – the total numbers of items correct – are sufficient for the ability parameters θ_i. The conditional distributions of $Y_{ij} \mid Y_{i+}$ and $Y_{ij} \mid Y_{+j}$ are therefore free of the parameters θ_i and α_j, respectively, and so maximizing the conditional likelihoods in each case gives consistent estimates of the item easiness and person ability parameters. This model provides a very powerful justification for the use of the total test score as an estimate of person ability.

An alternative approach to the consistency difficulty is to model the ability parameters, for example by giving them a normal distribution:

$$
\begin{aligned}
Y_{ij} \mid \theta_i &\sim b(1, p_{ij}) \text{ independently}, \\
\text{logit } p_{ij} &= \alpha_j + \theta_i, \\
\theta_i &\sim N(0, \sigma_A^2).
\end{aligned}
$$

This model can be recognized immediately as a two-level variance component model, with persons as the upper level and item responses within person as the lower level. The items are different, rather than the same as in the obesity example, but the model structure is identical – items are differentiated by their easiness parameters, rather than through the age trend.

The model has an unidentifiability property as in the normal variance component model: the mean of the ability distribution is aliased with an item parameter: it is only *differences* of item parameters which are identified. A common convention in item response theory is to constrain the item parameters to sum to zero, and include a non-zero ability mean in the model. We follow the zero mean model convention above. The random effect form of the model allows a simple change to the probit link function; this cannot be used in the conditional fixed-effects approach since it does not have sufficient statistics.

We give an example shortly, but first note that this model depends on the very strong assumption that the items differ only in easiness, or difficulty.

9.10.2 *The two-parameter model*

A more general model which weakens this assumption is the *two-parameter logit (2PL) model*, in which the items have different slope parameters as well as different intercept parameters:

$$
\begin{aligned}
Y_{ij} &\sim b(1, p_{ij}), \\
\text{logit } p_{ij} &= \alpha_j + \beta_j \theta_i.
\end{aligned}
$$

The parameter β_j determines the discriminatory power of item j, in the sense that an item with a large value of β_j identifies ability over a narrower range than an item with a small value, as the probability of a correct answer changes more rapidly with small changes in ability. In this fixed-effect model, the parameters α_j, β_j and θ_i cannot be consistently estimated by conditioning, because the 'sufficient' statistics now depend on the other model parameters. A distributional assumption for the ability parameters is essential; for example,

$$
\begin{aligned}
Y_{ij} \mid \theta_i &\sim b(1, p_{ij}) \\
\text{logit } p_{ij} &= \alpha_j + \beta_j \theta_i \\
\theta_i &\sim N(0, \sigma_A^2).
\end{aligned}
$$

A probit link gives the *2PP model*.

The notational convention in IRT is unfortunately to reverse the usual notation for slopes and intercepts, to reverse the sign of the easiness parameter and to denote the resulting item difficulty parameter by b_j; the two-parameter logit model becomes

$$\text{logit } p_{ij} = a_j(\theta_i - b_j),$$

and the one-parameter Rasch model becomes

$$\text{logit } p_{ij} = \theta_i - b_j.$$

The parameter a_j under this convention is called the *item discrimination parameter*. At the risk of further confusion, we maintain the conventional regression model notation for the parameters.

The model is the analogue for binary items of the normal factor model. It has an unidentifiability issue similar to that of the normal factor model: we could re-parametrize the model as

$$
\begin{aligned}
Y_{ij} \mid \theta_i^* &\sim b(1, p_{ij}), \\
\text{logit } p_{ij} &= \alpha_j + \beta_j^* \theta_i^*, \\
\theta_i^* &\sim N(0, 1),
\end{aligned}
$$

where $\theta_i^* = \theta_i/\sigma_A$, and $\beta_j^* = \sigma_A \beta_j$. Since θ_i is unobservable, the standard deviation of the ability distribution is unidentifiable if the regression parameters are unrestricted. However, this unidentifiability does not prevent the EM algorithm converging (one of its great assets). In the analysis below, the standard deviation is *not* fixed, so the β_j are determined only up to an arbitrary scale factor; common scalings are to fix the geometric mean of the β_j to be 1, or β_1 to be 1.

It is an important feature of the Rasch model that it does not suffer this unidentifiability: in (the random effect version of) this model σ is identifiable because of the fixed equal slopes of 1.

An interesting question is whether the distribution of ability in the 2PL model is itself identifiable – is the normal distribution a convenient model assumption which can be contradicted by the data, or would any other model give the same estimates of the identifiable parameters? It is clear that the standard deviation of the ability distribution is not identifiable, as noted above.

A deficiency of the two-parameter model is that makes no provision for *guessing*: a person with very low ability ($\theta_i \rightarrow -\infty$) has essentially zero probability of getting any item correct. This deficiency is addressed in the three-parameter model.

9.10.3 *The three-parameter logit (3PL) model*

Guessing with multiple-choice items is modelled as a random choice amongst the possible answers to the item. If the item has m possible answers, then a random choice would suggest a probability of $1/m$ of a correct answer for a person regardless of ability. However some answers for some items may be clearly incorrect and so the probability of a correct answer by guessing may be different from $1/m$, and is taken as an item-specific third parameter in the model.

The three-parameter model is a discrete mixture, with a probability γ_j of a correct answer on item j for very low ability, increasing to 1 as ability $\rightarrow \infty$:

$$\begin{aligned}\Pr[Y_{ij} = 1 \mid t_i] &= \gamma_j + (1 - \gamma_j) p_{ij},\\ \text{logit}\, p_{ij} &= \alpha_j + \beta_j \theta_i,\\ \theta_i &\sim N(0, 1).\end{aligned}$$

The introduction of an additional mixture parameter further complicates both the identification of the model parameters and the analysis of the data. Since all the item response models are mixed Bernoulli models, these are identified only by the common person ability across different items. This person ability does not affect the guessing parameter, and so the guessing parameter may be unidentifiable (van der Linden and Hambelton, 1997) without external information; this may be provided by a prior distribution based on the plausibility of the alternatives to the correct responses.

We now discuss an example.

9.10.4 *Example – The Law School Aptitude Test (LSAT)*

Bock and Aitkin (1981) illustrated the Gaussian quadrature analysis of the two-parameter probit model with data from examinees on Sections 6 and 7 of the LSAT. We use below the data from Section 7 which had five binary items, giving 32 possible response patterns, all of which were observed.

The data are given below, and are in the file `lsat`. A correct item is scored 1 and an incorrect one zero. The number with each pattern is n.

```
> data(lsat, package = "SMIR")
Item     n Item     n Item     n Item     n
12345      12345      12345      12345
-----------------------------------------------
00000   12 01000   10 10000    7 11000    6
00001   19 01001    5 10001   39 11001   25
00010    1 01010    3 10010   11 11010    7
00011    7 01011    7 10011   34 11011   35
00100    3 01100    7 10100   14 11100   18
00101   19 01101   23 10101   51 11101  136
00110    3 01110    8 10110   15 11110   32
00111   17 01111   28 10111   90 11111  308
-----------------------------------------------
```

This section of the test was clearly quite easy (for this population of examinees): the total-score distribution is

```
> tapply(lsat$wt7, list("Number of items correct " = apply(lsat[,
+      1:5], 1, sum)), sum)

Number of items correct
  0   1   2   3   4   5
 12  40 114 205 321 308
```

We first estimate the item response probabilities directly from the sample means.

```
> sapply(lsat[, 1:5], function(x) sum(x * lsat$wt7)/1000)

   y1    y2    y3    y4    y5
0.828 0.658 0.772 0.606 0.843
```

Items 1 and 5 are very easy, item 4 the hardest. We repeat the Bock and Aitkin analysis of Section 7 with the two-parameter logit model, and fit the Rasch model first, as well.

The binary response data do not need to be read in, since they form a regular pattern, but we need to expand the dataset to individual binary responses.

```
> library(npmlreg)
> lsat.gq20.rasch <- allvc(y ~ Item - 1, data = Lsat,
+      family = binomial, random = ~1 | Person,
+      random.distribution = "gq",
+      k = 20, plot.opt = 0, verbose = FALSE)

> summary(lsat.gq20.rasch)

Coefficients:
       Estimate Std. Error  t value
Item1 1.868     0.092       20.219
Item2 0.791     0.074       10.734
Item3 1.461     0.083       17.513
```

```
Item4 0.521      0.071       7.302
Item5 1.993      0.096      20.834
z     1.010      0.042      23.999
Random effect distribution - standard deviation:   1.010

-2 log L:        5329.8      Convergence at iteration  12
```

The estimated standard deviation of ability is just over 1.0. How does the estimation of the ability distribution affect the estimated item parameters?

```
> lsat.np2.rasch <- allvc(y ~ Item, data = Lsat, family = binomial,
+      random = ~1 | Person, random.distribution = "np",
+      k = 2, plot.opt = 0, verbose = FALSE)

> summary(lsat.np2.rasch)
Coefficients:
        Estimate Std. Error     t value
Item2 -1.074      0.116        -9.252
Item3 -0.405      0.121        -3.349
Item4 -1.341      0.115       -11.640
Item5  0.123      0.128         0.962
MASS1  0.546      0.095         5.765
MASS2  2.374      0.099        24.081

Mixture proportions:
    MASS1      MASS2
0.3081     0.6919

Random effect distribution - standard deviation:   0.844

-2 log L:        5332.2     Convergence at iteration  92
> lsat.np3.rasch <- update(lsat.np2.rasch, k = 3)

> summary(lsat.np3.rasch)
Coefficients:
        Estimate Std. Error    t value
Item2 -1.077      0.116        -9.258
Item3 -0.407      0.121        -3.358
Item4 -1.345      0.115       -11.649
Item5  0.124      0.129         0.965
MASS1  0.265      0.103         2.578
MASS2  1.852      0.095        19.410
MASS3  3.465      0.163        21.253

Mixture proportions:
    MASS1      MASS2      MASS3
    0.1910     0.6126     0.1964

Random effect distribution - standard deviation:   0.996
```

```
-2 log L:        5330.6     Convergence at iteration  30

> lsat.np4.rasch <- update(lsat.np2.rasch, k = 4)
> summary(lsat.np4.rasch)

Coefficients:
        Estimate Std. Error    t value
Item2 -1.078    0.116        -9.262
Item3 -0.408    0.121        -3.362
Item4 -1.347    0.116       -11.652
Item5  0.125    0.129         0.969
MASS1 -1.105    0.267        -4.142
MASS2  0.905    0.093         9.684
MASS3  2.550    0.105        24.330
MASS4  2.702    0.207        13.034

Mixture proportions:
     MASS1        MASS2        MASS3        MASS4
0.0218      0.3955       0.5257      0.0571

Random effect distribution - standard deviation:   0.918

-2 log L:            5329     Convergence at iteration  417
```

The disparity remains constant, apart from small variations, as K is increased further. Comparison of the item parameters is made difficult by the different mass-point distributions.

Because it is differences of item parameters which are identifiable, we give in Table 9.8 the parameter estimates centred on the first item parameter, together with the corresponding mean and standard deviation of the estimated ability distributions, with their masses and mass-points, for K up to 6. The ability distributions are shown in Table 9.9. The item easiness parameters are correlated +1.0 with the logistic transformations of the proportion correct on the items.

Table 9.8. Item parameters for nonparametric and normal ability distributions

Item			K			μ	σ	Disparity
	1	2	3	4	5			
2	1.868	0.794	1.463	0.527	1.991	−0.057	0.844	5332.19
3	1.868	0.791	1.461	0.523	1.992	−0.002	0.996	5330.61
4	1.868	0.790	1.460	0.521	1.993	−0.040	0.918	5329.05
5	1.868	0.790	1.460	0.521	1.993	−0.034	0.981	5328.98
6	1.868	0.790	1.460	0.521	1.993	−0.042	0.983	5329.03
GQ20	1.868	0.791	1.461	0.521	1.993	0.000	1.010	5329.80

Table 9.9. Ability distributions

K							
2							
	Masses	0.308	0.692				
	Mass-points	0.546	2.374				
3							
	Masses	0.191	0.613	0.196			
	Mass-points	0.265	1.852	3.465			
4							
	Masses	0.022	0.396	0.526	0.057		
	Mass-points	−1.105	0.905	2.550	2.702		
5							
	Masses	0.007	0.277	0.457	0.245	0.014	
	Mass-points	−3.087	0.603	1.984	2.989	3.519	
6							
	Masses	0.005	0.114	0.346	0.428	0.104	0.003
	Mass-points	−4.086	0.255	1.191	2.538	2.978	3.384

Although the discrete distributions differ in shape, they have essentially the same means and variances, and the item parameters are almost invariant; further, the disparities hardly change from the 20-point normal model. This is to be expected since at most about $K/2$ mass-points are identified with K items; the normal model for ability is not contradicted by the data.

9.11 Spatial dependence

A popular use of variance component models is in *small-area estimation*, in which geographical or other spatially defined administrative regions or areas report counts of disease or accident cases, with explanatory variables for the cases and the regions, and we want to model the sources of variation over regions in the mean count (or the case proportion, if there is a population base for the count) while allowing for individual variations in risk. The small regions give individual area risk estimates with relatively large variability; the variance component model allows the among-region variability to improve the estimation of the local region risks. In the biostatistical literature this is known as *disease mapping*. The first example in Aitkin (1999) is of this kind; it is taken from Tsutakawa (1985), who modelled the variation in lung cancer mortality rates for 84 cities in Missouri.

A limitation of the NPML analysis in this example, and generally in single- or two-level random effect models, is that it does not allow spatial dependence between neighbouring regions, as used for example in Clayton and Kaldor (1987); Breslow and Clayton (1993) and other authors. A popular extension (e.g. Besag *et al.*, 1992) of the simple random effect model for disease mapping is to include an additional spatial random effect for each region whose (conditional) mean value is equal to the average of the random effects for neighbouring regions (appropriately defined).

Care is needed in such models as the resulting joint distribution of all the random effects is likely to be singular unless it incorporates a regression parameter to reduce the very high intra-area correlation implied by the construction of the conditional means. This problem is usually handled by including an overdispersion effect as well for each region; this inflates the variance of the spatial random effects and reduces their correlation, but this is still determined by the size of the overdispersion variance component.

Initial spatial examination of the posterior means from the model *without* spatial dependence may be carried out to establish whether such dependence actually exists; as Clayton and Kaldor (1987) noted, 'there is no a priori reason why geographic proximity should be reflected in correlated cancer rates'. It should be noted that the simple variance component model provides consistent estimates of the regression model parameters in the presence of spatial correlation, but these are inefficient if real spatial dependence exists, and extensions of the NPML approach to spatial modelling would be very useful.

9.12 Multivariate correlated responses

A remarkable extension of the two-level model allows a very general analysis of multivariate responses linked by a latent structure. A simple example is the case of correlated bivariate Bernoulli responses, of which we gave an example in Section 5.9, with two signs of toxaemia in pregnant women. In Chapter 5 these were treated by contingency table analysis, but they can also be analysed through the two-level model: we simply allow different models for the two responses while maintaining their correlation through the common upper-level random effect. We do not give details here; the final model is very similar to that in Chapter 5 and does not lead to different conclusions.

A current important application of this approach is to survival studies in which a quality-of-life measure is also available, and the analysis needs to take into account the correlation between the survival event time and the quality of life measure. More general models could allow, for example, a normal response, a Poisson count, and a Bernoulli indicator to be linked through their conditional independence given a latent variable. Such extensions allow the development of very rich and complex model analyses.

9.13 Discreteness of the NPML estimate

The discrete nature of the NPML estimate may be found unattractive, if one believes *a priori* in the existence of a continuous mixing distribution. An alternative approach assuming a smooth mixing density was described by Davidian and Gallant (1993); it requires numerical quadrature with library optimization algorithms. Magder and Zeger (1996) took as a mixing distribution a finite mixture of normals, guaranteeing a continuous mixing distribution estimate. They found that the likelihood for the model with equal variances approaches a maximum as the common variance tends to zero (reflecting the optimality of the NPML estimate over *all* mixing distributions), and is very flat in the variance parameter away from zero, so the information about distributional shape of the mixing distribution is inherently limited.

Bibliography

Abramowitz, M. and Stegun, I.A. (ed.) (1970). *Handbook of Mathematical Functions.* Dover, New York.

Aitchinson, J. and Aitken, C.G.G. (1976). Multivariate binary discrimination by kernel method. *Biometrika*, **63**, 413–420.

Aitkin, M. and Alfó, Marco (2003). Longitudinal analysis of repeated binary data using autoregressive and random effect modelling. *Statistical Modelling*, **3**, 291–303.

Aitkin, M., Anderson, D., Francis, B., and Hinde, J. (1989). *Statistical Modelling in GLIM.* Oxford Statistical Science Series. Oxford University Press.

Aitkin, M., Anderson, D., and Hinde, J. (1981). Statistical modelling of data on teaching styles (with Discussion). *Journal of the Royal Statistical Society, Series A: General*, **144**, 419–461.

Aitkin, M., Boys, R.J., and Chadwick, T.J. (2005). Bayesian point null hypothesis testing via the posterior likelihood ratio. *Statistics and Computing*, **15**(3), 217–230.

Aitkin, M. and Clayton, D. (1980). The fitting of exponential, Weibull and extreme value distributions to complex censored survival data using GLIM. *Applied Statistics*, **29**, 156–163.

Aitkin, M. and Foxall, R. (2003). Statistical modelling of artificial neural networks using the multi-layer perceptron. *Statistics and Computing*, **13**, 227–239.

Aitkin, M. and Francis, B. (1982). Reader reaction: Interactive regression modelling. *Biometrics*, **38**, 511–13.

Aitkin, M., Laird, N.M., and Francis, B. (1983). A reanalysis of the Stanford heart transplant data (with Comments). *Journal of the American Statistical Association*, **78**, 264–92.

Aitkin, M. and Rocci, R. (2002). A general maximum likelihood analysis of measurement error in generalized linear models. *Statistics and Computing*, **12**(2), 163–174.

Aitkin, M. and Wilson, G.T. (1980). Mixture models, outliers, and the EM algorithm. *Technometrics*, **22**, 325–331.

Aitkin, M. (1974). Simultaneous inference and choice of variable subsets in multiple regression. *Technometrics*, **16**, 221–7.

Aitkin, M. (1978). The analysis of unbalanced cross-classifications (with Discussion). *Journal of the Royal Statistical Society, Series A: General*, **141**, 195–211.

Aitkin, M. (1981). A note on the regression analysis of censored data. *Technometrics*, **23**, 161–163.

Aitkin, M. (1987). Modelling variance heterogeneity in normal regression using GLIM. *Applied Statistics*, **36**, 332–339.

Aitkin, M. (1991). Posterior Bayes factors (with Discussion). *Journal of the Royal Statistical Society B*, **53**, 111–142.

Aitkin, M. (1994). An EM algorithm for overdispersion in generalized linear models. In *Proceedings of the 9th International Workshop on Statistical Modelling, Exeter.*

Aitkin, M. (1995). Probability model choice in single samples from exponential families using Poisson log-linear modelling, and model comparison using Bayes and posterior Bayes factors. *Statistics and Computing*, **5**, 113–120.

Aitkin, M. (1996*a*). A general maximum likelihood analysis of overdispersion in generalized linear models. *Statistics and Computing*, **6**, 251–262.

Aitkin, M. (1996*b*). A short history of a Vietnam War attitude survey. *Statistics*, **17**, 1–9.

Aitkin, M. (1997). The calibration of P-values, posterior Bayes factors and the AIC from the posterior distribution of the likelihood (with Discussion). *Statistics and Computing*, **7**, 253–261.

Aitkin, M. (1999). A general maximum likelihood analysis of variance components in generalized linear models. *Biometrics*, **55**, 117–128.

Aitkin, M. (2000). The review is of Computer-Assisted Analysis of Mixtures and Applications: Meta-Analysis, Disease Mapping and Others. (D. Böhning) in *Biometrics*, **56**, 651–652.

Aitkin, M. (2001). Likelihood and Bayesian analysis of mixtures. *Statistical Modelling*, **1**, 287–304.

Allen, D.M. (1971). Mean square error of prediction as a criterion for selecting variables. *Technometrics*, **13**, 469–475.

Andersen, E.B. (1973). Conditional inference for multiple-choice questionnaires. *British Journal Mathematical Statistics in Psychology*, **26**, 31–44.

Andersen, E.B. (1980). *Discrete Statistical Models with Social Science Applications*. North-Holland, Amsterdam.

Anderson, D.A. and Aitkin, M. (1985). Variance component models with binary response: Interviewer variability. *Journal of the Royal Statistical Society, Series B: Methodological*, **47**, 203–210.

Anderson, J.A. (1984). Regression and ordered categorical variables (with Discussion). *Journal of the Royal Statistical Society B*, **46**, 1–30.

Anscombe, F.J. (1964). Normal likelihood functions. *Annals of the Institute of Statistical Mathematics*, **26**, 1–19.

Ashford, J.R. (1959). An approach to the analysis of data from semi-quantal responses in biological response. *Biometrics*, **15**, 573–581.

Atkinson, A.C. (1981). Robustness, transformations and two graphical displays for outlying and influential observations in regression. *Biometrika*, **68**, 13–20.

Atkinson, A.C. (1982). Regression diagnostics, transformations and constructed variables (with Discussion). *Journal Royal Statistical Society B*, **44**, 1–36.

Atkinson, A.C. (1985). *Plots, Transformations and Regression. An Introduction to Graphical Methods of Diagnostic Regression Analysis*. Oxford University Press, Oxford.

Atkinson, C. and Polivy, J. (1976). Effects of delay, attack and retaliation on state depression and hostility. *Journal of Abnormal Psychology*, **85**, 370–76.

Barnard, G.A., Jenkins, G.M., and Winsten, C.B. (1962). Likelihood inference and time series. *Journal of the Royal Statistical Society A*, **125**, 321–372.

Barnard, G.A. (1949). Statistical inference. *Journal of the Royal Statistical Society B*, **11**, 115–139.

Barnett, V. and Lewis, T. (1994). *Outliers in Statistical Data* (2 edn). Wiley, New York, Chichester.

Barnett, V. (1975). Probability plotting methods and order statistics. *Applied Statistics*, **24**, 95–108.

Bartlett, M.S. (1957). A comment on D.V. Lindley's statistical paradox. *Biometrika*, **44**, 533–534.

Bartlett, R.H., Roloff, D.W., Cornell, R.G., Andrews, A.F., Dillon, P.W., and Zwischenberger, J.B. (1985, Oct). Extracorporeal circulation in neonatal respiratory failure: a prospective randomized study. *Pediatrics*, **76**(4), 479–487.

Basu, D. (1977). On partial sufficiency: A review. *Journal of Statistical Planning and Inference*, **2**, 1–13.

Baxter, L.A., Coutts, S.M., and Ross, G.A.F. (1980). Applications of linear models in motor insurance. In *Proceedings of the 21st International Congress of Actuaries, Zurich*, pp. 11–29.

Begg, C.B. (1990). On inferences from Wei's biased coin design for clinical trials (with Discussion). *Biometrika*, **77**, 467–484.

Belsley, D.A., Kuh, E., and Welsch, R.E. (1980). *Regression diagnostics: Identifying Influential Data and Sources of Collinearity*. Wiley, New York.

Bennett, M.D. (1974). The emotional response of husbands to suicide attempts by their wives. Unpublished MD thesis, Sydney University.

Bennett, S. and Whitehead, J. (1981). Fitting logistic and log-logistic regression models to censored data using GLIM. *GLIM Newsletter*, **4**, 12–19.

Berger, J.O. and Wolpert, R. (1986). *The Likelihood Principle*. Institute of Mathematical Sciences, Hayward.

Besag, J., York, J., and Mollie, A. (1992). Bayesian image restoration, with two applications in spatial statistics. *Annals of the Institute of Statistical Mathematics*, **43**, 1–59.

Birnbaum, A. (1962). On the foundations of statistical inference. *Journal of the American Statistical Association*, **57**, 269–306.

Bishop, Y. M. M., (1969). Full contingencytables, logits and split contingency tables. *Biometrics*, **25**, 383–399.

Bishop, Y.M.M., Fienberg, S.E., and Holland, P.W. (1975). *Discrete Multivariate Analysis: Theory and Practice*. MIT Press, Cambridge Mass.

Bissell, A.F. (1972). A negative binomial model with varying element sizes. *Biometrika*, **59**, 435–441.

Blundell, R. and Bond, S. (1998). Initial conditions and moment restrictions in dynamic panel data models. *Journal of Econometrics*, **68**, 115–132.

Bock, R.D. and Aitkin, M. (1981). Marginal maximum likelihood estimation of item parameters: An application of an EM algorithm. *Psychometrika*, **46**, 443–459.

Bock, R.D. and Yates, G. (1973). *MULTIQUAL: Log-linear Analysis of Nominal or Ordinal Qualitative Data by the Method of Maximum Likelihood*. National Educational Resources, Chicago.

Bock, R.D. (1975). *Multivariate Statistical Methods in Behavioral Research*. McGraw-Hill, New York.

Böhning, D. (1999). *Computer-Assisted Analysis of Mixtures and Applications*. Chapman & Hall, Boca Raton.

Box, G.E.P. and Cox, D.R. (1964). An analysis of transformations (with Discussion). *Journal of the Royal Statistical Society B*, **26**, 211–52.

Box, G.E.P. and Tidwell, P.W. (1962). Transformations of the independent variable. *Technometrics*, **4**, 531–50.

Box, G.E.P. (1980). Sampling and Bayes inference in scientific modelling and robustness (with Discussion). *Journal of the Royal Statistical Society A*, **143**, 383–430.

Box, G.E.P. (1983). An apology for ecumenism in statistics. In *Scientific inferences, data analysis and robustness* (ed. G. Box, T. Leonard, and C. Wu). Academic Press, New York.

Breslow, N.E. and Clayton, D.G. (1993). Approximate inference in generalized linear mixed models. *Journal of the American Statistical Association*, **88**, 9–25.

Breslow, N. (1974). Covariance analysis of censored survival data. *Biometrics*, **30**, 89–99.

Brownlee, K.A. (1965). *Statistical Theory and Methodology in Science and Engineering* (2 edn). Wiley, New York.

Brown, P.J., Stone, J., and Ord-Smith, C. (1983). Toxaemic signs during pregnancy. *Applied Statistics*, **32**, 69–72.

Brown, P.J. (1982). Multivariate calibration (with Discussion). *Journal of the Royal Statistical Society B*, **44**, 287–321.

Butler, R. (1986). Predictive likelihood inference and applications (with Discussion). *Journal of the Royal Statistical Society B*, **48**, 1–38.

Carroll, R.J., Ruppert, D., and Stefanski, L.A. (1995). *Measurement Error in Nonlinear Models*. Chapman & Hall, London.

Chadwick, T.J. (2002). A general Bayes theory of nested model comparisons. Unpublished PhD thesis, The University of Newcastle upon Tyne.

Chambers, J.M. and Hastie, T.J. (1992). *Statistical Models in S*. Wadsworth & Brooks/Cole.

Chatfield, C. and Goodhardt, G.J. (1970). The beta-binomial model for consumer purchasing behaviour. *Applied Statistics*, **19**, 240–50.

Chernoff, H. (1954). On the distribution of the likelihood ratio. *Annals of Mathematical Statistics*, **25**, 573–578.

Clayton, D.G. and Hills, M. (1994). *Statistical Models in Epidemiology*. University Press, Oxford.

Clayton, D. and Kaldor, J. (1987). Empirical Bayes estimates of age-standardized relative risks for use in disease mapping. *Biometrics*, **43**, 671–681.

Cleveland, W.S. (1985). *The Elements of Graphing Data*. Monterey, CA:Wadsworth.

Collett, D. (2003). *Modelling Survival Data in Medical Research*. Chapman & Hall, Boca Raton.

Cook, R.D. and Weisberg, S. (1982). *Residuals and Influence in Regression*. Chapman & Hall, London.

Cook, R.D. (1977). Detection of influential observations in linear regression. *Technometrics*, **19**, 15–18.

Copas, J.B. (1983). Regression, prediction and shrinkage (with Discussion). *Journal of the Royal Statistical Society B*, **45**, 311–54.

Copas, J.B. (1997). Statistical inference for non-random samples. *Journal of the Royal Statistical Society B*, **59**, 55–95.

Cox, D.R. (1958). Some problems connected with statistical inference. *Annals of Mathematical Statistics*, **29**, 357.

Cox, D.R. (1972). Regression models and life tables (with Discussion). *Journal of the Royal Statistical Society B*, **34**, 187–220.

Cox, D.R. and Oakes, D. (1984). *Analysis of Survival Data*. Chapman & Hall, London.

Crouch, E.A.C. and Speigelman, D. (1990). The evaluation of integrals of the form $\int_{-\infty}^{+\infty} f(t)\exp(-t^2)dt$: application to logistic-normal models. *Journal of the American Statistical Association*, **85**, 464–469.

Crowder, M.J. (1978). Beta-binomial ANOVA for proportions. *Applied Statistics*, **27**, 34–7.

Crowder, M.J. (2001). *Classical Competing Risks*. Chapman & Hall, Boca Raton.

Crowley, J. and Hu, M. (1977). Covariance analysis of heart transplant survival data. *Journal of the American Statistical Association*, **72**, 27–36.

Davidian, M. and Gallant, A.R. (1993). The nonlinear mixed effects model with a smooth random effects density. *Biometrika*, **80**, 475–488.

Davies, R.B. (1987). *Longitudinal Data Analysis*, Chapter Mass Point Methods for Dealing with Nuisance Parameters in Longitudinal Studies. Avebury, Aldershot, Hants.

Day, N.E. (1969). Estimating the components of a mixture of two normal distributions. *Biometrika*, **56**, 463–474.

Dempster, A.P. (1997). The direct use of likelihood in significance testing. *Statistics and Computing*, **7**, 247–252.

Dempster, A.P., Laird, N.M., and Rubin, D.B. (1977). Maximum likelihood from incomplete data via the EM algorithm (with Discussion). *Journal of the Royal Statistical Society B.*, **39**, 1–38.

Dempster, A.P. (1974). The direct use of likelihood in significance testing. In *Proc. Conf. Foundational Questions in Statistical Epidemiology* (ed. O. Barndorff-Nielsen, P. Blaesild, and G. Sihon), pp. 335–352.

Dietz, E. and Böhning, D. (1995). Statistical inference based on a general model of unobserved heterogeneity. In *Statistical Modelling: Proceedings of the 10th International Workshop on Statistical Modelling, Innsbruck, Austria, 10-14 July, 1995*, Innsbruck, Austria. International Workshop on Statistical Modelling: Springer-Verlag, New-York.

Diggle, P.J., Hegarty, P., Liang, K.Y., and Zeger, S.L. (2002). *Analysis of Longitudinal Data* (2 edn). Clarendon Press, Oxford.

Dodge, Y. (1999). The guinea pig of multiple regression. In *Robust Statistics, Data Analysis, and Computer Intensive Methods: In Honor of Peter Huber's 60th Birthday*, Volume 109 of *Lecture notes in statistics*, pp. 91–117. Springer-Verlag, New York.

Draper, N.R. and Smith, H. (1998). *Applied Regression Analysis* (2nd edn). Wiley, New York.

Dunn, P.K. and Smyth, G.K. (2006). *dglm: Double generalized linear models*. R package version 1.3.

Edwards, A.W.F. (1972). *Likelihood*. Cambridge University Press, Cambridge.

Efron, B. (1977). The efficiency of Cox's likelihood function for censored data. *Journal of the American Statistical Association*, **83**, 414–425.

Efron, B. (1986). Double exponential families and their use in generalized linear regression. *Journal of the American Statistical Association*, **81**, 709–721.

Einbeck, J., Darnell, R., and Hinde, J. (2006). *npmlreg: Nonparametric maximum likelihood estimation for random effect models*. R package version 0.34.

Einbeck, J. and Hinde, J. (2005). A note on NPML estimation for exponential family regression models with unspecified dispersion parameter. Technical Report IRL-GLWY-2005-04, National University of Ireland, Galway.

Einbeck, J. and Hinde, J. (2006). A note on NPML estimation for exponential family regression models with unspecified dispersion parameters. *Austrian Journal of Statistics*, **35**, 233–243.

Erickson, B.H. and Nosanchuk, T.A. (1979). *Understanding Data*. Milton Keynes, UK, Open University Press.

Everitt, B.S. and Hand, D.J. (1981). *Finite Mixture Distributions*. Chapman & Hall, London.

Fay, R.E. (1996). Alternative paradigms for the analysis of imputed survey data. *Journal of the American Statistical Association*, **91**, 490–498.

Feigl, P. and Zelen, M. (1965). Estimation of exponential probabilities with concomitant information. *Biometrics*, **21**, 826–38.

Filliben, J.J. (1975). The probability plot correlation coefficient test for normality. *Technometrics*, **17**, 111–17. Correction. *17*, 520.

Finch, S.J., Mendell, N.R., and Thode, H.C. (1989). Probabilistic measures of adequacy of a numerical search for a global maximum. *Journal of the American Statistical Association*, **84**, 1020–1023.

Finney, D.J. (1947). The estimation from individual records of the relationship between dose and quantal response. *Biometrika*, **34**, 320–34.

Fisher, R.A. (1935). The logic of inductive inference (with Discussion). *Journal of the Royal Statistical Society*, **98**, 39–54.

Fitzmaurice, G.M., Laird, N.M., and Lipsitz, S.R. (1994). Analysing incomplete longitudinal binary responses: a likelihood-based approach. *Biometrics*, **50**, 601–612.

Follman, D.A. and Lambert, D. (1989). Generalizing logistic regression by nonparametric mixing. *Journal of the American Statistical Association*, **84**, 295–300.

Francis, B.J., Green, M., and Payne, C. (eds.) (1993). *The GLIM System: Release 4 Manual*. Clarendon Press, Oxford.

Freireich, Emil J, Gehan, Edmund, Frei, Emil, III, Schroeder, Leslie R., Wolman, Irving J., Anbari, Rachad, Burgert, E. Omar, Mills, Stephen D., Pinkel, Donald, Selawry, Oleg S., Moon, John H., Gendel, B. R., Spurr, Charles L., Storrs, Robert, Haurani, Farid, Hoogstraten, Barth, and Lee, Stanley (1963). The Effect of 6-Mercaptopurine on the Duration of Steroid-induced Remissions in Acute Leukemia: A Model for Evaluation of Other Potentially Useful Therapy. *Blood*, **21**(6), 699–716.

Fuller, W.A. (1987). *Measurement Error Models*. Wiley, New York.

Gehan, E.A. (1965). A generalized Wilcoxon test for comparing arbitrarily singly-censored samples. *Biometrika*, **52**, 203–23.

Gelman, A., Carlin, J.B., Stern, H.S., and Rubin, D.B. (1995). *Bayesian Data Analysis*. Chapman & Hall, London.

Goldstein, H. (2003). *Multilevel Statistical Models* (3 edn). Edward Arnold, London.

Goodman, L.A. (1979). Simple models for the analysis of ordered categorical data. *Journal of the American Statistical Association*, **74**, 537–52.

Gottschalk, L.A. and Gleser, G.C. (1969). *The Measurement of Psychological States through the Content Analysis of Verbal Behavior*. University of California Press, Berkeley.

Haberman, S.J. (1974). Log-linear models for frequency tables with ordered classifications. *Biometrics*, **30**, 589–600.

Hall, P. and La Scala, B. (1990). Methodology and algorithms of empirical likelihood. *International Statistical Review*, **58**, 109–127.

Harrison, G.A. and Brush, G. (1990). On correlations between adjacent velocities and accelerations in longitudinal growth data. *Annals of Human Biology*, **17**(1), 55–57.

Hartley, H.O. (1950). The maximum F-ratio as a short cut test for heterogeneity of variances. *Biometrika*, **37**, 308–12.

Heckman, J.J. and Singer, B. (1984). A method for minimizing the impact of distributional assumptions in econometric models of duration. *Econometrica*, **52**, 271–320.

Henderson, H.V. and Velleman, P.F. (1981). Building multiple regression models interactively. *Biometrics*, **37**, 391–411.

Higgins, J.E. and Koch, G. G. (1977). Variable selection and generalized chi-square analysis of categorical data applied to a large cross-sectional occupational health survey. *International Statistical Review*, **45**, 51–62.

Hinde, J. (1982). Compound Poisson regression models. In *GLIM82* (ed. R. Gilchrist). Springer-Verlag, New York.

Hoaglin, D.C. and Welsch, R.E. (1978). The hat matrix in regression and ANOVA. *The American Statistician*, **32**, 17–22. Corrigenda 32, 146.

Hodges, J.L., Krech, D., and Crutchfield, R.S. (1975). *StatLab: An Empirical Introduction to Statistics*. McGraw-Hill Ryerson, Toronto.

Hollander, M., McKeague, I.W., and Yang, J. (1997). Likelihood ratio-based confidence bands for survival functions. *Journal of the American Statistical Association*, **92**, 215–226.

Hope, A.C.A. (1968). A simplified Monte Carlo significance test procedure. *Journal of the Royal Statistical Society B*, **30**, 582–598.

Hosmer, D.W. and Lemeshow, S. (1999). *Applied Survival Analysis: Regression Modeling of Time to Event Data*. Wiley-Interscience, New York.

Jacobs, R.A., Jordan, M.I., Nowlan, S.J., and Hinton, G.E. (1991). Adaptive mixtures of local experts. *Neural Computation*, **3**, 79–87.

Jeffreys, H. (1961). *Theory of Probability*. Clarendon Press, Oxford.

Johnson, P.O. and Neyman, J. (1936). Tests of certain linear hypotheses and their applications to some educational problems. *Statistical Research Memoirs*, **1**, 57–93.

Jones, K. (1975). A geographical contribution to the aetiology of chronic bronchitis. Unpublished B.Sc. dissertation, University of Southampton.

Jordan, M.I. and Jacobs, R.A. (1994). Hierarchical mixtures of experts and the EM algorithm. *Neural Computing*, **6**, 181–214.

Jöreskog, K.G. and Sörbom, D. (1981). *LISREL V. Analysis of Linear Structural Relationships by Maximum Likelihood and Least Square Methods*. Department of Statistics, University of Uppsala.

Jorgensen, B. (1997). *The Theory of Dispersion Models*. CRC Press, Boca Raton.

Kalbfleisch, J.D. and Prentice, R.L. (1980). *The Statistical Analysis of Failure Time Data*. Wiley, New York.

Kalbfleisch, J.D. and Sprott, D.A. (1970). Application of likelihood methods to models involving large numbers of parameters (with Discussion). *Journal of the Royal Statistical Society B*, **32**, 175–208.

Kaplan, E.L. and Meier, P. (1958). Non-parametric estimation from incomplete observations. *Journal of the American Statistical Association*, **53**, 457–481.

Kass, R. and Raftery, A.E. (1995). Bayes factors. *Journal of the American Statistical Association*, **90**, 773–795.

Kiefer, J. and Wolfowitz, J. (1956). Consistency of the maximum likelihood estimator in the presence of infinitely many nuisance parameters. *Annals of Mathematical Statistics*, **27**, 887–906.

Ku, H.H. and Kullback, S. (1974). Loglinear models in contingency table analysis. *The American Statistician*, **28**, 115–22.

Kuhn, M. and Weaston, S. (2007). *odfWeave: Sweave processing of Open Document Format (ODF) files*. R package version 0.5.9.

Laird, N.M. (1978). Nonparametric maximum likelihood estimation of a mixing distribution. *Journal of the American Statistical Association*, **73**, 805–811.

Laird, N.M. and Ware, J.H. (1986). Random-effects models for longitudinal data. *Biometrics*, **38**, 963–974.

Lange, K.L., Little, R.J.A., and Taylor, J.M.G. (1989). Robust statistical modeling using the t distribution. *Journal of the American Statistical Association*, **84**, 881–896.

Larson, M.G. and Dinse, G.E. (1985). A mixture model for the regression analysis of competing risks data. *Applied Statistics*, **34**, 201–211.

Lawless, J.F. (1982). *Statistical Models and Methods for Lifetime Data*. Wiley, New York.

Lawless, J.F., Kalbfleisch, J.D., and Wild, C.J. (1999). Semiparametric methods for response-selective and missing data problems in regression. *Journal of the Royal Statistical Society B*, **61**, 413–438.

Lesaffre, E. and Spiessens, B. (2001). On the effect of the number of quadrature points in a logistic random-effects model: An example. *Applied Statistics*, **50**, 325–335.

Liang, K.Y. and Zeger, S.L. (1986). Longitudinal data analysis using generalized linear models. *Biometrika*, **73**, 13–22.

Lindley, D.V. (1957). A statistical paradox. *Biometrika*, **44**, 187–192.

Lindsay, B.G. (1983). The geometry of mixture likelihoods, part I: A general theory. *Annals of Statistics*, **11**, 86–94.

Lindsay, B.G. (1995). *Mixture Models: Theory, Geometry and Applications*. Institute of Mathematical Statistics, Hayward.

Lindsey, J.K. (1974). Construction and comparison of statistical models. *Journal of the Royal Statistical Society B*, **36**, 418–25.

Lindsey, J.K. (1996). *Parametric Statistical Inference*. Clarendon Press, Oxford.

Lindsey, J.K. (1999). Some statistical heresies. *Journal of the Royal Statistical Society D*, **48**, 1–40.

Lindsey, J.K. and Mersch, G. (1992). Fitting and comparing probability-distributions with log linear-models. *Computational Statistics and Data Analysis*, **13**, 373–384.

Little, R.J. and Rubin, D.B. (1987). *Statistical Analysis with Missing Data*. Wiley, New York.

Little, R.J.A. and Schluchter, M.D. (1985). Maximum likelihood estimation for mixed continuous and categorical data with missing values. *Biometrika*, **72**, 497–512.

Longford, N.T. (1993). *Random Coefficient Models*. Clarendon Press, Oxford.

Lord, F.M. (1952). A Theory of Test Scores. *Psychometric Monographs*, **7**.

Lord, F.M. and Novick, M.R. (1968). *Statistical Theories of Mental Test Scores*. Addison-Wesley, Reading Mass.

Louis, T.A. (1982). Finding the observed information matrix when using the EM algorithm. *Journal of the Royal Statistical Society B*, **44**, 226–233.

Magder, L.S. and Zeger, S.L. (1996). A smooth nonparametric estimate of a mixing distribution using mixtures of Gaussians. *Journal of the American Statistical Association*, **91**, 1141–1151.

Mantel, N. (1966). Evaluation of survival data and two new rank order statistics arising in its consideration. *Cancer Chemotherapy Reports*, **50**, 163–70.

Maritz, J.S. and Lwin, T. (1989). *Empirical Bayes Methods* (2 edn). Chapman & Hall, London.

McCullagh, P. (1980). Regression models for ordinal data (with Discussion). *Journal of the Royal Statistical Society B*, **42**, 109–42.

McCullagh, P. and Nelder, J.A. (1989). *Generalized Linear Models*. Chapman & Hall, London.

McCulloch, C.E. (1997). Maximum likelihood algorithms for generalized linear mixed models. *Journal of the American Statistical Association*, **92**, 162–170.

McGilchrist, C.A. (1994). Estimation in generalized mixed models. *Journal of the Royal Statistical Society B*, **56**, 61–69.

McLachlan, G.J. (1987). On bootstrapping the likelihood ratio test statistic for the number of components in a normal mixture. *Applied Statistics*, **36**, 318–324.

McLachlan, G.J. and Krishnan, T. (1997). *The EM Algorithm and Extensions*. John Wiley, New York.

McLachlan, G.J. and Peel, D. (1997). *Computing Science and Statistics*, Chapter on a resampling approach to choosing the number of components in normal mixture models. Interface Foundation, Fairfax Station, Virgina.

McLachlan, G.J. and Peel, D. (2000). *Finite Mixture Models*. John Wiley, New York.

McLachlan, G.J. and Basford, K.E. (1988). *Mixture Models: Inference and Applications to Clustering*. John Wiley, New York.

Meng, X.L. and van Dyk, D.A. (1997). The EM algorithm – an old folk song sung to a fast new tune. *Journal of the Royal Statistical Society B*, **59**, 511–567.

Miller, R.G. and Halpern, J. (1982). Regression with censored data. *Biometrika*, **69**, 521–31.

Minder, Ch.E. and Whitney, J.B. (1975). A likelihood analysis of the linear calibration problem. *Technometrics*, **17**, 463–71.

Murphy, S.A. and van der Vaart, A.W. (2000). On profile likelihood (with Discussion). *Journal of the American Statistical Association*, **95**, 449–485.

Nelder, J.A. (1977). A reformulation of linear models (with Discussion). *Journal of the Royal Statistical Society A*, **140**, 48–76.

Nelder, J. A. (1985) Quasi-likelihood and GLIM. In *Generalized Linear Models*. Springer-Verlag, Berlin.

Nelder, J.A. and Wedderburn, R.W.M. (1972). Generalized linear models. *Journal of the Royal Statistical Society A*, **135**, 370–84.

Neuhaus, J.M. (1992). Statistical methods for longitudinal and clustered designs with binary responses. *Statistical Methods in Medical Research*, **1**, 249–273.

Neuhaus, J.M. and Jewell, N.P. (1990). Some comments on Rosner's multiple logistic model for clustered data. *Biometrics*, **46**, 523–534.

Oakes, D. (1999). Direct calculation of the information matrix via the EM algorithm. *Journal of the Royal Statistical Society B*, **61**, 479–482.

Owen, A.B. (1988). Empirical likelihood ratio confidence-intervals for a single functional. *Biometrika*, **75**, 237–249.

Owen, A.B. (1995). Nonparametric likelihood confidence bands for a distribution function. *Journal of the American Statistical Association*, **90**, 516–521.

Owen, A.B. (2001). *Empirical Likelihood*. Chapman & Hall/CR, Boca Raton.

Patterson, H.D. and Thompson, R. (1971). Recovery of interblock information when block sizes are unequal. *Biometrika*, **58**, 545–554.

Pearson, E.S. and Hartley, H.O. (1966). *Biometrika Tables for Statisticians* (3rd edn), Volume 1. Cambridge University Press, Cambridge.

Pfefferman, D. (1993). The role of sampling weights when modelling survey data. *International Statistical Review*, **61**, 317–337.

Pinheiro, J.C. and Bates, D.M. (2000). *Mixed-Effects Models in S and S-PLUS*. Springer-Verlag, New-York.

Plackett, R.L. (1977). The marginal totals of a 2×2 table. *Biometrika*, **64**, 37–42.

Postman, M., Hunchra, J.P., and Geller, M.J. (1986). Probes of large-scale structures in the Corona Borealis region. *The Astronomical Journal*, **92**, 1238–1247.

Potthoff, R.F. (1964). On the Johnson-Neyman technique and some extensions thereof. *Psychometrika*, **29**, 241–56.

Pregibon, D. (1981). Logistic regression diagnostics. *Annals of Statistics*, **9**, 705–24.

Prentice, R.L. (1973). Exponential survivals with censoring and explanatory variables. *Biometrika*, **60**, 279–88.

Prentice, R.L. and Gloeckler, L.A. (1978). Regression analysis of grouped survival data with application to breast cancer data. *Biometrics*, **34**, 57–67.

Racine, A., Grieve, A.P., Flühler, H., and Smith, A.F.M. (1986). Bayesian methods in practice: Experiences in the pharmaceutical industry. *Applied Statistics*, **35**, 93–150.

Rasch, G. (1960). *Probabilistic Models for Some Intelligence and Attainment Tests*. Danish Institute for Educational Research, Copenhagen.

Reiersöl, O. (1950). Identifiability of a linear relation between variables which are subject to error. *Econometrika*, **18**, 375–89.

Richardson, S. and Green, P.J. (1997). On Bayesian analysis of mixtures with an unknown number of components (with Discussion). *Journal of the Royal Statistical Society B*, **59**, 731–792.

Ridout, M.S. (1990). Non-convergence of Fisher's method of scoring – a simple example. *GLIM Newsletter*, **20.06**.

Rodriguez, G. and Goldman, N. (1995). An assessment of estimation procedures for multilevel models with binary responses. *Journal of the Royal Statistical Society A*, **158**, 73–89.

Roeder, K. (1990). Density estimation with confidence sets exemplified by superclusters and voids in the galaxies. *Journal of the American Statistical Association*, **85**, 617–624.

Roger, J.H. and Peacock, S.B. (1982). Fitting the scale as a GLIM parameter for Weibull extreme value, logistic and log-logistic regression models with censored data. *GLIM Newsletter*, **6**, 30–37.

Royall, R. (1997). *Statistical Evidence: A Likelihood Paradigm*. Chapman & Hall, London.

Rubin, D.B. (1976). Inference and missing data. *Biometrika*, **63**, 581–92.

Rubin, D.B. (1987). *Multiple Imputation for Nonresponse in Surveys*. Wiley, New York.

Rubin, D.B. and Thayer, D.T. (1982). EM algorithms for factor analysis. *Pyschometrika*, **47**, 69–76.

Ryan, T., Joiner, B., and Ryan, B. (1976). *Minitab Students Handbook*. Duxbury Press, North Scituate, Mass.

Scallan, A., Gilchrist, R., and Green, M. (1984). Fitting parametric link functions in generalized linear models. *Computational Statistics and Data Analysis*, **2**, 37–49.

Schafer, D.W. (2001). Semiparametric maximum likelihood for measurement error regression. *Biometrics*, **57**, 53–61.

Schafer, J.L. (1997). *Analysis of Incomplete Multivariate Data*. Chapman & Hall, London.

Silvapulle, M.J. (1981). On the existence of maximum likelihood estimators for the binomial response model. *Journal of the Royal Statistical Society B*, **43**, 310–13.

Simpson, E.H. (1951). The interpretation of interaction in contingency tables. *Journal of the Royal Statistical Society B*, **13**, 426–431.

Skrondal, A. and Rabe-Hesketh, S. (2003). *Generalized Latent Variable Modeling: Multilevel, Longitudinal and Structural Equation Models*. Chapman & Hall/CRC Press, Boca Raton.

Smith, T.M.F. (1976). The foundations of survey sampling: a review (with Discussion). *Journal of the Royal Statistical Society A*, **139**, 183–204.

Smyth, G.K. (1986). Modelling the dispersion parameter in generalized linear models. In *Proc. Statistical Computing Section*, pp. 278–283. American Statistical Association, Alexandria.

Smyth, G.K. (1989). Generalized linear models with varying dispersion. *Journal of the Royal Statistical Society B*, **51**, 47–60.

Smyth, G.K. and Verbyla, A.P. (1999). Double generalized linear models: approximate REML and diagnostics. In *Proceedings of the 14th International Workshop on Statistical Modelling, Graz, July 19–23 1999*, pp. 66–80.

Sprott, D.A. (1973). Normal likelihoods and their relation to large sample theory of estimation. *Biometrika*, **60**, 457–465.

Sprott, D.A. (1980). Maximum likelihood in small samples: estimation in the presence of nuisance parameters. *Biometrika*, **67**, 515–523.

Sprott, D.A. (2000). *Statistical Inference in Science*. Springer-Verlag, New York.

Steele, B.M. (1996). A modified EM algorithm for estimation in generalized mixed models. *Biometrics*, **52**, 1295–1310.

Stevens, S.S. (1946). On the theory of scales of measurement. *Science*, **103**, 677–80.

Stone, M. (1974). Cross-validity choice and assessment of statistical predictions (with Discussion). *Journal of the Royal Statistical Society B*, **36**, 111–47.

Stroud, A.H. and Sechrest, D. (1966). *Gaussian Quadrature Formulas*. Prentice-Hall, Englewoods Cliffs, New Jersey.

Stukel, T.A. (1988). Generalized logistic models. *Journal of the American Statistical Association*, **83**, 426–431.

Tanner, M.A. (1996). *Tools for Statistical Inference: Methods for the Exploration of Posterior Distributions and Likelihood Functions* (3rd edn). Springer-Verlag, New York.

Tate, R.F. and Klett, G.W. (1959). Optimal confidence intervals for the variance of a normal distribution. *Journal of the American Statistical Association*, **54**, 674–682.

Thode, H.C., Finch, S.J., and Mendell, N.R. (1988). Simulated percentage points for the null distribution of the likelihood ratio test for a mixture of two normals. *Biometrics*, **44**, 1195–1201.

Thompson, R. (1981). Survival data and GLIM. Letter to the editor. *Applied Statistics*, **30**, 310.

Titterington, D.M., Smith, A.F.M., and Makov, U.E. (1985). *Statistical Analysis of Finite Mixture Distributions*. John Wiley, Chichester.

Tsutakawa, R.K. (1985). Estimation of cancer mortality rates: a Bayesian analysis of small frequencies. *Biometrics*, **41**, 69–79.

van der Linden, W.J. and Hambelton, R.K. (1997). *Handbook of Modern Item Response Theory*. Springer-Verlag, New York.

Venables, W.N. and Ripley, B.D. (2002). *Modern Applied Statistics with S* (4th edn). Springer, New York. ISBN 0-387-95457-0.

Walker, S. (1996). An EM algorithm for nonlinear random effects models. *Biometrics*, **52**, 934–944.

Wand, M.P. and Jones, M.C. (1994). *Kernel Smoothing*. CRC Press, Boca Raton.

Warnes, G.R. (2005). *gdata: Various R programming tools for data manipulation*. R package version 2.1.2.

Wechsler, D. (1949). *Wechsler Intelligence Scale for Children*. The Psychological Corporation, New York.

Wedderburn, R.W.M. (1974). Quasi-likelihood functions, generalized linear models and the Gauss-Newton method. *Biometrika*, **61**, 439–447.

White, H. (1982). Maximum likelihood estimation of misspecified models. *Econometrica*, **29**, 1–25.

Whitehead, J. (1980). Fitting Cox's regression model to survival data using GLIM. *Applied Statistics*, **29**, 268–75.

Wilson, E.B. and Hilferty, M.M. (1931). The distribution of chi-square. *Proceedings of the Natural Academy of Sciences*, **17**, 684–688.

Wolynetz, M.S. (1979). Maximum likelihood estimation in a linear model from confined and censored normal data. Algorithm AS139. *Applied Statistics*, **28**, 195–206.

Woolson, R.F. and Clarke, W.R. (1984). Analysis of categorical incomplete longitudinal data. *Journal of the Royal Statistical Society A*, **147**, 87–99.

Wrigley, N. (1976). *Introduction to the Use of Logit Models in Geography*. Geo. Abstracts Ltd, CATMOG 10, University of East Anglia, Norwich.

Yates, F. (1984). Tests of significance for 2 × 2 contingency tables (with Discussion). *Journal of the Royal Statistical Society A*, **147**, 426–63.

Yusuf, S., Peto, R., Lewis, J., Collins, R., and Sleight, P. (1984). Beta blockade during and after myocardial infarction: An overview of the randomized trials. *Progress in Cardiovascular Diseases*, **27**, 335–371.

Zelen, M. (1969). Play the winner rule and the controlled clinical trial. *Journal of the American Statistical Association*, **64**, 131–146.

R function and constant index

Dataset Index

Subject index